Growth Factors, Peptides, and Receptors

GWUMC Department of Biochemistry
Annual Spring Symposia
Series Editors:
Allan L. Goldstein, Ajit Kumar, and J. Martyn Bailey
The George Washington University Medical Center

Recent volumes in this series:

ADVANCES IN MOLECULAR BIOLOGY AND
TARGETED TREATMENT FOR AIDS
Edited by Ajit Kumar

BIOLOGY OF CELLULAR TRANSDUCING SIGNALS
Edited by Jack Y. Vanderhoek

BIOMEDICAL ADVANCES IN AGING
Edited by Allan L. Goldstein

CARDIOVASCULAR DISEASE
Molecular and Cellular Mechanisms, Prevention, and Treatment
Edited by Linda L. Gallo

CELL CALCIUM METABOLISM
Physiology, Biochemistry, Pharmacology, and Clinical Implications
Edited by Gary Fiskum

EUKARYOTIC GENE EXPRESSION
Edited by Ajit Kumar

GROWTH FACTORS, PEPTIDES, AND RECEPTORS
Edited by Terry W. Moody

NEURAL AND ENDOCRINE PEPTIDES AND RECEPTORS
Edited by Terry W. Moody

PROSTAGLANDINS, LEUKOTRIENES, AND LIPOXINS
Biochemistry, Mechanism of Action, and Clinical Applications
Edited by J. Martyn Bailey

PROSTAGLANDINS, LEUKOTRIENES, LIPOXINS, AND PAF
Mechanism of Action, Molecular Biology, and Clinical Applications
Edited by J. Martyn Bailey

THYMIC HORMONES AND LYMPHOKINES
Basic Chemistry and Clinical Applications
Edited by Allan L. Goldstein

Growth Factors, Peptides, and Receptors

Edited by

Terry W. Moody
The George Washington University Medical Center
Washington, D.C.

SPRINGER SCIENCE+BUSINESS MEDIA, LLC

Library of Congress Cataloging-in-Publication Data

International Washington Spring Symposium (12th : 1992 : George
  Washington University)
    Growth factors, peptides, and receptors / edited by Terry W.
  Moody.
       p.   cm. -- (GWUMC Department of Biochemistry annual spring
  symposia)
    "Proceedings of the Twelfth Washington International Spring
  Symposium at the George Washington University, held June 1-5, 1992,
  in Washington, D.C."--T.p. verso.
    Includes bibliographical references and index.
    ISBN 978-1-4613-6232-6   ISBN 978-1-4615-2846-3 (eBook)
    DOI 10.1007/978-1-4615-2846-3
    1. Growth factors--Congresses. 2. Growth factors--Receptors-
  -Congresses. 3. Peptide hormones--Congresses. 4. Peptide hormones-
  -Receptors--Congresses. 5. Second messengers (Biochemistry)-
  -Congresses.   I. Moody, Terry W. II. Title. III. Series.
  QP552.G76I58 1992
  599'.0192456--dc20
                                                              93-18563
                                                                  CIP

Proceedings of the Twelfth Washington International Spring Symposium
at The George Washington University, held June 1–5, 1992,
in Washington, D.C.

ISBN 978-1-4613-6232-6

© 1993 Springer Science+Business Media New York
Originally published by Plenum Press in 1993
Softcover reprint of the hardcover 1st edition 1993

PREFACE

The Twelfth Annual Washington Spring Symposium on Health Sciences attracted over 300 scientists from 20 countries. It was held at the Lisner Auditorium of the George Washington University in Washington, D.C. during June 1-5, 1992. The theme of the meeting was "Growth Factors, Peptides, and Receptors," and speakers emphasized both basic and clinical research in these areas.

The seven plenary sessions emphasized Peptides, Growth Factors, Peptide Receptors, Growth Factor Receptors, Second Messengers, Proliferation, and Clinical Correlations. The chapters in this volume are derived from each of these scientific sessions plus the poster and special sessions.

The Abraham White Distinguished Scientist Award was presented to Dr. Solomon H. Snyder for his numerous contributions to the field of neurochemistry. He presented the keynote address "Nitric Oxide: A Novel Neuronal Messenger." Dr. Snyder discussed the pathway of nitric oxide (NO) synthesis by the enzyme NO synthase. Released NO may be responsible for the neuronal toxicity associated with NMDA, an excitatory amino acid analogue. Dr. Snyder noted that NO may be the first of a new class of transmitters, with carbon monoxide being another candidate.

The Distinguished Public Service Award was presented to Senator Fritz Hollings in recognition of his leadership and outstanding achievements in the United States Senate and for his legislative support for biomedical research and education. In the symposium banquet address, Senator Hollings stressed the need for continued support of research to combat serious diseases such as cancer.

The field of peptide and growth factor research has been active for over three decades. It has been richly endowed with Nobel Laureates including Drs. Rita Levi Montalcini and Stanley Cohen (nerve growth factor isolation), Vincent DuVigneaud and Bruce Merrifield (peptide synthesis), Rosalyn Yalow (peptide radioimmunoassay), Roger Guillemin (TRH isolation), and Andrew Schally (somatostatin isolation). The symposium emphasized that important research contributions have been and continue to be made in peptides, growth factors, and receptors.

In its infancy, the field was concerned with the isolation and sequencing of peptides and growth factors. Radioimmunoassays and bioassays were developed to assay for biologically active peptides and growth factors. Subsequently, receptor binding and second messenger assays were developed to estimate the potency of various peptide and growth factor analogues. This resulted in the definition of both receptor agonists and antagonists. With the advance of molecular biology, genes for the growth factors and peptides were cloned. Agents were defined which alter peptide and growth factor gene expression. More recently, peptide and growth factor receptors have been cloned and receptor subtypes identified.

Peptides and growth factors are synthesized in the form of high-molecular-weight precursor proteins, and thiol proteases as well as enzymes with cathepsin D and subtilisin-like activity regulate posttranslational processing. The processed peptides and growth factors affect a multitude of biological processes. Endothelin is a potent vasoconstictor, but VIP, in contrast, is a potent vasodilator. FGF, which binds to receptors and can be translocated to the nucleus, causes angiogenesis and wound repair. Transforming growth factor β causes cellular differentiation reducing the growth of many cells. NGF, which binds to receptors and causes neurite outgrowth, increases cholinergic parameters such as choline uptake, acetylcholine synthesis, and choline acetyltransferase activity; cholinergic function is deficient in Alzheimer's disease. It may be possible to deliver growth factors such as NGF to the brain using genetic engineering techniques.

Peptide receptors, such as bombesin/gastrin releasing peptide and CCK, were cloned and found to be linked to G-proteins. The receptors contain approximately 400 amino acids and 7 transmembrane domains. The third cytosolic loop interacts with G-proteins, and bombesin/GRP as well as cholecystokinin cause phosphatidylinositol turnover. The inositol-1,4,5-trisphosphate released elevates cytosolic calcium whereas the diacylglycerol released activates protein kinase C. In contrast, VIP elevates cAMP and somatostatin inhibits the increase in cAMP caused by VIP. Potent receptor antagonists have been synthesized for bombesin/GRP, CCK, and VIP receptors. The EGF receptor is a large 170-kdalton protein which crosses the membrane once and contains a tyrosine kinase. When EGF binds to the extracellular domains, tyrosine kinase activity is stimulated causing phosphorylation of protein substrates and growth. Monoclonal antibodies have been elicited and found to function as EGF receptor antagonists. Also, EGF-toxin conjugates, which may serve as antitumor agents, have been defined.

TGFα, which interacts with the EGF receptor, causes proliferative changes in transgenic mice. EGF receptors are abundant in several cancers including breast, gastric, and non-small-cell lung cancer. Somatostatin receptors are present in several cancers including gastrointestinal, pituitary, medullary thyroid, and small-cell lung cancer. The somatostatin receptor is associated with tyrosine phosphatase activity resulting in the dephosphorylation of protein substrates. Potent somatostatin agonists such as octreotide can be radiolabeled and utilized to image tumors. In the future, knowledge obtained about peptides and growth factors may be important in the development of new cancer therapies.

The Washington Spring Symposium provided an international forum for the exchange of new ideas. This preface highlights only a few of the scientifically significant and exciting research results that were presented. This volume will be of interest to neuroscientists, gastroenterologists, psychiatrists, oncologists, and others interested in acquiring state-of-the-art information about peptides, growth factors, and receptors.

Terry W. Moody

CONTENTS

Part I—Peptides

Part II—Growth Factors

Part IV—Growth Factor Receptors

Part V—Second Messengers

Part I
Peptides

DISTRIBUTION OF ENDOTHELIN, A PUTATIVE GROWTH PROMOTING PEPTIDE, USING MODERN MICROSCOPICAL IMAGING METHODS

Julia M. Polak and Giorgio Terenghi

Department of Histochemistry
Royal Postgraduate Medical School
DuCane Road
London W12 0NN

INTRODUCTION

The concept that endothelial cells in culture can produce a vasoconstrictor substance and secrete it into their medium was followed by the remarkable discovery in 1988 by Yanagisawa and colleagues [1] of a very potent vasoconstrictory peptide which they termed endothelin. This original work is an example of scientific excellence, combining molecular biology, cellular biology and pharmacological expertise to demonstrate that this active peptide is composed of 21 amino acids with two disulfide bonds, derives from a larger precursor termed preproendothelin and produces very marked and sustained vasoconstriction. This initial discovery led to a large number of further publications dealing with the identification of several endothelin peptide isoforms and the genes that encode them, regional variations in endothelin distribution, synthetic pathways for endothelins, multiple endothelin receptor subtypes and their genes, diverse endothelin receptor-coupled signal transduction pathways, numerous physiological effect of endothelins and alterations in endothelin levels associated with disease [2].

The endothelins include at least three isopeptides; endothelin-1, -2, -3, each of which is derived from a separate gene. In humans, endothelin-1 mRNA codes for a 202-amino acid (preproendothelin-1) which undergoes proteolytic cleavage at 38-amino acid residue to generate pre- or "big-endothelin-1" through the action of a putative endothelin-converting enzyme. The endothelins are structurally and functionally homologous to another mammalian peptide, vasoactive intestinal contractor, and to sarafotoxins found in the venom of the burrowing asp, *Atractaspis engaddensis*. Endothelin-1 is produced by endothelial cells while endothelin-3 is likely to be produced by neural and pulmonary tissue. The site of production of endothelin-2 is still uncertain, but is likely to be the kidney. The existence of receptor subtypes for endothelins has been recognized for some time and two distinct types of receptors have been cloned, one exhibiting high affinity for endothelin-1 and low affinity for endothelin-3, designated as endothelin$_A$ receptor and the other displaying high affinity for both endothelin-1 and -3, classified as the endothelin$_B$ receptor subtype [3]. The selective endothelin$_A$ receptor which is expressed in vascular smooth muscle is thought to mediate vasoconstriction, whereas the non-selective endothelin$_B$ receptor subtype located in endothelial cells is thought to mediate vasodilation, via the release of nitric oxide.

Growth Factors, Peptides, and Receptors, Edited
by T.W. Moody, Plenum Press, New York, 1993

The endothelin literature is vast, comprising numerous original publications as well as extensive and detailed review articles. This contribution will deal with the distribution of endothelin immunoreactivity, of the mRNA and of receptor binding sites in a variety of tissues. The techniques that we have used include immunocytochemistry, *in vitro* receptor autoradiography with image analysis and *in situ* hybridization.

DISTRIBUTION

a) Central Nervous System and Hypothalamus

After the discovery of endothelin, our studies as well as those of others concentrated on the distribution of endothelin immunoreactivity, mRNA and receptor-binding sites. While investigating the latter, we were struck by the strong binding of endothelin to neural tissue and hence, we asked the question whether endothelin may indeed be produced by neural tissue. Our hypothesis was supported by neurophysiological findings demonstrating modulatory actions of endothelin on the spinal cord and direct effects on neuronal excitability[4]. Our first investigations focused on the localization of endothelin mRNA and endothelin-like immunoreactivity in samples of neurologically normal nervous system from adult humans using the techniques of *in situ* hybridization and immunocytochemistry[5-6]. We were able to demonstrate the mRNA and translated protein in small and large diameter neurones of the dorsal root ganglion as well as in neurones of the spinal cord (lamina IV-VI) and in many motor neurones (Fig 1). Co-localization studies demonstrated that endothelin transcripts were localized to all cells expressing ß-preprotachykinin, most of which also expressed CGRP mRNA. In the majority of the motoneurones, coexistence of endothelin and CGRP mRNA was also noted. In the human brain (Fig 2), endothelin mRNA and endothelin immunoreactivity were found to be present in cell bodies in laminae III and IV of the parietal, temporal and frontal cortex. Labelled cells were seen in many nuclei of the hypothalamus as well as the medulla, hippocampus, substantia nigra, Purkinje cell layer of the cerebellum and the dorsal motor nucleus of the vagus nerve. Co-localization studies again revealed coexistence with immunoreactivity for neuropeptides including NPY and, in the hypothalamus, neurophysin and galanin. We were also able to demonstrate the presence of these peptides in the posterior pituitary of the pig[7] and, subsequently, Bloom and colleagues demonstrated similar findings in the human hypothalamic-pituitary axis[8]. The presence of endothelin immunoreactivity and mRNA in the brain, hypothalamus and pituitary

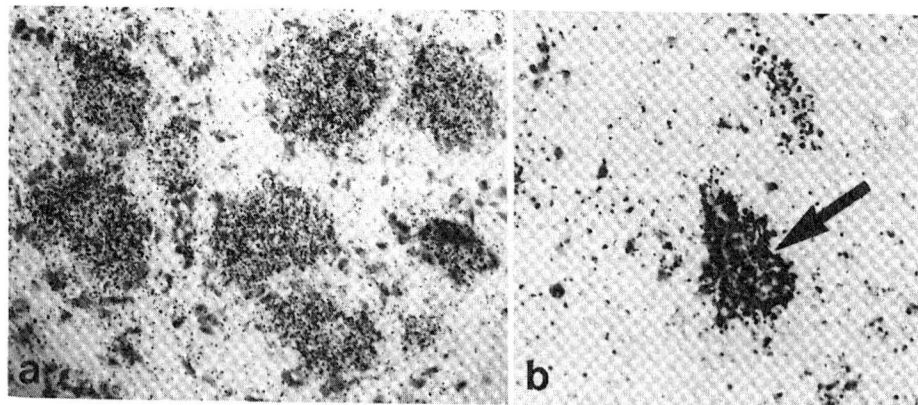

Figure 1. *In situ* hybridization for ET mRNA in (A) dorsal root ganglia cells, and (B) motoneurons in ventral horn of human lumbar spinal cord (A x 400, B x 340).

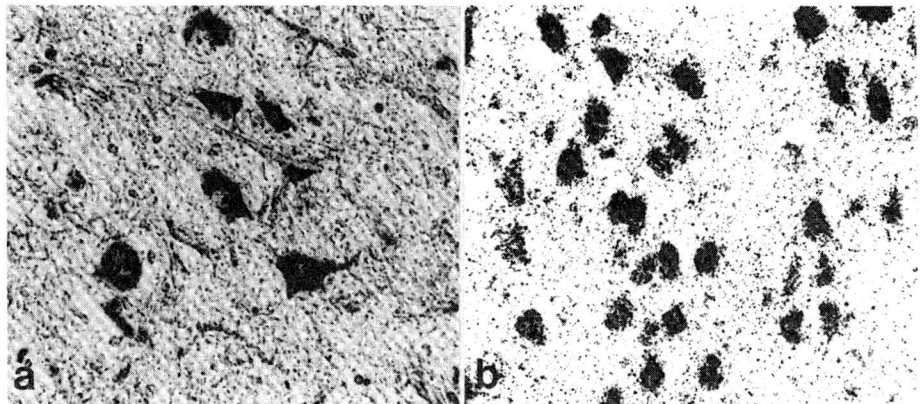

Figure 2. Cortex of human brain showing (A) ET immunoreactivity, and (B) mRNA for ET in several neuronal cells (x 250).

gland of different mammalian species is well accepted and supported by numerous subsequent publications as well as by demonstrations of many pharmacological and physiological actions of the endothelins in the nervous and endocrine systems[9-13]

b) Peripheral Nervous System

General consensus indicates that endothelin is constitutively produced and ?released from the cell of origin and hence, immunocytochemical localization in peripheral nerve terminals is unlikely unless the peptide is stored in considerable quantity. However, endothelin immunoreactivity can be demonstrated convincingly in the enteric nervous system of man and animals. We have been able to show the presence of endothelin immunoreactivity (Fig 3) and binding-sites (Fig 4) in the human adult and developing gastrointestinal tract[14-15].
By immunocytochemistry, we were able to localize endothelin to ganglion cells of the submucous and myenteric plexus and scattered nerves present, in particular, in the muscle layers. Coexistence studies demonstrated the presence of endothelin

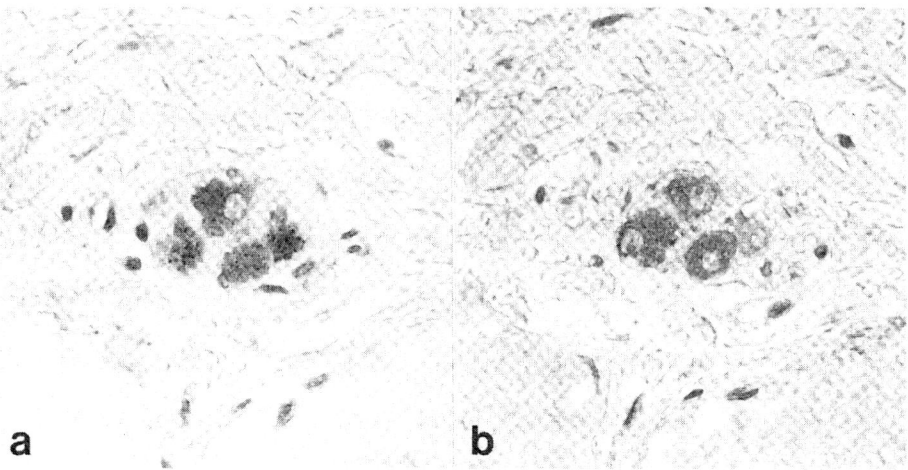

Figure 3. Serial sections of human colon showing a) endothelin immunoreactivity and b) vasoactive intestinal peptide immunostained in the same cells of the submucous plexus. (x 280)

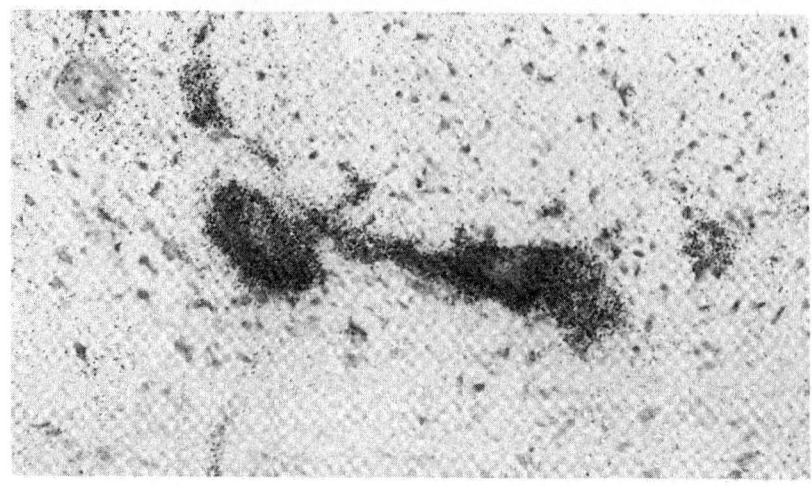

Figure 4. *In vitro* autoradiography of a section of human colon incubated with ^{125}I-endothelin-1 showing the myenteric plexus. Binding sites are visualized as silver grains in a covering layer of photographic emulsion. The binding is dense on the two central ganglion cells and weaker on the surrounding muscle layers. (x 440)

immunoreactivity and vasoactive intestinal polypeptide (VIP) in one and the same ganglion cells, in particular in the submucous plexus.

Investigations of endothelin-binding sites in the human bowel demonstrated high affinity binding sites for endothelin-1, in particular in the myenteric plexus, and lower affinity binding in the submucous plexus, mucosa, muscle layers and blood vessel walls. In the colonic myenteric plexusm, kinetic binding values were similar to those reported in the brain-stem and in porcine coronary arteries. The same pattern of endothelin immunoreactivity and binding sites was present throughout the gastrointestinal tract and was found in the fetal gut as early as 11 weeks (big-endothelin). Conversion to the 21-amino acid peptide endothelin-1 took place during late development (32 week gestation onwards). These studies provide anatomical support for numerous physiological studies demonstrating modulatory actions of endothelin in human and animal gastrointestinal tract[16].

c) Cardiovascular System

It is now widely accepted that endothelin-1 is the isoform produced by endothelial cells. Endothelin immunoreactivity, mRNA and binding sites have been demonstrated in the developing and adult human heart[17]. Our group originally determined the localization of specific, high affinity endothelin-1 binding sites in the human adult cardiovascular system[18-19] and other workers have demonstrated mRNA for endothelin-1 in the fetal atrium and ventricle[20] and in adolescent ventricular tissue, the latter by PCR techniques[21]. Our group subsequently demonstrated by *in vitro* autoradiography and immunohistochemistry, the localization of both endothelin binding sites and endothelin-like immunoreactivity in the human fetal heart. We demonstrated a tissue-specific localization of endothelin-1 and -3 with regional variations in both ligand binding density and affinity (Fig 5). High affinity endothelin-1 binding sites showed a tissue-specific distribution pattern with high density binding to the atria, ventricles and cardiac valve cusps. Specific, high density binding for endothelin-3 was also exhibited on valve cusps with lower density for this isoform being displayed in atria and ventricles. Microautoradiography examination demonstrated binding sites on the wall of the aorta, pulmonary and coronary arteries, myocardium, ventricular conduction system,

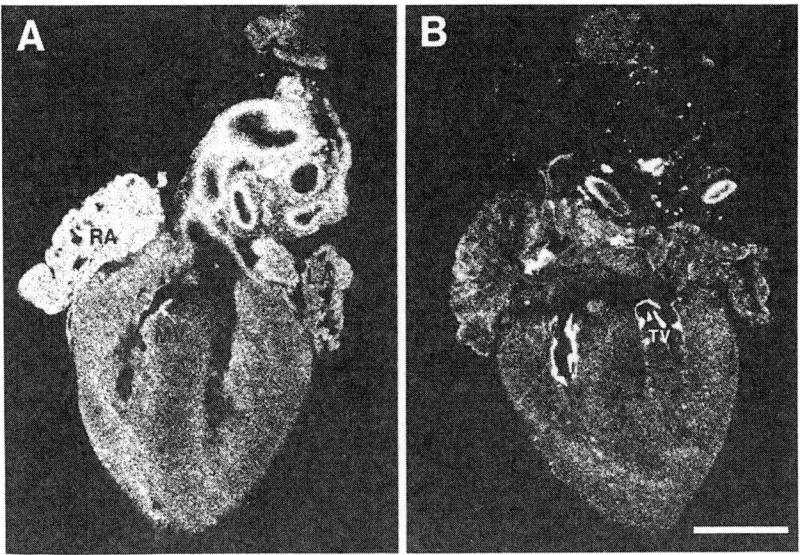

Figure 5. Dark-field images of autoradiographs demonstrating total ^{125}I-ET1 (A) and ^{125}I-ET3 (B) binding to adjacent sections of a human fetal heart. RA, right atrium; LA, left atrium; TV, tricuspid value; mv, mitral value. Bar = 2 mm.

endocardium and endothelial lining of valve cusps. Our studies are therefore indicative of the presence of receptor subpopulations in the human fetal heart. Immunoreactivity for endothelin was localized to a heterogeneous population of endothelial (Fig 6), endocardial and epicardial mesothelial cells.

The concordant distribution of specific binding sites and endothelin-immunoreactive cells suggest that endothelin may have a trophic role in the developing heart, acting locally in an autocrine or paracrine manner. Our studies give support to the reported actions of endothelin on the cardiovascular system, including positive inotrophic effects on myocardium, stimulation of c-fos and c-myc proto-oncogene expression, cardiac myocyte hypertrophy, atrial natriuretic peptide release and induction of fibroblast, vascular smooth muscle and endothelial cell proliferation [22].

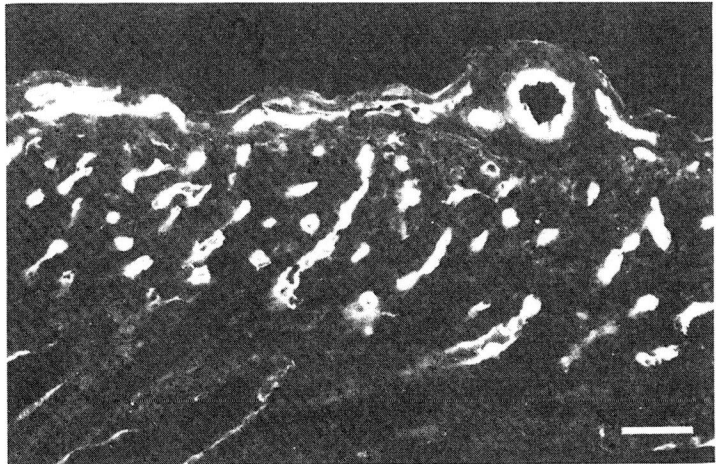

Figure 6. Immunofluorescence micrograph demonstrating the localization of endothelin-like immunoreactivity in human fetal coronary vessels. Bar = 50 μm.

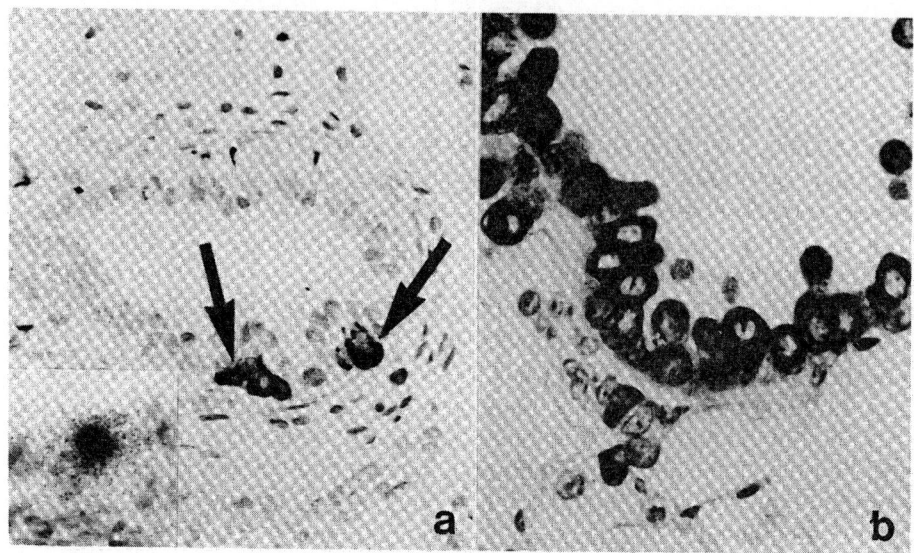

Figure 7. Immunoreactivity for endothelin in (a) endocrine cells (arrows) of human fetal lung and (b) bronchiolar epithelial cells in rat lung. A strong signal for *in situ* hybridization of endothelin mRNA to confirm endothelin synthesis in an endocrine cell in human fetal lung is shown in the inset in (a). (A x 240, B x 500)

d) Endothelin in Endocrine Cells: Respiratory Tract

Endothelin immunoreactivity is abundant in the respiratory tract and the localization is primarily non-endothelial. Endothelin immunoreactivity and mRNA is found in characteristic endocrine cells of the airway epithelium of the human respiratory tract [23] and in all cells of the airway epithelium in rats and mice [24] (Fig 7). This localization is interesting in view of the finding of relatively abundant, specific binding sites, not only in vascular smooth muscle, but also in the smooth muscle of airways [25] (Fig 8). Endothelin is known to exert potent vaso- and bronchoconstrictory properties [26]. In view of these findings and of the potent vasoconstrictory properties of endothelin, we asked the question whether endothelin

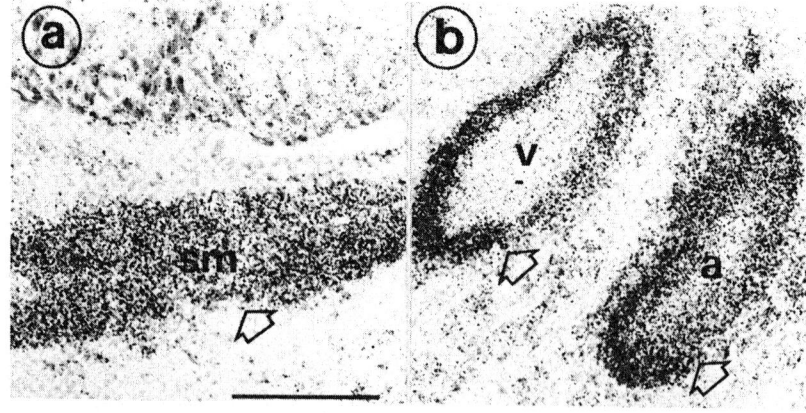

Figure 8. Autoradiogram to show strong binding of ^{125}I-endothelin-1 to smooth muscle (sm) in (a) bronchial and (b) vascular smooth muscle (A, artery; V, vein) in a section of adult human lung. (x 250)

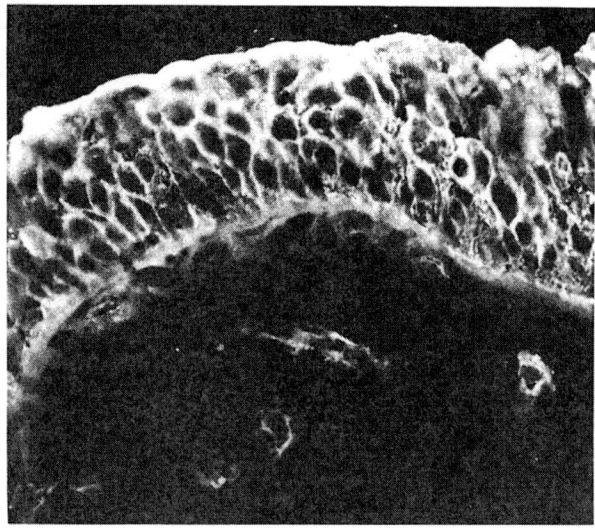

Figure 9. Endothelin immunoreactivity in bronchial epithelium in an endoscopic biopsy from an asthmatic patient. (x 450)

may be abnormally expressed in asthmatic patients. Indeed, this was the case[27]. In a series of endoscopic bronchial biopsies of mildly to moderately asthmatic patients, endothelin expression was dramatically upregulated and was found to be present in the entire airway epithelium, and not only confined to scattered epithelial cells as seen in human non-asthmatic airways (Fig 9). Whether enhanced endothelin expression occurs at the onset of the development of asthma or whether it is involved in tissue damage and repair remains to be elucidated. Interestingly, elevated levels of endothelin immunoreactivity have been reported recently in bronchoalveolar lavage of asthmatic patients[28]. The specific localization of endothelin immunoreactivity and binding sites in the respiratory tract of man and

Figure 10. ET-1-like immunoreactivity in endothelial cells of capillaries in the subepidermal papillae; indirect immunofluorescence technique. (x 325).

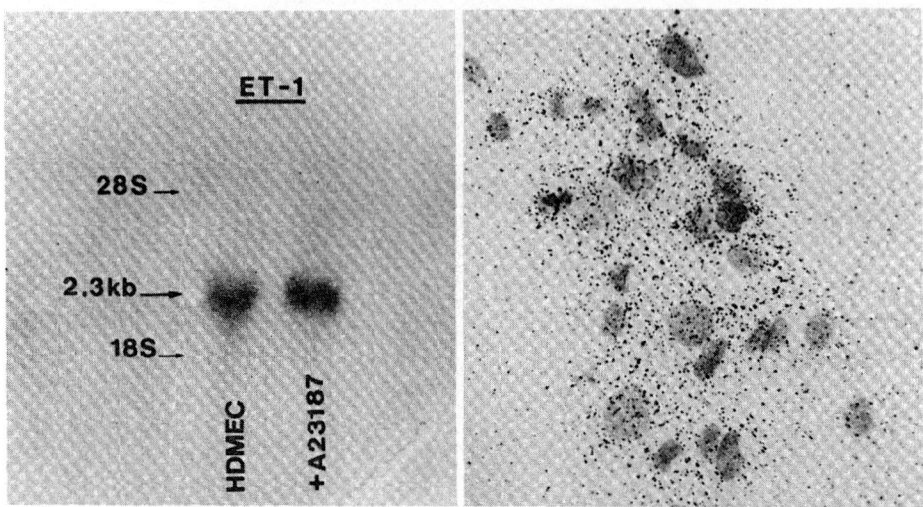

Figure 11. Northern blot analysis of unstimulated and A23187 ionophore stimulated HDMEC cells. Both extracts showed a single hybridization band corresponding to ET-1 mRNA.

Figure 12. ET-1 mRNA hybridization signal in human dermal microvessel endothelial culture (HDMEC) cells, using a ^{32}P-labelled probe. Emulsion dipped preparations, haematoxylin counterstaining (x 540).

mammals led us to investigate the presence of endothelin immunoreactivity and mRNA in a variety of pulmonary tumours[29]. Endothelin was found not only in endocrine tumours, but also in non-endocrine tumours of the respiratory tract including squamous cell carcinomas and adenocarcinomas. Specific binding studies in these tumours revealed that, contrary to expectations, binding sites were found not on tumour cells, but specifically on the wall of newly formed intratumour blood

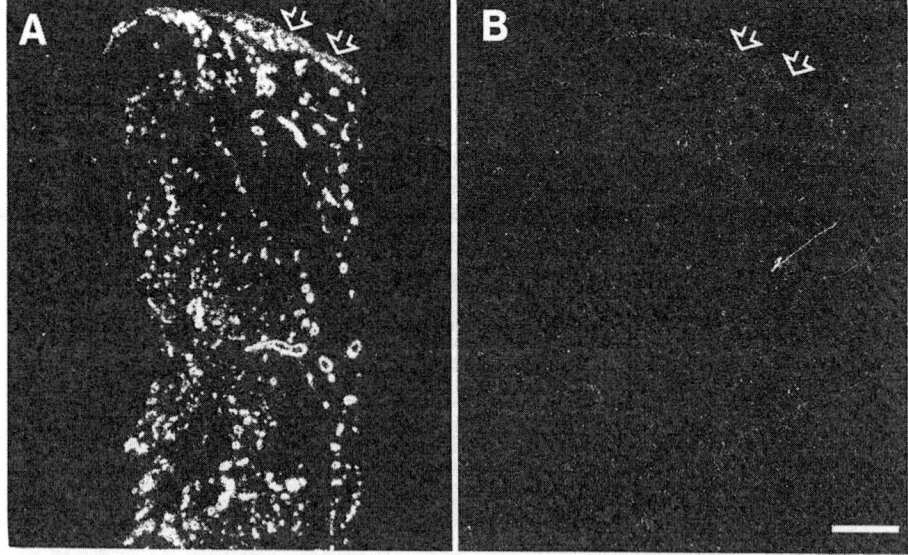

Figure 13. Dark-field images of autoradiographs demonstrating total (A) and non-specific ^{125}I-ET1 binding (B) to adjacent sections of osteoarthritic synovium. Arrows, synovial surface. Bar = 1 mm.

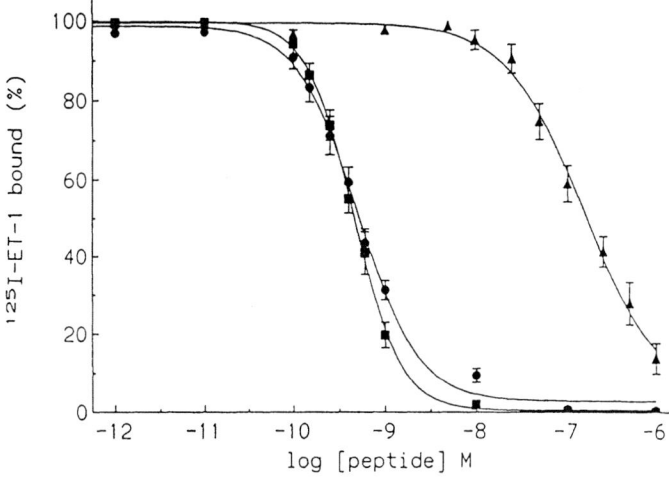

Figure 14. Competitive inhibition of specific ^{125}I-ET1 binding to blood vessels in rheumatoid synovial tissues by unlabelled ET1 (■), ET2 (•) and ET3 (▲), expressed as a percentage of ^{125}I-ET1 binding in the absence of unlabelled ± SEM of specific binding for 5 separate cases.

vessels. These studies suggest a paracrine loop whereby endothelin is produced and released from malignant tumour cells to activate receptors present in the wall of newly formed blood vessels. These studies are interesting and in keeping with the putative angiogenic and growth promoting properties of endothelin.

e) Endothelin in the Microvasculature

Endothelin immunoreactivity, mRNA and binding sites can be found in microvascular beds. We have undertaken a systematic study of human dermal [30] and synovial [31] microvessels. Endothelin immunoreactivity is found in endothelial cells of dermal (Fig 10) and synovial microvessels, the mRNA can also be demonstrated both by Northern blotting (Fig 11) and *in situ* hybridization (Fig 12) of cultured dermal microvessels. Specific binding sites are present on vascular smooth muscle and are particularly prominent in proliferating blood vessels during inflammation, as demonstrated by our group in human rheumatoid arthritis. Elevated circulating levels of endothelin have been demonstrated in a variety of microangiopathies including Raynaud's phenomenon [32] and diabetes [33].

In keeping with these findings, we have been able to demonstrate increased immunoreactivity in dermal microvessels of diabetic patients [34]. Inflammatory diseases such as rheumatoid arthritis also show changes in endothelin immunoreactivity and binding sites [13-14]. Elevated levels of endothelin have recently been reported in synovial fluid [35] and in the plasma of patients with rheumatoid arthritis [36].

CONCLUSIONS

The discovery of endothelin caused an almost unprecedented interest and the wealth of information achieved since its discovery, just over 4 years ago is remarkable. Distribution studies show that endothelin is widely distributed not only in vascular but also in non-vascular tissue and its expression is abnormal in a variety of disease states. Distribution of mRNA, immunoreactivity and binding sites strongly support pharmacological, physiological and molecular data and the changes in disease parallel and explain many functional disturbances. This is the exciting beginning of an understanding of vascular and non-vascular endothelin physiopathology.

REFERENCES

1. M. Yanagisawa, H. Kurihara, S. Kimura, Y. Tomobe, M. Kobayashi, Y. Mitsui, Y. Yazaki, K.Goto and T. Masaki, A novel potent vasoconstrictor peptide produced by vascular endothelial cells. *Nature* 332:411-415(1988).
2. G.M. Rubanyi, "Endothelin," Oxford University Press, New York, Oxford (1992).
3. D.J. Webb, Endothelin receptors cloned, endothelin converting enzyme characterized and pathophysiological roles for endothelin proposed. *TIPS.* 12:43-46(1991).
4. T. Yoshizawa, S. Kimura, I. Kanazawa, Y. Uchiyama, M. Yangisawa, T. Masaki, Endothelin localizes in the dorsal horn and acts on the spinal neurones: possible involvement of dihydropyridine-sensitive calcium channels and substance P release. *Neurosci Letts.* 102:179-184 (1989).
5. A. Giaid, S.J. Gibson, N.B.N.Ibrahim, S. Legon, S.R. Bloom, M. Yanagisawa, T. Masaki, I.M. Varndell and J.M. Polak, Endothelin 1, an endothelium-derived peptide, is expressed in neurons of the human spinal cord and dorsal root ganglia, *Proc Natl Acad Sci USA.* 86:7634-7638(1989).
6. A. Giaid, S.J. Gibson, M.T. Herrero, S. Gentleman, S. Legon, M. Yanagisawa, T. Masaki, N.B.N. Ibrahim, G.W. Roberts, M.L.Rossi and J.M. Polak, Topographical localisation of endothelin mRNA and peptide immunoreactivity in neurones of the human brain, *Histochemistry* 95:303-314(1991).
7. T. Yoshizawa, O. Shinmi, A. Giaid, M. Yangisawa, S.J. Gibson, S. Kimura, Y. Uchiyama, J.M. Polak, T. Masaki and I. Kanazawa, Endothelin: A novel peptide in the posterior pituitary system, *Science* 247:462-464(1991).
8. K. Takahashi, M.A. Ghatei, P.M. Jones, J.K. Murphy, H-C. Lam, D.J. O'Halloran, S.R. Bloom, Endothelin in human brain and pituitary gland: Presence of immunoreactive endothelin, endothelin messenger ribonucleic acid, and endothelin receptors, *J Clin Endocrinol & Metab.* 72:693-699(1991).
9. H. Matsumoto, N. Suzuki, H. Onda, M. Fujino, Abundance of endothelin-3 in rat intestine, pituitary gland and brain, *Biochem. & Biophys. Res. Comm.* 164:74-80(1989).
10. O. Shinmi, S. Kimura, T. Yoshizawa, T. Sawamura, Y. Uchiyama, Y. Sugita, I. Kanazawa, M. Yanagisawa, K. Goto, T. Masaki, Presence of endothelin-1 in porcine spinal cord: isolation and sequence determination, *Biochem & Biophys Res Comm.* 162:340-346(1989).
11. M.W. MacCumber, C.A. Ross, B.M. Glaser, S.H. Snyder, Endothelin: Visualization of mRNAs by *in situ* hybridization provides evidence for local action, *Proc Natl Acad Sci USA.* 86:7285-7289(1989).
12. M-E. Lee, S.M. de la Monte, S-C. Ng, K.D. Bloch, T. Quertermous, Expression of the potent vasoconstrictor endothelin in the human central nervous system, *J. Clin. Invest.* 86:141-147 (1990).
13. F. Fuxe, A. Cintra, B. Andbjer, E. Anggard, M. Goldstein, L.F. Agnati, Centrally administrated endothelin-1 produces lesions in the brain of the male rat, *Acta Physiol Scand.* 137:155-156 (1989).
14. C. Escrig, A.E. Bishop, H. Inagaki, G. Moscoso, K. Takahashi, I.M. Varndell, M.A. Ghatei, S.R. Bloom, J.M. Polak, Localisation of endothelin like immunoreactivity in adult and developing human gut, *Gut* 33:212-217(1992).
15. H. Inagaki, A.E. Bishop, C. Escrig, J. Wharton, T.G. Allen-Mersch and J.M. Polak, Localization of endothelin like immunoreactivity and endothelin binding sites in human colon, *Gastroenterol.* 101:47-54(1991).
16. N. Ishida, K. Tsukioka, M. Tomoi, K. Saida and Y. Mitsui, Differential activities of two distinct endothelin family peptides on ileum and coronary artery. *FEBS Lett.* 247:337-340(1989).
17. J. Wharton, R.A.D. Rutherford, L. Gordon, G. Moscoso, I. Scheimberg, J.A.R. Gaer, K.M. Taylor and J.M. Polak, Localization of endothelin binding sites and endothelin-like immunoreactivity in human fetal heart, *J Cardiovasc Pharmacol.* 17 (Suppl.7):S378-S384 (1991).
18. R.F. Power, J. Wharton, S.P. Salas, S. Kanse, M.A. Ghatei, S.R. Bloom and J.M. Polak, Autoradiographic localization of endothelin binding sites in human and porcine coronary arteries, *Eur J Pharmacol.* 160:199-200(1989).
19. R.F. Power, J. Wharton, S.R. Bloom and J.M. Polak, Autoradiographic localization of endothelin binding-sites in the cardiovascular and respiratory systems, *J Cardiovasc Pharmacol.* 63:550-556(1989).
20. K.D. Bloch, R.L. Eddy, T.B. Shows and T. Quertermous, cDNA cloning and chromosomal assignment of the gene encoding endothelin 3, *J Biol Chem.* 264:18156-61(1989).
21. D.J. Nunez, M.J. Brown, A.P. Davenport, C.B. Neylon, J.P. Schofield and R. Wyse, Endothelin-1 mRNA is widely expressed in porcine and human tissues, *J Clin Invest.* 85:1537-41 (1990).

22. V.J. Dzau, R.E. Pratt and J.P. Cooke, Endothelin as a growth factor in vascular remodeling and vascular disease, in: "Endothelin," G.M. Rubanyi ed., Oxford University Press, New York, Oxford (1992).
23. A. Giaid, J.M. Polak, V. Gaitonde, Q.A. Hamid, G. Moscoso, S. Legon, D. Uwanogho, M. Roncalli, O. Shinmi, T. Sawamura, S. Kimura, M. Yangisawa, T. Masaki and D.R. Springall, Distribution of endothelin-like immunoreactivity and mRNA in the developing and adult human lung. *Am J Respir Cell Mol Biol.* 4:50-58 (1991).
24. N. Rozengurt, D.R. Springall and J.M. Polak, Localization of endothelin-like immunoreactivity in airway epithelium of rats and mice, *J Pathol.* 160:5-8 (1990).
25. S.M. Kanse, M.A. Ghatei and S.R. Bloom, Endothelin binding sites in porcine aortic and rat lung membranes, *Eur J Biochem.* 182:175-179 (1989).
26. J.G. Filep, The endothelin family: novel mediators of bronchopulmonary diseases, *Drug of Today* 27:493-502 (1991).
27. D.R. Springall, P.H. Howarth, H. Counihan, R. Djukanovic, S.T. Holgate and J.M. Polak, Endothelin immunoreactivity of airway epithelium in asthmatic patients, *Lancet* 697-702 (1991).
28. S. Mattoli, M. Soloperto, M. Marini and A. Fasoli, Levels of endothelin in the bronchoalveolar lavage fluid of patients with symptomatic asthma and reversible airflow obstruction, *J Allergy Clin Immunol.* 88:376-384 (1991).
29. A. Giaid, Q.A. Hamid, D.R. Springall, M. Yangisawa, O. Shinmi, T. Sawamura, T. Masaki, S. Kimura, B. Corrin and J.M. Polak, Detection of endothelin immunoreactivity and mRNA in pulmonary tumours. *J Pathol.* 162:15-22 (1990).
30. G. Terenghi, H.A. Bull, C.B. Bunker, D.R. Springall, Y. Zhao, J. Wharton, P.M. Dowd and J.M. Polak, Endothelin-1 in human skin: immunohistochemical, receptor binding, and functional studies. *J Cardiovasc Pharmacol.* 17 (Suppl. 7):S467-S470 (1991).
31. J. Wharton, R.A.D. Rutherford, D.A. Walsh, P.I. Mapp, G.A. Knock, D.R. Blake and J.M. Polak, Autoradiographic localization and analysis of endothelin-1 binding sites in human synovial tissue, *Arthrit & Rheum.* (1992) in press.
32. K. Kanno, Y. Hirata, M. Shichiri, K. Emori, K. Ohta, N. Miyasaka and F. Marumo, Circulating endothelin in patients with Raynaud's phenomenon and collagen diseases, *J Vasc Med Biol.* 2:184 (1990).
33. K. Takahashi, M.A. Ghatei, H.C. Lam, D.J. O'Halloran and S.R. Bloom, Elevated plasma endothelin in patients with diabetes mellitus, *Diabetologia* 33:306-310 (1990).
34. G. Properzi, G. Terenghi, X.H. Gu, G. Poccia, R. Pasqua, S. Francavilla and J.M. Polak, Distribution of endothelin 1 and von Willebrand factor immunoreactivities in human diabetic skin. Correlation between quantitative immunohistochemistry and microangiopathy, (in preparation).
35. A.M. Nahir, A. Hoffman, M. Lorber and H.R. Keiser, Presence of immunoreactive endothelin in synovial fluid: Analysis of 22 cases, *J Rheumatol.* 18:678-680, (1991).
36. K. Kanno, Y. Hirata, M. Shirchiri, K. Emori, K. Ohta, N. Miyasaka and F. Marumo, Circulating endothelin in patients with Raynaud's phenomenon and collagen diseases, *J Vasc Med Biol.* 2:184, (1990).

VASOACTIVE INTESTINAL PEPTIDE: FROM MOLECULAR GENETICS TO NEUROTROPISM

Illana Gozes[1] and Douglas E. Brenneman[2]

[1]Department of Chemical Pathology, Sackler School of Medicine
Tel Aviv University, Tel Aviv, Israel
[2]Section on Molecular and Developmental Pharmacology
Laboratory of Developmental Neurobiology, NICHD, NIH
Bethesda, MD 20892, USA

INTRODUCTION

During development of the nervous system a large population of neurons die. It is hypothesized that this process is regulated by neuropeptides and growth factors, such as vasoactive intestinal peptide (VIP; Gozes and Brenneman, 1989). VIP is a major regulatory peptide in the mammalian brain. This 28-amino acid peptide exhibits neurotransmitter and hormonal roles and is involved in brain activity, neuroendocrine functions, cardiac activity, respiration, digestion and sexual potency. VIP gene expression in rat brain increases about 40-fold from birth to 14 days (Gozes et al., 1987), and decreases with maturation and aging (Gozes et al., 1988). Innervation controls neuropeptide expression, and lesions of the optic nerve during development resulted in changes in brain VIP mRNA (Levy Holtzman et al., 1989). Also, in the SCN (suprachiasmatic nucleus), in which the circadian clock resides, day-night cycles cause rhythmical alteration in the levels of VIP mRNA (Gozes et al., 1989c).

To assess VIP biological function *in vivo* and *in vitro* we have developed novel VIP antagonists e.g. a VIP-neurotensin hybrid antagonist (Gozes et al., 1989) which can differentiate between central and peripheral VIP receptors (Gozes et al., 1991a). This antagonist can induce neuronal cell death *in vitro* (Gozes et al, 1991) and inhibit the acquisition of milestones in the development of behavioral reflexes when chronically injected to developing rats (Hill et al., 1991). When administered intracerebrally to adult animals, this antagonist causes impairment in learning and memory acquisition (Glowa et al., 1991). The mechanism by which VIP-antagonists induce neuronal impairment may be mediated via novel growth factors associated with VIP (Brenneman et al., 1987; Brenneman, 1988, 1990; Brenneman et al., 1990; Brenneman et al., 1992).

In this short review we shall try to describe our recent studies on VIP gene expression, VIP receptor utilization and neurotropism: mitogenic properties and a secretagogue for neurotrophic factors.

MOLECULAR GENETICS

Studies on the VIP gene (Bodner et al., 1985; Gozes et al., 1986; Giladi et al., 1990) have indicated a developmentally regulated neuronal pattern of expression (Gozes, 1987, 1988; Gozes et al., 1987, 1988, Baldino et al., 1989) which may be associated with

Growth Factors, Peptides, and Receptors, Edited
by T.W. Moody, Plenum Press, New York, 1993

modulation of neuronal survival (Gozes and Brenneman, 1989). Immunoreactive VIP changes markedly with brain development (Said, 1982, 1984; McGregor et al., 1982; Nobou et al., 1985). In parallel, VIP mRNA in the frontal rat cortex increases about 40-fold from birth to 14 day of age and at 30 day of age the VIP mRNA decrease to about 60% of the peak levels (Gozes et al., 1987; Gozes, 1988). During rat embryonic development, significant amounts of VIP were undetectable in the brain (Emson et al., 1979; McGregor et al., 1982). In contrast, in the duodenum, as early as day 16 of gestation, limited amounts of VIP were detected (Emson et al., 1979), this is correlated with transient expression of VIP mRNA at this embryonic stage (Gozes et al., 1988). Moreover, preliminary results indicate transient expression of VIP mRNA on embryonic day 15.5 in the developing rat superior cervical ganglia (Davidson et al., 1992). With aging, VIP mRNA decreases further in content, especially in the aging cerebral cortex (Gozes et al., 1988). Factors regulating VIP gene activity may be diverse, such as hormonal regulation (Gozes and Shani, 1986; Gozes et al.,1989a,b; Lam et al., 1990; Chew et al., 1991; Rostene et al., 1992). Some hormonal regulation may be mediated via cAMP and protein kinase C (Fink et al., 1991; Gozes and Brenneman, 1989). Recent studies have suggested that cyclic-AMP and phorbol ester-induced transcriptional activation are mediated by the same enhancer element in the human vasoactive intestinal peptide gene (Fink et al., 1991). The same enhancer is found in the rat VIP gene (Giladi et al., 1990), implying stringent evolutionary constraints. In vivo, VIP expression may be partially controlled by synaptic activity, as lesions of the optic nerve during development result in changes in VIP mRNA in the suprachiasmatic nucleus-SCN (Levy Holtzman et al., 1989). Thus, following enucleation of the new-born rat, when VIP mRNA is measured at 30 days of age, an increase is observed in the SCN, which contrasts a decrease in the cerebral cortex. These results indicate different regulation by light and synaptic activity in the various brain regions. Additionally, in the SCN, in which the circadian clock resides, day-night changes are associated with rhythmical alteration in the levels of VIP mRNA (Gozes et al., 1989c; Albers et al., 1990; Okamoto et al., 1991) and higher levels of VIP mRNA are observed at night, paralleling the higher VIP mRNA levels observed in blind rats (Gozes et al., 1989c, Levy Holtzman et al., 1989). The lesion studies and day night alteration suggest the involvement of electrical activity in the regulation of VIP gene expression in vivo. Interestingly, preliminary studies indicate day-night alteration in the VIP mRNA, in the SCN, prior to the establishment of synaptic contacts with the optic nerve (Bardea et al., 1992), suggesting additional regulatory mechanisms. The involvement of electrical activity in the regulation of VIP gene expressions relationship has been directly addressed by in vitro experiments where blockade of synaptic activity with tetrodotoxin resulted in a reversible decrease in VIP mRNA and VIP immunoreactivity in dissociated spinal cord cultures (Foster et al., 1989; Agoston et al., 1991). The sensitivity of VIP expression to electrical blockade in the spinal cord cultures is dependent on the stage of development; i.e., the loss of VIP mRNA after tetrodotoxin treatment is greatest during a period where there is rapid growth (Agoston et al., 1991; Brenneman and Nelson, 1986). Thus, VIP is not only released during electrical activity (Brenneman et al., 1985a), but synaptic activity is also an important determinant regulating the expression of this growth-promoting peptide.

A MODEL SYSTEM TO STUDY VIP AND NEUROTROPISM

When spinal cord neurons are subjected to in vitro conditions, 50% of the neurons die within a critical period of development in culture (Brenneman et al., 1985) and electrical blockade accelerates neuronal cell death (Brenneman et al., 1983) if endogenous conditioning substances are removed (Brenneman et al., 1984). This enhanced cell death can be prevented by VIP (Brenneman and Eiden, 1986) and, indeed, electrical blockade inhibits secretion and synthesis of endogenous VIP (Brenneman et al., 1985a; Brenneman et al., 1987; Agoston et al., 1991). For VIP to exert its survival promoting activity, it is obligatory to have non-neuronal cells in the culture system (Brenneman et al., 1987; Brenneman et al., 1990). Glia are thought to have supportive roles for developing neurons by providing directional cues (Wessells et al., 1980; Silver et al., 1982; Noble et al., 1984) and growth-stimulating substances (Banker 1980; Muller et al., 1984; Eagleson et al., 1985).

Using a novel VIP-neurotensin hybrid antagonist (Gozes et al., 1989), two VIP receptor populations were identified on glial cells (Gozes et al., 1991). One low affinity binding site (Kd 80 nM), associated with cAMP production, and one high affinity (Kd 50 pM) binding site, which is associated with the capacity to induce the secretion of neuronal survival factor(s) (Brenneman et al., 1987; Brenneman et al., 1990; Gozes and Brenneman 1989; Gozes et al., 1991; Brenneman et al., 1992; Nielsen et al., 1990). In contrast to glial cells, neuronal cultures (e.g. hippocampal cell cultures) display only the low affinity VIP receptor (Gozes et al., 1991). Additionally, lymphoid cells display a third type of VIP receptor, which is not recognized by the VIP-neurotensin hybrid antagonist (Gozes et al., 1991a). Recently, a functional VIP receptor was isolated from a rat lung cDNA library. This protein was suggested to be Gi coupled and, when stimulated by VIP, caused an adenylate cyclase stimulation. In situ hybridization revealed a wide distribution of the receptor mRNA in the cerebral cortex and hippocampus (Ishihara et al., 1992).

The cAMP-associated VIP receptor may be involved, in part, in the mitogenic action of VIP. This was described for embryonic neurons in the superior cervical ganglion (Pincus et al., 1990), as well as for lung cancer cells (Moody et al., 1992) where VIP may act as an autocrine regulator (Gozes et al., 1992).

However, the proliferative and neurotrophic actions of VIP may be more complex in that indirect cellular interactions appear to be involved. The idea is that VIP released from neurons may diffuse to adjoining cells and stimulate high affinity receptors on astroglia, which have been shown to be sites of neurotrophic factors production (Banker, 1980; Brenneman et al., 1987). Although not yet fully characterized, the substances shown to be released by VIP include a variety of developmentally significant substances. For example, an interleukin-1-like (IL-1-like) substance is among the glia -derived substances released by VIP (Brenneman et al., 1992). This cytokine, IL-1 has been shown to increase glial proliferation (Giulian et al., 1988). VIP does produce a small but potent increase in the number of GFAP (glial fibrillary acidic protein) positive cells in spinal cord cultures (Brenneman et al., 1990). In addition, comparison of the proteins released from non-stimulated glial cells to the patterns of proteins released from VIP-stimulated astroglia in culture reveals quantitative as well as qualitative alterations (Brenneman et al., 1990; Brenneman, 1990). Some of these proteins may have neuronal survival activity and may act through different mechanisms to allow neuronal maintenance and survival (e.g. recent studies have indicated that VIP induces the secretion of protein nexin I; Festoff et al., 1990; which has been shown to be associated with neuronal survival).

In our most recent studies using sequential chromatographic procedures, a novel protein has been isolated which is secreted from glial cells upon stimulation by VIP. Since this factor was first discovered in a test paradigm which measured the rescue of neuronal cells from death induced by electrical blockade, the protein was called Activity Dependent Neurotrophic Factor (ADNF). Amino acid analysis reveals a protein which is relatively enriched in glycine residues as compared to other known neurotrophic molecules. Pharmacological studies suggest that ADNF can act at much lower concentrations than VIP. The optimal concentration of VIP which induces neuronal survival is 0.1 nM (Brenneman and Eiden, 1986), which is about 33 ng/ml peptide; however, recent studies with conditioned medium from VIP-stimulated astroglia show optimal survival activity at 0.1 ng/ml for the unfractionated protein mixture. This activity increases **200**-fold after DEAE-Sephacel chromatography. Thus, future studies on ADNF should provide insight into the mechanism of action of VIP in the central nervous system.

IN VIVO NEUROTROPHIC ACTIVITY OF VIP

As indicated above, studies have shown that VIP gene expression is developmentally determined reaching a peak concomitantly with synapse formation in the developing rat brain (Gozes et al., 1987) and decreasing in the aging brain (Gozes et al.,

1988). To investigate the biological roles of VIP *in vivo*, we have recently developed a potent VIP antagonist (Gozes et al., 1989). This antagonist was shown to block VIP activity *in vivo* (Gozes et al., 1989) and *in vitro* (Gozes et al., 1991). The antagonist can differentiate central nervous system and peripheral VIP receptors, preferring the central receptors (Gozes et al., 1991a). Our recent results suggest that blockade of VIP activity <u>in vivo</u> can result in neurological deficits and impairment of neuronal development, acquisition of reflexes and learning and memory mechanisms (Gozes et al., 1990; Hill et al., 1991; Panllilio et al., 1990, Glowa et al., 1992). These deficits are similar to deficits observed upon treatment with the HIV (human immunodeficiency virus) envelope protein gp120 (Glowa et al., 1992) which might be involved in the etiology of neurological deficits associated with AIDS (Brenneman et al., 1988; Brenneman et al., 1990a). Finally, the patterns of VIP gene expression in the brain are probably associated with the requirements for the biological functions of the peptide (Gozes and Brenneman, 1989).

REFERENCES

Agoston, D.V., Eiden, L.E., Brenneman, D.E., and Gozes, I., 1991, Spontaneous electrical activity regulates vasoactive intestinal peptide expression in dissociated spinal cord cultures. *Mol. Brain Res.* 10:235.

Alberts, H.E., Stopa, E.G., Zoeller, R.T., Kauer, J.S., King, J.C. Fink, J.S., Mobtaker, H., and Wolfe, H., 1990, Day-night variation in prepro VIP/PHI mRNA within the suprachiasmatic nucleus. *Mol. Brain Res.* 7:85.

Baldino, F., Fitzpatrick-McElligott, S., Gozes, I and Card, J.P., 1989, Localization of VIP and PHI-27 messenger RNA in rat thalamic and cortical neurons. *J. Mol. Neurosci.* 1:199.

Banker, G.A., 1980, Trophic interactions between astroglial cells and hippocampal neurons in culture. *Science* 209:809.

Bardea, A., Glazer, R., Liling, G., and Gozes, I., 1992, VIP influences electrophysiological and metabolic activities in vivo. *Soc. Neurosci. Abs.* In press.

Berg, D.K., 1982, Cell death in neuronal development: regulation by trophic factor. In: Neuronal Development, N.C. Spitzer, ed., Plenum Publishing Corp., New York. pp. 297-331.

Bodner, M. Fridkin, M. and Gozes, I., 1985, VIP and PHM-27 sequences are located on two adjacent exons in the human genome. *Proc. Natl. Acad. Sci. USA* 82:3548.

Brenneman, D.E., 1988, Regulation of activity-linked neuronal survival by vasoactive intestinal peptide. Annals of the New York Academy of Science, Vol. 527, S.I. Said and V. Mutt, eds., pp. 595-597.

Brenneman, D.E., 1990, Two-dimensional electrophoretic analysis of glial proteins released by VIP. *Soc. Neurosci. Abs.* 16:820.

Brenneman D.E., and Eiden,L.E., 1986, Vasoactive intestinal peptide and electrical activity influence neuronal survival. *Proc. Natl. Acad Sci. USA* 83:1159.

Brenneman, D.E., Eiden, L.E., and Siegel, R.E., 1985a, neurotrophic action of VIP on spinal cord cultures. *Peptides* 6 (suppl. 2):35.

Brenneman, D.E., Fitzgerald, S., and Litzinger, M.J., 1985, Neuronal Survival during electrical blockade is increased by 8-bromo cyclic adenosine 3′,5′ monophosphate. *J. Pharmacol. Exp. Therap.* 233:402.

Brenneman, D.E., Fitzgerald, S., and Nelson, P.G., 1984, Interaction between trophic action and electrical activity in spinal cord cultures. *Dev. Brain Res.* 15:211.

Brenneman, D.E., McCune, S.K., and Gozes, I., 1990a, Acquired immune deficiency syndrome and the developing nervous system. *Intl. Rev. Neurobiol.* 32:305.

Brenneman D.E., Neale, E., Habig, W., Bowers, L.M., and Nelson, P.G., 1983, Developmental and neurochemical specificity of neuronal deficits produced by electrical impulse blockade in dissociated spinal cord cultures. *Dev. Brain Res.* 9:13.

Brenneman D.E., Neale, E.A., Foster, G.A., d'Autremont, S., and Westbrook, G.L., 1987, Nonneuronal cells mediate neurotrophic action of vasoactive intestinal peptide. *J. Cell Biol.* 104:1603.

Brenneman, D.E., and Nelson, P.G., 1986, Peptide modulation of neuronal differentiation in culture. In: Model systems in development and aging of the nervous system. A. Verndakis (ed.) Boston, Martinus Nijhoff, pp. 257-276.

Brenneman, D.E., Nicol, T., Warren, D., and Bowers, L.M., 1990, Vasoactive intestinal peptide: a neurotrophic releasing agent and an astroglial mitogen. *J. Neurosci. Res.* 25:386.

Brenneman, D.E., Schultzberg, M., Bartfai, T., and Gozes, I., 1992, Cytokine regulation of neuronal survival. *J. Neurochem.* 58:454.

Brenneman, D.E., Westbrook, G.L., Fitzgerald, S.P., Ennist, D.L., Elkins, K.L., Ruff,M.R., and Pert, C.B., 1988, Neuronal cell killing by the envelope protein of HIV and its prevention by vasoactive intestinal peptide. *Nature* 335:639-642.

Chew, L.J., Murphy, D., and Carter, D.A., 1991, Differential use of 3' poly (A) addition sites in vasoactive intestinal peptide messenger ribonucleic acid of the rat anterior pituitary gland. *J. Neuroendocrinol.* 3:351-355.

Davidson, A., Black, I., Draoui, M., Zia, F., Liling, G., Fridkin, M., Brenneman, D.E., Moody, T.W. and Gozes, I., 1992, The neuropeptide VIP: an autocrine regulator of cell growth. *Soc. Neurosci. Abs.* In press.

Eagleson, K.L., Raju, T.R., and Bennett, M.R., 1985, Motoneurone survival is induced by immature astrocytes from developing avian spinal cord. *Dev. Brain Res.* 17:95.

Emson, P.C., Gilbert, R.F.T., Loren, I., Fahrenkrug, J., Sundler, F., and Schaffalitzky de Muckadell, O.B.,1979, Development of vasoactive intestinal polypeptide (VIP) containing neurones in the rat brain. *Brain Res.* 177:437.

Festoff, B.W., Rao, J.S., and Brenneman, D.E., 1990, Vasoactive intestinal peptide (VIP) is a secretagogue for protease nexin I (PNI) release from astrocytes. *Soc. Neurosci. Abs.* 16:909.

Fink, J.S., Verhave, M., Walton, K., Mandel, G., and Goodman,R.H., 1991, Cyclic AMP- and Phorbol ester-induced transcriptional activation are mediated by the same enhancer element in the human vasoactive intestinal peptide gene. *J. Biol. Chem.* 266:3882.

Foster, G.A., Eiden, L.E., and Brenneman, D.E., 1989, Regulation of discrete sub-populations of transmitter-identified neurones after inhibition of electrical activity in cultures of mouse spinal cord. *Cell Tissue Res.* 256:543.

Giladi, E., Shani, Y., and Gozes, I., 1990, The complete structure of the rat VIP-gene. *Mol. Brain Res.* 7:261.

Giulian, D., Woodward, J., Young, D.G., Krebs, J.F., and Lachman, L.B., 1988, Interleukin-1 injected into mammalian brain stimulates astrogliosis and neovascularization. *J. Neurosci.* 8:2485.

Glowa, J.R., Panlilio, L.V., Brenneman, D.E., Gozes, I., Fridkin, M., and Hill, J.M., 1992, Learning impairment following intracerebral administration of the HIV envelope protein gp120 or a VIP antagonist. *Brain Res.* 570:49.

Gozes, I., 1987, VIP gene expression. Brain Peptides Update-volume 1. J.B. Martin, M.J. Brownstein and D. Krieger, eds., John Wiley and Sons Ltd. Chapter 10, pp. 141-162.

Gozes,I., 1988, Biosynthesis and molecular biology: The VIP-gene. Vasoactive intestinal peptide and related peptides. *Annals NY Acad. Sci.*, Vol. 527, S.I. Said and V. Mutt, eds., pp. 77-86.

Gozes, I., Avidor, R., Biegon, A., and Baldino, F. Jr., 1989a, Lactation elevates vasoactive intestinal peptide messenger ribonucleic in rat suprachiasmatic nucleus. *Endocrinology* 124: 181-186.

Gozes, I., Bodner, M., Shani, Y., and Fridkin, M., 1986, Structure and expression of the vasoactive intestinal peptide (VIP) gene in a human tumor. *Peptides* 7:1.

Gozes, I., and Brenneman, D.E., 1989, VIP: Molecular biology and neurobiological function. *Mol. Neurobiol.* 3:201.

Gozes, I., Davidson, A., Draoui, M., and Moody, T.W., 1992, The VIP gene is expressed in non-small cell lung cancer cell lines. *Biomed. Res.* In press.

Gozes, I., Hill, J.M., Mervis, R.F., Fridkin, M., and Brenneman, D.E., 1990, VIP antagonist produces neuronal damage and retardation of behavioral development in neonatal rats. *Soc. Neurosci. Abs.* 16:1292.

Gozes, I., McCune, S.K., Jacobson, L., Warren, D., Moody, T.W. Fridkin, M., and Brenneman, D.E., 1991, An antagonist to vasoactive intestinal peptide affects cellular functions in the central nervous system. *J. Pharmacol. Experm. Therap.* 257:959.

Gozes, I., Meltzer, E., Rubinrout, S., Brenneman, D.E., and Fridkin, M., 1989, Vasoactive intestinal peptide potentiate sexual behavior: Inhibition by novel antagonist. *Endocrinology* 125:2945.

Gozes, I., Shachter, P., Shani, Y., and Giladi, E., 1988, Vasoactive intestinal peptide gene expression from embryos to aging rats. *Neuroendocrinology* 47:27.

Gozes, I., and Shani, Y., 1986, Hypothalamic VIP-mRNA is increased in lactating rats. *Endocrinology* 119:2497.

Gozes, I., Shani, Y., Liu, B., and Burbach, J.P., 1989c, Diurnal variation in vasoactive intestinal peptide messenger RNA in the suprachiasmatic nucleus of the rat. *Neurosci. Res. Comm.* 5:83.

Gozes, I., Shani, Y., and Rostene, W.H., 1987, Developmental expression of the VIP-gene in brain and intestine. *Mol. Brain Res.* 2:137.

Gozes, I., Werner, H. Fawzi, M.A.A., Shani, Y., Fridkin, M., and Koch, Y., 1989b, Estrogen regulation of vasoactive intestinal peptide mRNA in the rat hypothalamus. *J. Molec. Neurosci.* 1:55.

Gozes, Y., Brenneman, D.E., Fridkin, M., Asofsky, R., and Gozes, I., 1991a, A VIP antagonist distinguishes VIP receptors on spinal cord cells and lymphocytes. *Brain Res.* 540: 319.

Hill, J.M., Gozes, I., Hill, J.L., Fridkin, M., and Brenneman, D.E., 1991, Vasoactive intestinal peptide antagonist retards the development of neonatal behaviors in the rat. *Peptides* 12:187.

Ishihara, T., Shigemoto, R., Mori, K., Takahashi, K. and Nagata, S., 1992, Functional expression and tissue distribution of a novel receptor for vasoactive intestinal peptide. *Neuron* 8:811.

Lam, K.S.L., Srivasta, G., Lechan, R.M., Lee, T., and Reichlin, S., 1990, estrogen regulates the gene expression of vasoactive intestinal peptide in the anterior pituitary. *Neuroendocrinology* 52:417.

Levy Holtzman, R., Malach, R., and Gozes, I., 1989, Disruption of the optic pathway during development affects vasoactive intestinal peptide mRNA expression. *The New Biologist* 1:215.

McGregor, G.P., Woodhams, P.L., O'Shaughnessy, D.J., Ghatei, M.A., Polak, J.M., and Bloom, S.R., 1982, Developmental changes in bombesin, substance P, somatostatin and vasoactive intestinal polypeptide in the rat brain. *Neurosci. Lett.* 28:21.

Moody, T.W., Zia, F., Goldstein, A.L., Naylor, P.H., Sarin, E., Brenneman, D.E., Koros, A.M.C., Reubi, J.C., Korman, L.Y., Fridkin, M., and Gozes, I., 1992, VIP analogues inhibit small cell lung cancer growth. *Biomed. Res.*, In Press.

Muller, H.W., Beckh, S., and Seifert, W., 1984, Neurotrophic factor for central neurons. *Proc. Natl. Acad. Sci. USA* 81:1248-1252.

Nielsen, F.C., Gammeltoft, S., Westermark, B. and Fahrenkrug, J., 1990, High affinity receptors for vasoactive intestinal peptide on a human glioma cell line. *Peptides* 11:1225.

Noble, M., Fok-Seang, J., and Cohen, J., 1984, Glia are a unique substrate for the *in vitro* growth of central nervous system neurons. *J. Neurosci.* 4:1892.

Nobou, F., Besson, J., Rostene, W., and Rosselin, G., 1985, Ontogeny of vasoactive intestinal peptide and somatostatin in different structures of the rat brain: effects of hypo- and hypercorticism. *Dev. Brain Res.* 20: 296.

Okamoto, S., Okamura, H., Miyake, M., Takahashi, Y., Takagai, S., Akagi, Y., Okamoto, H., and Ibata, Y., 1991, A diurnal variation of vasoactive intestinal peptide (VIP) mRNA under a daily light-dark cycle in the rat suprachiasmatic nucleus. *Histochem.* 95:525.

Panililio, L.V., Hill, J.M., Brenneman, D.E., Fridkin, M., Gozes, I., and Glowa, J.R., 1990, GP120 and a VIP antagonist impair Morris water maze performance in rats. *Soc. Neurosci. Abs.* 16:1330.

Pincus, D.W., DiCicco-Bloom, E.M., and Black, I., 1990, Vasoactive intestinal peptide regulates mitosis, differentiation and survival of cultured sympathetic neuroblasts. *Nature* 343:564.

Rostene, W., Dussaillant, M. Denis, P., Gozes, I., Montagne, M.-N. and Berod, A., 1992, Distribution and regulation of vasoactive intestinal peptide mRNA and binding sites in the mammalian eye, brain and pituitary. *Biomed. Res.* In Press.

Said, S.I., ed., 1982, Vasoactive intestinal polypeptide, Advances in peptide hormone research, Raven Press, New York.

Said, S.I., 1984, Isolation, localization and characterization of gastrointestinal peptides. *Peptides* 5:143.

Silver, J., Lorenz, S.E., Wahlsten, D., and Coughlin, J., 1982, Axonal guidance during development of the great cerebral commissures: descriptive and experimental studies, in vivo, on the role of preformed glial pathways. *J. Comp. Neurol.* 210:10.

Wessells, N.K., Letourneau, P.C., Nuttall, R.P., Luduena Anderson, K., and Geiduschek, J.M., 1980, Responses to cell contacts between growth cones, neurites, and ganglionic non-neuronal cells. *J. Neurocytol.* 9:647.

DESIGN AND STRUCTURE/CONFORMATION-ACTIVITY STUDIES OF A PROTOTYPIC CORTICOTROPIN-RELEASING FACTOR (CRF) ANTAGONIST: MULTIPLE ALANINE SUBSTITUTIONS OF CRF$_{12-41}$

Tomi K. Sawyer[1],*, Douglas J. Staples[1], Carol A. Bannow[1],
John H. Kinner[1], Linda L. Maggiora[1], Dawna Evans[2],
Mark Prairie[3], William Krueger[3], and Robert A. Lahti[2]

[1]Biochemistry
[2]CNS Research
[3]Physical & Analytical Research
The Upjohn Company
Kalamazoo, MI 49001

*Current address:
Parke-Davis Pharmaceutical Research Division
Warner-Lambert Company
Ann Arbor, MI 48106

INTRODUCTION

The known physiological role(s) and proposed pathophysiological properties of the neuroendocrine peptide CRF have been previously described (for review, see 1), and CRF has been shown to exert a variety of CNS-mediated effects on behavior[2,3], cardiovascular system[4,5], reproduction[6,7], gastrointestinal secretion[8,9], motility[10], and transit[11]. Of particular significance is that CRF may, therefore, be involved in stress stimuli-induced activation of neural/humoral pathways leading towards anxiety and depressive disorders (*e.g.*, depression, panic and anorexia nervosa). Nevertheless, the molecular pharmacology and mechanisms which are involved in stress-induced behavioral, endocrine and metabolic activities are not well defined. The discovery and development of potent CRF antagonists may provide key molecular probes to investigate the biological activities of endogenous CRF in animal models as well as for studying the molecular pharmacology of CRF-receptor interactions. Such studies have been reported[12,13] and have been primarily based upon synthetic modification of CRF; yet the emergence of a high affinity analog of low molecular mass (*i.e.*, small peptide or peptidomimetic) has remained elusive to date. Nevertheless, studies[14-18] on the blockade of endogenous CRF using CRF antiserum or prototypic CRF antagonists have probed the possible role that endogenous CRF may have on the effects of stress in different animal models. Of noteworthy contribution to such CRF research has been both structure-activity and structure-conformation studies[12,13,19,20] to investigate CRF

SEEPPISLDL^{10}-TFHLLREVLE^{20}-MARAEQLAQQ^{30}-AHSNRKLMDIA^{41}-amide
human/rat CRF

phe^{12}-HLLREVLE-Nle^{21}-ARAEQLAQQAHSNR-Arg^{36}-L-Nle^{38}-DI-Ile^{41}-amide
CRF analog **I**

Figure 1. Chemical structure of human/rat CRF and CRF analog **I**. Abbreviations: S, serine; E, glutamic acid; P, proline; I, isoleucine; L, leucine; T, threonine; F, phenylalanine; H, histidine; R, arginine; V, valine; M, methionine; A, alanine; Q, glutamine; N, aspargine; K, lysine; D, aspartic acid; phe, D-phenylalanine; Nle, norleucine.

receptor binding and functional properties (agonism/antagonism). These studies have culminated in the identification of prototypic CRF antagonists (or partial agonists) which were modified fragment analogs of the native peptide. Specifically, compound **I** (Fig. 1) has been advanced[13] as a significant lead towards the development of high affinity CRF receptor antagonists. In this report we describe analogs of **I** to further explore the role of side-chain functionlization in CRF receptor binding using a strategy of multiple (iterative) Ala substitution with a particular focus on the central domain of this CRF analog corresponding to CRF_{22-31}. In addition, the structure-conformation properties of these analogs were investigated by circular dichroism spectroscopy.

CRF PEPTIDE ANALOG CHEMISTRY AND BIOLOGICAL TESTING

All CRF peptide analogs (Fig. 2) were prepared by total synthesis using solid-phase methods and an ABI Peptide Synthesizer Model-430. N-Boc-amino acids, DDC/HOBT coupling, and HF-labile benzhydrylamine resins were employed for the peptide synthesis.

phe-HLLREVLE-Nle-ARAE-<u>A-A</u>-A-<u>A</u>-QA-HSNRRL-Nle^{38}-DII-amide
CRF analog **II**
phe-HLLREVLE-Nle^{21}-A-<u>A</u>-A-<u>A-A-A</u>-A-<u>A-A</u>-A-HSNRRL-Nle^{38}-DII-amide
CRF analog **III**
phe-HLLREVLE-Nle^{21}-ARA-<u>A-A-A</u>-A-<u>A-A</u>-A-HSNRRL-Nle^{38}-DII-amide
CRF analog **IV**
phe-HLLREVLE-Nle^{21}-A-<u>A</u>-AE-<u>A-A</u>-A-<u>A-A</u>-A-HSNRRL-Nle^{38}-DII-amide
CRF analog **V**
phe-HLLREVLE-Nle^{21}-A-<u>A</u>-A-<u>A-A-A</u>-A-<u>A</u>-QA-HSNRRL-Nle^{38}-DII-amide
CRF analog **VI**

Figure 2. Chemical structures of CRF analogs **II-VI**. Abbreviations: see Fig. 1 legend. Ala modifications (relative to CRF analog **I**) are highlited by underlines.

Table I. Chemical and Analytical Properties of CRF Analogs **I-VI**

Analog	Molecular Formula	MW	FAB-MS [M· + H]+, m/z	HPLC* k'	Purity
I	$C_{158}H_{265}N_{51}O_{43}$	3567	3568	5.9	99%
II	$C_{151}H_{253}N_{49}O_{41}$	4208	4209	6.1	98%
III	$C_{144}H_{241}N_{45}O_{38}$	3211	3211	6.1	99%
IV	$C_{147}H_{248}N_{48}O_{38}$	3296	3297	6.4	98%
V	$C_{146}H_{243}N_{45}O_{40}$	3269	3270	6.3	98%
VI	$C_{144}H_{241}N_{45}O_{38}$	3211	3211	6.1	99%

*HPLC data determined using a Vydac C18 analytical column, flow rate at 1.5 mL/min, 220 nm detection, and a linear gradient of 100% H_2O-100% CH_3CN (0.1 % TFA)/17 min.

Preparative reversed-phase HPLC was used to purify the peptides to homogeneity and the final compounds were analysed by FAB-MS (Table I).

CRF receptor binding of each peptide was determined using radioreceptor methods which employed ^{125}I-labeled ovine CRF and rat brain frontal cortex membranes (adapted from deSouza[21]). Rat frontal cortex tissue was Polytroned in buffer containing 50 mM TRIS, 10 mM MgCl2, 2 mM EGTA, 0.32 M sucrose, 0.1 % BSA, aprotinin (100 U/mL), and 0.1 mM bacitracin, pH. 7.0. The resulting homogenate was centrifuged at 10,000xg for 10 min, the supernatant discarded, and the resulting pellet resuspended (final tissue dilution was 3.33 mg/0.1 mL). The incubation mixture was prepared in 1.5 mL plastic microfuge tubes and consisted of 0.1 mL of vehicle or compound, 0.1 mL of ^{125}I-labeled ovine CRF (30,000 cpm; final concentration of 30 pM; specific activity of 2200 Ci/mmol; New England Nuclear), and the incubation was initiated by the addition of 0.1 mL of tissue homogenate. A two hour incubation with constant shaking was performed at room temperature and termination was attained by centrifugation (supernatants were aspirated, the pellet gently washed with 1.0 mL of PBS containing 0.01% Triton X-100 and re-centrifuged and the supernatant aspirated off). The final tissue pellets were then analyzed by gamma counter (triplicate assays, non-specific binding was determined by 1.0 μM ovine Tyr-CRF.

Circular dichroism spectroscopy of CRF analog **I** and its analogs was performed using a Jasco Model 720 Spectrometer. Molar ellipticity data for each compound was determined as a function of wavelength (178-260 nm). The individual CRF analogs were dissolved in 9% acetic acid followed by phosphate buffer (pH 7.2), and CD spectra measurements were performed on each peptide solution contained in a 0.1 mm cell.

CRF RECEPTOR BINDING STRUCTURE-ACTIVITY RELATIONSHIPS

Multiple (iterative) Ala substitution in the central domain (CRF_{22-31} sequence) of CRF analog **I** was examined by a series of compounds (**II-VI**) in which the Arg[23], Glu[25], Gln[26], Leu[27], Gln[29], and/or Gln[30] were modified. The structure-activity of these CRF

Table II. CRF Receptor Binding Affinities of CRF Analogs **I-VI**

Analog	Primary Structure of Central Domain	Receptor Binding Ki, nM
I	Ala-Arg-Ala-Glu-Gln-Leu-Ala-Gln-Gln-Ala	41.5
II	Ala-Arg-Ala-Glu-Ala-Ala-Ala-Ala-Gln-Ala	24.6
III	Ala-Ala-Ala-Ala-Ala-Ala-Ala-Ala-Ala-Ala	>10,000
IV	Ala-Arg-Ala-Ala-Ala-Ala-Ala-Ala-Ala-Ala	121.3
V	Ala-Ala-Ala-Glu-Ala-Ala-Ala-Ala-Ala-Ala	112.3
VI	Ala-Ala-Ala-Ala-Ala-Ala-Ala-Ala-Gln-Ala	277.5

derivatives were evaluated in terms of CRF receptor binding and the results are summarized in Table II. Analog **II** was found to be as potent as the lead compound (**I**) and the $Ala^{26,27,29}$ substitutions in this derivative were therefore considered as acceptable. Additional Ala substitutions for Arg^{23}, Glu^{25} and/or Gln^{30} were then evaluated as a triple or double residue modification (*cf.* compounds **III** and **IV-VI**, respectively). Triple substitution of CRF analog **II** by Ala resulted in a complete loss in biological activity (compound **III** elicited essentially no binding at 10 µM). However, double substitutions of **II** to further explore the role of the Arg^{23}, Glu^{25} or Gln^{30} residues resulted in only ~5- to 10-fold decrease in binding affinity (compounds **IV-VI**). Therefore, it may be concluded that at least one of these residues is critical to molecular recognition of the CRF analog **I** in binding to the receptor. The biological properties of these CRF analogs will be investigated in terms of their functional properties (*i.e.*, antagonist, partial agonist or agonist). It is noted that CRF analog **I** has been previously described as a partial agonist (<10% intrinsic activity) by Rivier and co-workers[13] and it will be interesting to evaluate the effect of multiple (iterative) Ala substitution of the CRF_{22-31} sequence of **I** in terms of the resulting agonist/antagonist profiles for CRF analogs **II-VI**.

CRF STRUCTURE-CONFORMATION RELATIONSHIPS

To further investigate the structure-conformation properties of the above CRF peptide analogs we evaluated their CD spectra as a probe of their secondary structure (i.e., degree of alpha-helicity). These studies extend earlier reports of CRF conformational analysis[19,20] and a prototypic CRF antagonist peptide referred to as "alpha-helical CRF(9-41)" which also provided impetus to us to design a new series of potential CRF antagonists emphasizing considerations of the 3-D structural properties and the identification of key substructural features (*i.e.*, amino acids or "domain" of amino acids as referred to in Fig. 1) which are important for biological activity at the CRF receptor. As shown in Fig. 3A and 3B, the CD spectra of CRF analogs **I-III** and **IV-VI** were determined.

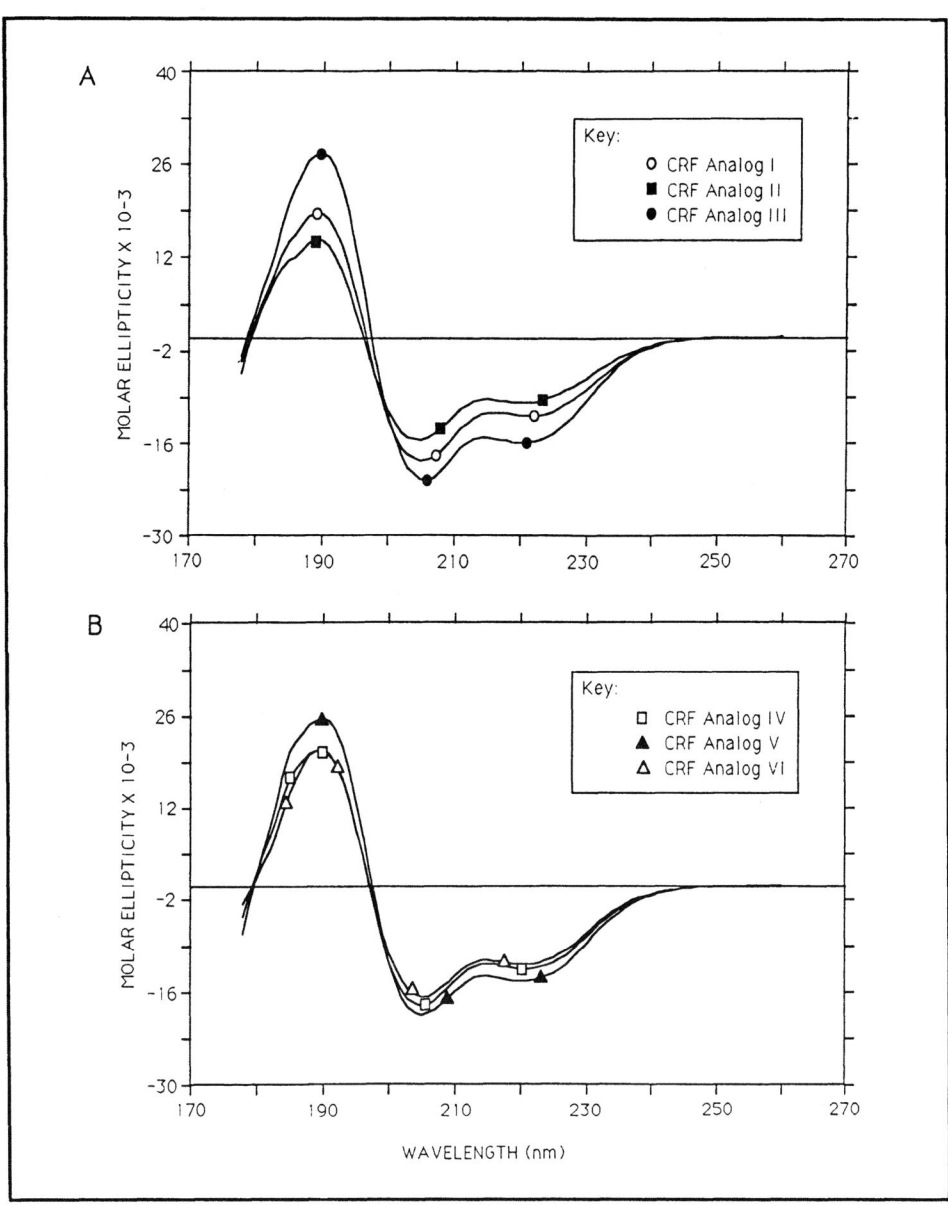

Figure 3. Circular dichroism of CRF analogs **I-III** (panel A) **and III-VI** (panel B).

The above data indicate that multiple (iterative) Ala modification of CRF analog **I** did result in secondary structural changes in solution which could be observed by circular dichroism spectroscopy. The degree of helicity was not measured in an absolute manner, but calculations at 222 nm by the methods of Chen and co-workers[22] did provide a measurement of this secondary structure to evaluate the structure-conformation properties of these peptides on a relative basis (Table III). The results of these studies indicate that CRF analogs **I-VI** all elicit alpha-helicity in aqueous solution (24-44% range), but there is no clear correlation between structure-activity and structure-conformation based on these studies.

Table III. % Helicity of CRF Analogs **I-VI**

Analog	Primary Structure of Central Domain	% Helicity*
I	Ala-Arg-Ala-Glu-Gln-Leu-Ala-Gln-Gln-Ala	31
II	Ala-Arg-Ala-Glu-Ala-Ala-Ala-Ala-Gln-Ala	24
III	Ala-Ala-Ala-Ala-Ala-Ala-Ala-Ala-Ala-Ala	44
IV	Ala-Arg-Ala-Ala-Ala-Ala-Ala-Ala-Ala-Ala	39
V	Ala-Ala-Ala-Glu-Ala-Ala-Ala-Ala-Ala-Ala	31
VI	Ala-Ala-Ala-Ala-Ala-Ala-Ala-Ala-Gln-Ala	33

*% Helicity = [Molar Ellipticity @ 222 nM + 2.34] + -30,300

SUMMARY AND FUTURE STUDIES

The discovery and development of CRF antagonists (peptide or peptidomimetic) remains a challenging area of research. Deletion of amino acids at the C-terminus of CRF results in complete loss in CRF receptor recognition, however, N-terminal deletion is tolerated to some extent. Based on a previous investigation by Salk researchers and colleagues[13] describing a potent CRF_{12-41} antagonist analog (**I**) we were interested to further study its structure-activity and structure-conformation properties. In evaluation of the primary structure of **I** (D-Phe[12]-His-Leu-Leu-Arg-Glu-Val-Leu-Glu-Nle-Ala[22]-Arg-Ala-Glu-Gln-Leu-Ala-Gln-Gln-Ala[31]-His-Ser-Asn-Arg-Arg-Leu-Nle-Glu-Ile-Ile[41]-NH2) we noted that the internal 22-31 sequence contained four Ala residues, three Gln, one Glu, one Arg and one Leu. To further explore the role of side-chain functionlization in CRF receptor binding, we designed a series of **I** analogs having multiple (iterative) Ala substitutions within the central domain and discovered high affinity derivatives. Specifically, the [Ala[26,27,29]]-substituted analog of **I** exhibited an IC_{50} of 25 nM (displacement of ^{125}I-CRF binding to rat frontal cortex). Additional (double) Ala substitution were also tolerated albeit with ~5- to 10-fold decreased binding affinity, but per-Ala substitution throughout the entire 22-31 sequence lead to an apparent complete loss ($IC_{50} > 10$ μM) in CRF receptor recognition. In conclusion, these results demonstrate the critical role of key amino acid residues within the central decapeptidyl domain of a potent CRF antagonist. Interestingly, the structure-conformation properties of these CRF analogs as evaluated by circular dichroism showed no obvious correlation between to secondary structure (*i.e.*, alpha-helicity) to biological activity (*i.e.*, receptor binding). Therefore, it may be proposed that the side-chain moieties throughout the 22-31 sequence are exquisitely important for CRF receptor recognition as opposed to maintaining conformational stability of the peptide.

A recent report by Kornreich and co-workers[23] which impacts on this work involves a elegant structure-activity investigation of ovine CRF by single Ala substitution (i.e., Ala scanning). Their work supported the importance of the N-terminal region, residues 5-19, for

CRF receptor binding and activation, and that the C-terminal region may be particularly important for structure-conformation properties. Our future studies will be focused on determining their comparative CRF receptor binding and functional properties (agonism/antagonism) to advance the structure-activity relationships of ligand-receptor recognition and signal transduction of this series of putative CRF antagonists analogs. Such work will be of significance to the rational design and discovery of high affinity CRF receptor antagonists for application in both *in vitro* and *in vivo* model systems to advance our understanding of the biological properties of this neuroendocrine peptide.

REFERENCES

1. C. Rivier, M. Smith, and W. Vale, Regulation of adrenocorticotropic hormone (ACTH) secretion by corticotropin-releasing factor (CRF) *in:* Corticotropin-releasing Factor: Basic and Clinical Studies of a Neuropeptide. E. DeSouza, and C.B. Nemeroff, Eds., CRC Press Inc., Boca Raton, FL, pp. 175-189 (1990).

2. R.E. Sutton, G.F. Koob, M.L. Moal, J. Rivier, and W. Vale, Corticotropin-releasing factor produces behavioral activation in rats. *Nature (London)*, 297: 331 (1982).

3. D.R. Britton, G.F. Koob, J. Rivier, and W. Vale, Intraventricular corticotropin-releasing factor enhances behavioral effects of novelty. *Life Sci.,* 31: 363 (1982).

4. L.A. Fisher, J. Rivier, C. Rivier, J. Speiss, W. Vale, and M.R. Brown, Corticotropin-releasing factor (CRF): Central effects on mean arterial pressure and heart rate in rats. *Endocrinology*, 110: 2222 (1982).

5. L.A. Fisher, Corticotropin-releasing factor: Central nervous system effects on baroreflex control of heart rate, *Life Sci.*, 42: 2645 (1988).

6. C. Rivier, J. Rivier, and W. Vale, Stress-induced inhibition of reproductive functions: Role of endogenous corticotropin-releasing factor, *Science,* 231: 607 (1986).

7. A. Barbarino, L.D. Marinis, A. Tofani, S.D. Casa, C. D'Amico, A. Mancini, S.M. Corsello, R. Sciuto, and A. Barini, Corticotropin-releasing hormone inhibition of gonadotropin release and effect of opioid blockade, *J. Clin. Endocrinol. Metab.*, 68: 5623 (1989).

8. Y. Tache, Y. Goto, M.W. Gunion, W. Vale, J. Rivier, and M. Brown, Inhibition of gastric acid secretion in rats by intracerebral injection of corticotropin-releasing factor, *Science,* 222: 935 (1983).

9. R.L. Stevens Jr., H. Yang, J. Rivier, and Y. Tache, Intracisternal injection of corticotropin-releasing factor blocks surgical stress-induced inhibition of gastric acid secretion in the rat, *Peptides,* 9: 1067 (1988).

10. L. Bueno, and J. Fioramonfi, Effects of corticotropin-releasing factor, corticotropin and cortisol on gastrointestinal motility in dogs, *Peptides,* 7: 73 (1986).

11. C.L.Williams, J.M. Peterson, R.G. Villar, and T.F. Burks, Corticotropin-releasing factor directly mediates colonic response to stress, *Am. J. Physiol.*, 153: G582 (1987).

12. J. Rivier, C. Rivier, and W. Vale, Synthetic competitive antagonists of corticotropin-releasing factor: Effect on ACTH secretion in the rat, *Science ,* 224: 889 (1984).

13. J. Rivier, C. Rivier, R. Galyean, G. Yamamoto, and W. Vale, Corticotropin-releasing factor: Characterization of new analogs, *in:* Peptide Chemistry, T. Shiba and S. Sakakibara, Eds., Protein Research Foundation, Osaka, Japan, pp. 597-600 (1988).

14. C. Rivier and W. Vale, Modulation of stress-induced ACTH release by corticotropin-releasing factor, catecholamines and vasopressin, *Nature* , 305: 325 (1983).

15. C. Rivier, J. Rivier, and W. Vale, Inhibition of adrenocorticotropin hormone secretion in the rat by immunoneutralizaton of corticotropin-releasing factor,*Science* , 218: 377 (1982).

16. K.T. Britton, G. Lee, W. Vale, J. Rivier, and G.F. Koob, Corticotropin-releasing factor (CRF) receptor antagonist blocks activating and 'anxiogenic' actions of CRF in the rat, *Brain Res.*, 369: 303 (1986).

17. M.R. Brown, T.S. Gray, and L.A. Fisher, Corticotropin-releasing factor antagonist: Effects on the autonomic nervous system and cardiovascular function, *Regul. Peptides,* 16: 321 (1986).

18. D.D. Krahn, B.A. Gosnell, M. Grace, and A.S. Levine, CRF antagonist partially reverses CRF- and stress-induced effects on feeding, *Brain Res. Bull.,* 17: 285 (1986).

19. S.H. Lau, J. Rivier, W.Vale, E.T. Kaiser, and F.J. Kezdy, Surface properties of an amphiphilic peptide hormone and of its analog, *Proceedings of the National Academy of Sciences USA* , 80: 7070 (1983).

20. P.V. Pallai, M. Mabilia, M.Goodman, W. Vale, and J. Rivier, Structureal homology of corticotropin-releasing factor, sauvagine , and urotensin I: Circular dichroism and prediction studies, *Proc. Natl. Acad. Sci., U.S.A.,* 80: 6770 (1983)

21. E.B. DeSouza, Corticotropin-releasing factor receptors in the rat central nervous system: Characterization and regional distribution, *J. Neuroscience,* 7: 88 (1987).

22. Y.-H. Chen, J.T. Yang, and H. Martinez, Determination of the secondary structures of proteins by circular dichroism and optical rotatory dispersion, *Biochemistry* , 22: 4120 (1972).

23. W.D. Kornreich, R. Galyean, J.-F. Hernandez, A. Grey Craig, C.J. Donaldson, G. Yamamoto, C. Rivier, W. Vale, and J. Rivier, Alanine series of ovine corticotropin releasing factor (oCRF): A structure-activity relationships study, *J. Med. Chem.* , 35: 1870 (1992).

RETINOIC ACID REGULATES preproVIP EXPRESSION
IN THE HUMAN NEUROBLASTOMA CELL LINE NB-1

Birgitte Georg, Birgitte S. Wulff, and Jan Fahrenkrug

Dept. of Clinical Chemistry
Bispebjerg Hospital
University of Copenhagen
DK-2400 Copenhagen NV

INTRODUCTION

Vasoactive intestinal polypeptide (VIP) is a widely distributed 28 amino acid neuropeptide with a broad spectrum of biological activities. In the brain, VIP seems to synchronize blood flow, energy metabolism, and cortical activity besides controlling pituitary hormone secretion. In the gastrointestinal-, respiratory-, and urogenital tract, the peptide plays a transmitter role controlling motility, blood flow, and secretion (Fahrenkrug, 1989). The actions are mediated through binding to G-protein coupled receptors and activation of adenylate cyclase (Amiranoff and Rosselin, 1982; Laburthe and Couvineau, 1988).

VIP is derived from a 170 amino acid precusor (preproVIP) by posttranslational processing. Seven exons are present in the VIP gene and each of these gives rise to a domain also characterized in the preproVIP peptide. The predominant human VIP mRNA is 1.8 kb but occasionally, a VIP mRNA of about 1.6 kb is found in addition. Two VIP mRNA species differing in the 3'untranslated region are detected in some rat tissues. It has been revealed that they arise by polyadenylation at the two potential sites in the rat VIP mRNA molecule (Chew et al., 1991). The human VIP mRNA contains three potential polyadenylation sites (Tsuskada et al., 1985) so polyadenylation at one of the more upstream sites is a probable mechanism for formation of the smaller human VIP mRNA molecule.

A number of peptides are formed by mono- and dibasic cleavage of preproVIP. These are: preproVIP 22-79; PHM (**P**eptide with N-terminal **H**istidine and C-terminal **M**ethionine-amide; PHV (**P**eptide with N-terminal **H**istidine and C-terminal **V**aline, which is PHM C-terminally extended with preproVIP 111-122); preproVIP 111-122; VIP, which exists as both amidated and glycine-extended peptide; and preproVIP 156-170 (Fahrenkrug

Growth Factors, Peptides, and Receptors, Edited
by T.W. Moody, Plenum Press, New York, 1993

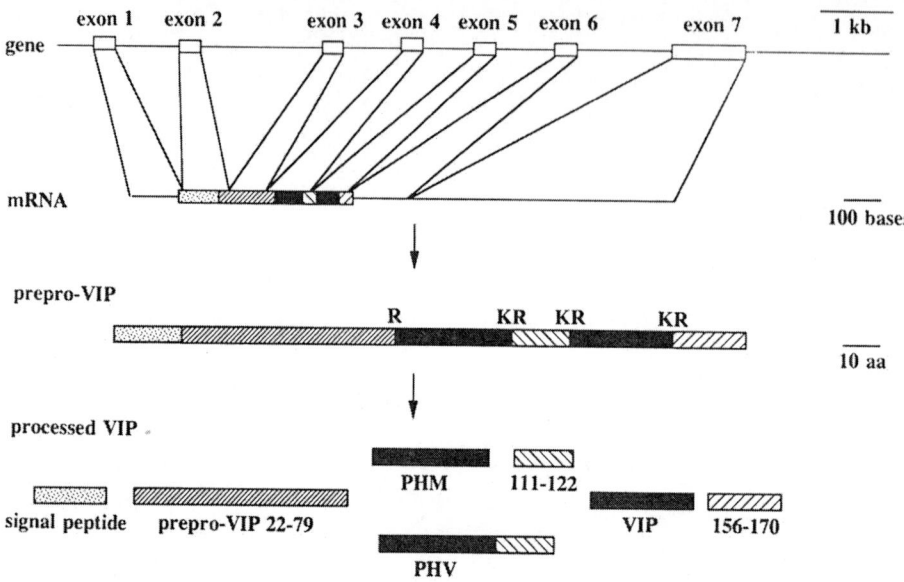

Figure 1. Schematic representation of the structure of the VIP gene and mRNA, preproVIP, and preproVIP derived peptides. Mono– and dibasic cleavage sites in preproVIP are indicated with single letter amino acid codes (K=lysine, R=arginine).

and Emson, 1989; Fahrenkrug, 1991). In figure 1, the structure of the VIP gene, mRNA, and preproVIP derived peptides are illustrated.

Both VIP and PHM are C–terminally amidated peptides, and both are like PHV and the immediate precursor of amidated VIP (glycine–extended VIP) biologically active (Fahrenkrug et al., 1989; Palle et al., 1989; Palle et al., 1992).

The expression of VIP is known to be regulated at the transcriptional level by cAMP and the phorbol–ester Phorbol 12–myristate 13–acetate (PMA) (Hayakawa et al., 1984; Ohsawa et al., 1985). The enhancer sequence responsible for the cAMP mediated induction have been identified (Tsukada et al., 1987) and, it has been shown that this cAMP responsive element also mediates the PMA response (Fink et al., 1991). Physiological mediators causing induction of cAMP and protein kinase C activity have however not been disclosed and presently, knowlegde of regulation of VIP gene expression mediated by other substances is sparse. Besides regulation at the transcriptional level, it is possible that VIP expression is regulated at the RNA level either through alternative splicing/polyadenylation and/or differential stability of mRNA species. In addition, a posttranslational regulation of the VIP expression is probable as tissue specific processing of preproVIP has been shown (Fahrenkrug and Emson, 1989; Bredkjær et al., 1991).

In the present study, the effect of the vitamin A metabolite retinoic acid (RA) on the VIP expression has been examined. RA is known to influence growth and differentiation of various cell types both in vivo and in vitro and, neuroblastoma cells are among those often seen to differentiate. Effects of RA are exerted by binding to nuclear receptors belonging to the steroid/thyroid hormone superfamily of receptors (Sidell et al., 1983; Glass et al., 1991).

MATERIALS AND METHODS

Cell Culture

The human neuroblastoma cell line NB-1 was obtained from Professor N. Yanaihara (Shizuoka University School of Phamacy, Japan). The cells were grown to confluency in RPMI 1640 containing: 10% FCS, penicillin (50 U/ml), and streptomycin (0.05 mg/ml). After confluency was reached, the cultures were kept in serum–free media containing 0.2% human serum albumin for 48 hours. At onset of experiments, fresh serum–free medium was added to parallel dishes which were further incubated with or without RA (All–trans retinoic acid was obtained from Sigma, dissolved to 10 mM in dimethyl sulfoxide, and kept in aliquots at −20°C until use) for another 48 hours period or as indicated.

In time–course experiments, RA was added to a final concentration of 10 μM RA to parallel dishes at different time–points before harvest of the cells. To minimize the serum–free period in long time–course experiments, RA was added to the first dish simultaneous with the cultures being set under serum–free conditions.

At the end of experiments, conditioned media were removed and if necessary centrifuged to remove dead cells. The media were kept on ice and frozen at −20°C until analysis. After removal of the medium, the cells were scraped of the plate in ice-cold phosphate–buffered saline by use of rubber-policeman, counted in a haemocytometer, and frozen at −20°C until analysis.

Radioimmunosassays

The peptides were extracted from cell pellets by boiling water/acetic acid extraction. The peptide concentrations in cell extracts and conditioned media were measured by radioimmunoassays (RIAs) directed against the following sequences: preproVIP 22–79, PHM, PHM+PHV, preproVIP 111–122, the N–terminal part of VIP, VIP (amidated), glycine–extended VIP, and preproVIP 156–170 (Fahrenkrug, 1987; Fahrenkrug, 1991; Fahrenkrug and Emson, 1989; Fahrenkrug and Pedersen, 1984; Fahrenkrug and Pedersen, 1986; Fahrenkrug and Schaffalitzky de Muckadell, 1977; Fahrenkrug and Schaffalitzky de Muckadell, 1978).

In all cell extracts and media, PHM and PHV were measured. In selected experiments, extract and medium were analyzed with all seven RIAs. The peptide concentrations were related to the number of cells.

RNA preparation and analysis

Total RNA was extracted by the acid guanidinium thiocyanate–phenol–chloroform extraction method (Chomczynski and Sacchi, 1987).

Slot blots containing dilutions of total RNA were subsequently hybridized to random primed ^{32}P–labeled plasmids containing VIP (obtained from professor dr. H. Okamoto, Tohoku University School of Medicine, Japan) and actin cDNA in sodium–phosphate buffer (0.5 M, pH: 7.2; 7% SDS; 1 mM EDTA) at 65°C. The blots were washed to a stringency of 0.02 M sodium–phosphate; 0.5% SDS; 65°C. To avoid the GC–tailing, the VIP cDNA was subcloned before used as probe in hybridizations.

Autoradiograms of the blots were scanned using a LKB 2222–020 laser densitometer. The slopes of the lines relating μg RNA added to the filter and absorbance were taken as an estimate of the relative amount of mRNA. The relative amount of actin mRNA was used to correct for variations in the total amount of RNA applied to the filter.

For northern blots 20 μg of total RNA was applied to gels. The blots were hybridized and washed as described for the slot blots.

RESULTS AND DISCUSSION

Expression of both VIP mRNA and preproVIP derived peptides were found to be induced by RA treatment of the cells. Many neuroblastoma cell lines are seen to differentiate in response to RA (Sidell et al., 1983) induction of preproVIP expression does however not seem to be due to differentiation of the cells as these become both more adherent and confluent in response to RA (Georg et al., 1992).

10–15 μM RA results in maximal induction of PHM+PHV immunoreactivity. An effect on the PHM+PHV expression is also observed using a concentration of about 0.1 μM, while high doses (50–100 μM) cause the amount of PHM+PHV to decrease to below the basal concentration probably due to a toxic effect. 10 μM RA leads to an approximately 4 fold induction in both conditioned medium and cell extract.

The concentrations of the various preproVIP derived peptides vary considerably. The basal peptide concentration of amidated PHM and VIP, and also glycine–extended VIP is low especially in the conditioned medium pointing toward ineffective processing and amidation activity in NB–1 cells. 10 μM RA causes however a rise in the concentration of these amidated and glycine–extended peptides to at least the same extend as the other preproVIP derived peptides in the cells. In the medium, the induction is on the contrary found to be most evident using the RIAs directed against: preproVIP 22–79, preproVIP 111–122, N–terminal VIP, and preproVIP 156–170. These results indicate selective storage of the biologically active peptides in the cells (Georg et al., 1992).

A 20 hours lag period is observed before induction of the peptides (PHM+PHV) in both cell extract and conditioned medium. In the extract, the level is stabilized after 48 hours treatment while in the medium, the level continue to rise for at least 96 hours. In both, the induced PHM+PHV level is stable during the 6 days experimental period.

As shown in figure 2, the VIP mRNA increases in response to RA. The effect of RA on the VIP mRNA level is first seen after a lag period of 8–20 hours after which, the VIP mRNA level increases until about 24 hours of exposure. The level is then stabilized and remain high for at least 96 hours. The effect of RA is evident on the most aboundant 1.8 kb VIP mRNA but occasionally the 1.6 kb VIP mRNA can be seen after RA treatment of the cells (Georg et al., 1992).

Because of the long lag period of the RA mediated VIP mRNA induction compared with for instance the one observed after cAMP stimulation, we decided to examine the possible involvement of newly synthesized proteins by addition of the translational inhibitor cycloheximide (CHX). Figure 3 shows a northern blot demonstrating a slight induction of the 1.8 kb mRNA after treatment with 10 μg/ml CHX alone for 24 hours (lane 2), and the

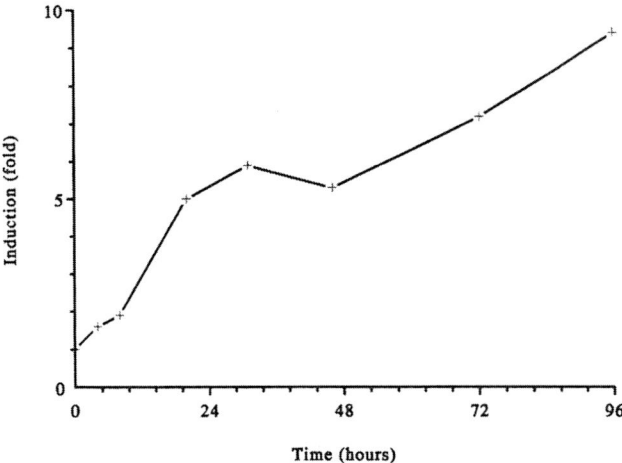

Figure 2. Induction of VIP mRNA at different time–points as estimated by scannings of autoradiograms of slot blot subsequently hybridized to VIP and actin cDNA probes.

more pronounced induction mediated by 10 μM RA (lane 3). Abolishment of the RA mediated induction is seen by concomitant treatment with CHX and RA (lane 4).

The RA receptors belong to the steroid hormone receptor superfamily. These receptors are intracellular/nuclear proteins that after binding to the ligand act as transcription factors able to either enhance or decrease the transcription of specific target genes. The effect of

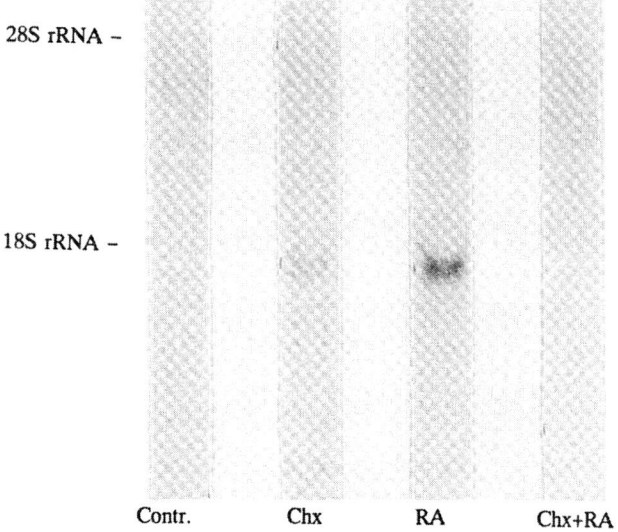

Figure 3. A northern blot hybridized to VIP cDNA. 20 μg total RNA was applied to each lane of the gel. The positions of the 28S and 18S rRNA bands are indicated to the left. Lane 1: RNA from NB–1 control cells. Lane 2: RNA from cells treated with 10 μg/ml cycloheximide (CHX) for 24 hours. Lane 3: RNA from cells treated with 10 μM retinoic acid (RA) for 24 hours. Lane 4: RNA from cells treated with both CHX and RA.

RA on the VIP gene expression does however not seem to be mediated directly. Both the slow mRNA induction and the abolishment by CHX indicate an indirect effect of RA. The effect could then be exerted by a yet undefined RA regulated transcription factor influencing the VIP gene expression or RA could affect the expression of factors influencing the stability of the VIP mRNA. Experiments evaluating the transcriptional activity of the VIP gene and/or the VIP mRNA stability have to be performed to reveal these possibilities.

The physiological significance of our observations is presently unknown but in contradiction to previous studies of regulation of VIP expression, the induction of the VIP expression mediated by RA does not seem to be due to differentiation of the cells toward a more neuronal phenotype. A more neuronal morphology of cultured cells has been observed concomitant with increases in the VIP expression mediated by cAMP and Nerve Growth Factor and a less differentiated state of the cells was seen when the expression of VIP was decreased by dexamethasone (Yanaihara et al., 1981; Beinfeld et al., 1988). The ability of RA to mediate induction of the VIP expression and simultanously affect the morphology of the NB-1 cells towards a less differentiated state shows that differentiation of cultured cells to a more neuron like cell type is not a prerequisite for induction of VIP expression.

ACKNOWLEDGMENT

The study was supported by grants from the Danish Biotechnology Center for Signal Peptide Research and the Danish Medical Research Council (f.no. 12–0819–1). The technical assistence of Lene Linnet Larsen and Anita Hansen is gratefully appreciated.

REFERENCES

Amiranoff, B. and Rosselin, G., 1982, VIP receptors and control af cyclic AMP production, *In*: Peptide Hormone Research Series. Vasoactive Intestinal Peptide, S. Said, ed., Raven Press, New York.

Beinfeld, M.C., Brick, P.L, Howlett, A.C., Holt, I.L., Pruss, R.M., Moskal, J.R., and Eiden, L.E., 1988, The regulation of vasoactive intestinal peptide synthesis in neuroblastoma and chromaffin cells, *In*: Ann. N.Y. Acad Sci. 527:68, S.I. Said and V. Mutt, eds..

Bredkjær, H.E., Rønnov-Jessen, D., Fahrenkrug, L., Ekblad, E., and Fahrenkrug, J., 1991, Expression of preproVIP-derived peptides in the human gastrointestinal tract: a biochemical and immunocyto-chemical study, Regul. Pept. 33:145.

Chew, L.-J., Murphy, D., and Carter, D.A., 1991, Differential use of 3'poly(A) addition sites in vasoactive intestinal peptide messenger ribonucleic acid of the rat anterior pituitary gland, J. Neuroendocrinol. 3:351.

Chomczynski, P. and Sacchi, N., 1987, Single-step method of RNA isolation by acid guanidinium thiocyanate-phenol-chloroform extraction, Anal. Biochem. 162:156.

Fahrenkrug, J., 1987, Co-existence and co-secretion of the structurally related peptides VIP and PHM, Scand. J. Clin. Lab. Invest. 47 (Suppl. 186):43.

Fahrenkrug, J., 1989, VIP and autonomic transmission, Pharmacol. Ther. 41:515.

Fahrenkrug, J., 1991, Glycine-extended processing intermediate of preproVIP: a new form of VIP in the rat, Biochem. Biophys. Res. Commun. 178:173.

Fahrenkrug, J. and Emson, P., 1989, Characterization and regional distribution of peptides derived from the vasoactive intestinal peptide precursor in the normal human brain, J. Neurochem. 53:1142.

Fahrenkrug, J., Ottesen, B., and Palle, C., 1989, Non–amidated forms of VIP (glycine–extended VIP and VIP–free acid) have full bioactivity on smooth muscle, Regul. Pept. 26:235.

Fahrenkrug, J. and Pedersen, J.H., 1984, Development and validation of a specific radioimmunoassay for PHI in plasma, Clin. Chem. Acta 143:183.

Fahrenkrug, J. and Pedersen, J.H., 1986, Co–secretion of peptide histidine methionine (PHM) and vasoactive intestinal peptide (VIP) in patients with VIP–producing tumours, Peptides 7:717.

Fahrenkrug, J. and Schaffalitzky de Muckadell, O.B., 1977, Radioimmunoassay of vasoactive intestinal poly-peptide in plasma, J. Lab. Clin. Med. 89:1379.

Fahrenkrug, J. and Schaffalitzky de Muckadell, O.B., 1978, Distribution of vasoactive intestinal polypeptide in the porcine central nervous system, J. Neurochem. 31:1445.

Fink, J.S., Verhave, M., Walton, K., Mandel, G., and Goodman, R., 1991, Cyclic AMP- and phorbol ester-induced transcriptional activation are mediated by the same enhancer element in the human vasoactive intestinal peptide gene, J. Biol. Chem. 266:3882.

Georg, B., Wulff, B.S., and Fahrenkrug, J., 1992, Regulation of prepro-VIP expression in the human neuroblastoma cell line NB-1, Biomed. Res. 13 (suppl. 2), in press.

Glass, C.K., DiRenzo, J., Kurokawa, R., and Han, Z., 1991, Regulation of gene expression by retinoic acid receptors, DNA and Cell Biology 10:623.

Hayakawa, Y., Obata, K.-I., Itoh, N., Yanaihara, N., and Okamoto, H., 1984, Cyclic AMP regulation of Pro-vasoactive intestinal polypeptide/PHM-27 synthesis in human neuroblastoma cells. J. Biol. Chem. 259:9207.

Laburthe, M. and Couvineau, A., 1988, Molecular analysis of vasoactive intestinal peptide receptors, Ann. N.Y. Acad. Sci. 527:296, S. Said and V. Mutt, eds..

Ohsawa, K., Hayakawa, Y., Nishizawa, M., Yamagami, T., Yamamoto, H., Yanaihara, N., and Okamoto, H., 1985, Synergistic stimulation of VIP/PHM-27 gene expression by cAMP and phorbol esters in human neuroblastoma cells, Biochem. Biophys. Res. Commun. 132:885.

Palle, C., Ottesen, B., and Fahrenkrug, J., 1992, Peptide histidine valine (PHV) is present and biological active in the human female genital tract, Peptides 38:101.

Palle, C., Ottesen, B., Jørgensen, J., and Fahrenkrug, J., 1989, Peptide histidine methionine and vasoactive intestinal peptide: occurence and relaxant effect in the human female reproductive tract, Biol. Reprod. 41:1103.

Sidell, N., Altman, A., Haussler, M.R., and Seeger, R.C., 1983, Effects of retinoic acid (RA) in the growth and phenotypic expression of several human neuroblastoma cell lines, Exp. Cell Res. 148:21.

Tsukada, T., Fink, J.S., Mandel, G., and Goodman, R.H., 1987, Identification of a region in the human vasoactive intestinal polypeptide gene responsible for regulation by cyclic AMP, J. Biol. Chem. 262:8743.

Tsukada, T., Horovitch, S.J., Montminy, M.R., Mandel, G., and Goodman, R.H., 1985, Structure of the human vasoactive intestinal polypeptide gene, DNA 4:293.

Yanaihara, N., Suzuki, T., Sato, H., Hoshino, M., Okaru, Y., and Yanaihara, C., 1981, Dibutyryl cAMP stimulation of production and release of VIP-like immunoreactivity in a human neuroblastoma cell line, Biomed. Res. 2:728.

STRUCTURE-DEPENDENT INHIBITORY EFFECT OF GALANIN ON INSULIN, GASTRIC ACID AND PANCREATIC AMYLASE SECRETION IN RAT

Wojciech J. Rossowski and David H. Coy

Peptide Research Laboratories and Gastroenterology Section
Department of Medicine
Tulane University School of Medicine
New Orleans, Louisiana

INTRODUCTION

Galanin was originally isolated as a 29-amino acid peptide from extracts of porcine small intestine on the basis of its amidated C-terminus (Sundstrom & Melander, 1988) . Subsequently rat (Kaplan et al., 1988), bovine (Rokaeus & Carlquist, 1988), sheep (Sillard et al., 1991) and human (Bersani et al., 1991; Evans & Shine, 1991) galanin primary structures were determined to be:

1

Porcine : Gly - Trp - Thr - Leu - Asn - Ser - Ala - Gly - Tyr - Leu -
Rat : Gly - Trp - Thr - Leu - Asn - Ser - Ala - Gly - Tyr - Leu -
Bovine : Gly - Trp - Thr - Leu - Asn - Ser - Ala - Gly - Tyr - Leu -
Sheep : Gly - Trp - Thr - Leu - Asn - Ser - Ala - Gly - Tyr - Leu -
Human : Gly - Trp - Thr - Leu - Asn - Ser - Ala - Gly - Tyr - Leu -

11

Porcine : Leu - Gly - Pro - His - Ala - Ile - Asp - Asn - His - Arg -
Rat : Leu - Gly - Pro - His - Ala - Ile - Asp - Asn - His - Arg -
Bovine : Leu - Gly - Pro - His - Ala -_Leu_- Asp -_Ser_- His - Arg -
Sheep : Leu - Gly - Pro - His - Ala - Ile - Asp - Asn - His - Arg -
Human : Leu - Gly - Pro - His - Ala -_Val-Gly_- Asn - His - Arg -

21

Porcine : Ser - Phe - His - Asp - Lys - Tyr - Gly - Leu - Ala - NH_2
Rat : Ser - Phe -_Ser_- Asp - Lys -_His_- Gly - Leu -_Thr_- NH_2
Bovine : Ser - Phe -_Gln_- Asp - Lys -_His_- Gly - Leu - Ala - NH_2
Sheep : Ser - Phe - His - Asp - Lys -_His_- Gly - Leu - Ala - NH_2
Human : Ser - Phe -_Ser_- Asp - Lys -_Asn_- Gly - Leu -_Thr-Ser_

Growth Factors, Peptides, and Receptors, Edited
by T.W. Moody, Plenum Press, New York, 1993

The first fifteen N-terminal amino acid residues are identical in all tested species. Species-specific differences in galanin amino acid composition are restricted to the C-terminal 14/15 amino acid residues, and the most extensive differences have been found in human galanin 1-30. Galanin-like immunoreactivity is widely distributed in the central and peripheral nervous systems (Rokaeus et al., 1984), especially in the hypothalamus (Bonnefond et al., 1990; Michener et al., 1990), hippocampus (Fisone et al., 1987; Fisone et al., 1989), adrenal medulla (Fried et al., 1991; Holst et al., 1991), gastrointestinal tract (Bauer et al., 1986; Bauer et al., 1989; Bishop et al., 1986), and pancreas (Ahren et al., 1988; Lindskog & Ahren, 1987; Lindskog, 1991). In the human pancreas, galanin-like immunoreactivity is localized in numerous nerve fibers around glandular acini, ductules and blood vessels, and in a few nerve fibres within islets (Shimosegawa et al., 1992). Immunostaining for galanin and for vasoactive intestinal peptide (VIP) showed the coexistence of the two immunoreactivities in a large proportion of nerve cells (Shimosegawa et al., 1992). Galanin-like immunoreactivity has also been shown to co-exist with VIP-like immunoreactivity in nerve fibres innervating human circular and longitudinal colonic muscle layers (Burleigh & Furness, 1990).

During basal conditions, the major source of circulating galanin-like immunoreactivity in pigs is the distal part of the gut, especially mid-colon (Harling & Holst, 1992). The mean half-life of galanin in plasma was 4.6 ± 0.3 min and most of endogenous and exogenous galanin was removed by the kidneys (Harling & Holst, 1992).

Galanin is a potent inhibitor of insulin and somatostatin secretion from perfused rat stomach and pancreas (Kwok et al., 1988; Madaus et al., 1988; Soldani et al., 1988) and also inhibits acetylcholine release from the ventral hippocampus (Fisone et al., 1987), dopamine secretion from pheochromocytoma cells (De Weille et al., 1989) and the rat median eminence (Nordstrom et al., 1987) and histamine release in rat hypothalamus and hippocampus (Arrang et al., 1991) as well as inhibiting histamine release from human basophils (Bergstrand et al., 1991). Potent inhibitory effects of galanin on serotonergic neurons in the rat brain have also been reported (Sundstrom & Melander, 1988) . Galanin strongly stimulates release of glucagon and growth hormone (Lindskog & Ahren, 1987; Maiter et al., 1990). Galanin also displays a potent, species-specific regulatory effect on gastrointestinal smooth muscles (Bauer et al., 1989; Ekblad et al., 1985; Fox et al., 1988; Katsoulis et al., 1990) and probably plays an important physiological role in the regulation of gastrointestinal sphincter functions (Allescher et al., 1989; Chakder & Rattan, 1991).

The N-terminal fragments of porcine galanin 1-10; 1-11; 1-15; and 1-20 retain significant activity in the contraction of longitudinal smooth muscle from rat jejunum (Ekblad et al., 1985; Fox et al., 1988) and isolated rat fundus strips (Katsoulis et al., 1990). Similarly, porcine galanin 1-15 inhibits insulin secretion from the pancreatic tumour beta-cell line Rin m 5F when used at higher dose level (IC_{50} 21 ± 9 nM as compared to 0.7 ± 0.2 nM for galanin 1-29) (Amiranoff et al., 1989) and porcine galanin 1-16 inhibited insulin release from isolated rat pancreatic islets at μM concentration (Gregersen et al., 1991). However, it was also found that the C-terminal porcine galanin fragments

12-29; 18-29; and 21-29 could recognize galanin-binding sites in the ventral hippocampus (Fisone et al., 1989), and porcine galanin 3-29 was active in inhibiting insulin secretion (Gregersen et al., 1991; Gregersen et al., 1992). Moreover, the porcine galanin 9-29 retains residual inhibitory activity in a test of gastric acid secretion stimulated by pentagastrin (Rossowski & Coy, 1989) and is also active in the displacement of the radioiodinated porcine galanin 1-29 from rat gastric smooth muscle membrane preparations (Rossowski et al., 1990). In in vivo studies porcine galanin 21-29 and 15-29 produces weak inhibition of field -stimulated activity in the distal part of canine ileum, while those fragments were not active in rats and guinea pigs (Fox et al., 1988). Galanin fragment 1-16 is also a full agonist in the displacement of radio-iodinated galanin 1-29 from hypothalamic receptor-binding sites, however, further shortening to galanin 1-15 and 1-10 causes gradual drop of affinity (K_D = 0.2 and 25 µM respectively as compared to 0.8 nM for porcine galanin 1-29 and 6.9 nM for galanin 1-16) (Land et al., 1991). Substitutions of individual amino acid residues in porcine galanin fragment 1-16 with L-Ala residue show that amino acid residues Gly[1]; Trp[2]; Asn[5]; Tyr[9]; Leu[10,11]; and Gly[12] are particularly important for the high affinity binding in displacement experiments (Land et al., 1991).

STRUCTURAL REQUIREMENTS FOR GALANIN-INDUCED INSULIN INHIBITION

Galanin inhibits insulin secretion stimulated by glucose, hormones, neurotransmitters and pharmacological agents (Hramiak et al., 1988). To inhibit insulin release, galanin interacts with high affinity membrane receptors on pancreatic beta-cells that are coupled to a pertussis toxin-sensitive inhibitory G-protein (G_{i1} and/or G_{i2}) (Cormont et al., 1991). It is likely that the receptor-activated G_i directly inhibits the activity of adenylate cyclase (Cormont et al., 1991). Recent studies show also that galanin inhibits proinsulin gene expression stimulated by the insulinotropic glucagon-like hormone GLP-I(7-37) and forskolin and that this action of galanin is mediated by a pertussis toxin-sensitive (G_i) pathway (Fehmann et al., 1992).

The inhibitory activity of galanin on insulin secretion is structure-dependent and the whole molecule is required for full activity. Porcine galanin 1-29 inhibits insulin release from Rin m 5F cells dose-dependently with IC_{50} = 0.7±0.2 µM (Amiranoff et al., 1989). Elimination of the first N-terminal amino acid residue results in about a 10 fold decrease in activity while elimination of first two amino acid residues result in the formation of a molecule which is more than 1000 fold less active as compared to the unmodified molecule when tested in the pancreatic beta cell line Rin m 5F (Amiranoff et al., 1989). In the perfused dog pancreas, porcine galanin 3-29 and 10-29 at a concentration of 1 nM did not significantly inhibit insulin secretion (Hermansen et al., 1989.) and porcine galanin 3-29 and 10-29 were ineffective for inhibiting (^{125}I) galanin binding to membranes from Rin-m 5F cells and rat brain (Gallwitz et al., 1990; Lagny-Pourmir et al., 1989). However in another experimental model (Gregersen et al., 1991; Gregersen et al., 1992) and in our in vivo studies in rat (Rossowski et al., 1992), rat galanin 3-29 at a concentration of 30 nmol

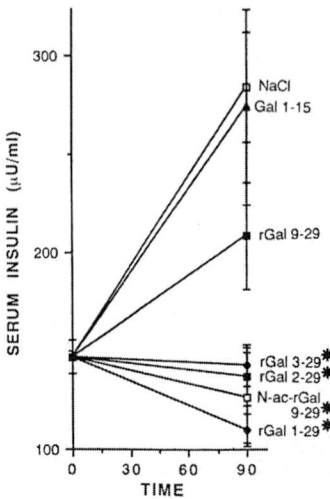

Fig 1. Effect of rat galanin (3nmol kg^{-1} h^{-1}) and rat galanin fragments: 2-29 (3nmol kg^{-1} h^{-1}); 3-29 (30nmol kg^{-1} h^{-1}); 9-29 (30nmol kg^{-1} h^{-1}); N^α-acetyl - 9-29 (30nmol kg^{-1} h^{-1}); and galanin 1-15 (30nmol kg^{-1} h^{-1}), on serum insulin level (μU/ml). After basal insulin determination rat galanin or galanin fragments were i.v. infused for 30 min. After infusion was stopped, blood samples were collected for RIA insulin determinations. Values are the means ±SEM (n=6-8), *p<0.05.

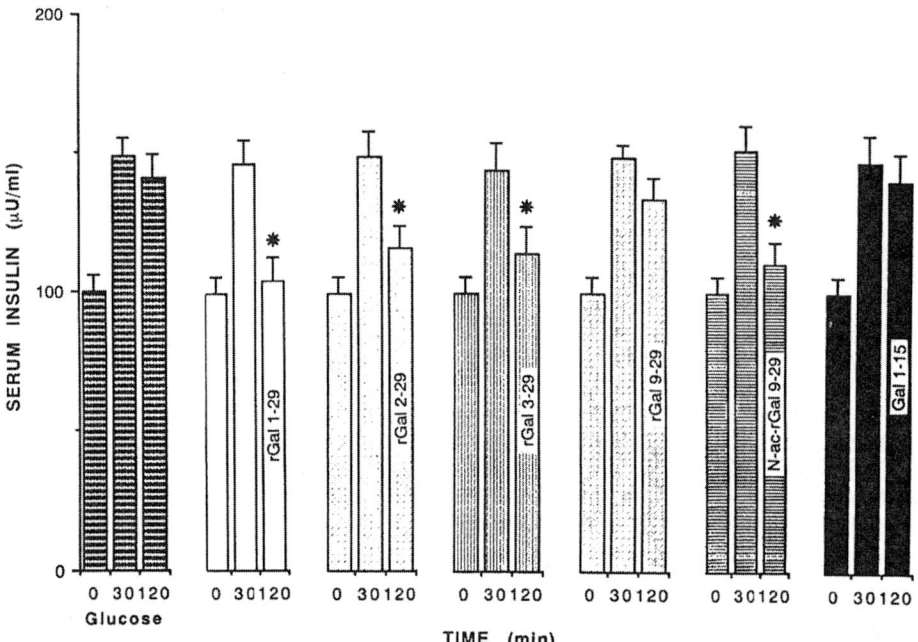

Fig. 2. Effect of rat galanin and rat galanin fragments on glucose-stimulated serum insulin level. After basal insulin determination, glucose at concentration of 2.8mmol kg^{-1} h^{-1} was i.v. infused for 120 min. Thirty minutes after beginning of glucose infusion, galanin or galanin fragments were i.v. infused into second femoral vein for the following 30 min. Blood samples for insulin determination were taken and 30 min. and 120 min. Galanin and galanin fragments doses are identical as indicated in Figure 1. Values are the means ±SEM (n=8), *p<0.05.

kg^{-1} h^{-1}, significantly inhibited both basal and glucose-stimulated insulin secretion (Fig. 1 and Fig. 2) which suggests that, at least in some species, the presence of the first two N-terminal amino acids is not absolutely required. Moreover, while rat galanin 9-29 was not active at dose 30 nmol kg^{-1} h^{-1}, the introduction of the acetyl group to the N-terminal amino acid residue, caused the reappearance of insulin inhibitory activity (Rossowski et al., 1992).

On the other hand galanin 1-15 (Amiranoff et al., 1989), galanin 1-16 and galanin1-11, were able to inhibit insulin secretion from isolated rat islets, but only when tested at significantly higher (10^{-6} M) concentrations (Gregersen et al., 1991). In our in vivo experiments (Rossowski et al., 1992), rat galanin 1-15 at doses 3 and 30 nmol kg^{-1} h^{-1} was not active in inhibiting both unstimulated and glucose-stimulated serum insulin concentration in pentobarbital-anesthetized rats (Fig. 1 and Fig. 2) (Rossowski et al., 1992).

STRUCTURAL REQUIREMENTS FOR GALANIN-INDUCED GASTRIC ACID INHIBITION IN RATS

Galanin-like immunoreactivity has been found in the nerve cell bodies and fibres in the myenteric plexi of the gastric antrum and corpus in several mammalian species (Bauer et al., 1986; Bishop et al., 1986). Galanin-positive nerve fibers were also found in rat gastric mucosa and submucosa (Bishop et al., 1986).

Similarly, high affinity galanin receptor binding sites have been identified on gastric and jejunal smooth muscle membrane preparations (K_D = 2.77±0.78 nM and 4.93±1.74 nM for gastric smooth muscle and jejunal smooth muscle membranes respectively) (Rossowski et al., 1990) and on dispersed smooth muscle cells from guinea pig stomach (K_D = 3 nM) (Gu et al., 1992). Recently, porcine gastric galanin was isolated from the gastric corpus and the amino acid composition and sequence analysis showed the identity of the gastric and intestinal forms of galanin (McDonald et al., 1992). Porcine galanin at concentrations of 10^{-10} -10^{-8} M inhibits rat basal gastrin secretion at luminal pH 7.0 and at concentrations of 10^{-9} -10^{-8} M at luminal pH 2.0. Porcine galanin also inhibits neuromedin C-stimulated gastrin release from isolated perfused rat stomach (Madaus et al., 1988). In dogs equipped with chronic gastric fistula, galanin has no effect on basal gastric acid secretion or on the secretion stimulated by agents acting directly on parietal cell (bethanechol and histamine), but it significantly decreases acid secretion stimulated by agents that activate cholinergic neuronal pathways through a central or peripheral mechanisms (2-deoxy-D-glucose or bombesin) (Soldani et al., 1988). In the G-cell enriched fraction from rat gastric mucosa, galanin tested at concentrations of 10^{-10}- 10^{-7} M has no effect on basal gastrin release but reduces the responses to neuromedin C, carbachol, TPA (12-0-tetradecanoyl-phorbol 13-acetate), and DBcAMP (N^6, 2´-O-dibutyryladenosine 3´,5´-cyclic monophosphate) with an IC$_{50}$ ranged between 1 x 10^{-10} and 8.6 x 10^{-10} M (Scheep et al., 1990). These results suggests therefore, that galanin might exert a direct inhibitory effect on rat gastrin cells by interfering at the intracellular mechanisms which are activated by protein kinase C and cAMP (Scheep et al., 1990). Porcine galanin did not affect serum gastrin concentration when i.v. infused at a dose of 40 pmol kg min for one hour into healthy human volunteers (Bauer et al., 1989).

We have previously found that porcine galanin i.v. infused into pentobarbital-anesthetized rats at dose of 3 nmol kg^{-1} h^{-1} , significantly inhibits pentagastrin-stimulated gastric acid secretion (Rossowski & Coy, 1989). Porcine galanin however, did not inhibit bethanechol-stimulated gastric acid secretion in rats (Rossowski & Coy, 1989) as well as in conscious dogs (Soldani et al., 1988). In addition, a dose-dependent and biphasic inhibitory effect of galanin on gastric acid secretion in rats was suggested (Yagci et al., 1990). We have also found that porcine galanin fragment 9-29 required 20 times higher concentration to induce residual gastric acid inhibition while porcine galanin fragment 15-29 was inactive at a dose of 10^{-7} M (Rossowski & Coy, 1989). Further, we have investigated the effect of rat galanin and several the N-terminal and C-terminal rat galanin fragments on pentagastrin-stimulated gastric acid secretion in conscious, chronic gastric fistula equipped rats (Mungan et al., 1992). We have found, in agreement with previous studies using porcine galanin, that rat galanin is a potent inhibitor of pentagastrin-stimulated gastric acid secretion in conscious rats. We also found that the N-terminal rat galanin fragments 1-10 and 1-15 are active in inhibiting pentagastrin-stimulated gastric acid secretion (Fig. 3) (Mungan et al., 1992; Rossowski et al., 1992). Elimination of Gly[1] residue decreases galanin inhibitory activity in stimulated gastric acid secretion. Rat galanin fragments 3-29 and 9-29 are unable to inhibit pentagastrin-stimulated gastric acid secretion

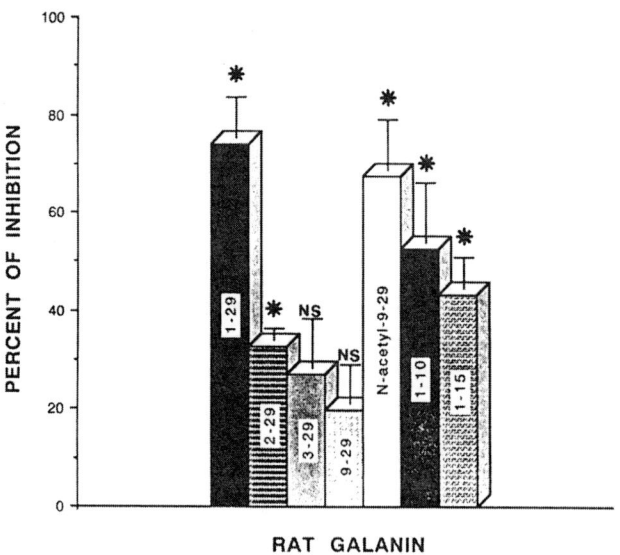

Fig. 3. Effect of rat galanin and rat galanin fragments on pentagastrin-stimulated gastric acid secretion in conscious, chronic gastric fistulae equipped rats. After basal collections, pentagastrin (24μg kg^{-1} h^{-1}) was i.v. infused and continued until the end of the experiment. Ninety minutes after pentagastrin infusion was started, i.v. rat galanin or galanin fragment infusion at a concentration of 3nmol kg^{-1} h^{-1} was added and continued for another 90 minutes. Values are the means ±SEM (n=6-7), *p<0.05.

significantly (Fig. 3.). However, acetylation of rat galanin 9-29 to give N^{α}-acetyl-rat galanin 9-29 results in conversion of essentially inactive peptide to an active molecule with high inhibitory activity in gastric acid secretion tests (Fig. 3) (Mungan et al., 1992; Rossowski et al., 1992). We speculate that this could be due to a favorable conformational effect on the carboxy-terminal region produced by α-acetylation. The fact that short N-terminal and C-terminal galanin sequences are biologically active in this system suggests that more than one receptor type could be involved having different peptide binding requirements.

STRUCTURAL REQUIREMENTS FOR GALANIN-INDUCED INHIBITION OF PANCREATIC AMYLASE SECRETION

Galanin is a potent inhibitor of basal and stimulated pancreatic amylase secretion in vivo in rats (Herzig et al., 1992; Rossowski et al., 1992; Yagci et al., 1991), pancreatic bicarbonate and protein secretion in dogs (Powers & Pappas, 1989) as well as stimulated amylase release from pancreatic acini in vitro (Ahren et al., 1988; Flowe et al., 1992; Herzig et al., 1992). Galanin inhibitory action on amylase release is not yet completely clear but it probably involve several different pathways such as inhibition of acetylcholine and insulin release, stimulation of glucagon secretion and/or direct inhibition of pancreatic acinar cell secretion. It was suggested that galanin acts as a sympathetic neurotransmitter in the dog pancreas (Dunning & Taborsky, 1988). It was found that electrical stimulation of the mixed autonomic pancreatic nerves in dogs (Dunning & Taborsky, 1989) or administration of the alpha$_2$-selective adrenoceptor antagonist yohimbine (Scheurink et al., 1992) markedly increases pancreatic output of both norepinephrine and galanin, suggesting that galanin might be a sympathetic cotransmitter in the endocrine pancreas. If this suggestion is correct, both cotransmitters e.g. norepinephrine and galanin at low concentrations can directly inhibit insulin release. However, species-specific differences have been found, e.g. in dogs, basal insulin secretion is not inhibited by pancreatic noradrenaline infusion (Scheurink et al., 1992). It was shown previously that in mice, rats and pigs both homologous and heterologous galanin significantly inhibits basal and stimulated insulin secretion in vivo and stimulated insulin secretion in vitro (for review see Lindskog, 1991). We have shown previously, that in rats, galanin significantly inhibits basal insulin and amylase secretion (Rossowski et al., 1992). There is strong evidence that insulin can affect pancreatic amylase secretion. High affinity insulin receptors have been found on pancreatic acinar cells (see review Williams & Goldfine, 1985) and dramatic reduction in pancreatic amylase secretion is accompanied by plasma insulin decrease in rats injected with streptozotocin. Infusion of exogenous insulin into diabetic rats restores pancreatic amylase content to control levels (Okabayashi et al., 1988). It was also found that i.v. administration of anti-insulin serum at a dose which almost completely eliminates circulating serum insulin, significantly inhibits pancreatic amylase secretion in a time-dependent manner (Lee et al., 1990; Rossowski et al, 1992). In diabetic patients, the pancreatic exocrine secretory response to exogeneous secretin and cholecystokinin is

43

impaired and a decrease in pancreatic enzyme secretion was correlated with residual insulin secretion (Newihi et al., 1988). Recent studies indicate, that insulin directly regulates pancreatic acinar cell function by binding to the high affinity insulin receptors, regulation of amylase messenger RNA (Korc et al., 1981), and by direct interaction with the clustered pancreatic amylase structural genes (Dranginis et al., 1984).

Porcine galanin at concentration of 10^{-8} - 10^{-9} M reduces CCK-8 and carbachol-stimulated amylase secretion from isolated rat pancreatic acini but does not affect basal amylase secretion (Ahren et al., 1988). In dogs equipped with chronic pancreatic fistula, galanin inhibits basal and meal-stimulated pancreatic protein and bicarbonate secretion (Powers & Pappas, 1989) in dose-dependent fashion. We have previously reported that porcine galanin inhibits basal and bombesin-, secretin-, and cholecystokinin-stimulated pancreatic protein and amylase secretion in pentobarbital-anesthetized rats (Yagci et al., 1991). This inhibitory effect of galanin is dose-dependent and biphasic with maximal inhibitory activity at 3 nmol kg^{-1} h^{-1}. Similar dose-dependent biphasic responses on amylase secretion have been observed with cerulein and carbamylcholine (Okabayashi et al., 1988) and they could indicate effects on the homeostasis of multiple hormone systems with competing biological functions, or may be a result of desensitization.

Continuous i.v. infusion of rat or porcine galanin 1-29 at a dose 3 nmol kg^{-1} h^{-1} into anesthetized rats results in a rapid decrease of basal and CCK-8-stimulated pancreatic amylase and protein secretion which is continued throughout the entire infusion period (Rossowski et al., 1992).

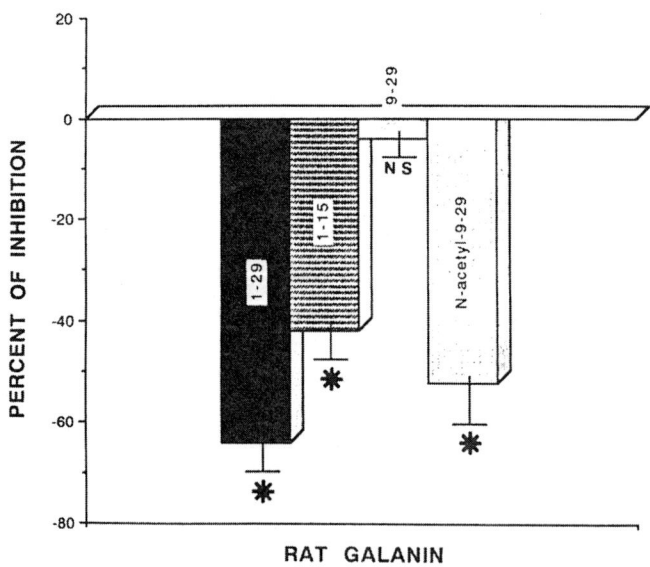

Fig. 4. Effect of rat galanin (3nmol kg^{-1} h^{-1}) and rat galanin fragments: 1-15 (3nmol kg^{-1} h^{-1}); 9-29 (30nmol kg^{-1} h^{-1}) and N^{α}-acetyl-9-29 (30nmol kg^{-1} h^{-1}) on basal (unstimulated) pancreatic amylase secretion. After basal (4x15 min) collection periods, galanin or galanin fragments were i.v. infused and continued for another 90 min. Values are the means ±SEM (n=9), *p<0.05.

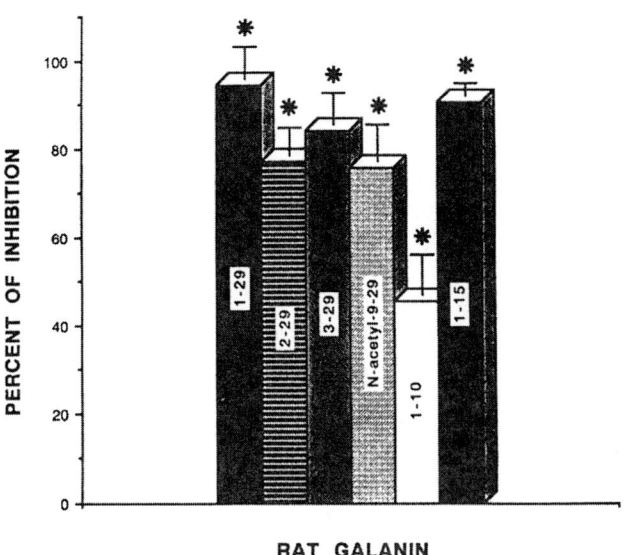

RAT GALANIN

Fig. 5. Effect of rat galanin (3nmol kg^{-1} h^{-1}) and rat galanin fragments: 2-29 (3nmol kg^{-1} h^{-1}); 3-29 (3nmol kg^{-1} h^{-1}); N$^{\alpha}$-acetyl-9-29 (30nmol kg^{-1} h^{-1}); 1-10 (3nmol kg^{-1} h^{-1}) and 1-15 (3nmol kg^{-1} h^{-1}) on CCK-8 stimulated pancreatic secretion. After basal collection, CCK-8 (0.5nmol kg^{-1} h^{-1}) was i.v. infused and continued until the end of the experiment. Sixty minutes after CCK-8 infusion was started, i.v. rat galanin or galanin fragment infusion was added and continued for another 90 min. Values are the means ±SEM (n=9-12), *p<0.05.

The N-terminal galanin fragment 1-15 when tested at the comparable dose of 3 nmol kg^{-1} h^{-1} was significantly less active in inhibiting basal but not CCK-8-stimulated pancreatic amylase secretion. Reduction of amino acid sequence to the first ten N-terminal residues in galanin 1-10, significantly reduced amylase inhibitory activity. Rat galanin C-terminal fragments 9-29 and 21-29 and porcine galanin fragments 9-29 and 15-29, when tested at doses of 3 and 30 nmol kg^{-1} h^{-1}, were not active in inhibiting basal pancreatic protein and amylase secretion. Introduction of the acetyl group onto the N-terminally deleted fragment rat galanin 9-29 to give N$^{\alpha}$-acetyl-rat galanin 9-29 resulted in the reappearance of inhibitory activity on the basal pancreatic amylase and protein secretion. The C-terminal rat galanin fragments: galanin 2-29, 3-29; N$^{\alpha}$-acetyl-galanin 9-29 significantly inhibited CCK-8-stimulated pancreatic amylase secretion in the in vivo rat model (Fig. 4 and Fig. 5), (Rossowski et al., 1992).

CONCLUSION

Rat galanin is a potent inhibitor of insulin, gastric acid and pancreatic amylase secretion in the homologous rat model in vivo. It is suggested that for full expression of inhibitory activity on individual secretory processes the full galanin amino acid sequence is certainly needed. However, there are tissue-specific differences in the recognition of specific modified galanin sequences by the galanin receptor binding sites. Rat galanin 2-29;

3-29; and N^α-acetyl-9-29 displayed similar activities, when tested at comparative doses, on insulin and pancreatic amylase secretion. The N-terminal galanin fragment 1-15 was practically not active in an insulin inhibition test when tested at doses 3 and 30 nmol kg^{-1} h^{-1}, but was fully active in inhibition of CCK-8-stimulated amylase secretion. Galanin fragment 1-10, which was least active in pancreatic amylase inhibition test, displayed strong inhibitory activity in inhibition of pentagastrin-stimulated gastric acid secretion in conscious chronic gastric fistula-equipped rats. Also, in contrast to the insulin and amylase test, rat galanin fragments: 2-29 and 3-29 displayed very weak inhibitory activity when tested in gastric acid secretion. The unexpected observation was reappearance or significant increase of the inhibitory galanin activity in the N^α-acetyl-rat galanin 9-29 fragment when tested in all studied secretory systems in vivo. The fact that N^α-acetyl-porcine galanin 9-29 fragment was less active than the corresponding rat fragment in inhibition of CCK-8-stimulated amylase secretion in rat, suggest that the C-terminal galanin sequences are biologically important. We have also shown that the presence of the first two N-terminal amino acid residues is not absolutely required for preservation of the galanin inhibitory activity at least in some biological systems.

ACKNOWLEDGMENT

We wish to thank Ms. Robyn Denenea for her excellent secretarial assistance.

REFERENCES

1. Ahren, B., Rorsman, P. and Berggren, P.O., 1988, Galanin and the endocrine pancreas. FEBS Lett. 229: 233.

2. Ahren, B., Andren-Sandberg, A., and Nilsson, A., 1988, Galanin inhibits amylase secretion from isolated rat acini. Pancreas, 3: 559.

3. Allescher, D. H., Daniel, E. E., Dent, J. and Fox, J. E. T., 1989, Inhibitory function of VIP-PHI and galanin on canine pylorus. Am. J. Physiol. 256: G789

4. Amiranoff, B., Lorinet, A.-M., Yanaihara, N. and Laburthe, M., 1989, Structural requirements for galanin action in the pancreatic cell line Rin m 5F. Eur. J. Pharmacol. 163: 205.

5. Arrang, J. M., Gulat-Marnay, C., Defontaine, N., and Schwartz, J.C., 1991, Regulation of histamine release in rat hypothalamus and hippocampus by presynaptic galanin receptors. Peptides, 12: 1113.

6. Bauer, F. E., Adrian, T.E., Christofides, N.D., Ferri, G.L., Yanaihara, N., Polak, J.M., Bloom, S.R., 1986, Distribution and molecular heterogeneity of galanin in human, pig, guinea pig, and rat gastrointestinal tracts. Gastroenterology, 91 : 877.

7. Bauer, F. E., Zintel, A., Kenny, M.J., Caldar, D., Ghatei, M.A., Bloom, S.R., 1989, Inhibitory effect of galanin on postprandial motility and gut hormonerelease in humans. Gastroenterology, 97: 260.

8. Bergstrand, H., Lundquist, B., and Ahren, B., 1991, Inhibition by galanin and by high K$^+$ of human basophil histamine release by calcium ionophoresbut not responses induced by anti-IgE, chemotactic peptide or phorbol ester. Br. J. Pharmacol., 103: 1381.

9.   Bersani, M., Johnsen, A.H., Hojrup, P., Dunning, B.E., Andersen, J.J., Holst, J.J., 1991, Human galanin: primary structure and identification of two molecular forms. FEBS Lett., 283:189.

10. Bishop, A. E., Polak, J.M., Bauer, F.E., Christofides, N.D., Carlei, F., Bloom, S.R., 1986, Occurence and distribution of a newly discovered peptide galanin, in the mammalian enteric nervous system. Gut, 27 :849.

11. Bonnefond, C., Palacios, J.M., Probst, A. and Mengod, G.,1990, Distribution of galanin mRNA containing cells and galanin receptor binding sites in human and rat hypothalamus. Eur. J. Neurosci., 2: 629.

12. Burleigh, D. E. and Furness, J.B., 1990, Distribution and action of galanin and vasoactive intestinal peptide in the human colon. Neuropeptides 16 :77.

13. Chakder, S. and Rattan, S., 1991, Effect of galanin on the opossum internal anal sphincter: structure-activity relationship. Gastroenterology,100:711.

14. Cormont, M., le Marchand-Brustel, Y., van Obberghen, E.E., Spiegel, A.M., Sharp, G.W.G., 1991, Identification of G protein a-subunits in RINm5F cells and their selective interaction with galanin receptor. Diabetes 40:1170.

15. De Weille, J. R., Fosset, M., Schmid-Antomarchi, H., Lazdunski, M., 1989, Galanin inhibits dopamine secretion and activates a potassium channel in pheochromocytoma cells. Brain Res., 485:199.

16. Dranginis, A., Morley, M., Nesbitt, M., Rosenblum, B.B., Meisler, M.H., 1984, Independent regulation of nonallelic pancreatic amylase genes in diabetic mice. J. Biol. Chem., 259:12216.

17. Dunning, B. E. and Taborsky, G.J., Jr., 1988, Galanin: a sympathetic neurotransmitter in the endocrine pancreas. Diabetes, 37:1157

18. Dunning, B. E. and Taborsky, G.J., Jr., 1989, Galanin release during pancreatic nerve stimulation is sufficient to influence islet function. Am. J. Physiol., 256 E191.

19. Ekblad, E., Hakanson, R., Sundler, F. and Wahlestedt, C., 1985, Galanin: neuromodulatory and direct contractile effect on smooth muscle preparations. Br. J. Pharmac., 86:241.

20. Evans, H. F. and Shine, J., 1991, Human galanin: molecular cloning reveals a unique structure. Endocrinology, 129:1682.

21. Fehmann, H.-C. and Habener, J.F., 1992, Galanin inhibits proinsulin gene expression stimulated by the insulinotropic hormone glucagon-like peptide-I(7-37) in mouse insulinoma TC-1 cells. Endocrinology, 130:2890.

22. Fisone, G., Wu, C.F., Consolo, S., Nordstrom, O., Brynne, N., Bartfai, T., Melander, T., Hokfelt, T., 1987, Galanin inhibits acetylcholine release in the ventral hippocampus of the rat: histochemical, autoradiographic, in vivo, and in vitro studies. Proc. Natl. Acad. Sci. USA, 84:7339.

23. Fisone, G., Langel, U., Carlquist, M., Bergman, T., Consolo, S., Hokfelt, T., Unden, A., Andell, S., Bartfai, T., 1989, Galanin receptor and its ligands in the rat hippocampus. Eur. J. Biochem. 181:269.

24. Flowe, K. M., Lally, K.M., and Mulholland, M.W., 1992, Galanin inhibits rat pancreatic amylase release via cholinergic suppression. Peptides, 13:487.

25. Fox, J. E. T., Brooks, B., McDonald, T.J., Barnett, W., Kostolanska, F., Yanaihara, C., Yanaihara, N., Rokaeus, A., 1988, Actions of galanin fragments on rat, guinea-pig, and canine intestinal motility. Peptides, 9:1183.

26. Fried, G., Wikstrom, L.M., Franck, J. and Rokaeus, A., 1991, Galanin and neuropeptide Y in chromaffin granules from the guinea-pig. Acta Physiol. Scand. 142:487.

27. Gallwitz, B., Schmidt, W.E., Schwarzhoff, R. and Creutzfeldt, W., 1990, Galanin: structural requirements for binding and signal transduction in Rin m 5F insulinoma cells. Biochem. Biophys. Res. Comm. 172:268.

28. Gregersen, S., Hermansen, K., Langel, U., Fisone, G., Bartfai, T. Ahren, B., 1991, Galanin-induced inhibition of insulin secretion from rat islets: effect of rat and pig galanin and galanin fragments and analogues. Eur. J. Pharmacol., 203:111.

29. Gregersen, S., Hermansen, K and Ahren, B., 1992, Galanin fragments and analogues: effects on glucose stimulated insulin secretion from isolated rat islets. Pancreas, 6:216.

30. Gu, Z. F., Pradhan, T.K., Coy, D.H., Jensen R.T., 1992, Dispersed smooth muscle cells from guinea pig stomach possess high affinity galanin receptors which mediate relaxation. Gastroenterology, 102:A453.

31. Harling, H., and Holst, J.J., 1992, Circulating galanin: origin, metabolism, and pharmacokinetics in anesthetized pigs. Am. J. Physiol. 262 :E52.

32. Hermansen, K., Yanaihara, N. and Ahren B., 1989, On the nature of the galanin action on the endocrine pancreas: studies with six galanin fragments in the perfused dog pancreas. Acta Endocrinologica (Copenh), 121:545.

33. Herzig, K. H., Brunhe, G., Schaffer, M., Gallwitz, B., Folsch, U.R., 1992, Galanin inhibits pancreatic enzyme secretion by inhibiting cholinergic transmission. Gastroenterology, 102:A269.

34. Holst, J. J., Ehrhart-Bornstein, M., Messel, T., Poulsen, S.S., Harling, H., 1991, Release of galanin from isolated perfused porcine adrenal glands: role of splanchnic nerves. Am. J. Physiol., 261:E31.

35. Hramiak, I. M., Dupre, J., . McDonald, T.J., 1988, Effects of galanin on insulin responses to hormonal, neuropeptidal, and pharmacological stimuli in conscious dogs. Endocrinology, 122:2486.

36. Kaplan,L. M., Spindel, E.R., Isselbacher, K.J., Chin, W.W., 1988, Tissue-specific expression of the rat galanin gene. Proc. Natl. Acad. Sci. USA, 85:1065

37. Katsoulis, S., Schmidt, W.E., Schworer, H., Creutzfeld, W., 1990, Effect of galanin, its analogues and fragments on rat isolated fundus strips. Br. J. Pharmacol., 101:297.

38. Korc, M., Owerbach, D., Quinto, C., Rutter, W.J., 1981, Pancreatic islet-acinar cell interaction: amylase messenger RNA levels are determined by insulin. Science, 213:351.

39. Kwok, Y. N., Verchere, C.B., McIntosh, C.H.S., Brown, J.C., 1988, Effect of galanin on endocrine secretions from the isolated perfused rat stomach and pancreas. Eur. J. Pharmacol., 145:49.

40. Land, T., Langel, U., Low, M., Berthold, M., Unden, A., Bartfai, T., 1991, Linear and cyclic N-terminal galanin fragments and analogs as ligands at the hypothalamic galanin receptor. Int. J. Peptide Protein Res., 38 :267.

41. Lagny-Pourmir, I., Lorinet, A.M., Yanaihara, N., Laburthe, M., 1989, Structural requirements for galanin interaction with receptors from pancreatic beta cells and from brain tissue of the rat. Peptides, 10:757.

42. Lee, K. Y., Zhou, L., Ren, X.S., Chang, T.-M., Chen, W.Y., 1990, An important role of endogenous insulin on exocrine pancreatic secretion in rats. Am. J. Physiol.,258:G268.

43. Lindskog, S. and Ahren, B., 1987, Galanin: effects on basal and stimulated insulin and glucagon secretion in the mouse. Acta Physiol. Scand.129:305.

44. Lindskog, S., 1991, Galanin in the regulation of insulin secretion. Doctoral Dissertation. University of Lund.

45. Madaus, S., Schuszdiarra, V., Seufferlein, T. and Classen M., 1988, Effect of galanin on gastrin and somatostatin release from the rat stomach.Life Sci., 42:2381.

46. Maiter, D. M., Hooi, S.C., Koenig, J.I. and Martin, J.B., 1990, Galanin is a physiological regulator of spontaneous pulsatile secretion of growth hormone in the male rat. Endocrinology, 126:1216.

47. McDonald, T. J., Krantis, A., Clarke, M., Mutt, V., Jornvall, H.E., 1992, Characterization of porcine gastric galanin. Peptides, 13:589.

48. Michener, S. R., Aimone, L.D., Yaksh, T.L., Go, V.L.W., 1990, Distribution of galanin-like immunoreactivity in the pig, rat and human central nervous system. Peptides, 11:1217.

49. Mungan, Z., Ozmen, V., Ertan, A., Coy, D.H., Baylor, L.M., Rice, J.C., Rossowski, W.J., 1992, Structural requirwmwnts for galanin inhibition of pentagastrin-stimulated gastric acid secretion in conscious rats. Eur. J. Pharmacol. 214:53.

50. Newihi el H., Dooley, C.P., Saad, C., Staples, J., Zeidler, A., Valenzuela, J.D., 1988, Impaired exocrine pancreatic function in diabetics with diarrhea and peripheral neuropathy. Dig. Dis. Sci. 33:705.

51. Nordstrom, O., Melander, T., Hokfelt, T., Bartfai, T., Goldstein, M., 1987, Evidence for an inhibitory effect of the peptide galanin on dopamine release from the rat median eminence. Neurosci. Lett. 73:21.

52. Okabayashi, Y., Otsuki, M., Ohki, A., Suehiro, I., Baba, S., 1988, Effect of diabetes mellitus on pancreatic exocrine secretion from isolated perfused pancreas in rats. Dig. Dis. Sci.,33:711.

53. Powers, M. A., and Pappas, T.N., 1989, Galanin is a potent inhibitor of pancreatic exocrine secretion. Gastroenterology, 96:A398.

54. Rokaeus, A., Melander, T., Hokfelt, T., Lundberg, J.M., Tatemoto, K., Carlquist, M., Mutt, V., 1984, A galanin-like peptide in the central nervous system and intestine of the rat. Neurosci Lett.,47:161.

55. Rokaeus, A., and M. Carlquist, M., 1988, Nucleotide sequence analysis of cDNAs encoding a bovine galanin precursor protein in the adrenal medulla and chemical isolation of bovine gut galanin. FEBS Lett. 234:400.

56. Rossowski, W. J., and Coy, D.H., 1989, Inhibitory action of galanin on gastric acid secretion in pentobarbital-anesthetized rats. Life Sci. 44:1807.

57. Rossowski, W. J., Rossowski, T.M., Zacharia, S., Ertan, A., Coy, D.H., 1990, Galanin binding sites in rat gastric and jejunal smooth muscle membrane preparations. Peptides, 11:333.

58. Rossowski, W. J., Zacharia, S., Mungan, Z., Mills, M., Ertan, A., Coy, D.H., 1992, Galanin: effect of new galanin agonists on pancreatic amylase secretion and jejunal smooth muscle contraction. Gastroenterology, 102:A754.

58a. Rossowski, W. J., Mungan, Z., Hammer, R.A. and Coy, D.H., 1992, Structural requirements for galanin-induced insulin, gastric acid and amylase inhibition in rat. XIIth Washington International Spring Symposium " Growth Factors, Peptides and Receptors ' 92", Washington, D.C. June 1-5, 1992, Abstract No. 63.

59. Scheep, W., Prinz, C., Tatge, C., Hakanson, R., Schusdziarra, V., Classen, M., 1990, Galanin inhibits gastrin release from isolated rat gastric G-cells. Am. J. Physiol., 258:G596.

60. Scheurink, A. J. W., Mundinger, T.O., Dunning, B.E., Veith, R.C., Taborsky, G.J., Jr., 1992, α2-adrenergic regulation of galanin and norepinephrine release from canine pancreas. Am. J. Physiol., 262:R819.

61. Shimosegawa, T., Moriizumi, S., Koizumi, M., Kashimura, J., Yanaihara, N., Toyota, T., 1992, Immunohistochemical demonstration of galaninlike immunoreactive nerves in the human pancreas. Gastroenterology, 102 :263.

62. Sillard, R., Langel, U. and Jornvall, H., 1991, Isolation and characterization of galanin from sheep brain. Peptides, 12:855.

63. Soldani, G., Mengozzi, G., Della Longa, A., Intorre, L., Martelli, F., Brown, D.R., 1988, An analysis of the effects of galanin on gastric acid secretion and plasma levels of gastrin in the dog. Eur. J. Pharmacol. 154:313.

64. Sundstrom, E. and Melander, T., 1988, Effects of galanin on 5-HT neurons in the rat CNS. Eur. J. Pharmacol., 148:327.

65. Tatemoto, K., Rokaeus, A., Jornvall, H., McDonald, T.M., Mutt, V., 1983, Galanin - a novel biologically active peptide from porcine intestine. FEBS Lett., 164:124.

66. Williams,J. A. and Goldfine, I.D., 1985, The insulin-pancreatic acinar axis. Review. Diabetes, 34:980.

67. Yagci, R. V., Alptekin, N., Rossowski, W.J., Brown, A., Coy, D.H., Ertan, A., 1990, Inhibitory effect of galanin on basal and pentagastrin-stimulated gastric acid secretion in rats. Scand. J. Gastroenterol. 25:853.

68. Yagci, R. V., Alptekin, N., Zacharia, S., Coy, D.H., Ertan, A., Rossowski W.J., 1991, Galanin inhibits pancreatic amylase secretion in the pentobarbital-anesthetized rat. Regulatory Peptides, 34:275.

AMPHIBIAN OPIOIDS: NOVEL DERMORPHIN-LIKE PEPTIDES

Lucia Negri

Institute of Medical Pharmacology
University "La Sapienza"
00185 Rome, Italy

INTRODUCTION

Among the fourteen peptide families so far described in amphibian skin, that of the opioid peptides is certainly one of the most interesting, as it raises a number of fascinating problems in the field of biochemistry, pharmacology and molecular biology. Opioid peptides have been so far found only in the skin of South American hylid frogs belonging to the subfamily Phyllomedusinae. These are heptapeptides with the common amino-terminal sequence Tyr-D-Xaa-Phe, where D-Xaa is either D-alanine or D-methionine.

The first peptide isolated from several species of these frogs was dermorphin, which was shown to have high affinity and selectivity for μ-type opioid receptors[1, 2]. The amino acid sequence of several dermorphin precursors was established by means of a cDNA library prepared from skin of *Phyllomedusa sauvagei*[3]. The sequence of one of these cDNAs indicated the existence of another peptide which contained methionine as the second amino acid[3]. This peptide was subsequently isolated from the skin of *Phyllomedusa sauvagei*[4, 5]. As demonstrated independently by three groups[4, 6, 7] this peptide had a higher affinity and selectivity for δ-opioid receptors than any other known natural compound. This compound is now referred to as met-deltorphin.

Subsequently, two additional peptides with even higher affinity for the δ-receptor were isolated from the skin of *Phyllomedusa bicolor* . Like dermorphin these peptides contain D-alanine as the second amino acid and they have been termed ala-deltorphin I and ala-deltorphin II[8].

The described peptides have the following sequences:

Tyr-D-Ala-Phe-Gly-Tyr-Pro-Ser-NH$_2$	dermorphin
Tyr-D-Met-Phe-His-Leu-Met-Asp-NH$_2$	met-deltorphin
Tyr-D-Ala-Phe-Asp-Val-Val-Gly-NH$_2$	ala-deltorphin I
Tyr-D-Ala-Phe-Glu-Val -Val -Gly-NH$_2$	ala-deltorphin II

Growth Factors, Peptides, and Receptors, Edited
by T.W. Moody, Plenum Press, New York, 1993

Recently, the structure of four precursors for ala-deltorphins I and II was deduced from cloned cDNAs derived from the skin of *Phyllomedusa bicolor*. These peptides contained one copy of ala-deltorphin II, and either one, three or no copies of ala-deltorphin I, respectively. In each case, a normal GCG codon for L-alanine was found in those positions where a D-alanine is present in the end product. Moreover, the cDNA sequences revealed that a glycine residue required for the formation of the COOH-terminal amide was present after the ala-deltorphin sequences, as it was after the met-deltorphin and dermorphin sequences determined earlier [9]

From the amino acid sequences of the ala-deltorphin precursors the existence of three additional peptides related to dermorphin could be predicted. Because of the absence of glycine at the carboxyl terminus of these peptides, they presumebly contained a free α-carboxyl group.

Tyr-D-Ala-Phe-Gly-Tyr-Pro-Lys -OH   [Lys^7-OH]dermorphin
Tyr-D-Ala-Phe-Trp-Tyr-Pro-Asn -OH   [Trp^4,Asn^7-OH]dermorphin
Tyr-D-Ala-Phe-Trp-Asn-OH   [Trp^4,Asn^5-OH]dermorphin 1-5

These dermorphin analogs have now been isolated from methanol extracts of this frog skin and their amino acid sequence confirmed. Like dermorphin, these peptides contain a D-alanine as the second amino acid and share the common amino terminal sequence Tyr-D-Ala-Phe. The D-alanine present in the final products is encoded in the skin mRNA by a normal codon for L-alanine. As predicted, the end products extracted from the skin were found to contain a free α-carboxyl group at the C-terminus.

DERMORPHIN-LIKE PEPTIDES FROM *Phyllomedusa bicolor* SKIN

Isolation and Identification

Fresh skin methanol extracts were fractionated by HPLC using a reverse-phase column (Aquapore RP-300, 7 x 250 mm, Brownlee Labs, Applied Biosystems) eluted with a 60 min gradient of 0-30% acetonitrile in 0.2% (by vol.) trifluoroacetic acid, at a flow of 2.0 ml/min. Elution of the peptides was monitored on a Beckman 160 spectrometer at 214 nm.

The eluate of the HPLC column was tested for inhibitory action on electrically evoked contraction in isolated preparations of myenteric plexus-longitudinal muscle obtained from the small intestine of the guinea-pig (GPI) and preparations of vas deferens of mice (MVD)[10]. The synthetic reference peptide was dermorphin. To characterize further the frog dermorphins, bioassays were performed in the presence or absence of naloxone and the δ receptor antagonist naltrindole.

The dermorphin-like activity was detected in two peaks, the first eluting with acetonitrile 24%, after 49 min and the second eluting with acetonitrile 30% after 65 min. Peptide sequences were determined by automated Edman degradation with an Applied Bioystems model 475A gas-phase sequencer and were found to be

Tyr-Ala-Phe-Gly-Tyr-Pro-Lys   (first peak)
Tyr-Ala-Phe-Trp-Tyr-Pro-Asn   (second peak)

The electrophoretic mobility of both peptides was consistent with the presence of a free α-carboxyl group. This result was also confirmed by fast-atom-bombardment mass spectral analysis (m/z 845 and 960, respectively, for the two peptides). Comparison of amino acid analyses performed on acid hydrolysates of the peptides before and after treatment with D-amino acid oxidase (Sigma) confirmed the presence of D-alanine in both peptides. Consequently, the two novel peptides were named [Lys^7-OH]dermorphin and [Trp^4,Asn^7-OH]dermorphin, respectively.

In the extract from the skin of *Phyllomedusa bicolor* the dermorphin pentapeptide Tyr-D-Ala-Phe-Trp-Asn-OH, also predicted from the sequence of the cloned cDNAs, could not be detected as a distinct chromatographic peak or by bioassay. This may be due to the low content of this peptide as well as to its relatively low biological activity (see below).

On the basis of amino acid analysis and bioassay we estimate that one gram of fresh skin of *Phyllomedusa. bicolor* contains about 25 µg of [Lys7-OH]dermorphin and 68 µg of [Trp4,Asn7-OH]dermorphin. The total amount of dermorphins present in the skin of *Phyllomedusa. bicolor* is thus similar to that found in other Phyllomedusinae.

Synthesis of Peptides

The three peptides predicted from the sequence of the two deltorphin precursors, their related C-terminal amides and some substituted analogs were synthesized using fluorenylmethoxycarbonyl-polyamide active-ester chemistry on a Biolynx automated peptide synthesizer (Pharmacia Biochrom, Cambridge, U.K.). The products were purified by preparative HPLC using a reverse-phase column (Vydac C_{18}, 10 µm, 300 °A, 22 x 250 mm) with a 60 min gradient of 10-90% acetonitrile in 0.2% (by vol.) trifluoroacetic acid, at a flow of 15.0 ml/min. The expected amino acid sequences were confirmed by automated Edman degradation performed with an Applied Biosystems 470A protein sequencer.

Opioid Receptor Binding

Binding of the peptides to µ, δ and k opioid sites was assayed in crude membrane preparations[11] from rat (µ , δ) and guinea pig (k) brain at pH 7.4 in 50 mM Tris-HCl buffer, at 35 °C for 90 min. The µ-binding site was selectively labeled with [^{3}H][D-Ala^2, Phe-(Me)^4, Gly-ol^5]enkephalin ([^{3}H]DAGO, Amersham, UK) (0.5 nM); the δ-binding site with (3, 5-^{3}H-tyrosyl)-D-Ala-Phe-Val-Val-Gly-NH$_2$, ([^{3}H][D-Ala^2]deltorphin I)[8] (0.3 nM); and the k-binding site with (5α, 7α, 8β)–(—)–N-methyl-N-(7-(1-pyrrolidinyl)-1-oxaspiro-(4, 5)dec-8-yl)-(3, 4-^{3}H)benzeneacetamide, ([^{3}H]U-69,593, Du Pont de Nemours, NEN Division, Dreiech, Germany) (1 nM). Non-specific binding was determined in the presence of 50 µM naloxone (S.A.L.A.R.S., Como, Italy) for the µ and k systems, or in the presence of 5 µM naltrindole (R.B.I., Natick, MA) for the δ system. The inhibition constants of the various non-radioactive peptides were calculated by fitting the displacement curves with the nonlinear regression programme LIGAND[12], using one-site or two-site models. The results are given as means ± S.E. of at least six separate determinations. IC_{50} is the concentration of the competing ligand at which specific binding of labeled ligand is reduced by half. K_d and K_i are the equilibrium dissociation constants for the radioligand and inhibitor respectively, η_H is the slope of the log-logit plot of competition data and B_{max} is the maximum binding capacity of the radioligand.

Table 1 shows the results of the inhibition of binding of [^{3}H]DAGO, [^{3}H][D-Ala^2]deltorphin I and [^{3}H]U-69,593 to brain membrane preparations by the new peptides. With the exception of [Lys^7-NH$_2$]dermorphin (peptide 2 in table 1), the peptides produced smooth competition curves that were fitted best by a one-site model.

Table 1. Affinities for opioid receptors and biological activities on guinea pig ileum (GPI) and mouse vas deferens (MVD) of dermorphin-like peptides

PEPTIDE	μ system		δ system		κ system		GPI	MVD
	K_i (nM)	η_H	K_i (nM)	η_H	K_i (nM)	η_H	IC_{50} (nM)	IC_{50} (nM)
1. Tyr-ala-Phe-Gly-Tyr-Pro-Ser-NH$_2$	0.540 ± 0.021	0.96	929 ± 41	0.93	8162 ± 979	1.00	1.29 ± 0.11	16.5 ± 1.3
2. Tyr-ala-Phe-Gly-Tyr-Pro-Lys-NH$_2$	0.090 ± 0.008	0.62	1105 ± 185	—	617 ± 66	0.87	1.15 ± 0.13	13.6 ± 1.5
high affinity site	0.007 ± 0.001	—			—	—		
low affinity site	0.250 ± 0.013	—	—	—	—	—		
3. Tyr-ala-Phe-Gly-Tyr-Pro-Lys-OH	5.700 ± 0.510	0.98	1150 ± 172	0.81	—	—	3.82 ± 0.45	56.3 ± 7.8
4. Tyr-ala-Phe-Gly-Tyr-Pro-Arg-NH$_2$	0.200 ± 0.012	0.90	391 ± 25	0.98	192 ± 19	0.85	0.90 ± 0.17	10.2 ± 1.7
5. Tyr-ala-Phe-Gly-Tyr-Pro-Asn-NH$_2$	0.444 ± 0.031	0.89	421 ± 42	0.82	5348 ± 1175	0.86	1.57 ± 0.23	17.0 ± 2.6
6. Tyr-ala-Phe-Trp-Asn-NH$_2$	0.900 ± 0.052	0.85	480 ± 45	0.98	177 ± 12	1.01	5.00 ± 0.52	73.7 ± 9.1
7. Tyr-ala-Phe-Trp-Asn-OH	4.440 ± 0.395	0.97	715 ± 73	1.01	2396 ± 351	0.89	13.10 ± 1.20	205.2 ± 25.0
8. Tyr-ala-Phe-Trp-Tyr-Pro-Asn-NH$_2$	0.320 ± 0.026	0.95	690 ± 57	1.00	427 ± 47	1.03	0.58 ± 0.06	6.6 ± 0.9
9. Tyr-ala-Phe-Trp-Tyr-Pro-Asn-OH	2.900 ± 0.208	0.84	865 ± 69	0.91	905 ± 61	0.97	1.30 ± 0.20	10.4 ± 1.3
10. Tyr-ala-Phe-Trp-Tyr-Pro-Ser-NH$_2$	0.390 ± 0.013	0.93	528 ± 48	0.89	154 ± 14	0.91	1.00 ± 0.15	8.7 ± 1.2

54

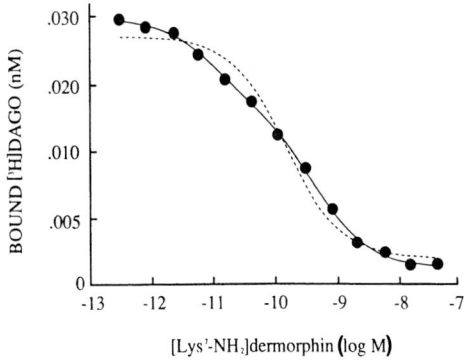

Figure 1. [Lys[7]-NH$_2$]dermorphin competition curve in the μ opioid system of the rat brain. Data were fitted by the non-linear regression program LIGAND: the solid line was the best fit for the two-site model, the broken line that for the one-site model. The apparent improvment in fit for the two-site model was significant (P <0.01 versus one-site).

Log-logit transformation of [Lys[7]-NH$_2$]dermorphin inhibition of the specific [^{3}H]DAGO binding gave a slope significantly lower than 1 (= 0.65) and an apparent K$_i$ of 0.09 $\pm$ 0.008 nM. Because the Hofstee plot was curvi-linear and the line of best fit assuming a one-site model was inadequate, a nonlinear regression analysis with a two-site model was used. This resulted in a shallow competition curve (Fig. 1), that fitted data significantly better than the monophasic curve (P < 0.01 versus one-site model). The computed K$_i$ values were 0.007 $\pm$ 0.001 nM for the high affinity site and 0.250 $\pm$ 0.013 nM for the low affinity site. However, for this analysis we assumed that the K$_d$ values of [^{3}H]DAGO for the two sites were identical (0.78 nM). If this was not the case, the computed K$_i$ values of the competing ligand were offset by an indeterminate amount.

To provide further evidence for two μ binding sites of different affinity, [^{3}H]DAGO binding was displaced by increasing concentrations of [Lys[7]-NH$_2$]dermorphin in presence of three different concentrations (0.1, 0.3 and 0.5 nM) of [Trp[4], Asn[7]-NH$_2$]dermorphin. Also in these cases, the experimental data were fitted significantly better by biphasic curves, but the low affinity site was progressively reduced as the concentration of [Trp[4], Asn[7]-NH$_2$]dermorphin increased from 0.1 nM to 0.5 nM (Table 2).

The peptides with a free C-terminal carboxyl group had a two hundred times lower affinity for μ sites than their amidated analogs. Because all peptides tested had very low

Table 2. Affinity for [^{3}H]DAGO binding sites of [Lys[7]-NH$_2$]dermorphin alone and in presence of three different concentrations of [Trp[4],Asn[7]]dermorphin

[^{3}H]DAGO 0.5 nM	high affinity site		low affinity site		total
	Ki	% sites	Ki	% sites	n° sites
	(nM)		(nM)		(fmol/mg)
[Lys[7]]derm	0.0050	35.5	0.25	64.5	98
+ [Trp[4],Asn[7]]derm 0.1 nM	0.0075	49.8	0.52	50.2	76
+ [Trp[4],Asn[7]]derm 0.1 nM	0.0068	65.0	1.33	35.0	64
+ [Trp[4],Asn[7]]derm 0.1 nM	0.0053	88.0	2.65	12.0	49

Ki, equilibrium dissociation constant of [Lys[7]-NH$_2$]dermorphin

affinity for δ and k sites, they behaved as highly selective ligands for μ sites (Table 1). Nevertheless, the new peptides showed higher affinity for k opioid receptors of guinea pig brain than did dermorphin.

Biological Activity on Isolated Preparations

Preparations of the myenteric plexus-longitudinal muscle obtained from the small intestine of male guinea pigs (400-500 g) and preparations of mouse vas deferens, were used for the study. The results were expressed as the IC_{50} values obtained from concentration-response curves.

The peptides tested all acted as potent μ opioid agonists on isolated organ preparations (Table 1). They were several times more active in inhibiting electrically evoked contractions in guinea pig ileum than in mouse vas deferens. The rank order of their biological potencies roughly paralleled that of the respective μ-binding affinities. However, there were two notable exceptions: i) [Trp^4,Asn^7-NH$_2$]dermorphin had twice the potency of [Lys^7-NH$_2$]dermorphin on guinea pig ileum, but one-third the affinity for μ-opioid receptors; ii) [Trp^4, Asn^7-OH]dermorphin had the same biological potency as [Lys^7-NH$_2$]dermorphin but a 1/30th the affinity for μ- opioid receptors.

Peptide Degradation

The peptides were degraded by crude brain homogenates (Table 3). An unexpected finding was that within few minutes of incubation with brain homogenates the biological activity of [Trp^4, Asn^7-NH$_2$] dermorphin and to a lesser extent that of [Trp^4]dermorphin was completely destroyed. Since both dermorphin and Tyr-D-Ala-Phe-Trp-Asn-NH$_2$ were degraded slowly, the rapid cleavage of [Trp^4,Asn^7-NH$_2$]dermorphin and [Trp^4]dermorphin probably occurred between Trp^4 and Tyr^5. The mixture of protease inhibitors we used (100 mM phenylmethanesulfonyl fluoride, 20 mg/ml soya bean trypsin inhibitor and 10 mg/ml bestatin) did not afford protection against this type of enzyme degradation.

Pharmacological Effects

Peptides, dissolved in saline, were delivered into the lateral ventricle of rat brain through a guide plastic cannula, with a Hamilton (Reno, NV) microliter syringe. All intracerebroventricular (i.c.v.) injections were made in a volume of 5 μl.

Table 3. Time-course of peptide degradation by rat brain homogenates

PEPTIDES	% Recovery				
	2 min	5 min	10 min	30 min	60 min
[Lys^7-NH$_2$]dermorphin	98	80	73	65	58
[Trp^4, Asn^7-NH$_2$]derm.	60	10	7	3	3
[Trp^4, Asn^5-NH$_2$]derm.1-5	99	90	80	65	57
[Trp^4]dermorphin	80	45	25	12	5
Dermorphin	98	80	60	35	23

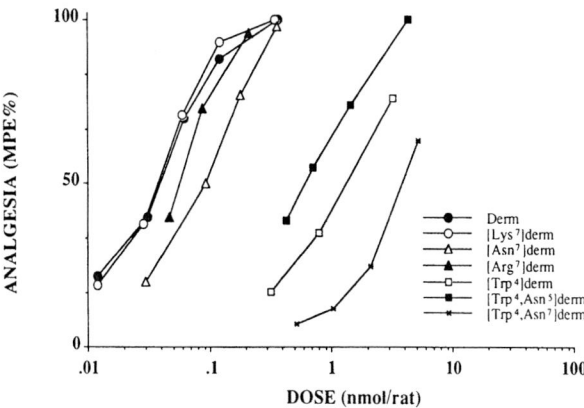

Figure 2. Dose-response curves for antinociception induced in rats after i.c.v. injections of the dermorphin amidated analogs

Each peptide dose was tested in a 10-rat group. In order to determine if antinociceptive effects were the result of activity at opioid receptors, naloxone (0.1 mg/kg, s.c.) was administered 15 min before i.c.v. injection of the peptides and naloxonazine (10 mg/kg, i.v.) was administered 24 hours before.

Antinociceptive responses to peptide administration were determined in rats by the tail-flick test [13]. The AD_{50} of each peptide was defined as the dose that produced an antinociceptive response in 50% of the animals tested. An antinociceptive response was considered to occur when an individual animal displayed a post-drug tail-flick latency value greater than its pre-drug tail-flick latency value plus three S.D.s of the control mean tail-flick latency of all animals in the group.

Catalepsy was evaluated by placing both forelimbs of the rat over a 10-cm-high horizontal bar and measuring the time the animal maintained the posture. Rats remaining more than 1 min on the bar were defined cataleptic (positive bar test). The CD_{50} of each peptide was defined as the dose that produced a positive bar test in 50% of the animals tested.

Peptides with a C-terminal amide group produced dose-related antinociceptive effects in rats (Fig. 2), whereas their analogs with a free α-carboxy group were about two hundred times less active (results not shown). By i.c.v. administration, [Lys^7-NH$_2$]dermorphin, [Arg^7-NH$_2$]dermorphin and [Asn^7-NH$_2$]dermorphin behaved as potent analgesic agents in the tail-flick test. The antinociception induced by [Lys^7-NH$_2$]dermorphin was long lasting (3-4 hrs), exceeding that of dermorphin itself. In contrast, [Trp^4, Asn^7-NH$_2$] dermorphin, and [Trp^4]dermorphin were about 1/80th and 1/20th as active as [Lys^7-NH$_2$]dermorphin, respectively. Their antinociceptive effects lasted less than one hour.

In addition to antinociception, i.c.v. injection of these peptides produced catalepsy (Table 4). With the exception of the tryptophan-containing analogs, the peptides tested showed CD_{50} values higher than the respective AD_{50} values. However, the ratio AD_{50}/CD_{50} varied widely among the different agonists. [Lys^7-NH$_2$]dermorphin showed the lowest ratio (0.073): at i.c.v. doses lower than 100 pmol per rat the peptide produced antinociception alone. In contrast, when given at i.c.v. doses that did not modify the reaction time in the tail-flick test, [Trp^4, Asn^7-NH$_2$] dermorphin induced catalepsy. For the latter peptide we calculated $(AD_{50}/CD_{50}) = 40$, a value 500 times higher than that calculated for [Lys^7-

Table 4. Antinociceptive and cataleptic effects of dermorphin-like peptides

PEPTIDES	M.W.	AD_{50}	CD_{50}	AD_{50}/CD_{50}
[Lys^7-NH$_2$]dermorphin	845	0.026 ± 0.009	0.364 ± 0.126	0.07
[Arg^7-NH$_2$]dermorphin	872	0.034 ± 0.008	0.149 ± 0.016	0.23
Dermorphin	804	0.035 ± 0.010	0.112 ± 0.037	0.31
[Asn^7-NH$_2$]dermorphin	830	0.034 ± 0.009	0.082 ± 0.024	0.41
[Trp^4, Asn^5-NH$_2$]derm.1-5	700	0.430 ± 0.044	0.210 ± 0.033	2.05
[Trp^4-NH$_2$]dermorphin	932	0.791 ± 0.048	0.130 ± 0.021	6.12
[Trp^4, Asn^7-NH$_2$]derm.	959	2.086 ± 0.311	0.052 ± 0.013	40.11

M.W., molecular weight; AD_{50}, ED_{50} (nmol/rat, i.c.v.) for antinociception in the tail-flick test; CD_{50}, ED_{50} (nmol/rat, i.c.v.) for catalepsy in the bar test.

NH$_2$]dermorphin. Both antinociception and catalepsy were reversed by naloxone (data not shown).

When injected intraperitoneally (i.p.) in rats, [Lys^7-NH$_2$]dermorphin produced a long-lasting antinociception that was dose-related (Fig. 3). In the range of doses tested, it did not induce catalepsy. By i.p. administration, the AD_{50} of [Lys^7-NH$_2$]dermorphin (36 ± 9 nmol/kg of body weight) was about 1/40th and 1/6th that of morphine (1500 ± 260 nmol/kg) and dermorphin (225 ± 38 nmol/kg), respectively.

DISCUSSION

The results obtained clearly demonstrate that the dermorphin-related peptides from the skin of *Phyllomedusa bicolor* and their amidated analogs are highly potent and selective μ-opioid agonists. In binding to opioid receptors, [Lys^7-NH$_2$]dermorphin showed an affinity for the preferred μ site at least 3 or 4 hundred times greater than its affinity for k and δ sites. Moreover, its competition curve for [^{3}H]DAGO binding was fitted best by a two-site model. At very low concentrations, in the picomolar range, this peptide reduced binding by some small amount, to produce a distinct change in the curve slope, followed at higher concentrations (nanomolar) by a steeper reduction in the remaining bound [^{3}H]DAGO. Previous data in the literature[14] show that DAGO binding to μ-opioid receptors sometimes reveals high and low affinity sites. The current explanation of this phenomenon is that the

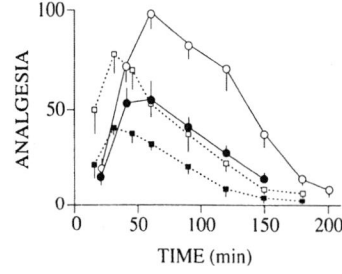

Figure 3 Time-course of the antinociception produced by i.p. injections of [Lys^7]dermorphin (●, 0.2 mmol/kg; O, 0.5 mmol/kg) and morphine HCl (■, 5 mmol/kg; □, 13 mmol/kg)

high affinity site represents a high affinity state of the receptor rather than a distinct μ receptor subtype. Our data are difficult to reconcile with this interpretation. The self-competition curves of DAGO were smooth and symmetrical whereas the heterologous competition curves with [Lys^7-NH$_2$]dermorphin clearly showed two binding sites. In addition, displacement of [^{3}H]DAGO binding by [Lys^7-NH$_2$]dermorphin in the presence of [Trp^4, Asn^7-NH$_2$]dermorphin at three different concentrations (0.1, 0.3 and 0.5 nM) produced three biphasic competition curves in which the density of [^{3}H]DAGO binding to the low affinity site was progressively shifted toward lower B_{max} values. A simple explanation for all these results might be that the two binding sites represent two distinct receptor subtypes of the μ system, one being the site preferred by [Lys^7-NH$_2$]dermorphin and the other the site preferred by [Trp^4, Asn^7-NH$_2$]dermorphin.

Evidence also exists for distinct functional roles of the two μ receptor subtypes. When [Lys^7-NH$_2$]dermorphin was injected i.c.v. in the rats, its AD_{50} in the test of antinociception was 1/10th its CD_{50} in the test of catalepsy, whereas in the case of [Trp^4, Asn^7-NH$_2$]dermorphin the AD_{50} was 40 times higher than the CD_{50}. However, until cloning and sequence analysis provide conclusive proof, the evidence provided here for the existence of two subtypes of the μ receptor must be regarded with circumspection.

Our experiments with brain homogenates demonstrated a large difference in the degradation rates of the tested peptides and no suitable protective cocktails could be devised. The possibility that in vivo enzyme degradation may generate cleavage products that can account in part for the difference observed in the ED_{50} values between the test of antinociception and that of catalepsy is not suggested by our data, but it cannot be rigorously excluded. However, both antinociception and catalepsy showed classical opioid behavior as evidenced by naloxone antagonism.

Of the peptides tested, [Lys^7-NH$_2$]dermorphin was the most potent μ-opioid agonist even when injected peripherally. When given i.p., its AD_{50} in the tail-flick test was 36 nmol/kg of body weight , which is a value 1/40th the AD_{50} of morphine (1500 nmol/kg). The duration of the antinociceptive effect also exceeded that of morphine. Thus its higher affinity and selectivity for μ sites and its greater antinociceptive potency, as compared with morphine, might well make [Lys^7-NH$_2$]dermorphin a superior analgesic agent.

REFERENCES

1. P.C. Montecucchi, R. de Castiglione, S. Piani, L. Gozzini, and V. Erspamer, Amino acid composition and sequence of dermorphin, a novel opiate-like peptide from the skin extracts of *Phyllomedusa sauvagei*, *Int. J. Peptide Prot. Res.* **17**, 275 (1981).

2. V. Erspamer, P. Melchiorri, G. Falconieri Erspamer, P.C. Montecucchi and R. de Castiglione, Phyllomedusa skin: a huge factory and store-house of a variety of active peptides, *Peptides* **6** (3): 7 (1985)

3. K. Richter, R. Egger and G. Kreil, D-Alanine in the frog skin peptide dermorphin is derived from L-Alanine in the precursors, *Science* **238**: 200 (1987)

4. G. Kreil, D. Barra, M. Simmaco, V. Erspamer, G. Falconieri Erspamer, L. Negri, C. Severini, R. Corsi, and P. Melchiorri, Deltorphin, a novel amphibian skin peptide with high selectivity and affinity for δ-opioid receptors, *Eur. J. Pharmacol.* **162**: 123 (1989).

5. A. Mor, A. Delfour, S. Sagan, M. Amiche, P. Pradelles, J. Rossier and P. Nicolas, Isolation of dermenkephalin from amphibian skin, a high affinity delta selective opioid heptapeptide containing a D-amino acid residue, FEBS *Lett.* **255**: 268 (1989).

6. L.H. Lazarus, W.E. Wilson, R. de Castiglione and A. Guglietta, Dermorphin gene sequence peptide with high affinity and selectivity for delta-opioid receptors, *J. Biol. Chem.* **264**: 3047 (1989).

7. M. Amiche, S. Sagan, A. Mor, A. Delfour and P. Nicolas, Dermenkephalin (Tyr-D-Met-Phe-His-Leu-Met-Asp-NH2): a potent and fully specific agonist for the delta-opioid receptors, *Mol. Pharmacol.* **35**: 774 (1989).

8. V. Erspamer, P. Melchiorri, G. Falconieri Erspamer, L. Negri, R. Corsi, C. Severini, D. Barra, M. Simmaco and G. Kreil, Deltorphins: A family of naturally occurring peptides with high affinity and selectivity for δ opioid binding sites, *Proc. Natl. Acad. Sci. USA* **86**, 5188 (1989).

9. K. Richter, R. Egger, L. Negri, R. Corsi, C. Severini and G. Kreil, cDNAs encoding [D-Ala2]deltorphin precursors from skin of *Phyllomedusa bicolor* also contain genetic information for three dermorphin-related opioid peptides, *Proc. Natl. Acad. Sci. USA* **87**, 4836 (1990).

10. M. Broccardo, V. Erspamer, G. Falconieri Erspamer, G. Improta, G. Linari, P. Melchiorri and P.C. Montecucchi, Pharmacological data on dermorphins, a new class of potent opioid peptides from amphibian skin, *Br. J. Pharmac.* **73**, 625 (1981).

11. H.W. Kosterlitz, J.A.H. Lord, S.J. Paterson and A.A. Waterfield, Effects of changes in the structure of enkephalins and of narcotic analgesic drugs on their interactions with μ- and δ-receptors, *Br. J. Pharmacol.* **68**, 333 (1980).

12. P.J. Munson and D. Rodbard, Ligand: a versatile computerized approach for characterization of LIGAND-binding systems, *Anal. Biochem.* **107**, 220-239 (1980)

13. F.E. D'Amour and D.L. Smith, Method for determining loss of pain sensation, *J. Pharmacol. Exp. Ther.* 72, 74 (1941).

14. A. Goldstain and A. Naidu, Different opioid receptors *Mol. Pharmacol.* **36**, 265 (1989).

PROTEASES FOR NEUROPEPTIDE PRECURSOR PROCESSING IN BOVINE ADRENAL MEDULLARY CHROMAFFIN GRANULES

Vivian Y.H. Hook, Anahit V. Azaryan, and Timothy J. Krieger

Department of Biochemistry
Uniformed Services University of the Health Sciences
Bethesda, MD 20814

INTRODUCTION

Neuropeptides acting as neurotransmitters and endocrine hormones are synthesized as protein precursors[1,2]. These precursors require specific proteolytic processing at dibasic (Lys-Arg, Lys-Lys, Arg-Arg, and Arg-Lys) and, less frequently, at monobasic arginine cleavage sites to generate the smaller active peptides. Processing of the prohormone occurs primarily in secretory vesicles[1,2] where the physiologically relevant peptide is stored for secretion. It is also noted that some processing has been found to occur in the Golgi apparatus[3,4].

Our laboratory has been concerned with elucidating the proteolytic processing steps that occur in secretory vesicles[5-9]. We utilize chromaffin granules (secretory vesicles) of bovine adrenal medulla as a model system to define the proteolytic enzymes required to convert neuropeptide precursors to active peptides. Chromaffin granules contain several neuropeptides including enkephalins[10-12], substance P[13], neuropeptide Y[14], and galanin[15]; thus, these secretory vesicles (chromaffin granules) should contain the respective processing enzymes. Our approach takes advantage of the high yield of isolated homogeneous chromaffin granules that allows purification of adequate amounts of processing proteases required for future microsequencing, molecular cloning, and antibody production to generate specific probes for cell biological studies of processing enzymes.

A second important aspect of our strategy to identify processing enzymes is the use of authentic recombinant precursors generated by expression of the respective cDNAs[5,6,8]. To identify relevant processing enzymes, it is important to assay putative processing activities with full-length prohormone substrates, instead of short peptides, to find proteases that are required for the initial steps of processing. We have chosen as model substrate, recombinant ^{35}S-(Met)-preproenkephalin, generated from its cDNA[16] by in vitro transcription and translation, since enkephalin peptides are the major neuropeptides present in adrenal medulla[10-12].

In addition to high levels of enkephalins, bovine adrenal medulla also contains other neuropeptides[13-15] such as neuropeptide Y, substance P, and galanin that are synthesized as precursors requiring proteolytic processing. Although these precursors possess similar dibasic cleavage sites, it is not known whether they are processed by identical or different proteases. Therefore, during enzyme purification we also assayed column fractions with the substance P precursor, ^{35}S-(Met)-ß-preprotachykinin, to detect proteases that may prefer the tachykinin over the enkephalin precursor.

Interestingly, the enkephalin and tachykinin precursor substrates have resulted in identification of proteases[6,8,9] of different mechanistic classes. The enkephalin precursor has resulted in identification and purification of a novel 'prohormone thiol protease' (PTP)[6], and the tachykinin precursor has resulted in identification of a chromaffin granule aspartyl protease (CGAP) that resembles cathepsin D[8]. Moreover, only peptide-MCA (methylcoumarin amide) substrates detected serine protease(s)[9] that are immunologically related to mammalian PC1/PC3 and PC2 (PC=prohormone convertase) genes[17-23], homologues of the yeast Kex2 protease[24], that are likely to be involved in neuropeptide precursor processing.

PROHORMONE THIOL PROTEASE (PTP)

Enkephalins are the most abundant neuropeptides in bovine chromaffin granules (secretory vesicles). Therefore, these granules must contain the proenkephalin processing enzymes. To detect these enzymes, recombinant ^{35}S-(Met)-preproenkephalin (^{35}S-(Met)-PPE), generated from its cDNA by in vitro transcription and translation, was used as substrate. Purification from bovine chromaffin granules resulted in the isolation of a novel 'prohormone thiol protease' (PTP) that represents the major enkephalin precursor cleaving activity in this organelle[6].

Chromatography on concanavalin A-Sepharose, Sephacryl S-200, chromatofocusing, and thiopropyl-Sepharose resulted in an 88,000-fold purification with a recovery of 35% of enzyme activity. The 33 kDa PTP is a glycoprotein with a pI of 6.0. It cleaved ^{35}S-(Met)-PPE with a pH optimum of 5.5, indicating that it is functional at the intragranular pH of 5.5-6.0[25]. Interestingly, production of trichloroacetic acid-soluble products was optimal at a slightly different pH of 4.0, suggesting that processing of initial precursor and intermediates may require different pH conditions. The thiol dependence of PTP was indicated by its requirement for dithiothreitol and inhibition by thiol protease inhibitors iodoacetate, p-hydroxymercuribenzoate, mercuric chloride, and cystatin C. These biochemical properties (binding to concanavalin A, pH optimum, pI, molecular weight) distinguish PTP from other thiol proteases (cathepsins B, H, L, and S), indicating that a unique thiol protease has been identified.

Importantly, in vitro processing of the enkephalin precursor by PTP mimicks in vivo processing of proenkephalin (PE)[6]. Proenkephalin in vivo is converted to multiple intermediates that contain the NH$_2$-terminal segment of PE lacking (Met)enkephalin sequences[11,12,26]. To determine if PTP generates these intermediates, use of ^{35}S-(Cys)-PPE as substrate allowed detection of intermediates containing the NH$_2$-terminal segment of PPE since all cysteine residues of PPE are located in this NH$_2$-terminal region (fig.1). PTP converted ^{35}S-(Cys)-PPE to multiple 22.1, 21.6, 17.7, 17.3, and 15.0 kDa intermediates that were labeled with ^{35}S-Cys and which, therefore, possess the precursor's NH$_2$-terminal fragment. Thus, PTP in vitro converts PE to intermediate products similar to those detected in vivo, indicating processing of PE from its COOH-terminal region.

PTP was capable of generating (Met)enkephalin, the final product of enkephalin precursor processing[6,7]. In studies of PTP cleavage sites, peptide F, a (Met)enkephalin-containing peptide intermediate derived from proenkephalin, was used as substrate. Proteolytic products of peptide F were separated by HPLC and identified by microsequencing and amino acid compositional analyses. PTP converted peptide F to (Met)enkephalin by cleavage at Lys-Arg and Lys-Lys dibasic sites (fig. 2).

To determine whether PTP is capable of cleaving all sites needed for proenkephalin processing, further studies assessed PTP's ability to cleave the dibasic and the monobasic sites within the enkephalin-containing peptide BAM-22P[7] (fig. 2). Cleavage products were separated by HPLC and subjected to microsequencing to determine their identity. PTP cleaved BAM-22P at the Lys-Arg site between the two basic residues. The Arg-Arg site was cleaved at the NH$_2$-terminal side of the paired basic residues to generate (Met)enkephalin. Furthermore, the monobasic arginine site was cleaved at its NH$_2$-terminal side by PTP. These findings show that PTP prefers arginine at the P1' position of the cleavage site. Overall, these cleavage studies[6,7] demonstrate that PTP possesses the necessary specificity for all the dibasic and monobasic cleavage sites required for converting proenkephalin into the bioactive (Met)enkephalin .

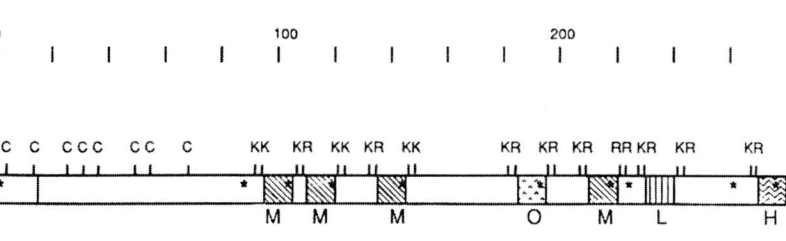

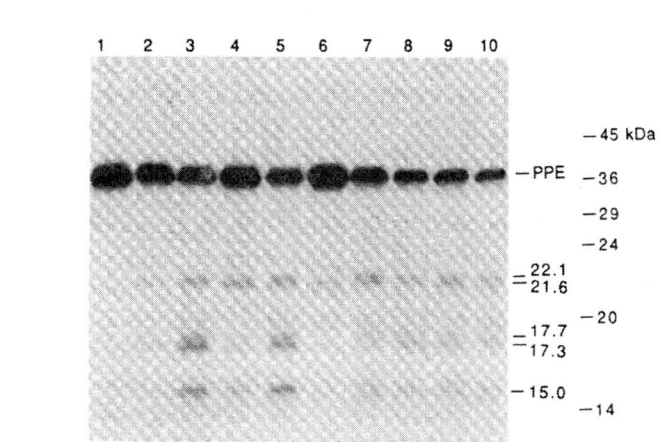

Figure 1. Cleavage of ^{35}S-(Met)- and ^{35}S-(Cys)-preproenkephalin (^{35}S-(Met)-PPE and ^{35}S-(Cys)-PPE) by 'prohormone thiol protease' (PTP). (a) Relative positions of the methionine (*) and cysteine (C) residues within the enkephalin precursor are indicated. K = Lys; R = Arg; M = (Met)enkephalin; L = (Leu)enkephalin; O = M-Arg^6-Gly^7-Leu^8; H = M-Arg^6-Phe^7. (b) ^{35}S-(Met)-PPE (lanes 1-5) and ^{35}S-(Cys)-PPE (lanes 6-10) were incubated with purified PTP for 0 (lanes 1 and 6), 30 (lanes 2 and 7), 60 (lanes 3 and 8), 90 (lanes 4 and 9), and 120 (lanes 5 and 10) min. Products were analyzed by SDS-PAGE gels and autoradiography.

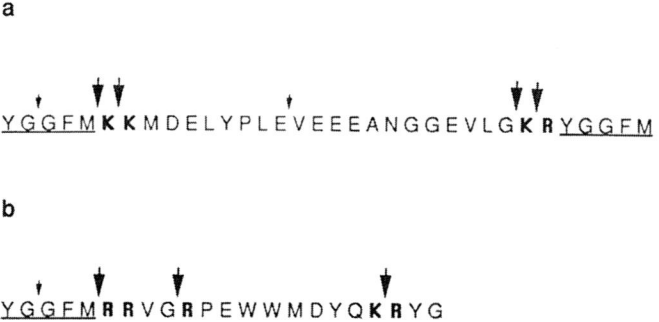

Figure 2. Major PTP cleavage sites within peptide F (a) and BAM-22P (b) are indicated by arrows. Peptide F and BAM-22P correspond to residues #107-140 and #212-236 within PPE (fig. 1), respectively.

Chromaffin granules contain several neuropeptides, and, therefore, PTP may be involved in cleaving other peptide precursors such as the substance P precursor, ß-protachykinin. To examine PTP's preference for different substrates, cleavage of ^{35}S-(Met)-ß-preprotachykinin (^{35}S-(Met)-ß-PPT) was examined. Interestingly, PTP demonstrated selectivity for the enkephalin over the tachykinin precursor since a 5-fold higher concentration of PTP was required to generate equivalent cleavage of ^{35}S-(Met)-ß-PPT[6]. It is noted, however, that the substrate concentration in these in vitro assays was 10^{-10} M, whereas, the estimated in vivo concentration of precursors within secretory vesicles is estimated at approximately 10^{-5} to 10^{-4} M[27]. Proper assessment of precursor selectivity should be conducted near in vivo substrate levels, requiring milligram amounts of precursors instead of the microgram amounts produced by in vitro translation. Such studies should be possible with high level expression of recombinant proenkephalin and ß-protachykinin in E. coli.

Overall, PTP's secretory vesicle localization, acidic pH optimum, appropriate proteolytic products, and cleavage site specificity provide strong support for a role of PTP in neuropeptide precursor processing. Novel specificity for cleavage at the NH_2-terminal side of arginine at dibasic or monobasic processing sites distinguish PTP from other proteases cleaving at the COOH-terminal side of basic residues including trypsin[28] and the yeast Kex2 protease[29,30]. Thus, experimental evidence shows that PTP is a novel protease that participates in neuropeptide precursor processing.

CHROMAFFIN GRANULE ASPARTYL PROTEASE (CGAP)

Our initial study showing PTP preference for the enkephalin over the tachykinin precursor suggests that use of different neuropeptide precursors as substrates may detect precursor selective proteases. Therefore, purification of potential tachykinin precursor cleaving enzymes from bovine chromaffin granules was undertaken using ^{35}S-(Met)-ß-preprotachykinin (^{35}S-(Met)-ß-PPT) as model substrate. Purification by concanavalin A-Sepharose, Sephacryl S200, and chromatofocusing resulted in a chromaffin granule aspartyl protease (CGAP) that clearly preferred the tachykinin over the enkephalin precursor[8].

CGAP was optimally active at pH 5.0-5.5, indicating that it would be active within the acidic intragranular environment. Cleavage at basic residues was suggested by HPLC and HVE (high voltage electrophoresis) identification of the tachykinin intermediate ^{35}S-(Met)-NKA-Gly-Lys (NKA = neurokinin A) as the major acid soluble product generated from ^{35}S-(Met)-ß-PPT. Neuropeptide K, an intermediate of protachykinin processing, was cleaved at a Lys-Arg basic residue site (fig. 3), as assessed by identification of proteolytic products by microsequencing and amino acid composition analyses. CGAP's localization to the chromaffin granule fraction, its optimum activity near the acidic intragranular pH, and cleavage at Lys-Arg sites support a role for CGAP in neuropeptide precursor processing.

Purified CGAP on a denaturing SDS-PAGE gel was composed of 47, 30, and 16.5 kDa polypeptides. However, on a non-denaturing native gel, these three polypeptides migrated as a single band, indicating that they are associated with one another. On gel filtration column chromatography, CGAP activity represented by the three polypeptides eluted with an apparent M_r of 45-55 kDa. These results suggest that two CGAP forms exist: a single 47 kDa polypeptide and a complex of 30+16.5 kDa associated subunits. The three CGAP polypeptides were similar to bovine cathepsin D in NH_2-terminal sequences and amino acid composition, indicating that CGAP appears to be a cathepsin D-related protease or cathepsin D itself[8]. The 47 and 16.5 kDa polypeptides of CGAP possesssed identical NH_2-terminal sequences, suggesting that the 16.5 kDa polypeptide may be derived from the 47 kDa form by proteolysis.

CGAP cleavage specificity also resembled cathepsin D, as shown by cleavage of neuropeptide K (NPK) at hydrophobic residues between Leu-Tyr and Phe-Val residues (fig. 3). Combined with evidence of CGAP cleavage at the Lys-Arg dibasic site of NPK, CGAP could potentially be involved in processing adrenal medullary proendothelin to endothelin[31,32] which requires cleavages at both hydrophobic and dibasic residue sites that

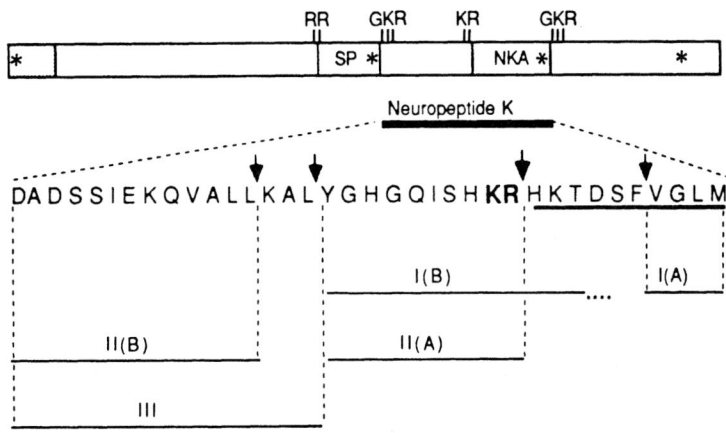

Figure 3. CGAP cleavage of neuropeptide K. Neuropeptide K (NPK) within ß-preprotachykinin is illustrated. SP represents the tachykinin substance P with the sequence Arg-Pro-Lys-Pro-Gln-Gln-Phe-Phe-Gly-Leu-Met-NH$_2$. NKA represents the tachykinin neurokinin A with the sequence His-Lys-Thr-Asp-Ser-Phe-Val-Gly-Leu-Met-NH$_2$. Asterisks indicate methionine residues. Arrows indicate CGAP cleavage sites within NPK.

would be compatible with CGAP's cleavage site specificity. CGAP's cathepsin D-like specificity for hydrophobic residues, however, also suggests that it may be involved in degrading precursor segments that do not become part of active peptide sequences. Immunoelectron microscopic studies (preliminary observations) showing colocalization of CGAP with (Met)enkephalin in chromaffin granules confirm the proposed secretory vesicle function of CGAP.

In summary, CGAP is localized in secretory vesicles, prefers the tachykinin over the enkephalin precursor as substrate, is active at the acidic intravesicular pH of 5.5-5.8, and cleaves mostly at hydrophobic residues but also demonstrates cleavage at a dibasic Lys-Arg site. These properties are compatible with participation of CGAP in the proteolytic processing of neuropeptide precursors with regard to possible proendothelin processing or proteolysis of segments of precursors that do not become active peptides.

CHROMAFFIN GRANULE SUBTILISIN-RELATED PROHORMONE CONVERTASES (PC's)

Among several candidate endoproteolytic processing enzymes, only the yeast KEX2 gene product has been characterized at both the molecular and enzymatic levels[24,29,30]. The Kex2 enzyme is a Ca^{2+}-dependent subtilisin-like serine protease that is required for in vivo processing of pro-α-mating factor at the COOH-terminal side of Lys-Arg and Arg-Arg sites. Recently, mammalian Kex2 homologues have been cloned[17-23] and proposed as prohormone convertases PC1 (also known as PC3) and PC2.

Isolation and microsequencing of a chromaffin granule membrane-bound glycoprotein, known as glycoprotein H (GpH), demonstrated that GpH represents the bovine forms of PC1/PC3 and PC2[33]. PC1/PC3 and PC2 in chromaffin granules have also been demonstrated by immunoblots[34]. Whereas the ability of PC1/PC3 and PC2 to function as putative processing enzymes of POMC (proopiomelanocortin) was demonstrated by cotransfection of PC1/PC3 and PC2 genes with the POMC gene in BSC-40 cells[21,22], it will be necessary to characterize the endogenous PC1/PC3 and PC2 activities purified from mammalian tissues with recombinant prohormones near in vivo substrate concentrations. An advantage of studying PC1/PC3 and PC2 activities expressed in vivo over expression in cell lines by DNA transfection is that authentic post-translational processing occurs in vivo which may be important for enzyme activity.

It is important to determine the relative roles of PC1/PC3 and PC2 in proenkephalin processing in chromaffin granules. The first step is to identify PC1/PC3- and PC2-like

Table 1. Effect of protease inhibitors and modulators on Boc-Gln-Arg-Arg-MCA cleaving activity in the membrane fraction of chromaffin granules.

Inhibitor	Concentration	Inhibition, %
None	--	0
CaCl$_2$	10 mM	-17
EGTA	2.5 mM	20
DFP	1.5 mM	70
α$_1$-Antitrypsin	0.1 mg/ml	59
Soybean trypsin inhibitor	0.1 mg/ml	65
PMSF	0.1 mM	3
Leupeptin	2 mM	96
Antipain	1 mM	91
TLCK	0.1 mM	73
[D-Tyr]-Glu-Phe-Lys-Arg-CK	20 µM	90
Z-Arg-Leu-Val-Gly-CHN$_2$	1 mM	0
Z-Leu-Val-Gly-CHN$_2$	1 mM	0
Cystatin C	1 mM	0
Iodoacetamide	5 mM	35
HgCl$_2$	5 mM	30
DTT	1 mM	-37
Pepstatin A	0.1 mM	0
1,10-O-Phenanthroline	0.1 mM	10

Activity was assayed with boc-Gln-Arg-Arg-MCA (100 µM) at pH 8.0.

activities in chromaffin granules. We have accomplished this by demonstrating that chromaffin granules contain boc-Gln-Arg-Arg-MCA (MCA= methylcoumarin amide) cleaving activity[9] that resembles PC2 expressed in Xenopus oocytes[35] and yeast Kex2 activities[29,30].

The chromaffin granules contained boc-Gln-Arg-Arg-MCA cleaving activity that was stimulated by Ca^{++}, and the majority (approximately 70%) of this granule activity was membrane-bound with some activity in the soluble fraction of granules. This boc-Gln-Arg-Arg-MCA membrane-bound cleaving activity, like human PC2[35] and yeast Kex2[29,30], was inhibited by high (millimolar) concentrations of serine protease inhibitors (DFP, α$_1$-antitrypsin, and soybean trypsin inhibitor (Table I). Potent inhibition by the active site-directed inhibitor {D-Tyr}-Glu-Phe-Lys-Arg-CK (20 µM) suggested selectivity for paired basic residue cleavage sites; this inhibitor corresponds to the Glu-Phe-Lys-Arg sequence at the junction of ACTH and ß-lipotropin peptides within proopiomelanocortin that is cleaved by PC2, PC1/PC3, and Kex2 in DNA cotransfection experiments[21,22,36]. The effects of protease inhibitors indicates that the chromaffin granule membrane boc-Gln-Arg-Arg-MCA cleaving activity represents a serine protease with specificity for paired basic residues.

Cleavage site studies (Table 2) showed that boc-Gln-Arg-Arg-MCA, boc-Arg-Val-Arg-Arg-MCA, boc-Gly-Lys-Arg-MCA, and boc-Gly-Arg-Arg-MCA were effectively cleaved, with boc-Gln-Arg-Arg-MCA as the best substrate. These results indicate cleavage on the carboxyl side of Arg-Arg and Lys-Arg pairs. Higher activity with substrates containing Arg-Arg or Lys-Arg pairs and low activity with boc-Glu-Lys-Lys-MCA indicates preference for arginine in the P1 position. These cleavage site studies provide further evidence for specificity of this serine proteolytic activity for paired basic residues.

Table 2. Hydrolysis of MCA-peptide substrates by chromaffin granule membranes.

Substrate	Proteolytic Activity (pmol AMC/hr/mg protein)
Boc-Gln-Arg-Arg-MCA	1326
Boc-Arg-Val-Arg-Arg-MCA	1031
Boc-Gly-Lys-Arg-MCA	844
Boc-Gly-Arg-Arg-MCA	562
Boc-Glu-Lys-Lys-MCA	375

Assays were performed in 0.1 M Tris-HCl, pH 8.0, with substrates at a final concentration of 100 μM.

The boc-Gln-Arg-Arg-MCA cleaving activity in chromaffin granule membranes was shown to be immunologically related to PC1/PC3 and PC2[37]. Specific antisera raised to COOH-terminal peptide sequences of PC1/PC3 and PC2 (antisera were a generous gift from Dr. Y. Peng Loh, NICHD, NIH) immunoprecipitated 40% and 36%, respectively, of boc-Gln-Arg-Arg-MCA cleaving activity in deoxycholate solubilized chromaffin granule membranes. Thus, the activity detected with boc-Gln-Arg-Arg-MCA is indeed related to PC1/PC3 and PC2. These immunodepletion results, together with substrate specificity and inhibitor profile data[9], indicate that the boc-Gln-Arg-Arg-MCA cleaving activity in chromaffin granule membranes represents PC1/PC3 and PC2 subtilisin-like proteases that are involved in prohormone processing[17-23].

It is of interest to note that PC1/PC3- and PC2-like activities were detected with the peptide-MCA substrates, but such subtilisin-like activities in chromaffin granules were not readily detected with the radiolabelled enkephalin and tachykinin precursors. These results show that proteases may possess selectivity towards certain substrates. Thus, fluorogenic peptide substrates with dibasic sites that correspond to cleavage sites of a precursor molecule are useful for assays of prohormone convertases[29,30,35]. It will be important to define the major proteolytic step(s) in proenkephalin processing catalyzed by each protease.

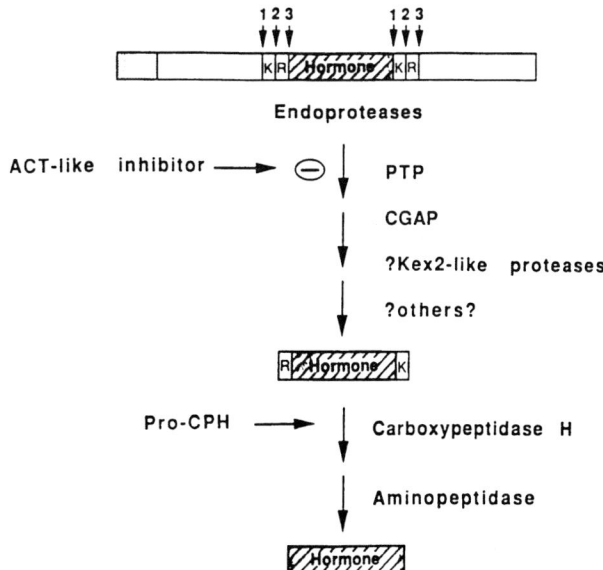

Figure 4. Involvement of PTP, CGAP, and PC1- and PC2-like proteases in chromaffin granule neuropeptide precursor processing.

REMARKS

We have identified three proteases of different mechanistic classes in chromaffin granules -- the 'prohormone thiol protease'[6,7], a chromaffin granule aspartyl protease (CGAP)[8], and PC1/PC3- and PC2-like serine proteases[9] -- that are involved in neuropeptide precursor processing (fig. 4). It is also important to consider that other proteases[38-43] studied in adrenal medulla and various tissues are likely to be present in chromaffin granules for prohormone processing. It is evident from these studies that choice of substrate was crucial in detecting each of these activities, since each protease appears to possess selectivity for particular prohormone substrates. These findings lead to the hypothesis that each processing enzyme may possess different affinities for various precursors. Further characterization of processing enzyme selectivity for prohormones will be important for understanding the molecular mechanisms responsible for the production of peptide hormones and neurotransmitters with diverse physiological actions.

ACKNOWLEDGEMENTS

This work was supported by grants from NIH, NIDA, and NMRDC. The authors thank Dr. Y. Peng Loh (NICHD, NIH) for the gift of anti-PC1/PC3 and PC2 antisera.

REFERENCES

1.   K. Docherty and D. Steiner. Posttranslational proteolysis in polypeptide hormone biosynthesis. Ann. Rev. Physiol. 44: 6266(1982).

2.   H. Gainer, J.T. Russell, and Y.P. Loh. The enzymology and intracellular organization of peptide precursor processing: the secretory vesicle hypothesis. Neuroendocrin. 40, 1717(1985).

3.   E. Schnabel, R.E. Mains, and M.G. Farquhar. Proteolytic processing of pro-ACTH/endorphin begins in the Golgi complex of pituitary corticotropes and AtT-20 cells. Mol. Endocrinol. 3:1223(1989).

4.   A. Lepage-Lezin, P. Joseph-Bravo, G. Devilliers, L. Benedetti, J.M. Launay, S. Gomez, and P. Cohen. Prosomatostatin is processed in the Golgi apparatus of rat neural cells. J. Biol. Chem. 266:1679(1991).

5.   V.Y.H. Hook, D. Hegerle, and H.U. Affolter. Cleavage of recombinant enkephalin precursor by endoproteolytic activity in bovine adrenomedullary chromaffin granules. Biochem. Biophys. Res. Commun. 167: 722(1990).

6.   T.K. Krieger, and V.Y.H. Hook. Purification and characterization of a novel thiol protease involved in processing the enkephalin precursor. J. Biol. Chem., 266, 8376(1991).

7.   T.K. Krieger and V.Y.H. Hook. Prohormone thiol protease and enkephalin precursor processing: cleavage at dibasic and monobasic sites. J. Neurochem. 59:26(1992).

8.   T.K. Krieger, and V.Y.H. Hook. Purification and characterization of a cathepsin D protease from bovine chromaffin granules. Biochemistry 31: 4223(1992).

9.   A.V. Azaryan, and V.Y.H. Hook. Kex2-like proteolytic activity in adrenal medullary chromaffin granules. Biochem. Biophys. Res. Commun. 185: 398(1992)

10.   A.S. Stern, R.C. Lewis, S. Kimura, K.L. Kilpatrick, B.N. Jones, K. Kojima, S.Stein, and S. Udenfriend. Biosynthesis of the enkephalins: precursors and intermediates, in: Biosynthesis, Modification,and Processing of Cellular and Viral Polypeptides, H.A. Blough and J.M. Tiffany, ed., Academic Press, 99(1980).

11.   N.P. Birch, A.D. Davies, D.L. Christie. An investigation of the molecular properties and stability of intermediates of proenkephalin in isolated bovine adrenal medullary chromaffin granules. J. Biol. Chem. 262: 3383(1987).

12. B.A. Spruce, Jackson, P.J. Lowry, D.P. Lane, and D.M. Glover. Monoclonal antibodies to a proenkephalin A fusion peptide synthesized in Escherichia coli recognize novel proenkephalin A precursor forms. J. Biol. Chem. 263: 19788(1988).

13. A. Saira, S.P. Wilson, A. Molnar, O.H. Viveros, F. Lembeck. Substance P and opiate-like peptides in human adrenal medulla. Neurosci. Lett. 20:195(1980).

14. A. Higuchi, H.Y.T. Yang, and S.L. Sabol. Rat neuropeptide Y precursor gene expression. J. Biol. Chem. 263:6288(1988)

15. A. Rokeaus and M.J. Brownstein. Construction of a porcine adrenal medullary cDNA library and nucleotide sequence analysis of two clones encoding a galanin precursor. Proc. Natl. Acad. Sci. USA 83: 628(1986).

16. K. Yoshikawa, C. Williams, and S.L. Sabol. Rat brain preproenkephalin mRNA, cDNA cloning, primary structure, and distribution in the central nervous system. J. Biol. Chem. 259: 14301(1984).

17. S.P. Smeekens, and D.F. Steiner. Identification of a human insulinoma cDNA encoding a novel mammalian protein structurally related to the yeast dibasic processing protease Kex2. J. Biol. Chem. 265, 2997(1990).

18. N.G. Seidah, L. Gaspar, P. Mion, M. Marcinkiewicz, M. Mbikay, and M. Chretien. cDNA sequence of two distinct pituitary proteins homologous to kex2 and furin gene products: tissue-specific mRNAs encoding candidates for pro-hormone processing proteinases. DNA and Cell Biol. 9: 415(1990).

19. N. G. Seidah, M. Marcinkiewicz, S. Benjannet, L. Gaspar, G. Beaubien, M.G. Mattei, C. Lazure, M. Mbikay, and M. Chretien, M. Cloning and primary sequence of a mouse candidate prohormone convertase PC1 homologous to PC2, furin, and kex2: distinct chromosomal localization and messenger RNA distribution in brain and pituitary compared to PC2. Mol. Endocrinol. 5: 111(1991).

20. S.P. Smeekens, A.S. Avruch, J. LaMendola, S.J. Chan, and D.F. Steiner. Identification of a cDNA encoding a second putative prohormone convertase related to PC2 in AtT20 cells and islets of Langerhans. Proc. Natl. Acad. Sci. USA 88: 340(1991).

21. S. Benjannet, N. Rondeau, R. Day, M. Chretien, and N. G. Seidah. PC1 and PC2 are proprotein convertases capable of cleaving proopiomelanocortin at distinct pairs of basic residues. Proc. Natl. Acad. Sci. USA 88: 3564(1991).

22. L. Thomas, R. Leduc, B.A. Thorne, S.P. Smeekens, D.F. Steiner, and G. Thomas. Kex2-like endoproteases PC2 and PC3 accurately cleave a model prohormone in mammalian cells: evidence for a common core of neuroendocrine processing enzymes. Proc. Natl. Acad. Sci. USA 88: 5297(1991).

23. K. Nakayama, M. Hosaka, K. Hatsuzawa, and K. Murakami. Cloning and functional expression of a novel endoprotease involved in prohormone processing at dibasic sites. J. Biochem. 109: 803(1991).

24. D. Julius, A. Brake, L. Blair, R. Kunisawa, and J. Thorner. Isolation of the putative structural gene for the lysine-arginine-cleaving endopeptidase required for processing of yeast prepro-α-factor. Cell 37: 1075(1984).

25. H.B. Pollard, H. Shinko, C.E., Creutz, C.J. Pazoles, and J.S. Cohen. Internal pH and state of ATP in adrenergic chromaffin granules determined by ^{31}P nuclear magnetic resonance spectroscopy. J. Biol. Chem. 255: 1170(1979).

26. D.L. Liston, G. Patey, J. Rossier, P. Verbanck, and J.J. Vanderhaeghen. Processing of proenkephalin is tissue-specific. Science 225: 734(1984).

27. A. Ungar and J.H. Phillips. Regulation of the adrenal medulla. Physiol. Rev. 63: 787(1983).

28. A.J. Barrett and J.K. McDonald. In: Mammalian Proteases, a Glossary and Bibliography, vol.1, Academic Press. (1980).

29.   R.S. Fuller, A. Brake, and J. Thorner. Yeast prohormone processing enzyme (KEX2 gene product) is a Ca^{++}-dependent serine protease. Proc. Natl. Acad. Sci. USA 86: 1434(1989).

30.   C. Brenner and R.S. Fuller. Structural and enzymatic characterization of a purified prohormone-processing enzyme: secreted, soluble Kex2 protease. Proc. Natl. Acad. Sci. USA 89: 22(1992).

31.   M. Yanagisawa, H. Kurihara, S. Kumura, Y. Tomobe, M. Kobayashi, Y. Mitsui, Y. Yazaki, K. Goto, and T. Masaki. A novel potent vasoconstrictor peptide produced by vascular endothelial cells. Nature 332: 411(1988).

32.   T. Sawamura, S. Kimura, O. Shinmi, Y. Sugita, M. Yanagisawa, K. Goto, and T. Masaki. Purification and characterization of putative endothelin converting enzyme in bovine adrenal medulla: evidence for a cathepsin D-like enzyme. Biochem. Biophys. Res. Commun. 168: 1230(1990)

33.   D.L. Christie, D.C. Batchelor, and D.J. Palmer. Identification of kex2-related proteases in chromaffin granules by partial amino acid sequence analysis. J. Biol. Chem. 266: 15679(1991).

34.   R. Kirchmair, P. Gee, R. Hogue-Angelletti, A. Laslop, R. Fischer-Colbrie, and H. Winkler. Immunological characterization of the endoproteases PC1 and PC2 in adrenal chromaffin granules and in the pituitary gland. FEBS Lett. 297:302(1992).

35.   K.I.J. Shennan, S.P. Smeekens, D.F. Steiner, and K. Docherty. Characterization of PC2, a mammalian Kex2 homologue, following expression of the cDNA in microinjected Xenopus oocytes. FEBS Lett. 284: 277(1991).

36.   G. Thomas, B.A. Thorne, L. Thomas, R.G. Allen, D.E. Hruby, R.S. Fuller, and J. Thorner. Yeast Kex2 endopeptidase correctly cleaves a neuroendocrine prohormone in mammalian cells. Science 241: 226(1988).

37.   A.V. Azaryan and V.Y.H. Hook. Dibasic residue specific endopeptidase in adrenal medullary chromaffin granules is immunologically related to PC1/PC2 enzymes. Society for Neuroscience Abstracts 18:269(1992).

38.   F.S. Shen, S.F. Roberts, and I. Lindberg. A putative processing enzyme for proenkephalin in bovine adrenal chromaffin granule membranes. J. Biol. Chem. 264: 15600(1989).

39.   Y.P. Loh, D.C. Parish, and R. Tuteja. Purification and characterization of a paired basic residue-specific pro-opiomelanocortin converting enzyme from bovine pituitary intermediate lobe secretory vesicles. J. Biol. Chem. 260: 7194(1985).

40.   D.C. Parish, T. Tuteja, M. Altstein, H. Gainer, and Y.P. Loh. Purification and chartacterization of a paired basic residue-specific prohormone-converting enzyme from bovine pituitary neural lobe secretory vesicles. J. Biol. Chem. 261: 14392(1986).

41.   F.E. Estivariz, T.C. Friedman, T. Chikuma and Y.P. Loh. Processing of adrenocorticotropin by two proteases in bovine intermediate lobe secretory vesicle membranes. J. Biol. Chem. 267: 7456(1992).

42.   J. Bourdais, A.R. Perotti, H. Boussetta, N. Barre, G. Devilliers, and P. Cohen. Isolation and functional properties of an arginine-selective endoprotease from rat intestinal mucosa. J. Biol. Chem. 266: 23386(1991).

43.   H.W. Davidson, C.R. Rhodes, and J.C. Hutton. Intraorganellar calcium and pH control proinsulin cleavage in the pancreatic ß cell via two distinct site-specific endopeptidases. Nature 333: 93(1988).

AMYLIN AND NEUROMEDIN U - TWO NOVEL REGULATORY PEPTIDES OF THE GASTROENTEROPANCREATIC SYSTEM

S R Bloom

Endocrinology Unit, Royal Postgraduate Medical School
Hammersmith Hospital, Du Cane Road, London W12 ONN, UK

INTRODUCTION

Superficially neuromedin U and amylin have little in common. They certainly bear no sequence relationship to each other and their pharmacology is entirely different. Yet they both illustrate the ubiquity of the regulatory peptide system. They both are present in neural tissue and endocrine cells. In the latter they both appear to have a predominantly paracrine role.

AMYLIN

Islet amyloid polypeptide (IAPP) or amylin is a 37 amino acid peptide which was first identified in islet amyloid deposits in patients with insulinomas and non-insulin dependent diabetes mellitus (NIDDM)[1,2]. Human and cat amylin, but not rat and most other species examined, cause islet amyloid formation[3,4]. Amino acids 20 to 29 are the amyloidogenic region of the molecule in cat and man, spontaneously forming beta pleated structures in vitro[5]. There is a clear relationship with calcitonin gene related peptide (CGRP) as the position of the only sulphur cross-link, between cystines at position two and seven, is identical and both peptides are of identical length and have C terminal amides. The human amino acid sequence itself has a 43% and 46% homology to human αCGRP and βCGRP respectively. Amylin is derived from an 89 amino acid precursor[6] and appears to be the only active peptide produced by this pro-hormone. Unlike CGRP the genomic sequence (on chromosome 12) does not encode another active peptide homologous to calcitonin[7].

AMYLIN LOCALISATION

The majority of amylin is produced by the beta cells in the Islets of Langerhans. Here it is co-stored and co-released with insulin[8]. The amounts of amylin present in the beta cells are between a 10th and a 100th that of insulin. Attempts to dissociate release of insulin from that of amylin, for example in response to particular stimuli, have shown only minor differences in the pattern of release[8]. Thus it seems that synthesis and release of insulin and amylin are closely coupled. Radioimmunoassay and Northern Blotting of tissue extracts, and in situ hybridisation and immunocytochemistry of tissue sections demonstrate the presence of small amounts of amylin in the central nervous system with a particular localisation in the dorsal root ganglion. A small amount is also found in the

Growth Factors, Peptides, and Receptors, Edited
by T.W. Moody, Plenum Press, New York, 1993

lung and stomach[9]. Significant amounts have been found in small cell carcinoma of the lung cell lines and also in osteoblast cell lines[10,11].

AMYLIN RECEPTORS

It was early found that amylin could displace CGRP from its receptor[12]. Indeed much of its pharmacology (see below) may be the pharmacology of CGRP. On the other hand CGRP has powerful actions in relaxing vascular smooth muscle (vasodilation) whereas, although active in this respect, amylin is perhaps a 100 fold weaker. There seem to be at least two types of CGRP receptor, one showing a high affinity for amylin and one a low affinity. Until recently there was no amylin preferring receptor. However it has been reported that, particularly in the lung, an amylin preferring receptor does occur and that the binding is reduced by the non-metabolisable GTP-analogue GTP-γ-S, suggesting the coupling of the binding site to a G protein. Scatchard analysis of rat amylin binding in the lung gave it a dissociation constant of 10.4 ± 2.63 nmol and a maximal binding of 3.1 ± 0.97 pmol/mg with only a single class of binding site. Chemical cross-linking analysis showed binding of amylin to sites of M_r 67,000, 64,000 and 38,000. The affinity for amylin was 150 x greater than that of rat alpha CGRP. Stomach, spleen and some regions of the brain also show very high concentrations of specific amylin binding sites[13].

PHARMACOLOGY

As soon as amylin was synthesised and therefore available in significant quantities its pharmacology was established. The first action to be shown was a marked inhibition of the release of the insulin from the beta cells. However not all workers were able to demonstrate this[14]. There were two potential reasons for failure. First, certain methods of synthesis of amylin led to the presence of a small amount of contaminating mercury which attached itself to the sulphur residue and rendered the peptide inactive[15]. Initial confusion resulted from the fact that amino acid sequence and standard mass spectroscopy apparently showed the peptide to be of entirely correct sequence as the mercury was not detected by these means. Electron spray mass spectroscopy, on the other hand, did allow detection of the mercury and many synthetic batches were found to be significantly affected. The second reason for failure to demonstrate inhibition of insulin release was interference by endogenous amylin. Thus incubated islets contain large amounts of amylin in the medium from endogenous release. It is therefore likely that the receptors were already saturated before addition of exogenous amylin.

Amylin was also shown to dramatically increase hepatic gluconeogenesis by blocking the inhibition by insulin[16-19]. Indeed one report claimed it was more powerful than glucagon in increasing hepatic output of glucose[20]. As with the inhibition of insulin there has been controversy and a number of groups have failed to find any effect of amylin on the output of glucose from the liver[21]. Perhaps the most interesting pharmacological action was on skeletal muscle where amylin greatly inhibited the uptake of glucose[22,23]. The combination of inhibition of insulin release, stimulation of hepatic gluconeogenesis and inhibition of peripheral muscle uptake of glucose, suggested amylin to be a diabetogenic peptide[16]. It thus appeared to increase peripheral insulin resistance and diminish insulin output, both phenomena observed in NIDDM (type II diabetes). To exploit the possibilities of inhibiting amylin, and thereby improving or reversing diabetes, a venture capital company, the Amylin Corporation, was set up and has been responsible for a number of the seminal studies in the field. In addition to the effects of amylin on carbohydrate metabolism, amylin has also been shown to have a calcitonin-like effect on bone, decreasing the activity of the osteoclast and thereby lowering plasma calcium[10,24]. Since, as mentioned above, amylin can be produced by osteoblast cell lines there is the possibility that it has a paracrine role within bone[10].

PHYSIOLOGY

The physiology of substances working in a paracrine fashion is usually difficult to evaluate. In contrast it is fairly straightforward to demonstrate whether a regulatory peptide acts as a circulating hormone. The hypothesis has been advanced that amylin is released into the circulation from the beta cell and to act hormonally on liver and peripheral muscle[3,16]. Indeed there has been speculation that in early NIDDM there might be increased release of amylin (insulin levels are raised initially) and raised amylin concentrations might cause increased peripheral insulin resistance and accelerate the progress of the disease. It is even possible that excessive release of amylin in some individuals could be the actual cause of the onset of NIDDM. However examination of the doses needed in the pharmacological studies suggested these are always well above those likely to occur in the circulation[3,4]. There has been unanimity that an amylin rise of about 10 pmol/l plasma occurs after natural stimuli to the beta cell (for example a lunch meal)[3,25]. The observed circulating concentrations of exogenously infused biologically active amylin that will have any influence on insulin release are at least 1000pmol/l[3,26,27], and to affect hepatic output of glucose, or peripheral muscle glucose uptake, still higher concentrations are required. Thus it seems very unlikely that amylin does act as a circulating hormone. The effect on muscle and liver has been ascribed to the cross-reaction of amylin with CGRP receptors[28]. CGRP is delivered locally through the innervation and, in skeletal muscle, may play an important part in the maintenance of acetyl choline receptor number. The 8-37 amylin fragment has been found to be a weak inhibitor of natural 1-37 amylin[29]. Thus 10^{-4} mol/l 8-37 amylin reduces by half the inhibition of insulin release resulting from addition of amylin at a concentration of 1.5×10^{-6} mol/l in isolated perfused rat islets. Furthermore 8-37 amylin at a concentration of 4×10^{-5} mol/l enhances insulin release in response to 8 mmol of glucose and 2×10^{-7} mol/l carbachol by 48% (Wang,Z.L. et al. Unpublished observations). This suggests amylin 8-37 is able to block a tonic inhibitory effect of endogenous amylin in the perifused islet preparation.

PATHOLOGY

In diabetes mellitus circulating amylin concentrations are decreased[4,30]. Plasma amylin is raised in conditions of excessive insulin secretion (for example after steroids or in obesity)[31]. Amylin is also produced by pancreatic endocrine tumours, particularly insulinomas[1,11,32,33]. Plasma levels are rarely significantly raised however. In mixed tumours it is usually the beta cell component that is responsible for amylin production. Amylin can also be produced, in small amounts, by other tumour types, for example small cell carcinoma of the lung, as mentioned above. There appears to be no clinical consequence of this amylin production. Of great interest is the possible pathological role of islet amyloid. Amyloid has been shown to be present in the islet in NIDDM but is also found in the islets of non-diabetic elderly people. Amyloid is prevalent in insulinomas and this was the original source for amylin's purification[1]. In other organs amyloid (from other proteins) has been shown to have a deleterious effect on organ function, perhaps by mechanically choking the organ and, for example, occluding the vascular supply. Although this crude mechanical explanation seems unsatisfactory it is nonetheless true that reversal of amyloid deposition in, for example, the kidney greatly improves renal function. It is therefore at least possible that amyloid deposition within the Islets of Langerhans has a similar deleterious affect and may cause or exacerbate existing beta cell failure. The reason for the deposition of amylin as amyloid under such circumstances is unknown. One hypothesis is that the post-translational enzymic processing becomes defective (this is known to be the case for insulin in NIDDM where partially processed proinsulin is released into the circulation in unusually large concentrations).

ROLE OF AMYLIN

Amylin is a member of the CGRP family and cross-reacts with CGRP receptors which explains much of its pharmacology. Thus pharmacologically it increases peripheral insulin resistance but it seems unlikely to act physiologically as a circulating hormone. There is some evidence that it may play a paracrine role within the islet and, theoretically, may thereby help prevent overactive beta cells from exhausting themselves by providing a local feedback inhibition, maintaining an even intra and inter islet beta cell function. In the failing islet a powerful amylin inhibitor could provide a new therapeutic means for increasing insulin secretion. The possible deleterious effects and importance of amylin deposition as amyloid in the islet is currently unclear.

NEUROMEDIN U

Neuromedin U receives its name because of potent contractile activity on the uterus. It was isolated in 1985 from the porcine spinal cord in two molecular forms neuromedin U 25 and neuromedin U 8, the latter forming the C-terminal 8 amino acids of the former[34]. Neuromedin U has no significant structural resemblance to any other known regulatory peptide.

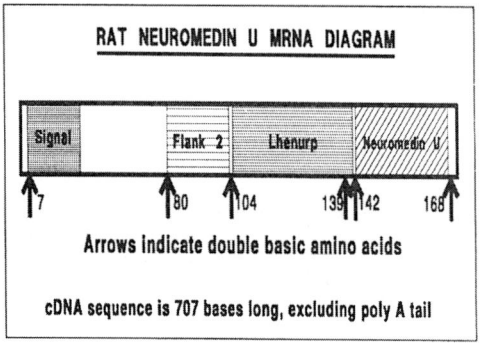

Neuromedin U has been isolated and sequenced from pig[34], rat[35], dog[36], rabbit[37], frog[38] and a partial sequence obtained from guinea pig[39] and chicken[40]. The C-terminal region of the molecule is very highly conserved, neuromedin U 8 having the full range of biological activities of neuromedin U 25 and being approximately a third as potent. In most species, including pig (the original source of isolation), the large form is the only form isolated, or is present in much larger amounts[41]. It may be, therefore, that the original isolation of neuromedin U 8 was an artifact due to degradation during the isolation procedure. Recently the cDNA of a precursor molecule encoding neuromedin U has been isolated from the rat and sequenced[42]. It shows the sequence coding neuromedin U to be at the C-terminal end. This sequence is flanked on the N-terminal side by a similar sized peptide bounded by double basic amino acids suggesting the possibility that it is separately cleaved and secreted. Our preliminary human sequence data indicates that only two amino acids are altered in this flanking peptide. This marked sequence conservation suggests the possibility of a biological role.

NEUROMEDIN U DISTRIBUTION

Radioimmunoassay of tissue extracts demonstrate neuromedin U immunoreactivity in the intestinal tract, with highest concentrations in the ileum, for example in the rat approaching 200 pmol/g[38,43,44]. Thus for this species neuromedin U is amongst the intestinal regulatory peptides found in highest concentration. The peptide is also found in the anterior pituitary[45], spinal cord, several regions of the brain (for example in the rat the nucleus accumbens has 40 pmol/g and the hypothalamus 30pmol/g) as well as lower concentrations in the thyroid and urogenital tract[43,46,47]. Immunocytochemistry of the gut demonstrates the neuromedin U to be localised exclusively within the nerves, particularly in the sub-mucous plexus[44,48-50]. The intestinal mucosa receives a dense innervation. Neuromedin U was frequently colocalised with CGRP[51] and, in the guinea pig, Furness et al demonstrated colocalisation with VIP, substance P, NPY and acetyl choline[50]. Extrinsic denervation does not affect NmU immunoreactivity. Within the CNS, NmU reactive cells were found particularly in the arcuate nucleus and immunoreactive fibres were found throughout the brain with the exception of the cerebellum[52]. Within the cord NmU-like immunoreactivity was mostly found in the dorsal horn with increasing concentrations towards the sacral region[53]. Significant levels of neuromedin U were also present in the dorsal root ganglion. Within the pituitary gland neuromedin U immunoreactivity was found in thyrotrophs and corticotrophs[46].

NEUROMEDIN U RECEPTORS

Little work has been done on the neuromedin U receptors so far. We have found high concentrations in the uterus and preliminary studies indicate an influence of GTP-γ-S suggesting a link with G proteins. Chemical cross linking analysis indicate the Mr of the receptor to be about 48,000. Receptor numbers are significantly altered by administration of steroids. The receptors are of high affinity and show no displacement by other regulatory peptides, present within, or known to have an action on, the uterus.

NEUROMEDIN U PHARMACOLOGY

Although concentrations of neuromedin U within the uterus are low, as mentioned above, receptor numbers are high and neuromedin U causes potent uterine smooth muscle contraction. During the initial isolation process, in which the uterine muscle contractility was first noted, neuromedin U was also shown to elevate blood pressure in the rat. The high concentrations within the gut led to examination of its effects on mesenteric blood flow[54]. In the dog it potently reduces superior mesenteric artery and portal vein flow without significantly affecting flow through other organ systems[48]. Neuromedin U has been shown to potently modify electrogenic ion transport in porcine distal jejunum[55] but in parallel with its low level of muscular innervation there was no detectable effect on smooth muscle contractility in this region in the gut. In contrast neuromedin U exerts a potent contractile effect on isolated human ileum and also urinary bladder[56]. We have demonstrated that circular muscle from a rat fundus is also contracted[57], hence suggesting that NmU may play a physiological role in mouth to caecum transit, although the site of action may vary between species.

A number of neuromedin U fragments have been prepared and the full range of biological activity, though relatively low potency, is demonstrated by the C-terminal 7 amino acids[58]. Indeed we have found that the last six amino acids had biological activity, though of very low potency.

PHYSIOLOGICAL CHANGES

The only studies so far reported on physiological alteration in neuromedin U content are in the anterior pituitary gland[45]. Here it has been demonstrated that systemic administration of thyroid releasing hormone (TRH) produced a five fold increase in anterior pituitary content. Both surgical and chemical thyroidectomy significantly reduced the pituitary content of neuromedin U. In contrast manipulation of the adrenal status with corticosterone or CRF, or the prolactin status with dopamine agonists and antagonists, did not affect neuromedin U concentrations.

CONCLUSION

Neuromedin U is a fairly numerous regulatory peptide in the rat (though not in man) but its physiological role is unclear. No antagonist has yet been developed and high affinity antibodies are difficult to obtain. Thus the standard means for exploring physiology (by inhibition) has been little exploited so far. The potent effect on uterine smooth muscle, but low concentrations of neuromedin U within this organ, are puzzling. The opposite phenomena is seen in the gut where there are high concentrations but pharmacological actions of neuromedin U are weak and idiosyncratic between species. The lack of biological actions has not encouraged research, but the considerable sequence conservation through from the amphibia implies a powerful evolutionary pressure.

It is interesting to contrast the effect of a possible role in human pathology for amylin (potentially a cause of maturity onset diabetes) with the absence of such, and consequentially far less research activity, for neuromedin U.

REFERENCES

1.   P. Westermark, C. Wernstedt, E. Wilander, D.W. Hayden, T.D. O'Brien and K.H. Johnson, Amyloid fibrils in human insulinoma and islets of Langerhans of the diabetic cat are derived from a neuropeptide-like protein also present in normal islet cells, *Proc Natl.Acad.Sci.U.S.A.* 84:3881 (1987).

2.   G.J. Cooper, A.C. Willis, A. Clark, R.C. Turner, R.B. Sim and K.B. Reid, Purification and characterization of a peptide from amyloid-rich pancreases of type 2 diabetic patients, *Proc Natl.Acad.Sci.U.S.A.* 84:8628 (1987).

3.   D. Bretherton-Watt and S.R. Bloom, Islet amyloid polypeptide, the cause of Type-2 Diabetes? *Trends Endo Metab* 2:203 (1992).

4.   D.F. Steiner, S. Ohagi, S. Nagamatsu, G.I. Bell and M. Nishi, Is islet amyloid polypeptide a significant factor in pathogenesis or pathophysiology of diabetes? *Diabetes* 40:305 (1991).

5.   K.H. Johnson, T.D. O'Brien and P. Westermark, Newly identified pancreatic protein islet amyloid polypeptide. What is its relationship to diabetes? *Diabetes* 40:310 (1991).

6.   T. Sanke, G.I. Bell, C. Sample, A.H. Rubenstein and D.F. Steiner, An islet amyloid peptide is derived from an 89-amino acid precursor by proteolytic processing, *J Biol.Chem.* 263:17243 (1988).

7.   S. Mosselman, J.W. Hoppener, C.J. Lips and H.S. Jansz, The complete islet amyloid polypeptide precursor is encoded by two exons, *FEBS Lett.* 247:154 (1989).

8.   H. Jamal, D. Bretherton-Watt, K. Suda, M.A. Ghatei and S.R. Bloom, Islet amyloid polypeptide-like immunoreactivity (amylin) in rats treated with dexamethasone and streptozotocin, *J Endocrinol.* 126:425 (1990).

9.   G.J. Ferrier, A.M. Pierson, P.M. Jones, S.R. Bloom, S.I. Girgis and S. Legon, Expression of the rat amylin (IAPP/DAP) gene, *J Mol.Endocrinol.* 3:R1 (1989).

10.   S.G. Gilbey, M.A. Ghatei, D. Bretherton-Watt, M. Zaidi, P.M. Jones, T. Perera, J. Beacham, S. Girgis and S.R. Bloom, Islet amyloid polypeptide: production by an osteoblast cell line and possible role as a paracrine regulator of osteoclast function in man, *Clin Sci* 81:803 (1991).

11.   D. Bretherton-Watt, M.A. Ghatei, S.E. Bloom, S. Williams and S.R. Bloom, Islet amyloid polypetide-like immunoreactivity in human tissue and endocrine tumours, *J Clin End* Submitted:(1992).

12. A. Chantry, B. Leighton and A.J. Day, Cross-reactivity of amylin with calcitonin-gene-related peptide binding sites in rat liver and skeletal muscle membranes, *Biochem.J* 277:139 (1991).

13. R. Bhogal, D.M. Smith and S.R. Bloom, Investigation and characterization of binding sites for islet amyloid polypeptide in rat membranes, *Endocrinology* 130:906 (1992).

14. C.L. Broderick, G.S. Brooke, R.D. DiMarchi and G. Gold, Human and rat amylin have no effects on insulin secretion in isolated rat pancreatic islets, *Biochem. Biophys. Res. Commun.* 177:932 (1991).

15. W.L. Cody, A.B. Giordani, S. Werness, M.D. Reily, J.A. Bristol, G. Zhu and D.T. Dudley, Analysis of rat amylin amide from commercial sources: identification of a mercury complex, *Biorg. Med. Chem. Let.* 1:415 (1991).

16. B. Leighton and G.J. Cooper, The role of amylin in the insulin resistance of non-insulin-dependent diabetes mellitus, *Trends. Biochem. Sci.* 15:295 (1990).

17. S.J. Koopmans, A.D. van Mansfeld, H.S. Jansz, H.M. Krans, J.K. Radder, M. Frolich, S.F. de Boer, D.K. Kreutter, G.C. Andrews and J.A. Maassen, Amylin-induced in vivo insulin resistance in conscious rats: the liver is more sensitive to amylin than peripheral tissues, *Diabetologia* 34:218 (1991).

18. R.O. Deems, R.W. Deacon and D.A. Young, Amylin activates glycogen phosphorylase and inactivates glycogen synthase via a cAMP-independent mechanism, *Biochem. Biophys. Res. Commun.* 174:716 (1991).

19. A.M. Gomez Foix, J.E. Rodriguez Gil and J.J. Guinovart, Anti-insulin effects of amylin and calcitonin-gene-related peptide on hepatic glycogen metabolism, *Biochem.J* 276:607 (1991).

20. M.W. Wang, P. Carlo, T.J. Rink and A.A. Young, Amylin is more potent and more effective than glucagon in raising plasma glucose concentration in fasted, anesthetized rats, *Biochem. Biophys. Res. Commun.* 181:1288 (1991).

21. A.A. Kassir, A.K. Upadhyay, T.J. Lim, A.R. Moossa and J.M. Olefsky, Lack of effect of islet amyloid polypeptide in causing insulin resistance in conscious dogs during euglycemic clamp studies, *Diabetes* 40:998 (1991).

22. A.A. Young, D.M. Mott, K. Stone and G.J. Cooper, Amylin activates glycogen phosphorylase in the isolated soleus muscle of the rat, *FEBS Lett.* 281:149 (1991).

23. S. Frontoni, S.B. Choi, D. Banduch and L. Rossetti, In vivo insulin resistance induced by amylin primarily through inhibition of insulin-stimulated glycogen synthesis in skeletal muscle, *Diabetes* 40:568 (1991).

24. H.K. Datta, M. Zaidi, S.J. Wimalawansa, M.A. Ghatei, J.L. Beacham, S.R. Bloom and I. MacIntyre, In vivo and in vitro effects of amylin and amylin-amide on calcium metabolism in the rat and rabbit, *Biochem Biophys Res Commun* 162:876 (1989).

25. D. Bretherton-Watt, S.G. Gilbey, M.A. Ghatei, S. Macrae and S.R. Bloom, Very high concentrations of islet amyloid polypeptide (IAPP) are necessary to alter the insulin response to intravenous glucose in man, *J Clin End* In Press:(1992).

26. D. Bretherton-Watt, S.G. Gilbey, M.A. Ghatei, J. Beacham and S.R. Bloom, Failure to establish islet amyloid polypeptide (amylin) as a circulating beta cell inhibiting hormone in man, *Diabetologia* 33:115 (1990).

27. M.A. Ghatei, H.K. Datta, M. Zaidi, D. Bretherton-Watt, S.J. Wimalawansa, I. MacIntyre and S.R. Bloom, Amylin and amylin-amide lack an acute effect on blood glucose and insulin, *J Endocrinol.* 124:R9 (1990).

28. G.C. Zhu, D.T. Dudley and A.R. Saltiel, Amylin increases cyclic AMP formation in L6 myocytes through calcitonin gene-related peptide receptors, *Biochem. Biophys. Res. Commun.* 177:771 (1991).

29. M.W. Wang, A.A. Young, T.J. Rink and G.J. Cooper, 8-37h-CGRP antagonizes actions of amylin on carbohydrate metabolism in vitro and in vivo,*FEBS Lett.* 291:195 (1991).

30. B. Ludvik, B. Lell, E. Hartter, C. Schnack and R. Prager, Decrease of stimulated amylin release precedes impairment of insulin secretion in type II diabetes,*Diabetes* 40:1615 (1991).

31. D. Bretherton-Watt, M.A. Ghatei, S.R. Bloom, H. Jamal, G.J. Ferrier, S.I. Girgis and S. Legon, Altered islet amyloid polypeptide (amylin) gene expression in rat models of diabetes,*Diabetologia* 32:881 (1989).

32. G. Rindi, G. Terenghi, G. Westermark, P. Westermark, G. Moscoso and J.M. Polak, Islet amyloid polypeptide in proliferating pancreatic B cells during development, hyperplasia, and neoplasia in humans and mice,*Am.J Pathol.* 138:1321 (1991).

33. H. Toshimori, R. Narita, M. Nakazato, J. Asai, T. Mitsukawa, K. Kangawa, H. Matsuo, K. Takahashi and S. Matsukura, Islet amyloid polypeptide in insulinoma and in the islets of the pancreas of non-diabetic and diabetic subjects,*Virchows Arch.A.Pathol.Anat.Histopathol.* 418:411 (1991).

34. N. Minamino, K. Kangawa and H. Matsuo, Neuromedin U 8 and U 25: Novel uterus stimulating and hypertensive peptides identified in porcine spinal cord,*Biochem.Biophys.Res.Commun* 130:1078 (1985).

35. J.M. Conlon, J. Domin, L. Thim, V. DiMarzo, H.R. Morris and S.R. Bloom, Primary structure of neuromedin U from the rat,*J Neurochem.* 51:988 (1988).

36. F. O'Harte, C.S. Bockman, P.W. Abel and J.M. Conlon, Isolation, structural characterization and pharmacological activity of dog neuromedin U,*Peptides* 12:11 (1991).

37. R. Kage, F. O'Harte, L. Thim and J.M. Conlon, Rabbit neuromedin U-25: lack of conservation of a posttranslational processing site,*Regul.Pept.* 33:191 (1991).

38. J. Domin, Y.G. Yiangou, R.A. Spokes, A. Aitken, K.B. Parmar, B.J. Chrysanthou and S.R. Bloom, The distribution, purification, and pharmacological action of an amphibian neuromedin U,*J Biol.Chem.* 264:20881 (1989).

39. R. Murphy, C.A. Turner, J.B. Furness, L. Parker and A. Giraud, Isolation and microsequence analysis of a novel form of neuromedin U from guinea pig small intestine,*Peptides* 11:613 (1990).

40. F. O'Harte, C.S. Bockman, W. Zeng, P.W. Abel, S. Harvey and J.M. Conlon, Primary structure and pharmacological activity of a nonapeptide related to neuromedin U isolated from chicken intestine,*Peptides* 12:809 (1991).

41. J. Domin, M.A. Ghatei, P. Chohan and S.R. Bloom, Characterization of neuromedin U like immunoreactivity in rat, porcine,guinea-pig and human tissue extracts using a specific radioimmunoassay,*Biochem Biophys Res Commun* 140:1127 (1986).

42. G. Lo, S. Legon, C. Austin, S.C. Wallis, Z.L. Wang and S.R. Bloom, Characterisation of complementary DNA encoding the rat neuromedin U precursor,*J.Mol.Endocrinol.* In Press:(1992).

43. J. Domin, M.A. Ghatei, P. Chohan and S.R. Bloom, Neuromedin U--a study of its distribution in the rat,*Peptides* 8:779 (1987).

44. J. Ballesta, F. Carlei, A.E. Bishop, J.H. Steel, S.J. Gibson, M. Fahey, R. Hennessey, J. Domin, S.R. Bloom and J.M. Polak, Occurrence and developmental pattern of neuromedin U-immunoreactive nerves in the gastrointestinal tract and brain of the rat,*Neuroscience* 25:797 (1988).

45. J. Domin, J.H. Steel, N. Adolphus, J.M. Burrin, U. Leonhardt, J.M. Polak and S.R. Bloom, The anterior pituitary content of neuromedin U-like immunoreactivity is altered by thyrotrophin-releasing hormone and thyroid hormone status in the rat,*J Endocrinol.* 122:471 (1989).

46. J.H. Steel, S. Van-Noorden, J. Ballesta, S.J. Gibson, M.A. Ghatei, J. Burrin, U. Leonhardt, J. Domin, S.R. Bloom and J.M. Polak, Localization of 7B2, neuromedin B, and neuromedin U in specific cell types of rat, mouse, and human pituitary, in rat hypothalamus, and in 30 human pituitary and extrapituitary tumors, *Endocrinology* 122:270 (1988).

47. J. Domin, A.M. Al-Madani, M. Desperbasques, A.E. Bishop, J.M. Polak and S.R. Bloom, Neuromedin U-like immunoreactivity in the thyroid gland of the rat, *Cell Tissue Res* 260:131 (1990).

48. S. Sumi, K. Inoue, M. Kogire, R. Doi, K. Takaori, T. Suzuki, H. Yajima and T. Tobe, Effect of synthetic neuromedin U-8 and U-25, novel peptides identified in porcine spinal cord, on splanchnic circulation in dogs, *Life Sci.* 41:1585 (1987).

49. M. Honzawa, T. Sudoh, N. Minamino, K. Kangawa and H. Matsuo, Neuromedin U-like immunoreactivity in rat intestine: regional distribution and immunohistochemical study, *Neuropeptides* 15:1 (1990).

50. J.B. Furness, S. Pompolo, R. Murphy and A. Giraud, Projections of neurons with neuromedin U-like immunoreactivity in the small intestine of the guinea-pig, *Cell Tissue Res.* 257:415 (1989).

51. J.P. Timmermans, D.W. Scheuermann, W. Stach, D. Adriaensen, M.H. De Groodt Lasseel and J.M. Polak, Neuromedin U-immunoreactivity in the nervous system of the small intestine of the pig and its coexistence with substance P and CGRP, *Cell Tissue Res.* 258:331 (1989).

52. M. Honzawa, T. Sudoh, N. Minamino, M. Tohyama and H. Matsuo, Topographic localization of neuromedin U-like structures in the rat brain: an immunohistochemical study, *Neuroscience* 23:1103 (1987).

53. J. Domin, J.M. Polak and S.R. Bloom, The distribution and biological effects of neuromedins B and U, *Ann.N.Y.Acad.Sci.* 547:391 (1988).

54. S.M. Gardiner, A.M. Compton, T. Bennett, J. Domin and S.R. Bloom, Regional hemodynamic effects of neuromedin U in conscious rats, *Am J Physiol* 258:R32 (1990).

55. D.R. Brown and F.L. Quito, Neuromedin U octapeptide alters ion transport in porcine jejunum, *Eur.J Pharmacol.* 155:159 (1988).

56. C.A. Maggi, R. Patacchini, S. Giuliani, D. Turini, G. Barbanti, P. Rovero and A. Meli, Motor response of the human isolated small intestine and urinary bladder to porcine neuromedin U-8, *Br.J Pharmacol.* 99:186 (1990).

57. M.A. Benito-Orfila, J. Domin, K.A. Nandha and S.R. Bloom, The motor effect of neuromedin U on rat stomach in vitro, *Eur J Pharmacol* 193:329 (1991).

58. M.A. Benito-Orfila, K.A. Nandha and S.R. Bloom, Actions of various neuromedin U fragments on the smooth muscle of the isolated rat uterus, *Brit Jour Pharm* 104:148 (1989).(Abstract)

Part II
Growth Factors

GENE TRANSFER INTO THE CENTRAL NERVOUS

SYSTEM: NEUROTROPHIC FACTORS

David M. Frim, [1,3] Julie K. Andersen,[1,2] James M. Schumacher,[3]
M. Priscilla Short,[1,2] Ole Isacson,[2,3] and Xandra Breakefield[1,2]

[1]Neuroscience Center, Neurology and Neurosurgery Services
Massachusetts General Hospital
Boston, Massachusetts 02114
[2]Program in Neuroscience
Harvard Medical School
Boston, Massachusettts 02115
[3]Neuroregeneration Laboratory
McLean Hospital
Belmont, Massachusetts 02178

INTRODUCTION

Neurotrophic factors, such as nerve growth factor (NGF), in addition to their role in neuronal development, have protective effects on neuronal survival and stimulatory effects on neuronal regeneration in adult animals (Barde, 1988; Snider and Johnson, 1989; Johnson et al, 1992). For this reason, treatment strategies for human neurodegenerative disease have focused on techniques by which neurotrophic factor gene products can be introduced into the adult central nervous system (CNS) (Hefti, 1986; Gage et al, 1987; Kromer, 1987; Dreyfus, 1989; Morgan, 1989; Gage et al, 1990; Olson, 1990; Lindvall, 1991).

Foreign gene products have been introduced into the mammalian CNS through a variety of methodologies. A secreted factor, such as a neurotrophic peptide, can be delivered by intracerebral implantation of an immortalized cell line altered by retroviral infection to secrete the desired product (Cepko, 1988; Rosenberg et al, 1988; Breakefield, 1989; Schumacher et al, 1991). Similarly, primary fibroblasts (Chang et al, 1990; Gage et al, 1990) or astrocytes (Cunningham et al, 1992) can be altered to secrete a desired product, then transplanted intracerebrally. Non-secreted products, such as synthetic enzymes (Wolff et al, 1989; Horellou et al, 1990), or neurotrophic molecules can be delivered to dividing CNS cells, such as embryonic neuroblasts or glia, by retrovirus infection, or to non-dividing adult neurons by recombinant herpes virus vectors (Breakefield and Deluca, 1991). These

Growth Factors, Peptides, and Receptors, Edited
by T.W. Moody, Plenum Press, New York, 1993

approaches allow an assessment of short- and long-term effects of neurotrophic factors in the CNS *in vivo*. Specifically, the effect of neurotrophic molecules in animal models of neurodegenerative disease can be investigated. This chapter will review studies conducted in our laboratories focused on the transfer of foreign genes, in particular neurotrophic factor genes, into the mammalian CNS.

GENE TRANSFER INTO THE CNS

Retroviral Vectors

The use of recombinant retroviral vectors for gene transfer into dividing cells has been well described (Cepko, 1988). In our laboratories, we have utilized recombinant retroviral plasmids, such as pL(X)RN, to construct a number of retroviruses capable of delivering genes of interest (Rosenberg et al, 1988; Wolf et al, 1988; Ezzedine, et al, 1991). Our studies have utilized two general approaches: the transfer of genetic material into the immortalized Rat 1 fibroblast cell line (Schumacher et al, 1991) and gene transfer into primary astocytes (Cunningham et al, 1992). Using these approaches, we have constructed fibroblasts capable of secreting NGF at high rates (Schumacher et al, 1991), and recovered primary astrocytes which secrete adequate NGF to support *in vivo* differentiation of co-transplanted adrenal medullary cells (Cunningham et al, 1992). These cell constructs utilize the viral LTR promoter, a presumed strong constitutive promoter, to drive transgene synthesis (Short et al, 1990).

Recently, we have had the opportunity to examine some aspects of the NGF-secreting fibroblasts over time *in vivo*. Seven and eighteen days after implantation into adult rat brain, we examined grafts of NGF-secreting fibroblasts histologically, immunocytochemically, and by in situ histohybridization techniques for overall morphological features, for NGF peptide accumulation, and for transgene mRNA expression (Figure 1). Fibroblast grafts, composed of either retrovirally altered or unaltered cells, survived well after intracerebral transplantation, forming fibromatous structures at the implantation site. Of note, the amount of immunoreactive NGF within NGF-producing grafts, as well as the transgene mRNA, were difficult to demonstrate 18 days after implantation. Cells actively synthesizing NGF appeared confined to peripheral areas of the grafts where continued graft growth was evident. These observations underscore a current concern about the use of secreting cell-lines: specifically, inability to maintain high levels of secretory product over time. Other investigators have found similar results in long term transplantation experiments with retrovirally constructed secreting cells (Palmer et al, 1991). Continued investigation into transgene promoter type and function should help overcome this difficulty.

Herpes Virus Vectors

The retrovirus approach to gene transfer into the CNS is limited to dividing cells (Cepko, 1988). Direct infection with genetically designed retrovirus vectors is incapable of delivering genetic information to non-dividing cells, such as neurons. Vectors derived from herpes-simplex virus, however, are an effective way of delivering foreign genetic information to neuronal cells of the CNS (Chiocca et al, 1990; Breakefield and Deluca, 1991; Geller et al, 1991). Herpes-virus particles infecting CNS neurons can induce a lytic infection or reside in a relatively benign latent state as an episomal unit within the cell nucleus. In the

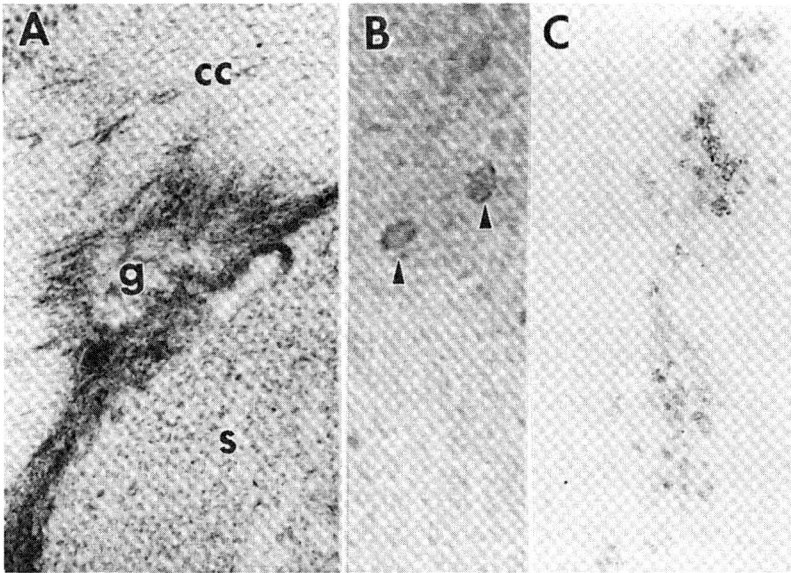

Figure 1. Eighteen day NGF-secreting fibroblast cell-line graft. A. Low power view of Nissl stained graft 18 days after implantation. Note tendrils of graft extension. Core of graft appears to be less dense than periphery. g, graft; s, striatum; cc, corpus callosum. B. High power view of NGF immunopositive fibroblasts 18 days after implantation located in periphery of graft shown in A. arrowheads, immunopositive NGF cells. C. In situ histohybridization (biotinylated-oligonucleotide-peroxidase method) for NGF-transgene mRNA 18 days after graft implantation. Section is adjacent to that shown in B.

latent state, the virus retains transcriptional activity, allowing stable expression of genes without compromising the cell's function (Stevens, 1989; Dobson et al, 1990; Breakefield et al, 1992). Infection with herpes-virus vectors rendered replication-deficient reduces the virulence of herpetic infection (Chiocca et al, 1990) and may increase the likelihood of viral latency and potential for long term foreign gene expression (Coen et al, 1989; Efstathiou et al, 1989; Leist et al, 1989; Kosz-Vnenchak et al, 1990).

We have demonstrated neuron-specific expression of a foreign gene by placement of a marker gene, LacZ, under the control of a neuron-specific promoter in a recombinant herpes-virus vector (Andersen et al, 1991). Specifically, using the promoter from the neuron specific enolase gene, the *E coli* LacZ gene, and a polyadenylation sequence from the human enkephalin gene, we constructed a recombinant herpes virus vector in which this gene cassette is placed within the thymidine kinase gene of the viral DNA (Figure 2). This renders the virus replication defective in neurons. This virus, after infection of cells in culture, promoted long term expression of bacterial β-galactosidase (Andersen et al, 1991). In addition, viral inoculation into rat frontal cortex yielded neurons able to produce the bacterial enzyme for up to one month. Long term neuron-specific expression may be due either to specificity of promoter or due to the fact that the virus can only enter latency in some neurons, not other cell types (Stevens, 1989). Whether promoter-dependent specificity of viral transgene expression can restrict expression to a neuronal sub-population remains to be shown.

The specific anatomical constraints of infection of CNS neurons with replication-deficient virus is unclear: accepted dogma predicts that only neurons that can directly take-up replication-deficient virus would be liable for viral infection with no transsynaptic transfer of

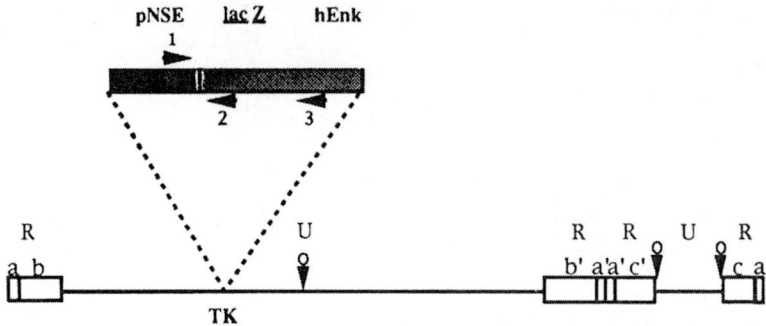

Figure 2. Schematic representation of replication deficient, neuron-specific herpes virus vector. Note insertion of NSE-LacZ-hEnk construct into herpes TK gene. pNSE, neuron specific enolase promoter; lac Z, E coli β–galactosidase gene; hEnk, human enkephalin gene polyadenylation site; TK, tyrosine kinase gene; o, origin of replication; R, HSV-1 repeat element; a, b, c, repeat elements; arrowheads, oligonucleotide primers, in sense and anti-sense directions, for PCR analysis (1,2) and in situ histohybridization assays (3, see Figure 3).

infection. Coupled with neuron-specific transgene expression, neuroanatomic specificity in extent of viral infection and gene delivery could provide a useful tool for treatment of neuroanatomically-defined neurological disease. Using a replication-deficient herpes virus containing the LacZ gene under the control of the neuron-specific enolase gene promoter (Figure 2), we found that herpes virus mediated gene delivery can be both neuronally- and neuroanatomically-specific (Andersen et al, submitted). When transgene delivery was examined for viral DNA by an in situ PCR technique, for transgene mRNA expression by in situ histohybridization, and for transgene protein product by immunocytochemistry and X-gal histochemistry, we found that after delivery of herpes vector into the rat striatum, transgene expression could be found only in neurons at the inoculation site and at sites reached through single-cell retrograde transport (Figure 3). This finding will be of value in designing strategies for neuroanatomically-specific neurotrophic factor gene delivery. Of note, we also documented an incidence of transgene expression seven days after inoculation of approximately one neuron expressing LacZ per 10^4 virions delivered. This low expression rate is likely a technical constraint that can be addressed by the use of different transgene promoters or different strains of recombinant herpes vectors.

NEUROTROPHIC FACTOR EFFECTS *IN VIVO*

Protective effects of NGF-secreting fibroblast grafts

The technology for delivery of NGF into the central nervous system has included systemic injection of agents to induce NGF synthesis (Furukawa and Furukawa, 1990), direct intracerebral infusion of NGF by minipump (Junard et al, 1990; Tusynski et al, 1990), and as described, biological delivery by genetically altered cell-lines (Gage et al, 1987; Breakefield, 1989) or primary autotransplants (Chang et al, 1990; Cunningham et al, 1992). Through these approaches, NGF has been shown to protect against cholinergic neuronal degeneration after axotomy (Hefti, 1986; Rosenberg et al, 1988; Gage et al, 1990; Koliatsos et al, 1991) and neuronal death after excitotoxic lesioning (Aloe, 1987; Schumacher et al, 1991). We have used the NGF-secreting immortalized rat fibroblast cell-line to confirm these observations and to further investigate the *in vivo* effects of NGF. Using a striatal

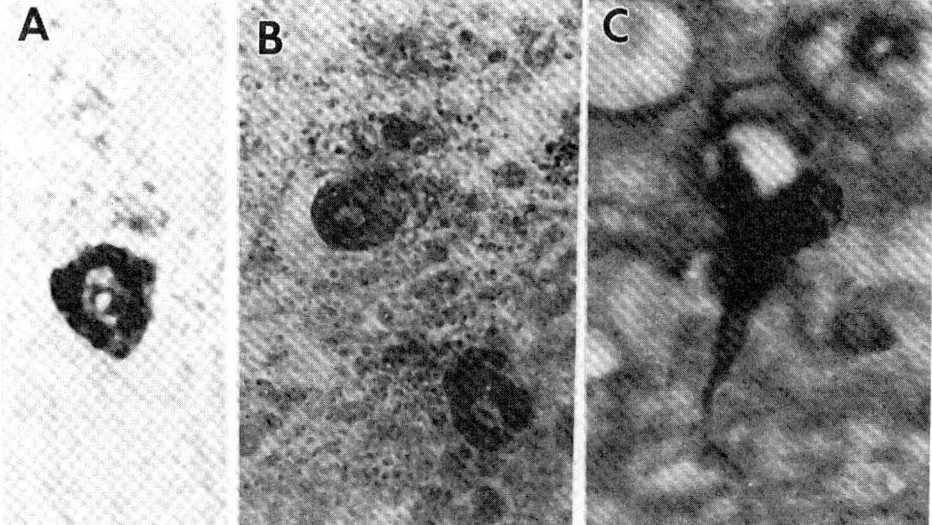

Figure 3. Neurons expressing the LacZ transgene seven days after herpes inoculation into rat striatum. A. In situ histohybridization (digoxigeninated-oligonucleotide-alkaline phosphatase method) for LacZ fusion mRNA in striatum. Note very strongly stained cell approximately 20 μ diameter. B. Two in situ histohybridization positive cells in frontal cortex, an area accessible from striatal injection site by single cell retrograde transport. C. Immunocytochemistry for *E coli* β-galactosidase in substantia nigra. Note strongly positive cell of neuronal morphology in another location accessible from striatal injection site by single cell retrograde transport.

excitotoxic lesion model of Huntington's disease (Schumacher et al, 1991), we have studied the local and distant effects of NGF-producing fibroblasts on lesion size, neuronal loss, and cholinergic neuronal survival. In our model, NGF-producing (NGF[+]) and non-NGF-producing (NGF[-]) fibroblasts are implanted unilaterally into the rat corpus callosum seven days before striatal infusion of quinolinate, a potent excitotoxin (Schwarcz et al, 1983).

We find that lesions located ipsilateral to NGF[+] implants are approximately 80% smaller in maximal cross-sectional area than lesions located ipsilateral to NGF[-] grafts (Schumacher et al, 1991; Frim et al, 1992). The distance from site of implant to lesion center is approximately one mm. Ipsilateral NGF[+] implants also cause marked cholinergic neuronal sparing, as seen with anti-cholineacetyltransferase immunostaining. When lesions are made contralateral to the NGF-producing implants (a distance of approximately four mm), no reduction in lesion size is noted (Figure 4). Additionally, there is no contralateral cholinergic cell sparing. These observations allow two conclusions: first, NGF secreted by the NGF[+] grafts is likely spread only locally through the neuropil, and not through the CSF pathways. This finding has been anticipated by previous studies of NGF directly injected into the striatum (Olson et al, 1991). Second, the protective cellular changes induced by the NGF[+] cells occur only locally, and investigation into NGF-mediated protective mechanisms can be concentrated in the area immediately adjacent to the grafts.

Mechanism of NGF-mediated excitotoxic protection

Building on the observations and conclusions described above, we have begun to investigate the mechanism of NGF-mediated neuronal protection after excitotoxic or knife-cut

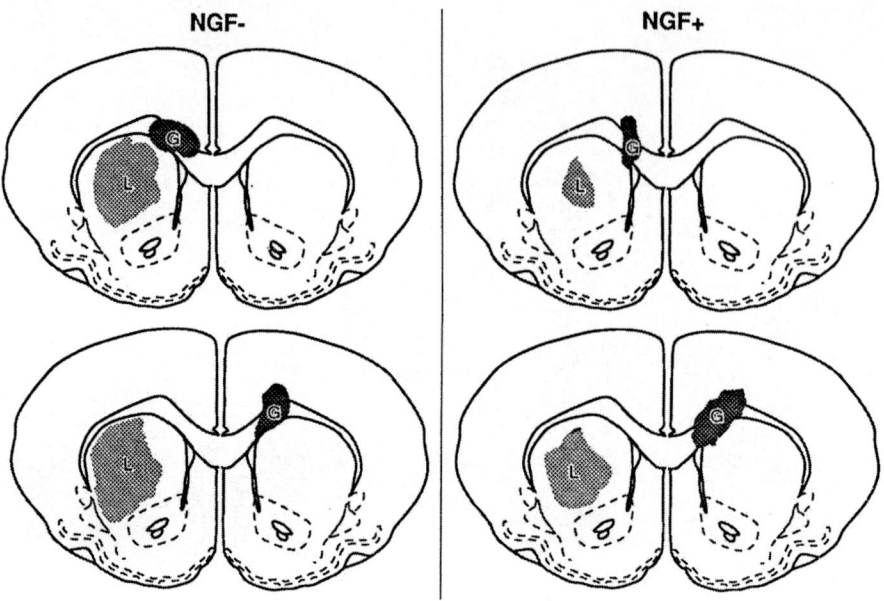

NGF- **NGF+**

Figure 4. Schematic representation of local and distant effects of NGF-producing cell implants. This figure represents a schematic of rat brain cross-sections taken at bregma +1.2 mm 11 days after striatal excitotoxic lesioning in the presence of NGF-producing (NGF+) and NGF-non-producing (NGF-) fibroblast cell-line implants. Note the proximity of the grafts (g) to the lesions (L) when implanted ipsilaterally versus contralaterally. The contralateral NGF+ graft has no effect on lesion size, as does the ipsilateral NGF+ graft.

lesions. Specifically, we examined the induction of protective enzymes and stress proteins in tissues surrounding the NGF[+] implants. Seven days after unilateral corpus callosum implantation of NGF[+] and NGF[-] grafts, we used histological techniques to evaluate acetylcholinesterase (AChE), and immunocytochemistry to visualize the stress proteins c-fos, c-myc, heat shock protein-72 kDa (hsp72), and ubiquitin (Ub), and the peroxidative enzyme catalase. We found no specific immunostaining in tissues surrounding either the NGF[+] or NGF[-] grafts using antibodies raised against c-fos, c-myc, or Ub, though both types of grafts stained faintly for c-fos. Both NGF[+] and NGF[-] grafts elicited a halo of hsp72 immunoreactivity in the tissue immediately surrounding the grafts. AChE staining was unchanged in areas adjacent to the grafts, and although there was no fibroblast AChE staining, multiple AChE positive fibers appeared to be growing into or traversing the NGF[+] grafts. In contrast to the above, NGF[+] cells caused increased catalase staining in astroglial cells adjacent to the grafts. These catalase positive astroglial cells were not seen in areas distant from the NGF[+] grafts, nor were they seen in brains implanted with NGF[-] cells.

These results suggest that the local protective effects of NGF[+] fibroblast grafts may be mediated through changes in peroxidative metabolism due to increased levels of catalase. This finding is consistent with previous observations of NGF-mediated protection against peroxide damage *in vitro* (Jackson et al, 1991). Inability to withstand free radical metabolism and lipid peroxidation is theorized to contribute to cell death after ischemic and traumatic insults (Hall and Braughler, 1989). Induction of catalase, an enzyme which reduces lipid peroxidation, may be a potent effector of neuronal protection. If so, NGF-mediated protection and neurotrophism in the adult brain may be caused in part by an increase in cell resistance to superoxide damage and lipid peroxidation.

The overall protective effects of NGF in the adult CNS are likely based on more than one mechanism. Other avenues of investigation include the evaluation of NGF-producing implant effects on local NMDA-receptor number and the ability of NGF implants to ameliorate the effects of metabolic toxins.

CONCLUSIONS AND FUTURE DIRECTIONS

The above described experiments portend the feasibility of the delivery of neurotrophic molecules into the CNS as treatment for neurological disease. Many technical and theoretical problems have yet to be addressed before a meaningful attempt at the clinical usage of these techniques is undertaken. Further recombinant engineering of viral vectors coupled with a more refined knowledge of neurotrophic factor protective mechanisms should yield significant progress towards useful neurotrophic factor gene transfer. Other neurotrophic molecules, for example, brain derived neurotrophic factor (Barde, 1988) need to be examined by secreting cell-line technology. In addition, neurotrophic molecule gene delivery in a herpes vector still awaits demonstration. Nevertheless, we continue to eagerly investigate each advance in the several model systems discussed in the hope of building upon the many observations contributed by our laboratories and others.

REFERENCES

Andersen JK, Garber D, Meaney CA, Breakefield XO. Gene transfer into cultured sensory neurons using herpes virus vectors: expression of bacterial LacZ using a mammalian neuronal promoter. Neurosci Abstr 17:1438, 1991.

Andersen JK, Frim DM, Isacson O, Breakefield XO. Stereotactic delevery of a neuron-specific herpes-virus vector into the rat striatum results in neuroanatomically specific transgene expression. Submitted for publication, 1992

Barde Y-A. What, if anything, is a neurotrophic factor? TINS 11:343-346, 1988.

Breakefield XO. Combining CNS transplantation and gene transfer. Neurobiol Aging 10:647-648, 1989.

Breakefield XO, Deluca NA: Herpes simplex virus for gene delivery to neurons. New Biologist 3:203-218, 1991.

Breakefield XO, Huang Q, Andersen JK, Kramer MF, Bebrin WR, Davar G, Vos B, Garber DA, Difiglia M, Coen D. Gene transfer into neurons using recombinant herpes virus vectors. Foundation IPSEN Proc, Paris, 1992, in press.

Cepko, C. Retroviral vectors and their applications in neurobiology. Neuron 1:345-353, 1988.

Chang PL, Capone JP, Brown GM. Autologous fibroblast implantation: Feasibility and potential problems in gene replacement. EMol Biol F Med F7:461-470, 1990.

Chiocca EA, Choi BB, Cai W, Deluca NA, Schaffer PA, DiFiglia M, Breakefield XO, Martuza RL. Transfer and expression of the lacZ gene in rat brain neurons mediated by herpes simplex virus insertion mutants. New Biologist 2:739-746, 1990.

Coen DM, Kosz-Vnenchak M, Jacobson JG, Leib DA, Board CL, Schaffer PA, Taylor KL, Knipe DM. Thymidine kinase-negative herpes simplex virus mutants establish latency in mouse trigeminal ganglia but do not reactivate. Proc Natl Acad Sci USA 86:47836-47840, 1989.

Cunningham LA, Hansen JP, Short MP, Bohn MC. The use of genetically altered astrocytes to provide nerve growth factor (NGF) to adrenal chromaffin cells grafted into the striatum. Brain Res, in press, 1992.

Dobson AT, Margolis TP, Sederati F, Stevens JG, Feldman LT. A latent, non-pathogenic HSV-1-derived vector stably expresses β-galactosidase in mouse neurons. Neuron 5:353-360, 1990.

Dreyfus CF. Effects of nerve growth factor on cholinergic brain neurons. Trends Pharmacol Sci 10:145-148, 1989.

Efstathiou S, Kamp S, Darby G, Minson AC. The role of herpes simplex type 1 thymidine kinase in pathogenesis. J Gen Virol 70:869-879, 1989.

Ezzedine ZD, Martuza RL, Short MP, Platika D, Malick A, Choi B, Breakefield XO. Selective killing of glioma cells in culture and in vivo by retrovirus transfer of the herpes simplex virus thymidine kinase gene. New Biologist 3:1-7, 1991.

Frim DM, Short MP, Rosenberg WS, Simpson J, Breakefield XO, Isacson O. Nerve growth factor secreting fibroblasts protect only locally against excitotoxic lesions in the rat striatum. J Neurosurg, in press, 1992.

Furukawa S, Furukawa Y. Nerve growth factor synthesis and its regulatory mechanisms: an approach to the therapeutic induction of nerve growth factor synthesis. Cerebrovasc Brain Metab Rev 2:32-344, 1990.

Gage FH, Wolf JA, Rosenberg MB, Xu L, YeeJK, Shults C, Friedmann T. Grafting genetically modified cells to the brain: possibilities for the future. Neuroscience 23:795-807, 1987.

Gage FH, Rosenberg MB, Tuszynski MH, Yoshida K, Armstrong DM, Hayes RC, Friedmann T: Gene therapy in the CNS: intracerebral grafting of genetically modified cells. Prog Brain Res 86:205-217, 1990

Geller AI, During MJ, Neve RL. Molecular analysis of neuronal physiology by gene transfer into neurons with herpes sinplex virus vectors. Trend Neurosci 14:428-432, 1991.

Hall ED and Braughler JM. Central nervous system trauma and stroke. II. Physiological and pharmacological evidence for involvement of oxygen radicals and lipid peroxidation. Free Rad. Biol. Med. 6: 303-311, 1989.

Hefti F. Nerve growth factor promotes survival of septal cholinergic neurons after fimbrial transections. J Neurosci 6:2155-2162, 1986.

Horellou P, Marlier L, Privat A, Mallet J. Behavior of engineered cells that synthesize L-Dopa or dopamine after grafting into the rat neostriatum. Eur J Neurosci 2:116-119, 1990.

Jackson GR, Apffel L, Werrbach-Perez K, Paerez-Polo JR. Role of nerve growth factor in oxidant-antioxidant balance and neuronal injury. I. Stimulation of hydrogen peroxide resistance. J Neurosci Res 25:360-368, 1990.

Johnson EM, Tatsuro K, Franklin J. A Calcium set-point hypothesis of neuronal dependence on neurotrophic factors. Exp Neurol 115:163-166, 1992.

Junard EO, Montero CN, Hefti F. Long-term administration of mouse nerve growth factor to adult rats with partial lesions of the cholinergic septohippocampal pathway. Exp Neurol 110:25-38, 1990.

Koliatsos VE, Applegate MD, Knussel B, Junard EO, Burton LE, Mobley WC, Hefti FF, Price DL. Recombinant human nerve growth factor prevents retrograde degeneration of axotomized basal forebrain cholinergic neurons in the rat. Exp Neurol 112:161-173, 1991.

Kosz-Vnenchak M, Coen DM, Knipe DM. Restricted expression of herpes simplex virus lytic genes during establishment of latent infection by thymidine kinase negative viruses. J Virol 64:5396-5402, 1990.

Kromer LF. Nerve growth factor treatment after brain injury prevents neuronal cell death. Science 235:214-216, 1987.

Leist TP, Sandri-Goldin RM, Stevens JG. Latent infections in spinal ganglia with thymidine kinase-deficient herpes simplex virus. J Virol 63:4976-4978, 1989.

Lindvall O. Prospects of transplantation in human degenerative diseases. Trends Neurol Sci 14:377-384, 1991.

Morgan DG. Considerations in the treatment of neurological disorders with trophic factors. Neurobiol Aging 10:547-549, 1989.

Olson L. Grafts and growth factors in CNS: basic science with clinical promise. Stereotact Funct Neurosurg 54-55:250-267, 1990.

Olson L, Wetmore C, Stromberg I, Ebendal T. Endogenous and exogenous nerve growth factor in the central nervous system. In Fuxe and Agnati (eds), *Volume Transmission in the Brain: Novel Mechanisms for Neural Transmission.* Raven Press; New York, 1991.

Palmer TD, Rosman GJ, Osborne WR, Miller AD. Genetically modified skin fibroblasts persist long after transplantation but gradually inactivate introduced genes. Proc Natl Acad Sci USA 88:1330-1334, 1991.

Rosenberg MB, Friedman T, Robertson RC, Tuszynski M, Wolff JA, Breakefield XO, Gage FH. Grafting genetically modified cells to the damaged brain: restorative effects of NGF expression. Science 242:1575-1578, 1988.

Schumacher JM, Short MP, Hyman BT, Breakefield XO, Isacson O. Intracerebral implantation of nerve growth factor-producing fibroblasts protects striatum against neurotoxic levels of excitatory amino acids. Neuroscience 45:561-570, 1991.

Schwarcz R, Whetsell WO, Mangano RM. Quinolinic acid: an endogenous metabolite that produces axon sparing lesions in rat brain. Science 219:316-319, 1983.

Short MP, Rosenberg MB, Ezzedine ED, Gage, J.H., Friedmann T, Breakefield X.O. Autocrine differentiation of PC12 cells mediated by retroviral vectors. Devel Neurosci 12:34-45, 1990.

Snider WD, Johnson EM. Neurotrophic molecules. Ann Neurol 26:486-506, 1989.

Stevens JG. Human herpesviruses: a consideration of the latent state. Microbiol Rev 53:318-332, 1989.

Tuszynski MH, U HS, Amaral DG, Gage FH. Nerve growth factor infusion into the primate brain reduces lesion-induced cholinergic neuronal degeneration. J Neurosci 10:3604-3614, 1990.

Wolf D, Richter-Landsberg C, Short MP, Cepko C, Breakefield, XO. Retrovirus mediated gene transfer of beta-nerve growth factor into mouse pituitary line AtT-20. Mol Biol Med 5:43-59, 1988.

Wolff JA, Fisher LJ, Xu L, Jinnah HA, Langlais PJ, Iuvone MP, O'Malley KL, Rosenberg, MB, Shimohara, S, Friedmann T, Gage FH. Grafting fibroblasts genetically modified to produce L-dopa in a rat model of Parkinson's Disease. Proc Natl Acad Sci USA 86:9011-9014, 1989.

THE STRUCTURAL AND BIOLOGICAL PROPERTIES OF HEPARIN-BINDING EGF- LIKE GROWTH FACTOR (HB-EGF)

Shigeki Higashiyama[1,2] and Michael Klagsbrun[2,3]

[1]Department of Biochemistry
Osaka University Medical School, Osaka 565 Japan
[2]Department of Surgical Research
Children's Hospital, Boston, MA, 02115
[3]Departments of Biological Chemistry and Molecular Pharmacology
Harvard Medical School, Boston, MA 02115

INTRODUCTION

The epidermal growth factor (EGF) protein family encompasses a number of structurally homologous mitogens including EGF, transforming growth factor-α (TGF-α), vaccinia growth factor (VGF), amphiregulin (AR), heparin-binding EGF-like growth factor (HB- EGF) and heregulin/Neu differentiation factor (HRG/NDF) (Carpenter and Wahl, 1990; Derynck et al., 1985; Brown et al., 1985; Shoyab et al., 1989; Higashiyama et al., 1991; Wen et al., 1992; Holmes et al., 1992). HB-EGF was originally identified in our laboratory in the conditioned medium of human monocytes, macrophages and the macrophage-like cell line, U-937. It was named heparin-binding EGF-like growth factor because of its strong affinity for immobilized heparin (elution with 1.0-1.2 M NaCl), a property not shared by EGF or TGF-α. It is speculated that the heparin-binding properties of HB-EGF may be of biological significance. In recent years, a growing number of heparin-binding growth factors have been identified. Some examples are members of the fibroblast growth factor (FGF) family, platelet-derived growth factor (PDGF), and vascular endothelial growth factor/vascular permeability factor (VEGF/VPF) (see review; Klagsbrun, 1992). Interaction with cell surface heparan sulfate proteoglycan has been demonstrated to facilitate the binding of basic FGF (bFGF) to its high affinity receptor and to enhance mitogenic activity. (Yayon et al., 1991; Rapraeger et al., 1991; Klagsbrun and Baird, 1991; Ornitz et al., 1992). We speculate that the ability of HB-EGF to bind to cell surface heparan sulfate proteoglycan might also modulate its biological activity.

This article reviews some of the structural and biological properties of HB-EGF.

PURIFICATION OF HB-EGF FROM U-937 CELL CONDITIONED MEDIUM

Heparin affinity chromatography of human macrophage and U-937 cell conditioned media revealed heterogeneous profiles of growth factor activity, mitogenic for BALB/c 3T3 cells, eluting with 1.0-1.2 M NaCl (Figure 1).

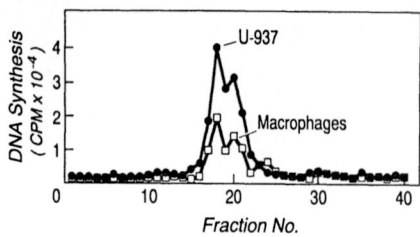

Figure 1. TSK-heparin column chromatography of cultured human macrophage and U-937 cell conditioned media. Isolated macrophages or U-937 cells were plated (2 x 10⁸ cells per T-150 flask) in RPMI 1640 containing 10% fetal calf serum, penicillin (100 U/ml), and streptomycin (100 ug/ml) (RPMI/10%FCS/PS). U-937 cells were stimulated for 24 hours with 60 nM TPA and transferred to serum-free medium (RPMI/PS). Macrophages were transferred to 5% serum medium (RPMI/5% FCS/PS). Conditioned media were collected after 24 hours and analyzed on a TSK-heparin 5PW column (8 x 75 mm) using fast performance liquid chromatography (FPLC). Bound protein was eluted with a 40- ml linear gradient of 0.2 to 2.0 M NaCl in 0.01 M Tris-HCl, pH 7.4, at 1 ml/min, and fractions were tested for the ability to stimulate DNA synthesis in BALB/c 3T3 cells.

A four-step protocol that included Bio-Rex 70 cation exchange chromatography, Cu-chelating Sepharose chromatography, TSK-heparin chromatography, and C_4 reversed phase-high performance liquid chromatography (RPLC) was developed to purify HB-EGF from the conditioned medium of U-937 cells treated with 12-O-tetradecanoylphorbol-13-acetate (TPA). The C_4 RPLC step revealed a complex profile of growth factor activity consisting of two major peaks and two smaller peaks. Each of these peaks was further purified by a separate second cycle of C_4 RPLC. The end result was that four apparently homogeneous and biologically active species could be resolved and isolated. These multiple forms were named HB-EGF Peak-1, -2, -3 and -4 in order of increasing retention time on C_4 RPLC. Their apparent molecular masses on SDS-PAGE under non-reducing conditions were determined to be 22, 19, 23, and 22.5 kDa respectively. All 4 forms were equally mitogenic for SMC and BALB/c 3T3 cells and were equally able to inhibit the binding of ^{125}I-EGF to the EGF receptor (Higashiyama et al., 1992).

That these four peaks of activity represented forms of HB- EGF was ascertained by N-terminal and tryptic fragment sequence analysis and comparison of the results to the predicted HB-EGF sequence based on the cDNA nucleotide sequence (Table 1). HB-EGF Peak-3 and HB-EGF Peak-4 appeared to contain the longest forms of HB-EGF. HB-EGF Peak-1, the predominant species in U-937 conditioned medium, appeared to be a truncated form missing the first 10 N-terminal amino acids. HB-EGF Peak-2 was a mixture of three truncated species missing the first 10, 14 and 19 N-terminal amino acid residues respectively.

Table 1. N-terminal amino acids of multiple HB-EGF forms

Predicted sequence from cDNA	DLQEA	DLDLL	RVTLS	SKPQA	LATPN	KEEHG
HB-EGF Peak-1			*			
HB-EGF Peak-2			*	*	*	
HB-EGF Peak-3		*				
HB-EGF Peak-4		*				

The predicted amino acid sequence shown corresponds to amino acids 63 to 92 in the 208-amino acid precursor of HB-EGF (Higashiyama et al. 1991). The asterisks denote the N-terminal amino acid of the various forms of HB-EGF. Note that HB-EGF Peak-2 contains three species of truncated HB-EGF. The species in HB-EGF Peak-4 is possibly N-terminally extended upstream of the N-terminal residue shown.

Table 2. Purification of HB-EGF from U-937 conditioned medium

Column	Protein (ug)	Mitogenic activity (ng/ml)	Purification (fold)	Yield (%)
Bio-Rex 70	450,000	7,500	1	100
Cu-chelating Sepharose	90,000	7,400	1	23
TSK-Heparin	295	32.8	229	18
C₄ RPLC			7,500	15
HB-EGF Peak-1	1.2	1.0		
HB-EGF Peak-2	1.0	1.0		
HB-EGF Peak-3	0.27	1.0		
HB-EGF Peak-4	0.21	1.0		

It can not be excluded at present that the four HB-EGF forms differ in their C-termini or in post-translational modification-spliced forms.

The purification of the various HB-EGF forms was approximately 7,500-fold, the yield was 15%, and about 2.7 ug of HB-EGF was recovered from 10 l of the conditioned medium (Table 2).

STRUCTURAL PROPERTIES OF HB-EGF

Tryptic digests were used to determine the full amino acid sequence of mature HB-EGF (Higashiyama et al., 1992). The results indicated that the longest form of secreted mature HB-EGF consists of at least 86 amino acids corresponding to positions 63 to 148 of the sequence predicted from the HB-EGF cDNA nucleotide sequence (Figure 2). The predicted threonine residues 13 and 23 amino acids from the N-terminus were not detected by the sequencer suggesting that these may be sites of O-glycosylation. That HB-EGF is a glycoprotein is supported by results indicating that endo-N-acetylgalactosaminidase (O-glycanase) lowers the apparent molecular mass of HB-EGF on SDS-PAGE from 18-20 kDa to about 14- 16 kDa (Higashiyama et al., 1992).

HB-EGF has six cysteine residues in its C-terminal domain with a spacing characteristic of members of the EGF family. Between the first and the sixth cysteine, the homologies of HB-EGF to EGF and to TGF-α are 41% (15/37) and 42% (15/36) respectively (Fig. 2). The high homology with EGF and TGF-α suggests that HB-EGF

HB-EGF, EGF, TGF-α Homologies

```
                              *        * * * * *   *    * *  ** ****   *  ( vs. EGF )
HB-EGF                        *        * * * *          * * ** ** *  **   *  ( vs.TGF-α )

DLQEADLDLLRVTLSSKPQALATPNKEEHGKRKKKGKGLGKKRDPCLRKYKDFCIH-GECKYVKELRAPSCICHPGYHGERCHGLSL...   86

             EGF          NSDSECPLSHDGYCLHDGVCMYIEALDKYACNCVVGYIGERCQYRDLKWWELR   53

             TGF-α   VVSHFNDCPOSHTQPCFH-GTCRFLVQEDKPACVCHSGYVGARCEHADLLA   50
```

Figure 2. Amino acid sequences of HB-EGF, EGF and TGF-α. The asterisks indicate amino acid homologies between HB-EGF and EGF, or HB-EGF and TGF-α (Carpenter and Wahl, 1990; Higashiyama et al., 1991). A putative heparin- binding sequence of HB-EGF is indicated by a solid line. The positions of the cysteine residues are shaded.

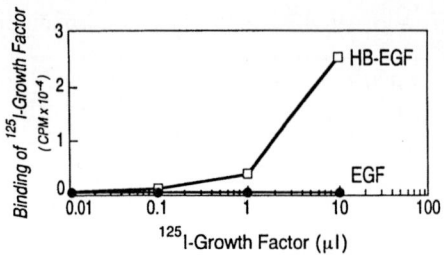

Figure 3. ^{125}I-HB-EGF binding to immobilized heparin. Ten ul of heparin-Sepharose equilibrated with 10 mM Tris-HCl, pH 7.4, 0.2 M NaCl, were incubated in microtubes with either ^{125}I-HB-EGF or ^{125}I-EGF (10000 cpm /ul). After a 5 min incubation at room temperature, the heparin-Sepharose beads were washed with equilibration buffer and counted in a gamma counter.

should bind to the EGF receptor. Indeed, it can be shown that HB-EGF competes with EGF for binding to A-431 cells and that it triggers phosphorylation of the EGF receptor (Higashiyama et al., 1992).

HB-EGF appears to be an N-terminally extended member of the EGF family relative to EGF and TGF-α possessing 35-40 additional amino acids. Since EGF (Figure 3) and TGF-α do not bind to heparin, it is likely that the heparin-binding domain of HB-EGF is located in its N-terminal region. Analysis of the HB-EGF sequence reveals a cationic stretch rich in Lys and Arg residues (Figure 2). This hydrophilic sequence is a possible candidate for being a heparin-binding domain.

HB-EGF is resistant to exposure to acidic pH as low as 2.5. Heating to 90 °C for 5 minutes results in a 36% loss in activity. On the other hand, HB-EGF is totally inactivated by exposure to 5 mM dithiothreitol suggesting that intact intrachain disulfide bonds are neccessary for mitogenic activity.

Chromatofocusing demonstrates that the isoelectric point (pI) of HB-EGF is between 7.2 and 7.8. Thus HB-EGF is considerably more cationic than EGF which has a pI of about 5.5

Some of the structual and biochemical properties of HB-EGF are summarized in Table 3.

BIOLOGICAL PROPERTIES OF HB-EGF

The source of HB-EGF has been to date primarily monocytes, macrophages and macrophage-like cells. Not surprisingly, HB-EGF is found in wound fluid, possibly as a

Table 3. Structural properties of HB-EGF

---

* Heparin-binding (elution with 1.1 M NaCl)
* Homologous to EGF (40%)
* Single chain polypeptide
* Longest mature protein form has at least 86 amino acids
* 20-23 kDa (as determined by SDS-PAGE)
* O-glycosylated but not N-glycosylated
* exists in multiple forms
* N-terminal truncations occur
* Processed from 208 amino acid precusor
* pI of 7.8 (EGF - pI of 5.5)
* DTT sensitive, heat resistant
* 2.5 kb mRNA

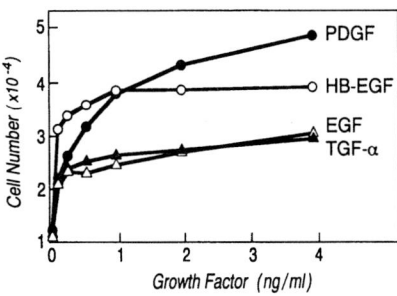

Figure 4. Smooth muscle cell proliferation. Bovine aortic SMC (BASMC) were plated (5 x 10³ cells/well, 24-well plate) in Dulbecco's modified Eagle's medium, 10% calf serum, penicillin (100 U/ml), and streptomycin (100 ug/ml) (DMEM/10%CS/PS). After attachment overnight, the medium was replaced with DMEM/2%CS/PS. HB-EGF, PDGF, EGF or TGF-alpha were added and BASMC were subsequently counted after 3 days.

macrophage product. HB-EGF is also expressed by vascular cells. HB-EGF can be detected in the conditioned medium of fetal human smooth muscle cells (SMC) and HB-EGF message is expressed by endothelial cells exposed to tumor necrosis factor (Yoshizumi et al., 1992).

In terms of target cells, HB-EGF has been shown to stimulate proliferation of SMC, fibroblasts and keratinocytes. Interestingly, HB-EGF is a more potent SMC mitogen than EGF or TGF-α (Figure 4) even though all three mitogens bind to the EGF receptor. HB-EGF at 100 pg/ml stimulates SMC proliferation to the same extent as does EGF or TGF-alpha at 4 ng/ml. HB-EGF potency is more comparable to PDGF AB with approximately a 5-fold greater specific activity for bovine aortic SMC (Figure 4) and BALB/c 3T3 cells (Figure 5).

On the other hand, HB-EGF is not mitogenic for endothelial cells or myoblasts, nor does it stimulate neurite outgrowth in PC 12 cells. In this regard the target cell specificities of HB-EGF are quite different than those of bFGF. The sources of HB-EGF and its target cell specificities are summarized in Table 4.

DISCUSSION

HB-EGF has some interesting structural and biological properties as follows: 1) HB-EGF, unlike EGF and TGF-α, binds to heparin and to cell surface heparan sulfate proteoglycan. Thus HB-EGF might be concentrated on the surfaces of cells that are rich in heparan sulfate proteoglycan. Growth factor-heparan sulfate interactions have been shown to modulate the activities of other heparin-binding growth factors such as bFGF and VEGF/VPF (Yayon et

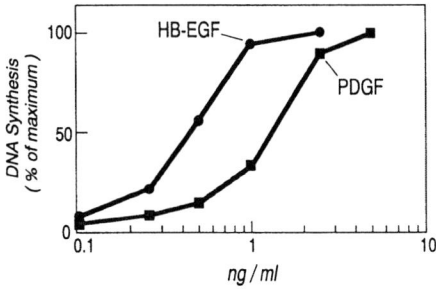

Figure 5. HB-EGF and PDGF mitogenicity for BALB/c 3T3 cells. Purified HB-EGF or recombinant PDGF-AB were tested for the ability to stimulate DNA synthesis in BALB/c 3T3 cells. Growth factor samples were added along with ³[H]-thymidine (1 uCi/200 ul/well) to confluent monolayer cells. After 36-48 hours, ³[H]-thymidine incorporated into BALB/c 3T3 DNA was counted in a scintillation counter.

Table 4. Biological properties of HB-EGF

Source
* Monocytes/Macrophages (U-937)
* Wound fluid (possible macrophage product)
* Smooth muscle cells
* Endothelial cells (induced by TNF-alpha)
* Breast carcinoma cells
* Vero cells

Targets
* Fibroblasts
* Smooth muscle cells (more potent than EGF, TGF-α)
* Epithelial cells (Keratinocytes)

Not a target
* Endothelial cells
* Myoblasts
* PC 12 cells

---

al., 1991; Rapraeger et al., 1991; Klagsbrun and Baird, 1991; Ornitz et al., 1992; Gitay-Goren et al., 1992). 2) Structural analysis of HB-EGF suggests that it may have a heparin-binding domain made up of contiguous amino acids upstream of the C-terminal EGF-like domain responsible for binding to the EGF receptor. To date, identification of heparin-binding domains in heparin- binding growth factors such as bFGF have been elusive possibly because these domains consist of tertiary structures. 3) HB-EGF is processed from a larger transmembrane precursor. It has been shown that the TGF-α precursor is mitogenic while still anchored to the cell membrane allowing direct cell-to-cell mitogenic signalling in what has been termed juxtacrine regulation (Wong et al., 1989; Anklesaria et al., 1990; Brachmann et al., 1989). Whether the HB-EGF precursor is biologically active has not been yet determined. A recent report has demonstrated at least one novel function for a membrane-associated form of HB-EGF, the ability to bind diphtheria toxin (Naglich et al.,1992). Thus the HB-EGF precursor appears to be multifunctional, acting as the source of a mature growth factor on one hand and as a source of the receptor for a totally different type of ligand on the other hand. 4) HB-EGF is a secreted product of monocytes and macro-phages. Since it is also a potent mitogen for fibroblasts and keratinocytes, a role in wound repair is suggested. The observation that HB-EGF is found in wound fluid produced after production of partial thickness excisional wounds in the skin of a pig, reinforces this notion. HB-EGF levels in pig wound fluid are detectable within a day after wounding suggesting that HB-EGF production is an early response to injury (Marikovsky et al., submitted). 5) HB-EGF is a potent SMC mitogen, more so than the homologous EGF family members EGF and TGF-α. The enhanced mitogenic activity might be due to the ability of HB-EGF, but not EGF or TGF-α, to bind to SMC heparan sulfate proteoglycan which could act as low affinity receptors that would modulate the response of the high affinity EGF receptor. Given its potent SMC mitogenic and chemotactic activity, HB-EGF might be considered as a possible regulator of smooth muscle cell hyperplasia as occurs in restenosis following angioplasty and in atherosclerosis. SMC also synthesize HB-EGF (Higashiyama et al, submitted). That SMC are targets for HB-EGF and also produce HB-EGF suggests a possible role for this growth factor in autocrine growth regulation. We speculate that following injury to SMC as occurs in angioplasty, injured SMC might induce HB-EGF synthesis which would then result in SMC proliferation.

In summary, HB-EGF has interesting structural properties that need to be explored more fully including elucidation of heparin- binding sites, diphtheria toxin binding sites and processing sites. There are numerous biological questions to pursue including the biological activity of the HB-EGF precursor and the role of HB-EGF in wound repair and smooth muscle cell hyperplasia. It is hoped that further investigation will illuminate these areas of inquiry.

ACKNOWLEDGEMENT

We thank Dr Judith A. Abraham (Scios, Mountain View, CA) for her helpful discussions.

REFERENCES

Anklesaria, P., Teixido, J., Laiho, M., Pierce, J.N., Greenberger, J.S. and Massague, J. 1990, Cell-cell adhesion mediated by binding of membrane-anchored transforming growth factor-α to epidermal growth factor receptors promotes cell proliferation, *Proc. Natl. Acad. Sc. USA.* 87:3289

Besner, G., Higashiyama, S., and Klagsbrun, M. 1990, Isolation and characterization of a macrophage-derived heparin-binding growth factor, *Cell Regulation.* 1:811.

Brachmann, R., Lindquist, P.B., Nagashima, M., Kohr, W., Lipari, T., Napier, M. and Derynck, R. 1989, Transmembrane TGF-α precursors activate EGF/TGF-α receptors, *Cell.* 56:691

Brown, J.P., Twardzik, D.R., Marquardt, H., and Todaro, G.J. 1985, Vaccinia virus encodes a polypeptide homologous to epidermal growth factor and transforming growth factor, *Nature.* 313:491.

Carpenter, G. and Wahl, M.I. 1990, The epidermal growth factor family, in:"Peptide Growth Factors and Their Receptors, Vol. I," M.B. Sporn and A.B. Roberts, ed., Springer-Verlag pp. 69, New York

Derynck, R., Roberts, A.B., Eaton, D.H., Winkler, M.E. 1985, Human transforming growth factor alpha: Precursor sequence, gene structure, and heterologous expression, *Cancer Cells.* 3:79.

Gitay-Goren, H., Soker, S., Vlodavsly, I. and Neufeld, G. 1992, The binding of vascular endothelial growth factor to its receptors is dependent on cell surface-associated heparin-like molecules, *J. Biol. Chem.* 267:6093

Higashiyama, S., Abraham, J.A., Miller, J., Fiddes, J.C., and Klagsbrun, M. 1991, A heparin-binding growth factor secreted by macrophage-like cells that is related to EGF, *Science.* 251:936.

Higashiyama, S. Lau, K., Besner, G., Abraham, J.A., and Klagsbrun, M. 1992, Structure of heparin-binding EGF-like growth factor: multiple forms, primary structure and glycosylation of the mature protein, *J. Biol. Chem.* 267:6206.

Higashiyama, S., Judith, A.A., Damm, D., Marikovsky, M., and Klagsbrun, M. Cultured fetal human smooth muscle cells synthesize heparin-binding EGF-like growth factor (HB-EGF), (submitted)

Holmes, W.E., Sliwkowski, M.X., Akita, R.W., Henzel, W.J., Lee, J. Park J.W., Yansura, D., Abadi, N., Raab, H., Lewis, G.D., Shepard, H.M., Kuang, W-J., Wood, W.I., Goeddel, D.V. and Vandlen, R.L. 1992, Identification of heregulin, a specific activator of p185^{erbB2}, *Science.* 256:1205.

Klagsbrun, M., and Baird, A. 1991, A dual receptor system is required for basic fibroblast growth factor activity, *Cell.* 67:1

Klagsbrun, M. 1992, Mediators of angiogenesis: the biological significance of basic fibroblast growth factor (bFGF)-heparin and heparan sulfate interactions, *Cancer Biology.* 3:81.

Marikovsky, M., Breuing, K., Liu, P.Y., Eriksson, E., Higashiyama, S., Farber, P.R., Abraham, J.A. and Klagsbrun, M. Appearance pf heparin-binding-EGF-like growth factor (HB-EGF) in wound fluid as a response to injury (submitted)

Naglich, J.G. Metherall, J.E., Russell, D.W., and Eidels, L. 1992, Expression coning of a diphtheria toxin receptor: identity with a heparin-binding EGF-like growth factor precursor, *Cell.* 69:1051.

Ornitz, D.M., Yayon, A., Flanagan, J., Svahn, C.M., Levi, E., and Leder, P. 1991, Heparin is required for cell-free binding of basic fibroblast growth factor to a soluble receptor and for mitogenesis in whole cells, *Mol. Cell. Biol.* 92:6564.

Rapraeger, A.C., Krufka, A., and Olwin, B.B. 1991, Requirement of heparan sulfate for bFGF-mediated fibroblast growth and myoblast differntiation, *Science.* 252:1705.

Shoyab, M., Plowman, G.D., McDonald, V.L., Brabley, J.G. and Todaro G.J. 1989, Structure and function of human amphiregulin: A member of the epidermal growth factor family, *Science.* 2433:1074.

Wen, D. Peles, E., Cupples, R., Suggs, S.V., Levy, R.B., Koski, R.A., Lu, H.S. and Yarden, Y. 1992, *neu* differentiation factor: a transmembrane glycoprotein containing and EGF domain and an immunoglobulin homology unit, *Cell.* 69:559.

Wong, S.T., Winchell, L.F., McCune, B.K., Earp, H.S., Teixido, J., Massague, J., Herman, B., and Lee, D.C. 1989, The TGF-α precursor expressed on the cell surface binds to the EGF receptor on adjacent cells, leading to signal transduction. *Cell.* 56:495.

Yayon, A., Klagsbrun, M., Esko, J.D., Leder, P., and Ornitz, D.M. 1991, Cell surface, heparin-like molecules are required for the binding of basic fibroblast growth factor to its high affinity receptor, *Cell.* 64:841.

Yoshizumi, M., Kourembanas, S., Temizer, D.H., Cambria, R.P., Quertermous, T., and Lee, M-E. 1992, Tumor necrosis factor increases transcription of the heparin-binding epidermal growth factor-like growth factor gene in vascular endothelial cells, *J. Biol. Chem.* 267:9467.

STRUCTURAL ORGANIZATION OF THE MULTIPLE TGF-ß GENES

Sonia B. Jakowlew

National Cancer Institute
Division of Cancer Prevention and Control
Biomarkers and Prevention Research Branch
Kensington, MD 20895
and
The George Washington University School of
Medicine and Health Sciences
Department of Biochemistry and Molecular Biology
Washington, DC 20037

INTRODUCTION

Transforming growth factor-beta (TGF-ß), a peptide first identified by its ability to cause phenotypic transformation of rodent fibroblasts[1,2], and isolated from a variety of natural sources[3-5], has been shown to be a general regulator of cellular activities in both normal and neoplastic cells[6-8]. Our understanding of the chemistry and biology of TGF-ß has increased substantially in recent years as our comprehension of the number of different cell types and physiological processes affected by TGF-ß continues to expand. Many different types of cells have been demonstrated to synthesize TGF-ß and essentially all cells have been shown to have a specific set of receptors for this peptide[9-18]. TGF-ß is a multifunctional molecule, since it can either stimulate or inhibit proliferation, differentiation and other critical processes in cell function; the effect that predominates is dependent on the particular cell type and the other growth factors present[19,20]. It is now well established that TGF-ß plays a major role an adult physiology in processes such as inflammation and tissue repair, control of hematopoiesis and control of steroidogenesis[21-26]. Moreover, it has been demonstrated that TGF-ß also controls differentiation and morphogenesis in embryonic development[27-30]. Studies have shown that TGF-ß may play an important role in such processes as myogenesis, chondrogenesis, osteogenesis, epithelial cell differentiation and adipogenesis, as well as in wound healing[31-52]. In addition to these processes, TGF-ß also regulates the turnover of extracellular matrix. For instance, TGF-ß stimulates production by fibroblasts of extracellular matrix components, such as type I, III and IV procollagen, fibronectin, and proteoglycan, and of cellular receptors for these proteins[53-63]. At the same time, TGF-ß inhibits production of enzymes catalyzing degradation of the extracellular matrix and stimulates production of inhibitors of these enzymes[64-70]. TGF-ß is thus a fundamental regulatory molecule, acting by both autocrine and paracrine mechanisms. As is true for most polypeptide growth factors, the basic molecular mechanism of action of TGF-ß is at present unknown. In spite of this, so many varied functions have been descibed for TGF-ß and are continuing to be demonstrated, that TGF-ß must be considered to be a general mediator of cell regulation, and one of importance for both positive and negative control of cell growth. In this brief review, our current understanding of the structural organization of the TGF-ß genes will be discussed.

MULTIPLE TGF-ß ISOFORMS

The TGF-ß superfamily is composed of several closely related polypeptides and an increasing number of structurally related, but functionally distinct, polypeptides. Five distinct isoforms of TGF-ß have been identified and characterized from mammalian, avian and amphibian species. TGF-ß1 was first purified to homogeneity from human platelets, human placenta and bovine kidney[3-5]. The assay used to monitor purification utilized the ability of TGF-ß to induce the growth of colonies of normal rat kidney fibroblasts in soft agar in the presence of epidermal growth factor[1,71]. TGF-ß2, the second isoform to be isolated, has also been purified from cells and tissues including bovine bone, porcine platelets, human glioblastoma cells and simian kidney BSC-1 cells[12,35,36,72,73]. So far, TGF-ßs 3, 4 and 5 have not been purified from any natural tissue source. cDNAs for TGF-ßs 1-5 have been cloned from a wide range of species. TGF-ß1 cDNAs have been isolated from human, simian, rodent, porcine, bovine and chicken[74-80]. Recent experiments using Southern blotting and restriction enzyme analyses suggest that chicken TGF-ß1 cDNA may actually be porcine TGF-ß1 cDNA, and studies utilizing the polymerase chain reaction suggest that the chicken may not contain a TGF-ß1 gene[81]. TGF-ß2 cDNAs have been isolated from human, simian, rodent, amphibian and chicken[73,82-86]. TGF-ß3 cDNAs have been isolated from human, murine, porcine and chicken[87-91]. So far, TGF-ßs 4 and 5 cDNAs have been isolated from only chicken and amphibian species, respectively[92,93].

A summary of the main features of TGF-ßs 1-5 is presented in Figure 1. TGF-ß1, the prototype of the TGF-ß family and a 25,000 molecular weight disulphide-linked homodimer, is encoded as a 391 amino acid precursor that contains a 29 amino acid signal peptide, a 250 amino acid pre-pro segment called a latency associated peptide (LAP) and an 112 amino acid mature peptide[74]; the TGF-ß1

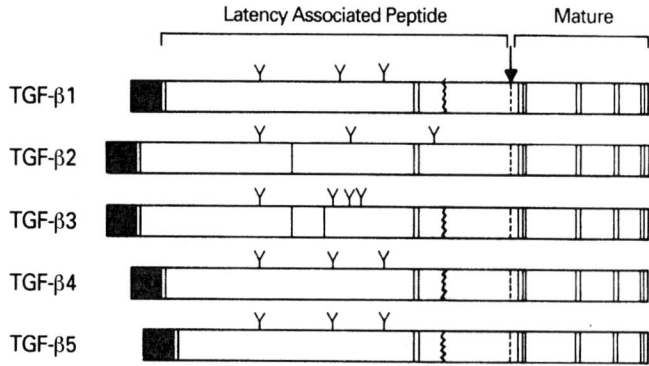

Figure 1. Comparison of the TGF-ß isoforms. The main features are indicated as follows: filled boxes, signal peptides; vertical bars, cysteine residues; dashed bar, proteolytic cleavage sites; jagged bars, Arg-Gly-Asp recognition sites; tree-like structures, N-linked glycosylation sites.

precursor must undergo proteolytic processing to release the mature peptide[94-97]. Like TGF-ß1, TGF-ßs 2-5 are also encoded as larger precursors which also must be similarly processed. Proteolytic processing is thought to occur at a four amino acid cleavage site which is identical in TGF-ßs 2, 3 and 5 (Arg-Lys-Lys-Arg); in TGF-ßs 1 and 4, the proteolytic processing site is modified to be Arg-His-Arg-Arg in TGF-ß1 and Arg-Arg-Arg-Arg in TGF-ß4. Unlike TGF-ß1, both TGF-ßs 2 and 3 are encoded as 412 amino acid precursors, each containing a 300 amino acid LAP and signal peptide, while TGF-ß4, originally thought to lack a portion of the LAP and a signal peptide, has been recently shown to contain a LAP region that is typical of the other TGF-ß isoforms and a signal peptide[92,98]. TGF-ß5 is the shortest of the TGF-ß isoforms, containing a 270 amino acid LAP and signal peptide[93]. The processed, mature TGF-ß1, 2, 3 and 5 peptides each contain 112 amino acids, while TGF-ß4 contains two additional amino acids, aspartic acid and glycine, thus making TGF-ß4 the longest processed TGF-ß isoform. Two other notable features of the TGF-ß isoforms are the conservation of nine cysteine residues in the processed peptide and conservation of the C-terminal sequence Cys-Lys-Cys-Ser-COOH which is found in all the TGF-ß isoforms except TGF-ß5, where an asparagine residue replaces the lysine residue. The degree of amino acid identity between the five TGF-ß isoforms ranges from 64-82% in the mature peptide and 28-45% in the LAP region. The high degree of conservation of the mature TGF-ß isoforms suggests the existence of evolutionary pressure to conserve distinct biological properties in each isoform.

Although the mature portions of the TGF-ß isoforms show significantly higher identity with one another than do their LAP regions, the LAP regions show several features that are also noteworthy. For example, all five TGF-ß LAPs share a region near the N-terminus that shows significant identity. All five LAP regions also show conservation of three cysteine residues, one cysteine located very near the signal peptide and two other cysteines separated by just one residue located in the C-terminal half of the molecule (see Figure 1). Also to be noted in the TGF-ß LAPs are sites for N-linked glycosylation, several which are conserved, as well as the cellular recognition site for fibronectin/vitronectin, Arg-Gly-Asp[99]. This Arg-Gly-Asp recognition site is found in all the TGF-ßs with the exception of TGF-ß2. Integrity of the TGF-ß LAP has been shown to be necessary for effective secretion. Several mutations have been introduced in the LAP region of TGF-ß1, all resulting in inhibition of secretion of not only TGF-ß1, but also of TGF-ßs 2 and 3 as well[94,95,100].

Examination of TGF-ßs 1, 2 and 3 from different species shows remarkable conservation. For example, cDNAs for murine, bovine, porcine, and simian TGF-ß1 have been sequenced and the derived amino acid sequences shown to have a high degree of identity with human TGF-ß1 cDNA[74-80]. All of the mature, processed TGF-ß1 peptides are identical, with the exception of murine TGF-ß1 which differs from human TGF-ß1 only in the substitution of a serine residue for an alanine residue. The mature, processed TGF-ß2 peptides of human and simian species are also identical, while that of chicken TGF-ß2 differs from human TGF-ß2 in the substitution of a histidine residue for an asparagine residue[73,82-86]. Mature human and chicken TGF-ß3 differ only in the substitution of a phenylalanine residue for a tyrosine residue, and human and porcine TGF-ß3 have only two amino acid differences[87-91]. Comparison of the identities of the LAP regions of these TGF-ß isoforms from different species shows identities greater than 86% for TGF-ß1, 82% for TGF-ß2 and 84% for TGF-ß3. It is not possible to compare the identity of TGF-ßs 4 and 5 in this fashion since each has been identified in only one species thus far. The significance of identity between species in the LAP region is not understood. The conservation of identity suggests that the LAP region might have an important biological function itself.

MULTIPLE TGF-ß GENES

Conservation of identity between the TGF-ß isoforms is also seen at the genomic level. The human TGF-ß1 precursor is encoded by seven exons, with the mature peptide being divided between exons five, six and seven[101]. The intron/exon splice junctions have been demonstrated to be conserved in the bovine and porcine genes[77,78]. The positions of the intron/exon splice junctions of human TGF-ß1 are also conserved in human TGF-ßs 2 and 3, with the exception of the first intron which differs by three nucleotides[87,102-104]. The genes for human and chicken TGF-ß2 have been shown to have seven exons at homologous positions[105]. The position of the first intron/exon splice junction in human and chicken TGF-ß3 has also been shown to be conserved. Chromosomal assignments of the genes for human and murine TGF-ßs 1, 2 and 3 indicate that these genes have become widely dispersed during evolution[93-95]. This suggests that the various TGF-ß isoforms originated from a series of duplications of a common ancestral gene, and that these genes were separated by chromosomal translocations.

PROMOTERS IN THE MULTIPLE TGF-ß GENES

Recent studies comparing developmental expression of TGF-ßs 1, 2 and 3 by immunohistochemical and in situ hybridization analyses indicate a complex pattern of expression for each isoform in developing mammals[27,83,88,109-113]. Although these three TGF-ß isoforms are sometimes co-expressed, the temporal pattern of specific cell types involved are often distinct. Because the TGF-ßs can act on nearly all cells, many mechanisms may exist for precise control of their activity. Based on immunohistochemical and in situ hybridization analyses, it has been hypothesized that the differential expression of these genes may be of greater importance than the functional specificity of each isoform.

To investigate the mechanistic basis for the differential regulation of expression of the TGF-ß isoforms and to define the DNA sequences essential for constitutive and inducible expression, the 5'-flanking regions of the human genes for TGF-ßs 1, 2 and 3 have been cloned. Analysis of the transcriptional start sites of human TGF-ß1 mRNAs by S1 nuclease and primer extension analyses shows two major start sites 271 nucleotides away from one another, with several additional minor start sites[114]. In contrast, analysis of the transcriptional start sites of human TGF-ßs 2 and 3 mRNAs shows only one major start site in each mRNA and multiple minor start sites[102-104].

DNA sequence analysis of the 5'-flanking regions of the human TGF-ß genes has identified significant differences in the regulatory regions of these genes. A summary of the main features of these promoter sequences is presented in Figure 2. Unlike the promoters of the human TGF-ß2 and 3 genes which contain consensus TATAA box sequences 20 to 30 nucleotides upstream of the transcriptional start sites, the human TGF-ß1 promoter does not contain a consensus TATAA box sequence and is characterized by the presence of a very GC-rich region that contains several potential Sp1 binding sites just upstream of the first transcriptional start site[114]. Several AP-1, AP-2, cAMP responsive elements (CRE), as well as other Sp1 sites, can be identified in each promoter. The TGF-ß1 promoter contains eleven putative Sp1 binding sites and three potential AP-1 sites, while both the TGF-ßs 2 and 3 promoters each contain substantially fewer potential Sp1 and AP-1 binding sites[102-104]. A variety of experimental approaches have identified the AP-1 sites in the TGF-ß1 gene as targets for phorbol ester control of gene expression and as the major positive regulatory sequences responsible for induction of expression of TGF-ß1[114-117]; similar experiments have shown that the consensus AP-1 sequences in TGF-ßs 2 and 3 do

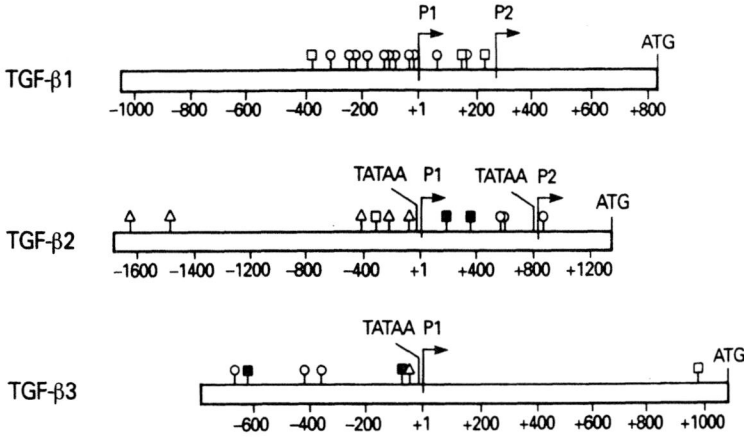

Figure 2. Comparison of the major features of the 5'-flanking regions of the human TGF-ß1, 2 and 3 genes. Transcriptional start sites are designated as P1 and P2. The positions of TATAA boxes are indicated. The consensus transcription factor binding sites are indicated as follows: open squares, AP-1 sites; closed squares, AP-2 sites; open triangles, CRE sites; open circles, Sp1 binding sites.

not confer responsiveness to phorbol ester in these genes[102-104]. Alternatively, in the TGF-ß2 and 3 genes, control seems to be mediated by CRE sequences and possibly by AP-2 sites. Transcription from the TGF-ß2 promoter has been demonstrated to be dependent upon the CRE binding site located 5' of the TATA box[118], while in the TGF-ß3 promoter, the CRE has been shown to be important for both basal and forskolin induction of promoter activity[103]. A unique feature of the TGF-ß2 promoter is the presence of approximately 50 nucleotides of homopurine-homopyrimidine sequence in the 5'-untranslated region, the significance of which is presently unknown[102]. Considering all these differences, it is possible that the very different character of the 5'-flanking regions of the TGF-ßs could be responsible for their differential regulation of expression of each of the isoforms. In addition, the long 5'-untranslated regions of each of the TGF-ß isoforms contain stretches of nucleotides which are very GC-rich, thus contributing secondary structure. Moreover, the 5'-untranslated regions also contain sequences which are capable of coding for other open reading frames. This suggests that translational regulation may also be important in regulation of expression of the TGF-ß isoforms. Several studies suggest that post-transcriptional regulation of TGF-ß1 is another important control point in regulating the activity of this growth factor. For example, expression of TGF-ß1 mRNA has been shown in unstimulated monocytes and in monocytes activated to become macrophages[119]; however, TGF-ß1 protein is secreted only by activated macrophages, suggesting that expression is controlled at the level of translation. In addition, activation of both B and T lymphocytes results in a significant increase in expression of TGF-ß1 mRNA as early as 2 hr following stimulation, whereas the rate of secretion of TGF-ß1 gradually increases over a period of 4 days, again suggesting post-transcriptional regulation[44]. Because the 5'-untranslated region of the human TGF-ß1 gene is very GC-rich, it has been postulated that potential secondary stem-loop structures may form and control expression. This has been examined by inserting DNA fragments of a highly GC-rich region of the 5'-untranslated region of the human TGF-ß1 gene into the corresponding region of the structural gene for human growth hormone driven by the simian virus 40 promoter; expression was shown to be significantly inhibited in a cell type-specific manner, suggesting that the secondary structure of TGF-ß1 mRNA plays an important role in post-transcriptional regulation of TGF-ß1 expression[120].

Two promoter regions have been identified in the human TGF-ß1 gene using chloramphenicol acetyltransferase activity in transfected human cells. Characterization of these promoter regions shows one extending about 1400 nucleotides upstream of the first transcriptional start site and the second promoter located between the two start sites[114] (see Figure 2). Sequences responsible for both promotion and inhibition of transcription of TGF-ß1 have been identified in the promoter upstream of the first transcriptional start site. In addition, both promoters of the TGF-ß1 gene are responsive to autoregulation and activation by phorbol ester, and in both promoters, autoinduction is mediated by binding of the AP-1 (Jun-Fos) complex with autoactivation of c-jun transcription by AP-1[117]. Moreover, induction of jun expression by TGF-ß1, as well as jun autoinduction, may amplify the action of TGF-ß1 during normal development and oncogenesis. The identification of a phorbol ester responsive element immediately downstream of the last TGF-ß1 exon suggests that a 3' enhancer element may play a role in TGF-ß1 transcription, and also suggests a basis for growth factor-mediated regulation of TGF-ß1 expression by activation of protein kinase C[121].

More is currently known about regulation of transcription of human TGF-ß1 than of TGF-ßs 2 and 3 because the gene for TGF-ß1 was characterized first. However, recent studies have focused on characterization of the genes for human TGF-ßs 2 and 3 and their regulation. Like TGF-ß1, analysis of the human TGF-ß2 gene has shown two promoter regions[102], while only one promoter region has been identified in the human TGF-ß3 gene[103,104] (see Figure 2). Sequences responsible for potential enhancer and silencer regions have been identified in the first promoter of the human TGF-ß2 gene[118]. Studies examining the activity of the first promoter in the human TGF-ß2 gene in embryonal carcinoma cells and their differentiated cells have shown that the human TGF-ß2 promoter contains at least two positive regulatory elements that are separated by a negative regulatory element; a CRE/ATF-like site appears to be responsible for part of the positive regulation[122]. In addition, fusion of the 5'-flanking region of the TGF-ß2 gene, including the first and second promoters, to the chloramphenicol acetyltransferase gene shows induction of the level of promoter activity, while sequences farther upstream in the first promoter modulate this promoter activity in a manner that is dependent on the cell line used. This suggests that regulation of TGF-ß2 transcription may be dependent on the cellular background.

Like TGF-ß2, the activity of the human TGF-ß3 promoter has also been shown to be highly variable in different cell types. Although the TGF-ß3 promoter contains a proximal CRE sequence that is critical for both basal and forskolin-induced promoter activity, this element has been shown not to be responsible for the variable cell-specific regulation of the promoter. Instead, regulation is thought to result from binding of a as yet unidentified protein that interacts with a repeated TCCC motif in the TGF-ß3 promoter[123]; this DNA binding protein may also regulate aspects of developmental and tissue-specific expression of TGF-ß3. Yet another level of control of regulation of TGF-ß3 is exerted through inhibition of translation by the 5'-untranslated region. It has been shown that the 5'-untranslated region of human TGF-ß3 mRNA contains as many as eleven potential open reading frames that are capable of inhibiting translation[124].

PROMOTERS IN OTHER VERTEBRATE TGF-ß GENES

In addition to the human TGF-ß genes, TGF-ß genes have also been cloned and sequenced from other vertebrate species including murine, porcine and avian. Characterization of the murine TGF-ß1 gene shows high sequence identity with the human TGF-ß1 gene; there is 66% nucleotide identity in 2.7 kb of DNA upstream

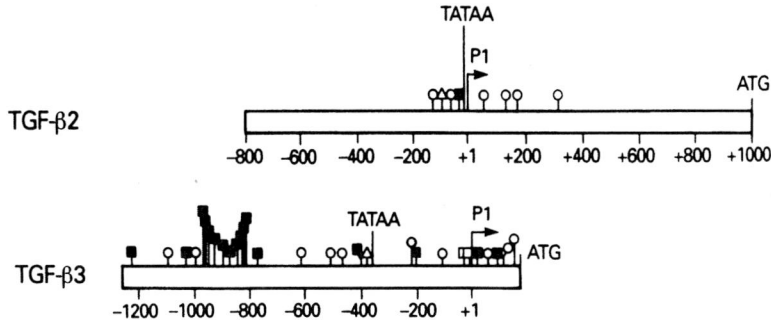

Figure 3. Comparison of the major features of the 5'-flanking regions of the chicken TGF-ß2 and 3 genes. Transcriptional start sites are designated as P1. The positions of TATAA boxes are indicated. The consensus transcription factor binding sites are indicated as follows: open squares, AP-1 sites; closed squares, AP-2 sites; open triangles, CRE sites; open circles, Sp1 binding sites.

of the translational start site suggesting conservation of transcriptional regulation for TGF-ß1[125]. Like the human TGF-ß1 gene, the murine TGF-ß1 gene does not contain a consensus TATAA box sequence, but does contain several Sp1 binding and AP-2-like sequences. Two transcriptional start sites separated by 290 nucleotides have been identified by S1 nuclease analysis. Also like the human TGF-ß1 gene, promoter activity can be demonstrated adjacent to both transcriptional start sites using chloramphenicol acetyltransferase fusion genes. In addition, the Harvey-ras oncogene has been shown to transcriptionally activate the murine TGF-ß1 promoter.

The chicken TGF-ß2 and 3 genes have also been cloned and characterized. A summary of the main features of their 5'-flanking regions is presented in Figure 3. Like the human TGF-ß1 gene, the chicken TGF-ß2 gene contains seven exons and six introns spanning about 70 kb[105]. Primer extension analyis has identified one major and two minor transcriptional start sites. A comparison of the chicken TGF-ß2 exon I sequence with the homologous region from the TGF-ß2 cDNAs of several mammalian species shows a high deegree of sequence conservation only in the coding region and no significant conservation in the 5'-untranslated region. Comparison of chicken and human TGF-ß2 promoter sequences shows homology limited to the region surrounding the transcriptional start site. The conserved sequence contains consensus TATAA box, CRE and AP-2-like sequence motifs. In contrast, the 3'-untranslated region of the TGF-ß2 gene is highly conserved; the overall identity between chicken and mammalian TGF-ß2 mRNAs in the 3'-untranslated region is about 66%. This suggests that the 3'-untranslated region may play a significant role in regulation of expression of the TGF-ß2 gene.

An analysis of chicken TGF-ß3 mRNA using S1 nuclease shows that chicken TGF-ß3 mRNA, like human TGF-ß3 mRNA, has only one transcriptional start site[126]. However, in contrast to human TGF-ß3, the chicken TGF-ß3 transcriptional start site is located 330 nucleotides farther downstream. Chracterization of the chicken TGF-ß3 gene shows TATAA box, CRE and AP-2-like consensus sequences upstream of the transcriptional start site[126,127] (see Figure 3). Moreover, 14 additional AP-2-like, 10 Sp1 and 2 AP-1-like sites can be identified. As in the human TGF-ß3 promoter, the CRE site shows activation by forskolin, an effect which could be shown by expression of TGF-ß3 mRNA in both cultured human and avain cells. Except for 32 bp of identity centered around the TATAA box and CRE site, the chicken TGF-ß3 promoter is structurally very different from the human TGF-ß3 promoter. This suggests that differences in the regulation of expression of the genes for mammalian and avian TGF ß3 may result in part from the unique structure of their promoters.

SUMMARY

The present characterization of the structural composition of the multiple TGF-ß genes, as well as demonstration of the control of regulation by specific sequence elements in several species, provides a basis for further study of the regulation of these genes. The identification and characterization of trans-acting factors that interact with cis-acting DNA elements in the TGF-ß promoters is beginning to clarify mechanisms by which the various TGF-ß isoforms have both distinct and overlapping patterns of expression. Further study will be needed to understand developmental and tissue-specific expression of the multiple TGF-ß genes.

REFERENCES

1. A.B. Roberts, M.A. Anzano, L.C. Lamb, J.M. Smith and M.B. Sporn. New class of transforming growth factors potentiated by epidermal growth factor. Proc. Natl. Acad. Sci. USA 78:5339(1981).

2. H.L. Moses, E.L. Branum, J.A. Proper and R.A. Robinson. Transforming growth factor production by chemically transformed cells. Cancer Res. 41:2842(1981).

3. R.K. Assoian, A. Komoriya, C.A. Meyers, D.M. Miller and M.B. Sporn. Transforming growth factor-beta in human platelets. J. Biol. Chem. 258:7155(1983).

4. C.A. Frolik, L.L. Dart, C.A. Meyers, D.M. Smith and M.B. Sporn. Purification and initial characterization of a type beta transforming growth factor from human placenta. Proc. Natl. Acad. Sci. USA 80:3676(1983).

5. A.B. Roberts, M.A. Anzano, C.A. Meyers, J. Wideman, R. Blacher, Y-C. Pan, S. Stein, S.R. Lehrman, J.M. Smith, L.C. Lamb and M.B. Sporn. Purification and properties of a type beta transforming growth factor from bovine kidney. Biochemistry 22:5692(1983).

6. A.B. Roberts, C.A. Frolik, M.A. Anzano and M.B. Sporn. Transforming growth factors from neoplastic and non-neoplastic tissues. Fed. Proc. 42:2621(1983).

7. J.R. Keller, G.K. Sing, L.R. Ellingsworth and F.W. Ruscetti. Transforming growth factor-ß: possible roles in the regulation of normal and leukemic hematopoietic cell growth. J. Cell. Biochem. 39:79(1989).

8. E. Van Obberghen-Schilling, N.S. Roche, K.C. Flanders, M.B. Sporn and A.B. Roberts. Transforming growth factor-ß1 positively regulates its own expression in normal and transformed cells. J. Biol. Chem. 263:7741(1988).

9. C.A. Frolik, L.M. Wakefield, D.M. Smith and M.B. Sporn. Characterization of a membrane receptor for transforming growth factor-ß in normal rat kidney fibroblasts. J. Biol. Chem. 259:10995(1984).

10. R.F. Tucker, E.L. Branum, G.D. Shipley, R.J. Ryan and H.L. Moses. Specific binding to cultured cells of ^{125}I-labeled type ß transforming growth factor from human platelets. Proc. Natl. Acad. Sci. USA 81:6757(1984).

11. L.M. Wakefield, D.M. Smith, T. Matsui, C.C. Harris and M.B. Sporn. Distribution and modulation of the cellular receptor for transforming growth factor-beta. J. Cell Biol. 105:965(1987).

12. S. Cheifetz, J.A. Weatherbee, M.L-S. Tsang, J.K. Anderson, J.E. Mole, R. Lucas and J. Massague. The transforming growth factor-beta system, a complex pattern of cross-reactive ligands and receptors. Cell 48:409(1987).

13. J. Massague, B. Kelly and C. Mottola. Stimulation by insulin-like growth factors is required for cellular transformation by type beta transforming growth factor. J. Biol. Chem. 260:4551(1985).

14. J. Massague. Subunit structure of a high-affinity receptor for type ß-transforming growth factor. J. Biol. Chem. 260:7059(1985).

15. B.O. Fanger, L.M. Wakefield and M.B. Sporn. Structure and properties of the cellular receptor for transforming growth factor type beta. Biochemistry 25:3083(1986).

16. J.L. Graycar, D.A. Miller, B.A. Arrick, R.M. Lyons, H.L. Moses and R. Derynck. Human transforming growth factor-ß3: recombinant expression, purification, and biological activities in comparison with transforming growth factors-ß1 and -ß2. Mol. Endocrinol. 3:1977(1989).

17. H.J. Schluesener and O. Lider. Transforming growth factor-beta-1 and factor-beta-2 cytokines with identical immunosuppressive effects and a potential role in the regulation of autoimmune T-cell function. J. Neuroimmunol. 24:353(1989).

18. G. Zugmaier, B.W. Ennis, B. Deschauer, D. Katz, C. Knabbe, G. Wilding, P. Daly, M.E. Lippman and R.B. Dickson. Transforming growth factors type-beta-1 and type-beta-2 are equipotent growth inhibitors of human breast cancer cell lines. J. Cell. Physiol. 141:353(1989).

19. A.B. Roberts, M.A. Anzano, L.M. Wakefield, N.S. Roche, D.F. Stern and M.B. Sporn. Type beta transforming growth factor: a bifunctional regulator of cellular growth. Proc. Natl. Acad. Sci. USA 82:119(1985).

20. R.F. Tucker, G.D. Shipley, H.L. Moses and R.W. Holley. Growth inhibitor from BSC-1 cells closely related to platelet type beta transforming growth factor. Science 226:705(1984).

21. T.A. Mustoe, G.F. Pierce, A. Thomason, P. Gramates, M.B. Sporn and T.F. Deuel. Transforming growth factor type beta induces accelerated healing of incisional wounds in rats. Science 237:1333(1987).

22. M. Hotta and A. Baird. Differential effects of transforming growth factor type ß on the growth and function of adrenocortical cells in vitro. Proc. Natl. Acad. Sci. USA 83:7795(1986).

23. J-J. Feige, C. Cochet, W.E. Rainey, C. Madani, and E.M. Chambaz. Type ß transforming growth factor affects adrenocortical cell-differentiated functions. J. Biol. Chem. 262:13491(1987).

24. J.R. Keller, C. Mantel, G.K. Sing, L.R. Ellingsworth, S.K. Ruscetti and F.W. Ruscetti. Transforming growth factor-ß1 selectively regulates early murine hematopoietic progenitors and inhibits the growth of IL-3 dependent myeloid leukemia cell lines. J. Exp. Med. 168:737(1988).

25. O. Avallet, M. Vigier, M.H. Perrard-Sapori and J.M. Saez. Transforming growth factor-ß inhibits Leydig cell functions. Biochem. Biophys. Res. Comm. 146:575(1987).

26. N.L. Thompson, K.C., Flanders, J.M. Smith, L.R. Ellingsworth, A.B. Roberts and M.B. Sporn. Expression of transforming growth factor-ß1 in specific cells and tissues of adult and neonatal mice. J. Cell Biol. 108:661(1989).

27. U.I. Heine, K.C. Flanders, A.B. Roberts, E.F. Munoz and M.B. Sporn. Role of transforming growth factor-ß in the development of the mouse embryo. J. Cell Biol. 105:2861(1987).

28. D. Kimelman and M. Kirschner. Synergistic induction of mesoderm by FGF and TGF-ß and the identification of an mRNA coding for FGF in the early Xenopus embryo. Cell 51:869(1987).

29. J.C. Smith. A mesoderm inducing factor is produced by a Xenopus cell line. Dev. 99:3(1987).

30. F. Rosa, A.B. Roberts, D. Danielpour, L.L. Dart, M.B. Sporn and I.B. Dawid. Mesoderm induction in amphibians: the role of TGF-ß2-like factors. Science 239:783(1988).

31. J.R. Florini, A.B. Roberts, D.Z. Ewton, S.L. Falen, K.C. Flanders and M.B. Sporn. Transforming growth factor-beta. A very potent inhibitor of myoblast differentiation, identical to the differentiation inhibitor secreted by Buffalo rat liver cells. J. Biol. Chem. 261:16509(1986).

32. J. Massague, S. Cheifetz, T. Endo and B. Nadal-Ginard. Type ß transforming growth factor is an inhibitor of myogenic differentiation. Proc. Natl. Acad. Sci. USA 83:8206(1986).

33. E.N. Olson, E. Sternberg, J.S. Hu, G. Spizz and C. Wilcox. Regulation of myogeneic differentiation by type beta transforming growth factor. J. Cell Biol. 103:1799(1986).

34. R.E. Allen and L.K. Boxhorn. Inhibition of skeletal muscle satellite cell differentiation by transforming growth factor-beta. J. Cell. Physiol. 133:567(1987).

35. P.R. Seyedin, P.R. Segarini, D.M. Rosen, A.Y. Thompson, H. Bentz and J. Graycar. Cartilage-inducing factor-B is a unique protein structurally and functionally related to transforming growth factor-beta. J. Biol. Chem. 262:1946(1987).

36. S.M. Seyedin, T.C. Thomas, A.Y. Thompson, D.M. Rosen and K.A. Piez. Purification and characterization of two cartilage-inducing factors from bovine demineralized bone. Proc. Natl. Acad. Sci. USA 82:2267(1985).

37. S.M. Seyedin, A.Y. Thompson, H. Bentz, D.M. Rosen, J.M. McPherson, A. Conti, N.R. Siegel, G.R. Galluppi and K.A. Piez. Cartilage-inducing factor-A. J. Biol. Chem. 261:5693(1986).

38. M. Centrella, T.L. McCarthy and E. Canalis. Transforming growth factor beta is a bifunctional regulator of replication and collagen synthesis in osteoblast-enriched cell cultures from fetal rat bone. J. Biol. Chem. 262:2869(1987).

39. M. Centrella, T.L. McCarthy and E. Canalis. Skeletal tissue and transforming growth factor-ß. FASEB J. 2:3066(1988).

40. J.L. Carrington, A.B. Roberts, K.C. Flanders, N.S. Roche and H.A. Reddi. Accumulation, localization and compartmentation of transforming growth factor-ß during endochondral bone development. J. Cell Biol. 107:1969(1988).

41. P.M. Petkovich, J.L. Wrana, A.E. Grigoriadis, J.N.M. Heersche and J. Sodek. 1,25 Dihydroxyvitamin D3 increases epidermal growth factor receptors and transforming growth factor-ß-like activity in a bone-derived cell line. J. Biol. Chem. 262:13424(1987).

42. J. Pfeilschrifter and G.R. Mundy. Modulation of type ß transforming growth factor activity in bone cultures by osteotropic hormones. Proc. Natl. Acad. Sci. USA 84:2024(1987).

43. J.H. Kehrl, A.B. Roberts, L.M. Wakefield, S.B. Jakowlew, M.B. Sporn and A.S. Fauci. Transforming growth factor beta is an important immunomodulatory protein for human B-lymphocytes. J. Immunol. 137:3855(1986).

44. J.H. Kehrl, L.M. Wakefield, A.B. Roberts, S.B. Jakowlew, M. Alvarez-Mon, R. Derynck, M.B. Sporn and A.S. Fauci. Production of transforming growth factor beta by human T lymphocytes and its potential role in the regulation of T cell growth. J. Exp. Med. 163:1037(1986).

45. J.H. Kehrl, A.S. Taylor, G.A. Delsing, A.B. Roberts, M.B. Sporn and A.S. Fauci. Further studies of the role of TGF-ß in human B cell function. J. Immunol. 143:1868(1989).

46. S.M. Wahl, D.A. Hunt, L.M. Wakefield, N. McCartney-Francis, L.M. Wahl, A.B. Roberts and M.B. Sporn. Transforming growth-factor beta (TGF-beta) induces monocyte chemotaxis and growth factor production. Proc. Natl. Acad. Sci. USA 84:5788(1987).

47. H.L. Moses, R.F. Tucker, E.B. Leof, R.J. Coffey, J. Halper and G.D. Shipley. Type beta transforming growth factor is a growth stimulator and a growth inhibitor. In: Feramisco, J., Ozanne, B. and Stiles, C. (eds) Cancer Cells 3, pp 65-71, Cold Spring Harbor Laboratory (1985).

48. G.D. Shipley, M.R. Pittelkow, J.J. Wille, R.E. Scott and H.L. Moses. Reversible inhibition of normal human prokeratinocyte proliferation by type ß transforming growth factor-growth inhibitor in serum-free medium. Cancer Res. 46:2068(1986).

49. M. Reiss and A.C. Sartorelli. Regulation of growth and differentiation of human keratinocytes by type ß transforming growth factor and epidermal growth factor. Cancer Res. 47:6705(1987).

50. R.J. Coffey, C.C. Bascom, N.J. Sipes, R. Graves-Deal, B.E. Weissman and H.L. Moses. Selective inhibition of growth-related gene expression in murine keratinocytes by transforming growth factor-ß. Mol. Cell. Biol. 8:3088(1988).

51. R.J. Coffey, N.J. Sipes, C.C. Bascom, R. Graves-Deal, C.Y. Pennington, B.E. Weissman and H.L. Moses. Growth modulation of mouse keratinocytes by transforming growth factors. Cancer Res. 48:1596(1988).

52. R.A. Ignotz and J. Massague. Type ß transforming growth factor controls the adipogenic differentiation of 3T3 fibroblasts. Proc. Natl. Acad. Sci. USA 82:8530(1985).

53. R.A. Ignotz and J. Massague. Transforming growth factor-beta stimulates the expression of fibronectin and collagen and their incorporation into the extracellular matrix. J. Biol. Chem. 261:4337(1986).

54. A.B. Roberts, M.B. Sporn, R.K. Assoian, J.M. Smith. N.S. Roche, L.M. Wakefield, U.I. Heine, L.A. Liotta, V. Falanga, J.H. Kehrl and A.S. Fauci. Transforming growth factor type-beta: rapid induction of fibrosis and angiogenesis in vivo and stimulation of collagen formation in vitro. Proc. Natl. Acad. Sci. USA 83:4167(1986).

55. T.I. Morales and A.B. Roberts. Transforming growth factor-ß regulates the metabolism of proteoglycans in bovine cartilage organ cultures. J. Biol. Chem. 263:12828(1988).

56. A.B. Roberts, K.C. Flanders, P. Kondaiah, N.L. Thompson, E. Van Obbergehen-Schilling, L.M. Wakefield, P. Rossi, B. deCrombrugghe, U.L. Heine and M.B. Sporn. Transforming growth factor ß: biochemistry and roles in embryogenesis, tissue repair and remodeling, and carcinogenesis. Recent Prog. Horm. Res. 44:157(1988).

57. J.L. Wrana, M. Maeno, B. Hawrylyshyn, K-L. Yao, C. Domenicucci and J. Sodek. Differential effects of transforming growth factor-ß on the synthesis of extracellular matrix proteins by normal fetal rat calvarial bone cell populations. J. Cell Biol. 106:915(1988).

58. J. Varga, J. Rosenbloom and S.A. Jimenez. Transforming growth factor-ß (TGF-ß) causes a persistent increase in steady-state amounts of type I and type III collagen and fibronectin mRNAs in normal human dermal fibroblasts. Biochem. J. 247:597(1987).

59. P. Rossi, G. Karsenty, A.B. Roberts, N.S. Roche, M.B. Sporn and B. de Crombrugghe. A nuclear factor 1 binding site mediates the transcriptional activation of a type I collagen promoter by transforming growth factor-ß. Cell 52:405(1988).

60. J.A. Madri, B.M. Pratt and A. Tucker. Phenotypic modulation of endothelial cells by transforming growth factor-ß depends upon the composition and organization of the extracellular matrix. J. Cell Biol. 106:1375(1988).

61. R.A. Ignotz, T. Endo and J. Massague. Regulation of fibronectin and type I collagen mRNA levels by transforming growth factor-ß. J. Biol. Chem. 262:6443(1987).

62. D.C. Dean, R.F. Newby and S. Bourgeois. Regulation of fibronectin biosynthesis by dexamethasone, transforming growth factor-ß and cAMP in human cell lines. J. Cell Biol. 106:2159 (1989).

63. R.P. Penttinen, S. Kobayashi and P. Bornstein. Transforming growth factor ß increases mRNA for matrix proteins both in the presence and in the absence of changes in mRNA stability. Proc. Natl. Acad. Sci. USA 85:1105(1988).

64. C.P. Chiang and M. Nilsen-Hamilton. Opposite and selective effects of epidermal growth factor and human platelet transforming growth factor-ß on the production of secreted proteins by murine 3T3 cells and human fibroblasts. J. Biol. Chem. 261:10478(1986).

65. M. Laiho, O. Saksela, P.A. Anderson and J. Keski-Oja. Enhanced production and extracellular deposition of the endothelial-type plasminogen activator inhibitor in cultured human lung fibroblasts by transforming growth factor-ß. J. Cell Biol. 103:2403(1986).

66. L.R. Lund, A. Riccio, P.A. Andreasen, L.S. Nielsen, P. Kristensen, M. Laiho, F. Blasi and K. Dano. Transforming growth factor-ß is a strong and fast acting positive regulator of the level of type-1 plasminogen activator inhibitor mRNA in WI-38 human lung fibroblasts. EMBO J. 6:1281(1987).

67. D.R. Edwards, G. Murphy, J.J. Reynolds, S.E. Whitman, A.J.P. Docherty, P. Angel and J.K. Heath. Transforming growth factor beta modulates the expression of collagenase and metalloproteinase inhibitor. EMBO J. 6:1899(1987).

68. C.M. Overall, J.L. Wrana and J. Sodek. Independent regulation of collagenase, 72 kDa-progelatinase, and metalloendoproteinase inhibitor (TIMP) expression in human fibroblasts by transforming growth factor-ß. J. Biol. Chem. 264:1860(1989).

69. L.M. Matrisian, P. Leroy, C. Ruhlmann, M-C. Gesnel and R. Breathnach. Isolation of the oncogene and epidermal growth factor-induced transin gene: complex control in rat fibroblasts. Mol. Cell. Biol. 6:1679(1986).

70. C.J. Roberts, T.M. Birkenmeier, J.J. McQuillan, S.K. Akiyama, S.S. Yamada, W.T. Chen, K.M. Yamada and J.A. McDonald. Transforming growth factor-ß stimulates the expression of fibronectin and of both subunits of the human fibronectin receptor by cultured human lung fibroblasts. J. Biol. Chem. 263:4586(1988).

71. J.E. De Larco and G.J. Todaro. Growth factors from murine sarcoma virus-transformed cells. Proc. Natl. Acad. Sci. USA 75:4001(1978).

72. M. Wrann, S. Bodmer, R. de Martin, C. Siepl, R. Hofer-Warbinek, K. Frei, E. Hofer and A. Fontana. T Cell suppressor factor from human glioblastoma cells is a 12.5 KD protein closely related to transforming growth factor-beta. EMBO J. 6:1633(1987).

73. S.K. Hanks, R. Armour, J.H. Baldwin, F. Maldonado, J. Spiess and R.W. Holley. Amino acid sequence of the BSC-1 cell growth inhibitor (polyergin) deduced from the nucleotide sequence of the cDNA. Proc. Natl. Acad. Sci. USA 85:79(1988).

74. R. Derynck, J.A. Jarrett, E.Y. Chen, D.H. Eaton, J.R. Bell, R.K. Assoian, A.B. Roberts, M.B. Sporn and D.V. Goeddel. Human transforming growth factor-beta cDNA sequence and expression in tumor cell lines. Nature 316:701(1985).

75. R. Derynck, J.A. Jarrett, E.Y. Chen and D.V. Goeddel. The murine transforming growth factor-beta precursor. J. Biol. Chem. 261:4377(1986).

76. K. Sharples, G.D. Plowman, T.D. Rose, D.R. Twardzik and A.F. Purchio. Cloning and sequence analysis of simian transforming growth factor-ß cDNA. DNA 6:239(1987).

77. R. Derynck and L. Rhee. Sequence of the porcine transforming growth factor-beta precursor. Nucleic Acids Res. 15:3187(1987).

78. E. Van Obberghen-Schilling, P. Kondaiah, R.L. Ludwig. M.B. Sporn and C.C. Baker. Complementary deoxyribonucleic acid cloning of bovine transforming growth factor-ß1. Mol. Endocrinol. 1:693(1987).

79. S-W. Qian, P. Kondaiah, A.B. Roberts and M.B. Sporn. cDNA cloning by PCR of rat transforming growth factor-ß1. Nucleic Acids Res. 18:3059(1990).

80. S.B. Jakowlew, P.J. Dillard, M.B. Sporn and A.B. Roberts. Nucleotide sequence of chicken transforming growth factor-beta 1 (TGF-ß1). Nucleic Acids Res. 16:8730(1988).

81. S.B. Jakowlew, M.B. Sporn and A.B. Roberts. Expression of transforming growth factor-ß isoforms in the developing chicken spleen. J. Cell Science (submitted).

82. L. Madisen, N.R. Webb, T.M. Rose, H. Marquardt, T. Ikeda, D. Twardzik, S. Seyedin and A.F. Purchio. Transforming growth factor-ß2: cDNA cloning and sequence analysis. DNA 7:1(1988).

83. R. de Martin, B. Haendler, R. Hofer-Warbinek, H. Gaugitsh, M. Wrann, H. Schlusener, J.M. Seifert, S. Bodmer, A. Fontana and E. Hofer. Complementary DNA for human glioblastoma-derived T cell suppressor factor, a novel member of the transforming growth factor-ß gene family. EMBO J. 6:3673(1987).

84. D.A. Miller, R.W. Pelton, E.Y. Chen, H.L. Moses and R. Derynck. Murine transforming growth factor-ß2 cDNA sequence and expression in adult tissues and embryos. Mol. Endocrinol. 3:1108(1989).

85. S.B. Jakowlew, P.J. Dillard, M.B. Sporn and A.B. Roberts. Complementary deoxyribonucleic acid cloning of an mRNA encoding transforming growth factor-ß2 from chicken embryo chondrocytes. Growth Factors 2:123(1990).

86. M.L. Rebert, N. Bhatia-Dey and I.B. Dawid. The sequence of TGF-ß2 from Xenopus laevis. Nucleic Acids Res. 18:2185(1990).

87. P. ten Dijke, P. Hanson, K.K. Iwata, C. Pieler and J.G. Foulkes. Identification of a new member of the transforming growth factor-ß gene family. Proc. Natl. Acad. Sci. USA 85:4715(1988)

88. R. Derynck, P.B. Lindquist, A. Lee, D. Wen, J. Tamm, G.L. Graycar, L. Rhee, A.J. Mason, D.A. Miller, R.J. Coffey, H.L. Moses and E.Y. Chen. A new type of transforming growth factor-ß, TGF-ß3. EMBO J. 7:3737(1988).

89. D.A. Miller, A. Lee, Y. Matsui, E.Y. Chen, H.L. Moses and R. Derynck. Complementary DNA cloning of the murine transforming growth factor-beta 3 (TGF-ß3) precursor and the comparative expression of TGF-ß3 and TGF-ß1 messenger mRNA in murine embryos and adult tissues. Mol. Endocrinol. 3:1926(1989).

90. F. Denhez, R. Lafyatis, P. Kondaiah, A.B. Roberts and M.B. Sporn. Cloning by polymerase chain reaction of a new mouse TGF-ß, mTGF-ß3. Growth Factors 3:135(1990).

91. S.B. Jakowlew, P.J. Dillard, P. Kondaiah. M.B. Sporn and A.B. Roberts. Complementary deoxyribonucleic acid cloning of a novel transforming growth factor-ß messenger ribonucleic acid from chick embryo chondrocytes. Mol. Endocrinol. 2:747(1988).

92. S.B. Jakowlew, P.J. Dillard, M.B. Sporn and A.B. Roberts. Complementary deoxyribonucleic acid cloning of an mRNA encoding transforming growth factor-beta 4 from chicken embryo chondrocytes. Mol. Endocrinol. 2:1186(1988).

93. P. Kondaiah, M.J. Sands, J.M. Smith, A. Fields, A.B. Roberts, M.B. Sporn and D.A. Melton. Identification of a novel transforming growth factor-ß mRNA in Xenopus laevis. J. Biol. Chem. 265:1089(1989).

94. L.E. Gentry, M.N. Lioubin, A.F. Purchio and H. Marquardt. Molecular events in the processing of recombinant type 1 pre-pro-transforming growth factor beta to the mature polypeptide. Mol. Cell. Biol. 8:4162(1988).

95. L.E. Gentry, N.L. Webb, G.J. Lim, A.M. Brunner, J.E. Ranchalis, D.R. Twardzik, M.N. Liobin, H. Marquardt and A.F. Purchio. Type 1 transforming growth factor beta: amplified expression and secretion of mature and precursor polypeptides in Chinese hamster ovary cells. Mol. Cell. Biol. 7:3418(1987).

96. L.M Wakefield, D.M. Smith, K.C. Flanders and M.B. Sporn. Latent transforming growth factor-ß from human platelets. J. Biol. Chem. 263:7646(1988).

97. L.M. Wakefield, D.M. Smith, S. Broz, M. Jackson, A.D. Levinson and M.B. Sporn. Recombinant TGF-ß1 is synthesized as a two component latent complex that shares some structural features with the native platelet latent TGF-ß1 complex. Growth Factors 1:203(1989).

98. D.W. Burt and S.B. Jakowlew. A new interpretation of a chicken transforming growth factor-ß4 complementary DNA. Mol. Endocrinol. 6:989(1992).

99. E. Ruoslahti and M.D. Pierschbacher. Arg-Gly-Asp: a versatile cell recognition signal. Cell 44:517(1986).

100. S-W. Qian, J.K. Burmester, J.R. Merwin, J.A. Madri, M.B. Sporn and A.B. Roberts. Identification of a structural domain that distinguishes the actions of the type-1 and type-2 isoforms of transforming growth factor-beta on endothelial cells. Proc. Natl. Acad. Sci. USA 89:6290(1992).

101. R. Derynck, L. Rhee, E.Y. Chen and A. Van Tilburg. Intron-exon structure of human transforming growth factor-ß precursor gene. Nucleic Acids Res. 15:3188(1987).

102. T. Noma, A.B. Glick, A.G. Geiser, M.A. O'Reilly, J. Miller, A.B. Roberts and M.B. Sporn. Molecular cloning and structure of the human transforming growth factor-ß2 gene promoter. Growth Factors 4:247(1991).

103. R. Lafyatis, R. Lechleider, S-J. Kim, S.B. Jakowlew, A.B. Roberts and M.B. Sporn. Structural and functional characterization of the transforming growth factor-ß3 promoter: a cAMP-responsive element regulates basal and induced transcription. J. Biol. Chem. 265:19128(1990).

104. A.R. Lopez, J. Cook, P.L. Deininger and R. Derynck. Dominant negative mutants of transforming growth factor-ß1 inhibit secretion of different transforming growth factor-ß isoforms. Mol. Cell. Biol. 12:1674(1992).

105. D.W. Burt and I.R. Paton. Molecular cloning and primary structure of the chicken transforming growth factor-ß2 gene. DNA Cell Biol. 10:723(1991).

106. D. Fujii, J.E. Brissenden, R. Derynck and U. Franke. Transforming growth factor-ß gene maps to human chromosome 19 long arm and to mouse chromosome 7. Somatic Cell Mol. Gen. 12:281(1986).

107. D.E. Barton, B.E. Foellmer, J. Du, J. Tamm, R. Derynck and U. Francke. Chromosomal locations of TGF-ß's 2 and 3 in the mouse and human. Oncogene Res. 3:323(1988).

108. P. ten Dijke, A.H.M. Geurts van Kessel, J.G. Foulkes and M.M. Le Beau. Transforming growth factor-beta type 3 maps to human chromosome 14, region q23-q24. Oncogene 3:721(1988).

109. S.A. Lehnert and R.J. Akhurst. Embryonic expression pattern of TGF-beta type 1 RNA suggests both paracrine and autocrine mechanisms of action. Dev. 104:263(1988).

110. R.W. Pelton, S. Nomura, H.L. Moses and B.L.M. Hogan. Expression of transforming growth factor-ß2 during murine embryogenesis. Dev. 106:759(1989).

111. R.W. Pelton, M.E. Dickinson, H.L. Moses and B.L.M. Hogan. In situ hybridization analysis of TGF-ß3 RNA expression during mouse development: comparative studies with TGF-ß1 and ß2. Dev. 110:609(1990).

112. R.W. Pelton, B. Saxena, M. Jones, H.L. Moses and L.I. Gold. Immunohistochemical localization of TGF-beta 1, TGF-beta 2 and TGF-beta 3 in the mouse embryo-expression patterns suggest multiple roles during embryonic development. J. Cell Biol. 115:1091(1991).

113. F.A. Millan, F. Denhez, P. Kondaiah and R.J. Akhurst. Embryonic gene expression patterns of TGF beta-1, beta-2 and beta-3 suggest different developmental functions in vivo. Dev. 111:131(1991).

114. S-J. Kim, A. Glick, M.B. Sporn and A.B. Roberts. Characterization of the promoter region of the human transforming growth factor-ß1 gene. J. Biol. Chem. 264:402(1989).

115. S-J. Kim, K-T. Jeang, A. Glick, M.B. Sporn and A.B. Roberts. Promoter sequences of the human TGF-ß gene responsive to TGF-ß1 autoinduction. J. Biol. Chem. 264:7041(1989).

116. S-J. Kim, F. Denhez, K-Y. Kim, J.T. Holt, M.B. Sporn and A.B. Roberts. Activation of the second promoter of the transforming growth factor-ß1 gene by transforming growth factor-ß1 and phorbol ester occurs through the same target sequences. J. Biol. Chem. 264:19373(1989).

117. S-J. Kim, P. Angel, R. Lafyatis, K. Hattori, K-Y. Kim, M.B. Sporn, M. Karin and A.B. Roberts. Autoinduction of transforming growth factor ß1 is mediated by the AP-1 complex. Mol. Cell. Biol. 10:1492(1990).

118. M.A. O'Reilly, A.G. Geiser, S-J. Kim, L.A. Bruggeman, A.X. Luu, A.B. Roberts and M.B. Sporn. Identification of an activating transcription factor (ATF) binding site in the human transforming growth factor-ß2 promoter. J. Biol. Chem., in press.

119. R.K. Assoian, B.E. Fleurdelys, H.C. Stevenson, P.J. Miller, D.K. Madtes, E.W. Raines, R. Ross and M.B. Sporn. Expression and secretion of type beta transforming growth factor by activated human macrophages. Proc. Natl. Acad. Sci. USA 84:6020(1987).

120. S-J. Kim, K. Park, D. Koeller, K-Y. Kim, L.M. Wakefield, M.B. Sporn and A.B. Roberts. Post-transcriptional regulation of the human transforming growth factor-ß1 gene. J. Biol. Chem. 267:13702(1992).

121. L. Scotto, P.I. Vaduva, R.E. Wager and R.K. Assoian. Type ß1 transforming growth factor gene expression: a corrected mRNA structure reveals a downstream phorbol ester responsive element in human cells. J. Biol. Chem. 265:2203(1990).

122. D. Kelly, M.A. O'Reilly and A. Rizzino. Differential regulation of the transforming growth factor type-ß2 gene promoter in embryonal carcinoma cells and their differentiated cells. Dev. Biol., in press.

123. R. Lafyatis, F. Denhez, T. Williams, M.B. Sporn and A.B. Roberts. Sequence specific protein binding to and activation of the TGF-ß3 promoter through a repeated TCCC motif. Nucleic Acids Res. 19:6419(1991).

124. B.A. Arrick, A.L. Lee, R.L. Grendell and R. Derynck. Inhibition of translation of transforming growth factor-ß3 mRNA by its 5'-untranslated region. Mol. Cell. Biol. 11:4306(1991).

125. A.G. Geiser, S-J. Kim, A.B. Roberts and M.B. Sporn. Characterization of the mouse transforming growth factor-ß1 promoter and activation by the Ha-ras oncogene. Mol. Cell. Biol. 11:84(1991).

126. S.B. Jakowlew, R. Lechleider, A.G. Geiser, S-J. Kim, T.A. Santa-Coloma, J. Cubert, M.B. Sporn and A.B. Roberts. Identification and characterization of the chicken transforming growth factor-ß3 promoter. Mol. Endocrinol., in press.

127. D.W. Burt, I.R. Paton and B.R. Dey. Comparative analysis of human and chicken transforming growth factor-ß2 and -ß3 promoters. J. Mol. Endocrinol. 7:175(1991).

DEXAMETHASONE MODULATES THE RESPONSE OF RAT STROMAL BONE MARROW DERIVED BONE-LIKE CELLS TO bFGF AND IGF-I

Naphtali Savion,[1,3] Anat Beit-Or,[1] Shlomo Kotev-Emeth,[1] and Sandu Pitaru,[2]

[1]The Maurice and Gabriela Goldschleger Eye Research Institute, Sackler Faculty of Medicine, Tel-Aviv University, Tel-Aviv, Israel
[2]Department of Oral Biology, The Maurice and Gabriela Goldschleger School of Dental Medicine, Tel-Aviv University, Tel-Aviv, Israel
[3]On Sabbatical at the Department of Molecular Biology, American Red Cross, Rockville, MD

INTRODUCTION

The formation of bone tissue during development and remodeling involves proliferation and differentiation of preosteoblastic cells and secretion and mineralization of the extracellular matrix by the differentiated osteoblasts. This complex process is regulated by systemic hormones and local factors.[1] The local factors synthesized and found in the skeletal tissue includes acidic fibroblast growth factor (aFGF), basic FGF (bFGF), insulin-like growth factor-I (IGF-I) and IGF-II, platelet-derived growth factor and transforming growth factor β.[2] Some of the growth factors primarily stimulate bone cell replication whereas others also affect the differentiated functions of the osteoblast.[3] The effect of FGFs on the proliferation and differentiation of bone cells was studied in different culture systems, all derived from fetal or neonatal origin.[4-9] There is general agreement that FGFs stimulate the proliferation of bone-derived cells[5-8] and that bFGF is more potent in this respect than aFGF.[7,8] However, the data regarding the effect of FGFs on different parameters that are considered to be expressed by the differentiated osteoblast are ambiguous: bFGF was reported to decrease the synthesis of type I collagen in rat calvaria cultures,[5] to reduce alkaline phosphatase, osteocalcin, type I collagen expression and parathyroid hormone (PTH)-stimulated adenylate cyclase in osteoblast-like cells,[7,8,10] and it was also reported that bFGF enhanced osteocalcin[8] and osteopontin[10] expression. IGF-I was shown to stimulate preosteoblastic cell proliferation in intact fetal rat calvariae, which increases a cell population capable of synthesizing type I collagen. Furthermore, in osteoblast-enriched cells from fetal rat parietal bone, IGF-I increased type I collagen synthesis independently of its effect on cell replication.[11] These studies clearly indicated that both FGFs and IGF-I stimulate the proliferation of fetal or neonatal bone-derived cells. However, the data regarding the effect of these growth factors on various differentiation parameters and their effect on cultures derived from adult animals, is not yet clear.

Bone metabolism is controlled not only by peptide hormones and growth factors but also by glucocorticoids. Prolonged treatment with glucocorticoids is known to induce osteoporosis and *in vitro* studies to explore this phenomenon using calvarial explants demonstrated that sustained exposure to high concentrations of glucocorticoids results in a decrease in bone cell replication and collagen synthesis.[12] Futhermore, high levels of cortisol were shown to decrease skeletal IGF-I synthesis by reducing IGF-I transcript levels, and this effect probably contributes to the inhibitory influence of cortisol on bone formation.[13] These studies may explain the observed steroid-induced osteoporosis, however, other studies using bone cell cultures demonstrated the opposite effect, namely, the osteogenic potential of glucocorticoids.[14-17] These conflicting data suggest that glucocorticoids affect bone cell metabolism by a complex mechanism that has not been completely clarified.

In adults, stromal bone marrow cells (SBMC) are believed to consist of a heterogeneous cell population that comprise the reticular, fibroblastic, adipocytic and osteogenic lineages of the endosseous bone system.[18] Isolated bone marrow cells from adult animals were shown to have osteogenic potential when cultured *in vivo* in diffusion chambers.[18-22] Furthermore, Maniatopoulos et al.[23] described a culture system in which SBMC obtained from young adult animals have the capacity to produce mineralized bone-like nodules *in vitro* when the culture medium is supplemented with dexamethasone, ascorbic acid and β-glycerophosphate.

The ultimate differentiation step of bone tissue is the mineralization of the extracellular matrix. Therefore, the model described by Maniatopoulos may serve as an optimal model to study the role of various growth factors in the proliferation and, more importantly, in the differentiation of bone marrow-derived bone cells in culture. Using this model, we have shown that bFGF stimulates the capacity of SBMC to produce mineralized bone-like tissue.[24] Further studies were undertaken to characterize the role of bFGF and IGF-I in bone formation by bone marrow-derived cells cultured in the presence of dexamethasone.

EFFECT OF bFGF ON CELL PROLIFERATION

SBMC cultured in the presence of 15% calf serum incorporated limited amounts of [³H]thymidine (Fig. 1). The addition of bFGF at increasing concentrations resulted in a dose-dependent stimulation of up to 50-fold at bFGF concentration of 1 ng/ml. This result indicated the significant effect of bFGF on the proliferation of SBMC cultured in the presence of 15% serum.

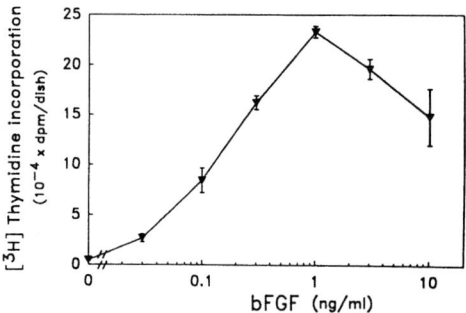

Figure 1. Effect of bFGF on [³H]thymidine incorporation in rat SBMC. SBMC cells were flushed out from femurs of adult rats (40-45 days old) and cultured in αMEM supplemented with 15% fetal calf serum, glutamine (2 mM), penicillin (100 U/ml), streptomycin (100 μg/ml), fungizone (0.25 μg/ml), Na-β-glycerophosphate (10 mM), ascorbic acid (50 μg/ml) (complete medium) and dexamethasone (10^{-7} M). Cultures were washed every second day and after 7 days were trypsinized and seeded in 24-well plates at a density of 1x10⁴ cells/well in the same complete medium supplemented with 10^{-8} M dexamethasone. bFGF was added every other day at the indicated concentrations. [³H]Thymidine (1 μCi/ml) was added on day 4 and cultures were further incubated for 24 h, washed, lysed in 0.1 M NaOH and the radioactivity of each well was determined. Each point represents the mean ± S.D. of 3 wells.

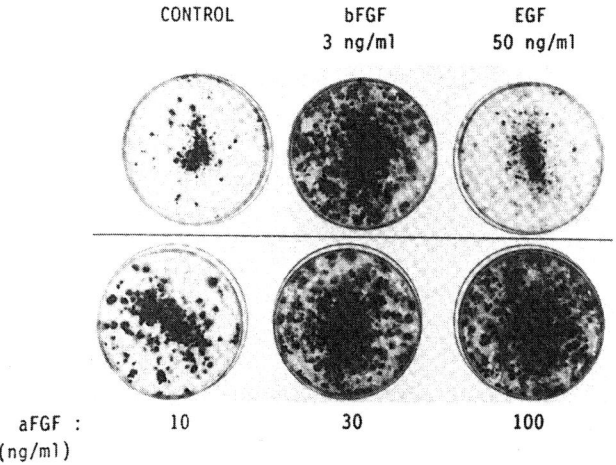

CONTROL   bFGF   EGF
            3 ng/ml   50 ng/ml

aFGF :   10   30   100
(ng/ml)

Figure 2. Mineralization of SBMC cultures in the presence of various growth factors. SBMC cultures were prepared and grown as described in Fig. 1 but trypsinized and seeded in 35 mm tissue culture dishes at a density of 4×10^4 cells/dish, exposed to a complete medium supplemented with dexamethasone (10^{-8} M). Various factors as indicated in the figure were added every other day, growth medium was changed twice a week and cultures were grown for 21 days, then fixed with formalin:methanol:water solution (1:1:1.5 v/v) and stained for 15 min with a saturated solution of Alizarin Red S (pH 4.0).

ROLE OF FGFs AND EPIDERMAL GROWTH FACTOR IN THE MINERALIZATION OF SBMC CULTURES

SBMC cultured in the presence of dexamethasone for 21 days differentiate and form bone-like nodules in tissue culture dishes. These bone-like nodules are mineralized and specifically stained by Alizarin Red S (Fig. 2). In this system we could study whether growth factors which induce the proliferation of SBMC may delay or induce the differentiation process. As shown in Fig. 2, both bFGF and aFGF significantly induced the formation of mineralized bone-like nodules which were heavily stained by Alizarin Red S. Maximum stimulation of bone-like tissue formation by bFGF was achieved at a concentration of 3 ng/ml[24] while the maximal effect of aFGF was observed at a concentration of 30 ng/ml. Epidermal growth factor (EGF) which was reported to induce rat calvaria bone cell proliferation with concomitant inhibition of differentiation into bone nodules,[25] had no effect on the mineralization of SBMC cultured in the presence of dexamethasone.

Previous studies showed that the stimulatory effect of growth factors on bone cell proliferation was associated with a decrease in the phenotypic expression of these cells.[5,7] bFGF was also shown to have a mitogenic effect on calvaria cells and to increase the osteocalcin content of the conditioned media, but mineralized bone-like nodule formation occurred only after removal of bFGF from the medium.[8] In the present study, bFGF and aFGF were shown to play a major inductive role in the proliferation and differentiation of bone marrow cells obtained from adult animals. As previously suggested,[5] FGFs stimulate the replication of osteoprogenitor cells which differentiate and manifest the osteoblastic phenotype. However, this is the first study to show that FGFs do not interfere in the differentiation process and that bone nodules were formed while FGFs were continuously present in the culture medium.

Transmission electron microscopic examination of 21-day-old control and bFGF treated cultures confirmed the formation of a multilayered structure of cells embedded

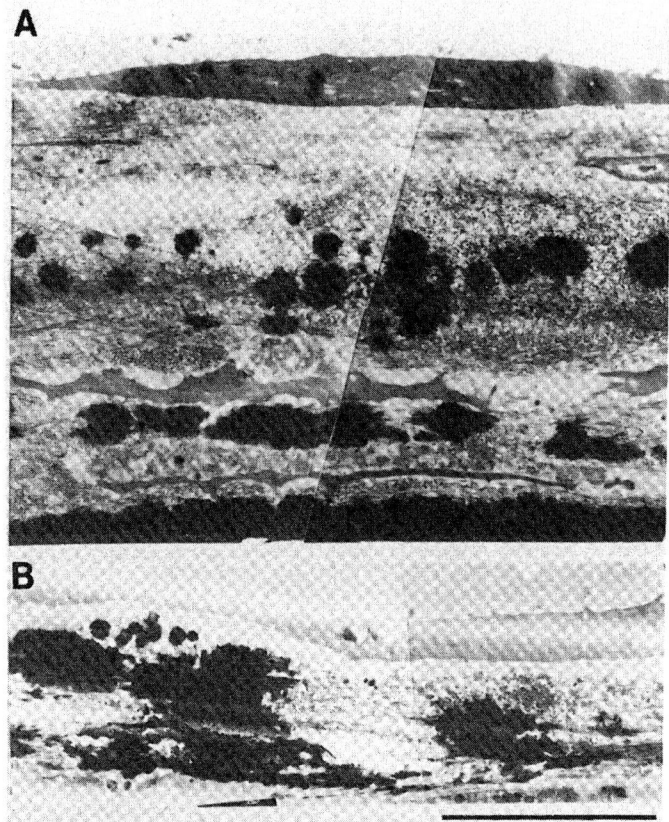

Figure 3. Transmission electron micrograph of mineralized SBMC cultures treated with or without bFGF. SBMC were prepared and cultured for 21 days as described in Fig. 2 in the presence of bFGF (3 ng/ml; A) or in it's absence (B). The micrographs depict a perpendicular section through a non-demineralized multilayered bone-like structure. The bFGF treated culture (A) demonstrates a higher number of cell layers than the control culture (B). At the top, an osteoblast-like cell lays on a collagenous non-mineralized matrix. Beneath this layer, initial mineral deposits associated with orthogonally arranged dense collagen bundles, were observed. Osteocyte-like cells entrapped within mineralized matrix developed on the floor of the dish. The bar represents 10 μm.

within extracellular matrix rich in collagen fibers (Fig. 3). Although the structural appearance of bFGF treated and untreated cultures were similar, the bFGF treated cultures exhibited a larger amount of extracellular matrix and a higher number of cell layers (5-6 versus 2-3 layers). Osteocyte-like cells could be identified as individual cells surrounded by collagen fibers that formed lacunae-like structures. Vesicular structures resembling those described for matrix vesicles[26] were observed within the extracellular space at higher magnification (not shown). The most prominent characteristic for bone tissue is the formation of mineralized matrix. In the superficial and middle layers, initial mineralization sites of spherical configuration associated with banded collagen fibers were noticed. Deeper, larger areas of the mineralized layer were observed on the floor of the dish.

This pattern of mineral deposition, the needle-shaped particles within deposits associated with banded collagen fibers, the flat appearance of the cells, and the orthogonal arrangement of collagen bundles were also described for bone-like structures produced *in vitro* by embryonal chicken and rat calvaria cells.[27-28] Examination by infrared spectrophotometry of mineral deposited in cultures of SBMC, pointed to its similarity with hydroxyapatite crystals deposited in the femur of 21-day-old rat *in vivo*.[29]

THE EFFECT OF bFGF ON THE SEQUENTIAL DEVELOPMENT OF BONE-LIKE TISSUE IN CULTURE

Since the development of mineralized structures *in vitro* is a progressive process that consists of cell proliferation, differentiation of the osteoblastic phenotype, matrix deposition, and mineralization of the extracellular matrix, the effect of bFGF on these parameters was determined throughout the culture period (Fig. 4). The rate of [³H]thymidine incorporation in SBMC cultures grown in the presence or absence of bFGF at various time points during the development of the bone-like tissue in culture is shown in Fig. 4A. A slight increase in the [³H]thymidine incorporation rate was observed in the control cultures throughout the experimental period. In the bFGF treated cultures, the rate of [³H]thymidine incorporation increased sharply starting from day 5, peaking on day 11 at values approximately 12 times higher than the controls and then declining to remain at levels 4 times higher than controls at the end of the experimental period. The bFGF-induced proliferation until day 11 of the cultures was associated with a concomitant increase in protein synthesis and matrix deposition[29]. This enhanced cell proliferation and matrix deposition during the first 10-11 days of culture is followed by a sharp decline of these parameters, accompanied by a marked increase in alkaline phosphatase activity (Fig. 4B).

Alkaline phosphatase activity was first observed in bFGF treated cultures at day 7 and in control cultures at day 10. In the bFGF treated cultures, this activity increased sharply until day 14 and then leveled off, whereas in the control cultures, it increased linearly between days 10-21. At all time points, the alkaline phosphatase activity was higher in the treated cultures (Fig. 4B). In order to further characterize the differentiation of the SBMC cultures, we measured the secretion of osteocalcin as a specific marker for bone tissue (Fig. 4C). Low levels of osteocalcin were found in the bFGF treated cultures during the first 10 days of culture. In these cultures, osteocalcin levels increased sharply between days 10-12, then leveled off until day 18 and decreased considerably during the last 3 culture days. In control cultures, osteocalcin expression increased gradually between days 10-18 and then declined slightly. Between days 10-18, osteocalcin expression was approximately 4 times higher in bFGF treated cultures than in controls. The ultimate step of the sequential process of SBMC differentiation is the formation of mineralized bone-like tissue in culture. To test the effect of bFGF on the development of mineralized structures, the areas stained by Alizarin Red S in each culture dish were determined by image analysis at the time points indicated in Fig. 4D. The results indicated that formation of mineralized nodules started earlier (day 8-9) and was considerably higher in the experimental cultures than in the controls. Although there was already some evidence of mineralized structures in control cultures at day 11, substantial mineral deposition was not detected until day 15. By this time, the mineralization values in bFGF treated cultures peaked and then leveled off at values 7 times higher than those of untreated cultures. Mineralized structure formation was most pronounced in control cultures between 15-20 days of culture.

In this study, we observed a biphasic sequence of events during the development of mineralized bone-like tissue in bFGF treated and control cultures. The first phase is characterized by cell proliferation and matrix accumulation until day 11. The subsequent second phase is characterized by a sharp decline in cell proliferation and matrix accumulation and concomitant expression of osteoblast differentiation as reflected by the progressive increase in alkaline phosphatase activity, osteocalcin expression and mineral deposition. Treatment of cultures with bFGF accentuated this biphasic sequence of events.

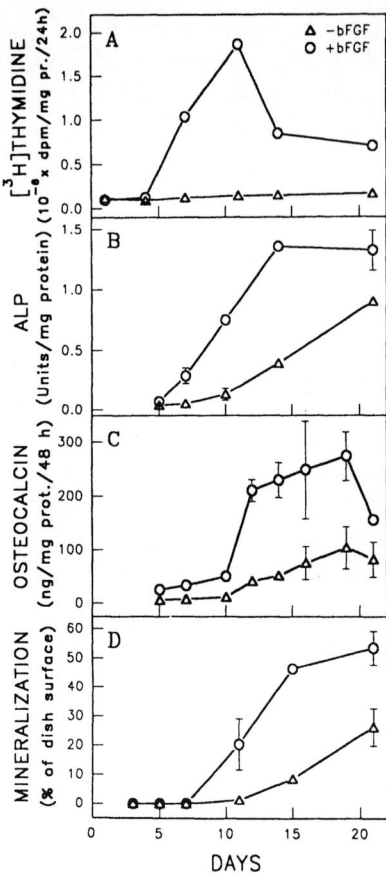

Figure 4. Effect of bFGF on [³H]thymidine incorporation (A), alkaline phosphatase activity (B), osteocalcin secretion (C) and formation of mineralized tissue (D) by SBMC in culture. Cultures of rat SBMC were prepared and grown in complete medium containing dexamethasone (10^{-8} M) as described in Fig. 2, in the presence (0) or absence (Δ) of bFGF (3 ng/ml). At the time points indicated in the figure the following parameters were studied; <u>A</u>. Cultures were washed and exposed to medium containing [³H]thymidine, further incubated for 24 h and [³H]thymidine incorporation was measured as described in Fig. 1; <u>B</u>. Cultures were washed, scraped and analyzed for alkaline phosphatase (ALP) activity using a specific kit and enzymatic activity was expressed as U/mg protein; <u>C</u>. A 48 h conditioned medium was collected and osteocalcin was determined by a RIA kit; <u>D</u>. Cultures were fixed and stained with Alizarin Red S as described in Fig. 2 and the dish area covered with heavy red stain representing calcified tissue was measured with an image analysis system. Each point represents the mean ± S.D. of three dishes and in cases were S.D. is not shown, it was smaller than the size of the symbols.

In contrast to our findings, other studies reported an inhibitory effect of bFGF on alkaline phosphatase expression by rat bone calvaria cells and on alkaline phosphatase, osteocalcin and type I collagen expression by ROS 17/2.8 cells. These findings suggest that bFGF may have opposing effects on bone cells in culture depending on the origin of the cells such as primary cultures versus cell lines, the age and species of the animal, the type of bone from which the bone cells were obtained, and the culture conditions such as the presence or absence of dexamethasone in the growth medium.

THE ROLE OF bFGF AND DEXAMETHASONE DURING THE PROLIFERATION AND DIFFERENTIATION PHASES

The biphasic sequence of events observed during the development of the mineralized bone-like tissue in culture, with proliferation in the first phase and differentiation in the second phase, provided us with a system to test the role of bFGF and dexamethasone in each phase. To examine this role, SBMC cultures were grown in a complete growth medium but either dexamethasone or bFGF was omitted from the medium for various time intervals during the culture period. The amount of mineralized bone-like tissue formation after 21 days under each condition is shown in Fig. 5.

Complete omission of either dexamethasone or bFGF from the complete culture medium for the entire culture period resulted in the formation of a limited number of small bone-like nodules. Exposure of cultures to dexamethasone for the first 7 days resulted in a significant increase in the number and size of bone nodules. Maximum bone-like tissue formation (90% of the dish area) was observed in cultures exposed to dexamethasone for either 14 or 21 days. Thus, additional exposure to dexamethasone for the last 7 days of culture did not further increase bone-like tissue formation. In the case of bFGF, exposure of the cultures to the mitogen for the first 7 days was enough to achieve maximal formation of bone-like tissue. Further, exposure of the cultures to bFGF up to day 14 or day 21 did not further increase the amount of bone-like tissue. However, exposure of cultures to either dexamethasone or bFGF-free medium for the first 7 or 14 days, followed by addition of the omitted component during the rest of the culture period, resulted in the formation of very limited amounts of bone-like tissue. These results indicate that exposure of SBMC to bFGF in combination with dexamethasone during the proliferative phase of the culture is essential and sufficient to obtain maximal formation of bone-like tissue in the culture. Continuous exposure of the cultures to both factors during the differentiative phase of the culture had neither stimulatory nor inhibitory effects on the formation of bone-like tissue.

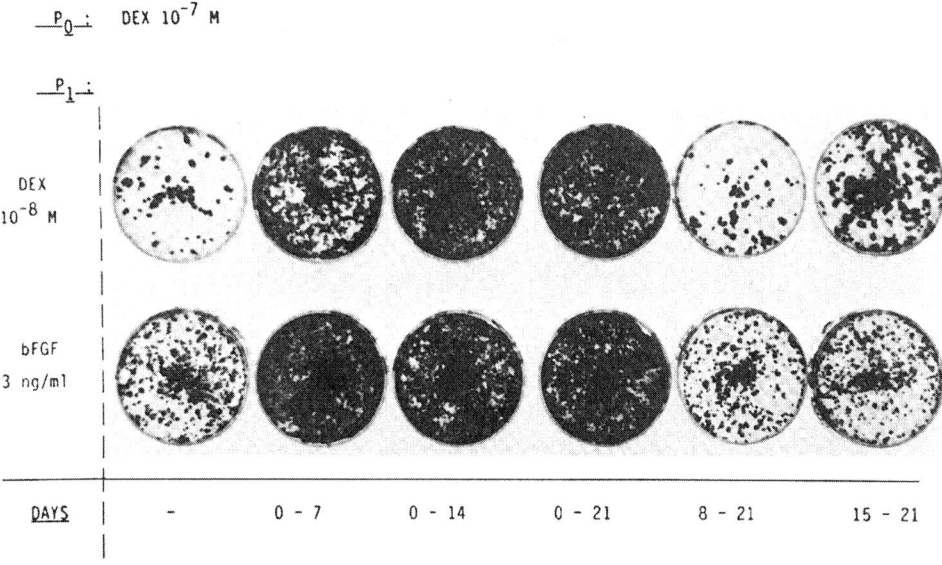

Figure 5: Effects of dexamethasone and bFGF omission from the growth medium on the formation of mineralized bone-like tissue. Cultures were prepared and grown as described in Fig. 2. Dexamethasone (DEX; upper row of dishes) and bFGF (lower row of dishes) were omitted from the complete growth medium and added to the cultures for various time intervals as indicated in the figure. After 21 days of culture, the dishes were fixed and stained as described in Fig. 2.

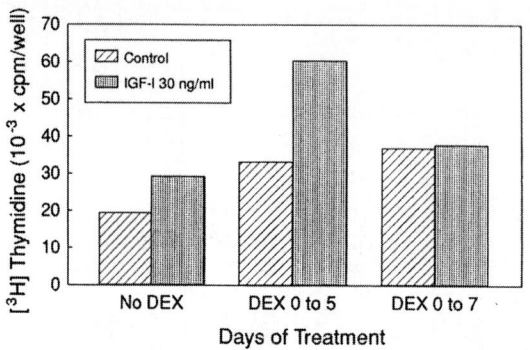

Figure 6: Effect of IGF-I on [³H]thymidine incorporation in SBMC treated with or without dexamethasone. SBMC were prepared and cultured as described in Fig. 1. Cultures were grown in the absence of dexamethasone (No DEX) or in the presence of dexamethasone (10^{-8} M) for the first 5 days of the culture (DEX 0 to 5) or for the entire culture period of 7 days (DEX 0 to 7). IGF-I (30 ng/ml) was added every other day. [³H]thymidine (1 μCi/ml) was added on day 6 and cultures were further incubated for 24 h, washed, lysed in 0.1 M NaOH and the radioactivity was determined. Each point represents the mean of 3 wells.

ROLE OF IGF-I IN THE PROLIFERATION OF SBMC IN CULTURE

Since IGF-I was suggested to be mitogenic for bone cells,[11] we studied the effect of IGF-I on the proliferation of SBMC (Fig. 6). SBMC were cultured for 7 days in a complete medium without dexamethasone in the absence or presence of IGF-I (30 ng/ml). Under these conditions, IGF-I increased [³H]thymidine incorporation by 50% above the control level. Addition of dexamethasone (10^{-8} M) to the culture medium for the first 5 days of the culture or until day 7 of the culture increased [³H]thymidine incorporation by about 80%. The addition of IGF-I to cultures exposed to dexamethasone for 7 days did not induce any further increase in [³H]thymidine incorporation measured on day 7 of the culture. However, when dexamethasone was omitted from the culture medium on day 5, the culture treated with IGF-I demonstrated an increase of 80% in [³H]thymidine incorporation. This data suggests that dexamethasone is inhibiting the mitogenic effect of IGF-I in SBMC cultures. To further study this observation, SBMC were cultured under various dexamethasone concentrations and the effect of IGF-I and bFGF on [³H]thymidine incorporation was studied (Fig. 7). Increasing concentrations of dexamethasone significantly inhibited [³H]thymidine incorporation by SBMC as indicated by more than an 8-fold decrease in the incorporated radioactivity in the presence of 10^{-7} M as compared to 10^{-9} M dexamethasone. The addition of IGF-I did not affect [³H]thymidine incorporation. However, the addition of bFGF-induced [³H]thymidine incorporation by about 6- or 4-fold in cultures exposed to 10^{-7} M or 10^{-8}-10^{-9} M, respectively. These results indicate that SBMC grown in the presence of dexamethasone do not proliferate in response to IGF-I but only in response to bFGF. Taking into account the observation that dexamethasone induces the formation of bone marrow-derived bone-like tissue in culture further suggests that dexamethasone inhibits the proliferation of most of the bone marrow-derived cell lineages except that of preosteoblast cells which then proliferate in response to bFGF, resulting in an increased osteoblast population capable of synthesizing and mineralizing bone matrix.

ROLE OF IGF-I IN THE MINERALIZATION OF SBMC TREATED WITH OR WITHOUT DEXAMETHASONE

Since IGF-I failed to stimulate the proliferation of SBMC treated cultures in contrast with the strong induction observed with bFGF, we studied the effect of IGF-I on the

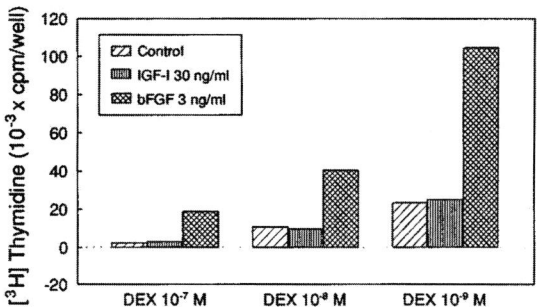

Figure 7: Effect of various concentrations of dexamethasone on [³H]thymidine incorporation by SBMC in the absence or presence of bFGF or IGF-I. SBMC cultures were prepared and cultured as described in Fig. 1 and exposed to various concentrations of dexamethasone (DEX) as indicated. The cultures were grown for 7 days in the absence or presence of bFGF (3 ng/ml) or IGF-I (30 ng/ml) as indicated in the figure. The growth factors were added every other day. [³H]thymidine (1 μCi/ml) was added on day 6 and the amount of incorporated radioactivity was determined as described in Fig. 6.

mineralization of SBMC in culture (Fig. 8). SBMC cultured in the absence of dexamethasone did not form any mineralized bone-like nodules in culture. Addition of IGF-I did not induce bone-like tissue formation. However, very limited amounts of small nodules appeared in the bFGF treated cultures. Exposure of SBMC cells to dexamethasone for the first 7 days of culture resulted in the formation of small amounts of mineralized bone-like tissue and this process was significantly enhanced by exposure of cultures to either IGF-I or bFGF. The continuous exposure of cultures to dexamethasone for 21 days resulted in a slight increase in the amount of mineralized bone-like tissue formation in the control

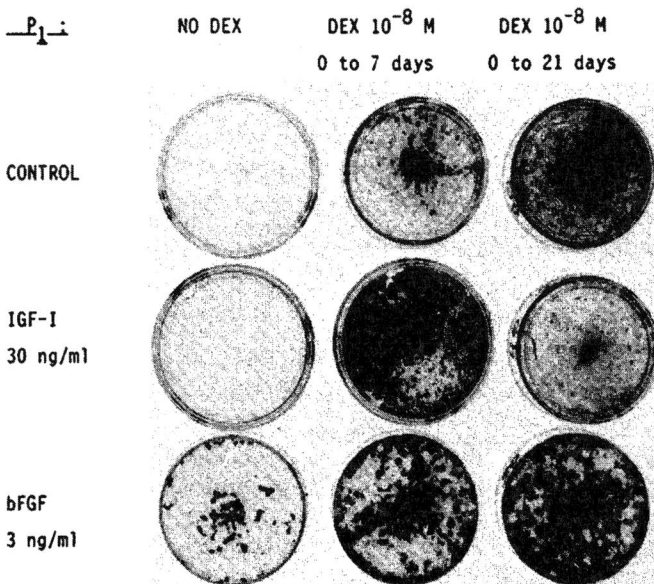

Figure 8. Effect of IGF-I on the mineralization of SBMC cultures treated with or without dexamethasone. SBMC were prepared and cultured as described in Fig. 2 and exposed to a complete medium and dexamethasone (DEX) was not added or dexamethasone (10^{-8} M) was added during the first 7 days of the culture or for the entire culture period of 21 days as indicated in the figure. All cultures were grown for 21 days in the absence of any growth factor (control) or in the presence of IGF-I (30 ng/ml) or bFGF (3 ng/ml) which were added to the cultures every other day. At the end of the culture period, cells were fixed and stained as described in Fig. 2.

and bFGF treated cultures, but in a complete inhibition in IGF-I treated cultures. These results indicated that IGF-I inhibited the dexamethasone-induced mineralized bone-like tissue formation. However, when SBMC were first exposed to dexamethasone for 7 days followed by exposure to IGF-I alone, then IGF-I significantly induced bone formation. It is suggested that dexamethasone modulates the differentiation of SBMC in culture and their response to various growth factors. On the one hand, dexamethasone potentiates the proliferation and differentiation of SBMC cultures in response to bFGF, while on the other hand, it blocks the proliferative response of SBMC to IGF-I and modulates IGF-I to act as an inhibitor of SBMC differentiation. Therefore, if IGF-I plays a major role in bone remodeling *in vivo* by stimulating bone cell replication and bone matrix synthesis, the present study may suggest a mechanism by which glucocorticoids induce osteoporosis, that is, by an indirect mechanism involving the modulation of IGF-I activity.

SUMMARY

Rat stromal bone marrow cells cultured in the presence of dexamethasone produced mineralized bone-like tissue. The addition of bFGF resulted in a significant increase in the formation of mineralized tissue. bFGF induced the proliferation, extracellular matrix synthesis and deposition, osteocalcin secretion and mineralized bone-like tissue formation. The bFGF treated cultures demonstrated the development of a multilayered structure resembling mineralized bone-like tissue consisting of cell layers embedded within bundles of collagen fibers associated with extensive mineral deposits. A biphasic sequence of events was observed during the development of mineralized bone-like tissue. The first phase is characterized by cell proliferation and matrix accumulation and the second phase which follows is characterized by a sharp decline in cell proliferation and matrix accumulation and a concomitant expression of osteoblast differentiation. Considering the possibility that dexamethasone enhances the entering of undifferentiated, multipotential stromal bone marrow cells into the osteogenic pathway,[23] then bFGF apparently increases mainly the proliferation of osteoprogenitor cells, resulting in an increased osteoblast population capable of synthesizing and mineralizing bone matrix. It is suggested that both dexamethasone and bFGF exert their full activity during the first phase and commit the cells to continue the differentiation process even in their absence. The role of both dexamethasone and bFGF in this process is not clear, but we suggest that dexamethasone may select the osteogenic lineage subpopulation of the bone marrow, and/or increases the response of this specific subpopulation to bFGF. Consequently, bFGF stimulates the proliferation of these osteoprogenitor cells which differentiate and manifest the osteoblastic phenotype. In contrast to bFGF, the addition of IGF-I to SBMC cultures in the presence of dexamethasone did not induce proliferation and mineralized bone-like tissue formation, but it resulted in the inhibition of mineralized bone-like formation compared with cultures treated with dexamethasone alone. However, the treatment of SBMC cultures with dexamethasone for 5 to 7 days followed by continuous exposure to IGF-I alone until the end of the culture period resulted in an increase in both cell proliferation and mineralized bone-like tissue formation.

Our studies indicate that dexamethasone modulates the response of SBMC to bFGF and IGF-I, and suggests that corticosteroids may modulate the activity of peptide growth factors within the skeletal system. bFGF is highly effective in the enhancement of mineralized bone-like tissue formation by SBMC cultured in the presence of dexamethasone. In contrast, IGF-I inhibits mineralized bone-like tissue formation by SBMC cultured in the presence of dexamethasone. If IGF-I is playing a major positive role in bone remodeling *in vivo* then these results may suggest that steroids induce osteoporosis via modulating the effect of IGF-I on bone cells, which in the presence of dexamethasone, acts as an inhibitor of bone tissue formation.

ACKNOWLEDGMENT

We wish to thank S. Young and J. Serejski for their invaluable assistance in the preparation of this manuscript.

REFERENCES

1) E. Canalis, The hormonal and local regulation of bone formation, Endocr. Rev. 4:62-77 (1983).

2) E. Canalis, T. McCarthy, and M. Centrella, Growth factors and the regulation of bone remodeling, J. Clin. Invest. 81:277-281 (1988).

3) E. Canalis, T.L. McCarthy, and M. Centrella, Growth factors and cytokines in bone cell metabolism, Annu. Rev. Med. 42:17-24 (1991).

4) P.V. Hauschka, A.E. Mavrakos, M.D., Iafrati, S.E. Doleman and M. Klagsbrun, Growth factors in bone matrix: Isolation of multiple types by affinity chromatography on heparin sepharose. J. Biol. Chem. 261:12665-12674 (1986).

5) E. Canalis, M. Centrella, and T. McCarthy, Effects of basic fibroblast growth factor on bone formation *in vitro*, J. Clin. Invest. 81:1572-1577 (1988).

6) E. Canalis, J. Lorenzo, W.H. Burgess, and T. Maciag, Effects of endothelial cell growth factor on bone remodelling *in vitro*, J. Clin. Invest. 79:52-58 (1987).

7) S.B. Rodan, G. Wesolowski, K. Thomas, and G.A. Rodan, Growth stimulation of rat calvaria osteoblastic cells by acidic fibroblast growth factor, Endocrinology 121:1917-1923 (1987).

8) R.K. Globus, P. Patterson-Buckendahl, and D. Gospodarowicz, Regulation of bovine bone cell proliferation by fibroblast growth factor and transforming growth factor, Endocrinology 123:98-105 (1988).

9) T.L. McCarthy, M. Centrella, and E. Canalis, Effects of fibroblast growth factors on deoxyribonucleic acid and collagen synthesis in rat parietal bone cells, Endocrinology 125:2118-2126 (1989).

10) S.B. Rodan, G. Weslowski, K. Yoon, and G.A. Rodan, Opposing effects of fibroblast growth factor and pertussis toxin on alkaline phosphatase osteopontin, osteocalcin, and type I collagen mRNA levels in ROA 17/2.8 cells, J. Biol. Chem. 264:19934-19941 (1989).

11) T.L. McCarthy, M. Centrella, and E. Canalis, Regulatory effects of insulin-like growth factors I and II on bone collagen synthesis in rat calvarial cultures, Endocrinology 124:301-309 (1989).

12) E. Canalis, Effect of glucocorticoids on type I collagen synthesis, alkaline phosphatase activity, and deoxyribonucleic acid content in cultured rat calvariae, Endocrinology 112:931-939 (1983).

13) T.L. McCarthy, M. Centrella, and E. Canalis, Cortisol inhibits the synthesis of insulin-like growth factor-I in skeletal cells, Endocrinology 126:1569-1575 (1990).

14) H.C. Tenenbaum and J.N.M. Heersche, Dexamethasone stimulates osteogenesis in chick periosteum *in vitro*, Endocrinology 117:2211-2217 (1985).

15) C.G. Bellows, J.E. Aubin, and J.N.M. Heersche, Physiological concentrations of glucocorticoids stimulate formation of bone nodules from isolated rat calvaria cells *in vitro*, Endocrinology 121:1985-1992 (1987).

16) A. Bennett, T. Chen, D. Feldman, R.L. Hintz, and R.G. Rosenfeld, Characterization of insulin-like growth factor I receptors on cultured rat bone cells: regulation of receptor concentration by glucocorticoids, Endocrinology 115:1577-1583 (1984).

17) T.J. Hahn, S.L. Westbrook, and L.R. Halstead, Cortisol modulation of osteoblast metabolic activity in cultured neonatal rat bone, Endocrinology 114:1864-1870 (1984).

18) M. Owen, Marrow stromal stem cells. J. Cell Sci. Suppl. 10:63-76 (1988).

19) A.J. Friedenstein, R.K. Chailakhyan, N.V. Latsinik, A.F. Panasyuk, and I.V. Keiliss-Borok, Stromal cells responsible for transferring the microenvironment of the hemopoietic tissue, Transplantation 17:331-340 (1974).

20) E.A. Luria, M.E. Owen, A.J. Friedenstein, J.F. Morris, and S.A. Kuznetsow, Bone formation in organ cultures of bone marrow, Cell Tissue Res. 248:449-454 (1987).

21) B.A. Ashton, T.D. Allen, C.R. Howlett, C.C. Eaglesom, A. Hattori, and M.E. Owen, Formation of bone and cartilage by marrow stromal cells in diffusion chambers *in vivo*, Clin. Orthop. Rel. Res. 151:294-307 (1980).

22) I. Bab, B.A. Ashton, D. Gazit, G. Marx, M.C. Williamson, and M.E. Owen, Kinetics and differentiation of marrow stromal cells in diffusion chambers *in vivo*, J. Cell Sci. 84:139-151 (1986).

23) C. Maniatopoulos, J. Sodek, A.H. Melcher, Bone formation *in vitro* by stromal cells obtained from bone marrow of young adult rats, Cell Tissue Res. 254:317-330 (1988).

24) D. Noff, S. Pitaru, and N. Savion, Basic fibroblast growth factor enhances the capacity of bone marrow cells to form bone-like nodules *in vitro*, FEBS Letters 250:619-621 (1989).

25) M.E. Antosz, C.G. Bellows, and J.E. Aubin, Biphasic effects of epidermal growth factor on bone nodule formation by isolated rat calvaria cells *in vitro*. J. Bone Mineral Res. 2:385-393 (1987).

26) H.C. Anderson, Mineralization by matrix vesicles, Scan. Electron Microsc. 2:953-964 (1984).

27) L.C. Gerstenfeld, S.D. Chipman, C.M. Kelly, K.J. Hodgens, D.D. Lee, and W.L. Landis, Collagen expression, ultrastructural assembly and mineralization of chicken embryo osteoblasts. J. Cell Biol. 106:979-989 (1988).

28) U. Bhargava, M. Bar-Lev, C.G. Bellows, and J.E. Aubin, Ultrastructural analysis of bone nodules formed *in vitro* by isolated fetal rat calvaria cells. Bone 9:155-163 (1988).

29) S. Pitaru, S. Kotev-Emeth, D. Noff, S. Kaffuler, and N. Savion, The effect of basic fibroblast growth factor on the growth and differentiation of adult bone marrow cells: enhanced development of mineralized bone-like tissue in culture, J. Bone Mineral Res., in press (1992).

BEHAVIORAL, HORMONAL AND NEUROCHEMICAL ALTERATIONS IN TRANSGENIC MICE OVEREXPRESSING TRANSFORMING GROWTH FACTORα

Leena Hilakivi-Clarke[1,2], Richard Goldberg[2], and Robert B. Dickson[1]

[1]Vincent T. Lombardi Cancer Research Center
[2]Department of Psychiatry
Georgetown University Medical School
Washington, D.C. 20007

INTRODUCTION

The presence of growth factors and their receptors in the brain, including nerve growth factor (NGF) (Levi-Montalcini, 1987; Shelton and Reichardt, 1986), transforming growth factor α (TGFα) (Wilcox and Derynck, 1988; Lazar and Blum, 1992), the insulin-like growth factors (Hasselbacher et al. 1985; Marks, Porte and Baskin, 1991), and basic fibroblast growth factor (Fallon et al. 1992), is well-established. However, the functional relevance of these observations is mostly unknown. We have been able to demonstrate that TGFα may be involved in the control of various behaviors.

TGFα is a 50 amino acid polypeptide that is cleaved from a larger integral membrane glycoprotein (proTGFα) (Derynck et al. 1987), and which interacts with cells through the EGF (epidermal growth factor) receptor. While TGFα expression is often associated with neoplastic transformation (Derynck et al. 1987; Bates et al. 1988), several studies have also implicated this growth factor in regulation of normal epithelial cells (Sandgren et al. 1990). Of particular interest is the observation that TGFα mRNA, protein and EGF/TGFα receptors have been observed in the normal rodent brain (Wilcox and Derynck, 1988; Lazar and Blum, 1992; Seroogy, Han and Lee, 1991). TGFα is detected in all brain regions of the mature mice and rats, including the brain stem, cerebellum, cerebral cortex, hippocampus, hypothalamus, thalamus and striatum (Lazar and Blum, 1992; Seroogy, Han and Lee, 1991). Most EGF/TGFα receptors are located in the glia cells (Ojeda et al. 1990), implicating TGFα action in glial-neuronal interactions. It has been postulated that TGFα acts as a neurotransmitter/modulator. This assumption is based on the observation that mRNA levels for TGFα in the brain are 15-170 times greater than the levels for EGF or NGF mRNA (Lazar and Blum, 1992; Shelton and Reichardt, 1986). Thus, TGFα gene expression is clearly higher than what is representative of a traditionally-defined neurotrophic agent.

To investigate the functional significance of TGFα in brain, we have used previously constructed TGFα transgenic mice of the CD-1 background (Jhappan et al. 1990). These animals overexpress human TGFα in multiple tissues, including the brain, throughout their life (Jhappan et al. 1990). Human TGFα binds to mouse EGF receptors

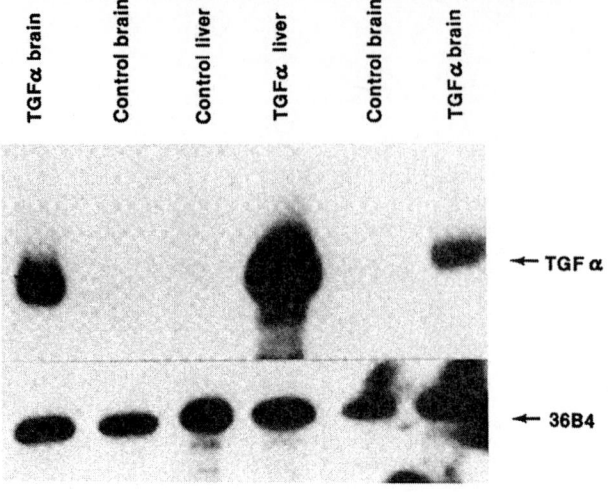

Figure 1. The expression of TGFα mRNA in the brain and liver in the male transgenic TGFα and non-transgenic control mice. TGFα expression was determined using a sensitive RNase protection assay. Total RNA (60 μg) was hybridized to a uniformly labelled TGFα riboprobe, subjected to RNase digestion, and electrophoresed in denaturing polyacrylamide gels. Relative RNA loading is indicated by the intensity of the 36B4 protected fragments. 36B4 is a constitutively expressed mRNA, which codes for a ribosomal protein.

with an equivalent affinity and induce comparable biological responses than mouse TGFα (Coffey et al. 1988). Figure 1 shows TGFα mRNA expression, measured by RNase protection assay, in whole brain and liver in the male TGFα transgenes and in the non-transgenic CD-1 controls. Our initial investigations using these transgenic mice indicate that TGFα may be involved in the regulation of depressive behavior, aggression, plasma 17β-estradiol (E2) levels, natural killer (NK) cell activity, and brain monoamines (Hilakivi-Clarke et al. 1992a; Hilakivi-Clarke et al. 1992b; Hilakivi-Clarke et al. 1992c).

TGFα AND BEHAVIOR

We have employed a variety of tests to monitor behavior in the TGFα mice. The data reveal that transgenic TGFα mice show altered behavioral patterns in the swim test of depressive behavior, in the resident-intruder paradigm of aggression, and in an open field test of locomotor activity (Hilakivi-Clarke et al. 1992a; Hilakivi-Clarke et al. 1992b; Hilakivi-Clarke et al. 1992c). In the swim test, the male TGFα mice spend a significantly longer time immobile than the male non-transgenic CD-1 controls (Table 1). Porsolt's swim test predicts the antidepressant efficacy of different drugs (Porsolt, Bertin and Jalfre, 1977) as well as the effects of a variety of stressors on behavior (Hilakivi et al. 1989; Weiss et al. 1981; Garcia-Marquez and Armario, 1987). In general, antidepressants shorten the time spent immobile in the water, whereas stressors lengthen it. The lengthened immobility in the male TGFα mice implies that their depressive behavior is increased and/or that their ability to cope with stress is impaired. In contrast, immobility in the swim test is significantly shorter in the female TGFα mice than in the female non-transgenic CD-1 mice (Hilakivi-Clarke et al. 1992a) (Table 1).

The data obtained in the resident-intruder test have shown that the male TGFα mice spend a significantly lengthened time in aggressive behavior (Table 1). The female TGFα mice appear less aggressive than the controls. Both male and female TGFα mice have higher locomotor activity scores in an open arena than the non-transgenic CD-1

Table 1. Alterations in behavior, plasma levels of E2, and NK cell activity in the male and female TGFα mice. The symbols indicate the significance of a change, higher than in the controls: + p<.05, + + p<.01, + + + p<.001; lower than in the controls: - p<.05, -- p<.01, --- p<.001; NS = non-significant result. The original data was presented in (Hilakivi-Clarke et al. 1992a) and (Hilakivi-Clarke et al. 1992b).

	Compared with male non-transgenic controls	Compared with female non-transgenic controls
Immobility in the swim test	+	--
Aggression in the resident-intruder test	+ + +	-
The index of anxiety in the plusmaze	NS	NS
Locomotor activity in an open field	+ +	+
Plasma E2 levels	+ + +	+ +
NK cell activity	-	---

control mice (Table 1). No differences in the plusmaze model of anxiety has been observed between the TGFα and control mice. These findings indicate that TGFα is involved in the processes controlling behaviors in the models of depressive, aggressive and locomotor activity. The results also suggest that this growth factor influences behavior in a gender-dependent manner, and, thus, influences sex-related differences in behavior.

TGFα and gender-related behavioral differences

A number of behavioral differences have been reported between male and female mice. It is well-established that female mice and rats are more active than males (Kennett et al. 1986; Archer, 1975). In support of these findings, the non-transgenic female CD-1 mice also exhibit higher locomotor activity scores than the CD-1 males (Hilakivi-Clarke, 1993). Aggressive tendencies in the resident-intruder test do not appear to significantly differ between various strains of female and male mice or rats (Scholtens et al. 1990; Barfield, 1984), including CD-1 mice. Our results also indicate that female CD-1 mice spend a longer time immobile in the swim test than male CD-1 mice. This is in line with the results obtained in rats suggesting that females are more sensitive to the depression-inducing effects of stress (Kennett et al. 1986).

The female transgenic TGFα mice spend significantly shorter times immobile in the swim test and exhibit less aggressive behavior than the TGFα males (Hilakivi-Clarke et al. 1992a). No differences in locomotor activity scores appear between female and male transgenic mice. Thus, insertion of an extra gene that regulates human TGFα expression markedly alters the gender-specific differences in the swim test, resident-intruder paradigm and open field in CD-1 mice. A number of systems are implicated to be responsible for the sex-related behavioral differences, including steroid hormones and monoamines (Reisert and Pilgrim, 1992; Goudsmit, Feenstra and Swaab, 1990). We are currently investigating the possibility that TGFα interacts with these systems in altering behavior in the male and female transgenic mice.

BIOLOGICAL MEDIATORS OF THE ASSOCIATION BETWEEN TGFα AND BEHAVIOR

Our data clearly indicate a functional role for TGFα in the brain. The results suggest that **(i)** TGFα influences behavior in a gender-specific manner, <u>and/or</u> **(ii)** the localization of TGFα mRNA expression and/or the levels of expression of TGFα are different in the male and female transgenic brain. To examine whether gender-related mechanisms mediate elevated TGFα expression on behavior, **plasma steroid** and **brain monoamine** levels have been measured. Both systems are known to be influenced by TGFα. Alexi et al. (1991) recently reported that TGFα stimulates dopamine uptake in rat fetal dopaminergic neurons in culture. It has been shown that in the hypothalamus of female rats TGFα induces release of luteinizing hormone-releasing hormone LHRH (Ojeda et al. 1990).

Estrogen

To study whether steroid hormones may mediate the interaction between TGFα and behavior, the plasma E2, testosterone and corticosterone levels were measured in the transgenic and non-transgenic mice. Our findings indicate that in male and female TGFα mice the plasma E2 levels are 4-6 times higher than in the controls (Hilakivi-Clarke et al. 1992a; Hilakivi-Clarke et al. 1992b) (Table 1). No significant differences have been noted in the plasma levels of testosterone and corticosterone. We

Table 2. The alterations in the concentrations of NE, DA and 5-HT, their metabolites, and in the ratio between monoamines and their respective metabolites in the frontal cortex, hypothalamus and brain stem in the transgenic male and female TGFα. The symbols indicate the significance of a change; higher than in the controls: + $p<.05$, + + $p<.01$; lower than in the controls: - $p<.05$, -- $p<.01$; NS = non-significant result. The data is presented in (Hilakivi-Clarke et al. 1992c).

	Frontal cortex		Hypothalamus		Brain Stem	
	Female	Male	Female	Male	Female	Male
NE	NS	NS	+ +	NS	NS	NS
DA	NS	NS	NS	NS	NS	NS
5-HT	NS	NS	NS	NS	+ +	NS
DOPAC	NS	NS	NS	NS	NS	NS
HVA	NS	NS	NS	NS	NS	NS
5-HIAA	NS	NS	NS	NS	NS	NS
DOPAC/ DA	NS	NS	NS	NS	NS	NS
HVA/DA	NS	NS	NS	NS	NS	NS
5-HIAA/ 5-HT	-	NS	NS	NS	NS	-

also found that only 20% of the female TGFα mice cycled regularly; 40% appeared to remain in estrus and 40% in anestrus. Thus, the mechanisms responsible for controlling reproductive cycle are clearly impaired in the TGFα mice.

The mechanisms causing an elevation in the plasma E2 levels remain open. It has been reported that TGFα induces release of luteinizing hormone-releasing hormone (LHRH) in the hypothalamus of female rats (Ojeda et al. 1990). LHRH stimulates the release of LH in the pituitary, which in turn stimulates estrogen release from the uterus in females and from the testes in males (Griffin and Ojeda, 1988). Thus, constitutive expression of TGFα may result in increased plasma estrogen levels via the hypothalamus. Another explanation for increased plasma levels of E2 in the transgenic mice is that TGFα modulates peripheral conversion of androgens to estrogen via aromatase (Clarke, Dickson and Lippman, 1991).

Monoamines

The monoamine analysis has revealed no significant alterations in the levels of monoamines or their metabolites in the male TGFα mice, as compared to male CD-1 controls (Table 2). In the female TGFα mice, significant elevations in the concentration of norepinephrine (NE) in the cerebral cortex and serotonin (5-HT) in the brain stem were measured (Hilakivi-Clarke et al. 1992c) (Table 2). The ratio between 5-hydroxyindoleacitic acid (5-HIAA) and 5-HT is significantly reduced both in males and females; males show a reduction in the brain stem and females in the frontal cortex. In the males, however, the 5-HIAA/5-HT ratio is reduced because of a low 5-HIAA concentration, whereas in the females 5-HT is significantly elevated, but 5-HIAA levels are normal. Therefore, the synthesis of 5-HT may be increased in the female TGFα mice, rather than their serotonin metabolism being slowed down.

We have failed to find any alterations in the concentration of dopamine (DA) or its metabolites 3,4-dihydroxyphenylacetic acid (DOPAC) and homovanillic acid (HVA) in the frontal cortex, hypothalamus, or brain stem (Table 3). The lack of changes in dopaminergic systems in the TGFα mice is perhaps surprising since both the male and female transgenic mice exhibit increased locomotor activity (Hilakivi-Clarke et al. 1992a). However, we did not analyze the monoamine levels in the striatum, which is one of the richest areas for dopaminergic neurons, and which is closely associated with the control of locomotor activity (Bradford, 1986).

Table 3. Effects of an acute administration of tryptophan and 5-HT uptake inhibitors on behavior in the swim test and resident-intruder test in transgenic TGFα mice and non-transgenic controls. The results are published in (Hilakivi-Clarke et al. 1992d).

	TGFα mice	Controls
Immobility in the swim test		
Tryptophan (100 mg/kg)	Decrease	No change
Zimelidine (12.5 mg/kg)	Decrease	Increase
Aggression in the resident-intruder test		
Tryptophan (100 mg/kg)	No change	No change
Zimelidine (12.5 mg/kg)	Decrease	No change
Clomipramine (10 mg/kg)	Decrease	No change

It remains to be determined whether the changes in the concentrations of 5-HT and NE result form a change in the transcription of the DNA encoding for rate limiting enzymes for monoamine biosynthesis or altered degradation caused by overexpression of TGFα.

TGFα expression

According to our preliminary observation, the expression of TGFα mRNA in the brain of the transgenic mice is similarly elevated in both sexes (Hilakivi-Clarke et al. 1992c). Further, no significant sex-associated differences are apparent in any region of the central nervous system for the TGFα mRNA expression in non-transgenic mice (Lazar and Blum, 1992). An exception is the pituitary, where significantly higher levels of TGFα mRNA have been observed in male mice (Lazar and Blum, 1992). Since gonadotrophins are released from the pituitary, and this gland contains a high number of gonadal steroid hormones, the sex-difference in TGFα mRNA expression in the pituitary may be due to differences in gonadal steroid hormone expression. These results strongly imply that TGFα expression is not linked to gender; however, TGFα may interact with brain monoamines and/or plasma E2 in inducing gender-specific behavioral alterations.

NEUROENDOCRINOLOGICAL SYSTEMS IN AFFECTING BEHAVIOR

Estrogen and testosterone are involved in regulating various behaviors. Maggi & Perez (Maggi and Perez, 1985) have presented both clinical and experimental evidence indicating that estrogen stimulates stereotyped locomotor behaviors. Thus, estrogen may be involved in the elevated locomotor activity seen in both the male and female transgenic TGFα mice.

The role of estrogen in depression is unclear. Non-transgenic female mice in estrus show longer immobility times in the swim test model than female mice in other stages of estrus cycle (Pare and Redei, 1991). Plasma levels of E2 peak immediately prior to estrus. However, estradiol does not itself affect immobility in the swim test in intact females, but they reverse the effects of ovariectomy (Bernardi et al. 1989). Ovariectomy increases depressive-like behavior in female mice. In male mice, castration lengthens immobility in the swim test, and testosterone reduces it to within normal limits (Bernardi et al. 1989). Similar to estrogen in female mice, testosterone does not affect immobility in control mice. It is to be noted that we failed to find a correlation between the time spent immobile in the swim test and plasma E2 levels in male CD-1 mice (unpublished observation). This would indicate that estrogen is not directly involved in the expression of depressive behavior in normal male mice.

High levels of testosterone are generally associated with aggressive behavior (Hilakivi et al. 1989; Miczek, 1987). The results of Brain et al. (Brain et al. 1988) and Hasan et al. (Hasan, Brain and Castano, 1988) imply that high plasma levels of estrogen may mediate social aggression in male rats and mice. Furthermore, experiments conducted by van de Poll et al. (1985) have shown that chronic treatment with estrogen induced high levels of aggression in male rats, whereas females showed virtually no aggression. It is also clear that female rats are the least aggressive during the estrus phase of their estrous cycle (Hood, 1984). The elevation in aggressive behavior in the male TGFα mice could, thus, result from the increased plasma levels of E2. High levels of plasma E2 in the female TGFα mice, on the other hand, may have led to a decrease in aggressive tendencies.

MONOAMINES AND BEHAVIOR

Monoamines are involved in the regulation of various behaviors. NE has been generally associated with behavioral activation requiring attention, such as learning (Matussek, 1988; Bradford, 1986). There is some evidence indicating that increases in the brain NE levels reduce depressive behavior. For example, acute treatments with inhibitors of monoamine uptake which increase the synaptic concentrations of NE, reduce depressive behavior in the swim test model (Porsolt, Bertin and Jalfre, 1977; Borsini and Meli, 1988). The reduced immobility in the swim test in the female TGFα mice may, thus, be a result of their elevated brain NE concentrations.

5-HT is clearly related to stress, depression and aggression. Low levels of 5-HT and its major metabolite 5-HIAA are typical to depressive individuals and depletion of 5-HT induces depressive behavior in humans and animals (Meltzer, 1990). On the other hand, elevation of plasma 5-HT levels by tryptophan is shown to relieve the symptoms of mild to moderate depression (Morand, Young and Ervin, 1983). Lowered 5-HT functions have been linked to increased aggressive tendencies and compounds which increase brain 5-HT levels generally inhibit aggression (Miczek, 1987). 5-HT is also involved with locomotor activity (Green and Backus, 1990). The role of 5-HT neurotransmission in the development and treatment of anxiety states, has recently been considered (Charney et al. 1990). However, the specific nature of the involvement of 5-HT in anxiety remains to be defined. In summary, the reduced ratio between 5-HIAA and 5-HT seen in the male TGFα mice may be associated with their increased immobility in the swim test and aggressive tendencies in the resident-intruder test. The low levels of depressive-like behavior and aggression exhibited by female transgenics may result from elevated brain 5-HT concentrations.

ROLE OF 5-HT IN INDUCING BEHAVIORAL CHANGES IN MALE TGFα MICE

To investigate whether the reduced ratio between 5-HT and its metabolite 5-HIAA noted in the male TGFα mice is associated with their increased immobility in the swim test or lengthened time spent in exhibiting aggression in the resident-intruder test, these animals were treated with serotonergic compounds (Hilakivi-Clarke et al. 1992d). The results showed that 5-HT precursor tryptophan and 5-HT uptake inhibitor zimelidine shortened immobility in the swim test in the male transgenic TGFα mice (Table 3). Tryptophan failed to influence immobility in the non-transgenic CD-1 mice. Further, an acute administration of zimelidine lengthened immobility in the male CD-1 mice. Increased aggressive behavior of the male TGFα mice was reversed with 5-HT uptake inhibitors zimelidine and clomipramine, but not with tryptophan (Table 3). Tryptophan, zimelidine and clomipramine all failed to significantly alter aggressive behavior in the male control mice.

The results suggest that the behavioral impairments seen in the male TGFα mice are not caused by lowered tryptophan content in the brain, or by reduced synthesis of 5-HT. Tryptophan is shown to increase the levels of 5-HT in the rodent brain; a maximal (approximately 80%) increase in brain 5-HT content is reported following administration of 100 mg/kg tryptophan (Fernstrom and Wurtman, 1972). Further, synthesis of 5-HT is significantly increased by tryptophan treatment (Fernstrom, 1988).

Similar to tryptophan, acute administration of 5-HT uptake inhibitors increases the cerebral concentrations of 5-HT (Fuller and Wong, 1990; Enna et al. 1981; Racagni, Ellison and Nelson, 1982), although the mechanism of increase is different in these compounds. Uptake inhibitors prevent 5-HT from being removed from the synaptic cleft (Fuller and Wong, 1990), whereas tryptophan increases the synthesis of 5-HT (Fernstrom and Wurtman, 1972). Further, 5-HT uptake inhibitors have a number of other effects on monoamine transmission (Enna, Malick and Richelson, 1981), any of which could contribute to the behavioral changes seen following an administration of these compounds. For example, clomipramine also inhibits uptake of NE in the brain (Raisman, Briley and Langer, 1980). Taken together, TGFα may influence behavior by affecting the synthesis and/or uptake of 5-HT in neurons.

CONCLUSIONS

In conclusion, our findings represent the first observation that TGFα may influence behavior. These results indicate that TGFα may prove to be a novel therapeutic target for the treatment of aberrant aggressive and depressive behaviors. Whether the behavioral alterations seen in the transgenic mice are the result of a direct action of TGFα, indicating that this growth factor acts as a neurotransmitter/ neuromodulator in the brain, or are indirectly caused through its effects on steroid hormones and/or brain monoamines, is currently under investigation.

A number of new research projects are expected to originate from our pilot work. At present, the results suggest that an increase in synaptic availability of 5-HT, either due to tryptophan administration or due to an uptake blockade of this neurotransmitter, shortens immobility in the swim test model of depression in the male TGFα mice. High levels of aggressive behavior are not reversed by tryptophan. However, an administration of 5-HT uptake inhibitors reverses elevated aggression, suggesting that alterations in the uptake mechanisms of this neurotransmitter may partially explain high aggressive tendencies in the male transgenic mice. Our unpublished observations indicate that a 5-HT2 receptor antagonist, ketanserin, also antagonizes behavioral alterations in these mice. Thus, TGFα may play a critical role in influencing affective behaviors through 5-HT transmission. Our future studies will further determine the interaction between 5-HT and TGFα and whether high levels of plasma estrogen participate to the expression of behavioral alterations in the transgenic TGFα mice.

Acknowledgements

We thank our collaborators Dr. P.K. Arora, Dr. T. Taira, Dr. R. Clarke, Dr. A. Lauber, Dr. M.E. Lippman, Ms. A. Wright, Ms. M.-B. Sabol, and Ms. V. Boulay for their contribution to this study. Our research was partly supported by grants from the Academy of Finland and American Cancer Society #BC 52754.

REFERENCES

Alexi, T., Denton, T.L., and Hefti, F., (1991), Effects of TGFα and TGFβ on ventral mesencephalic dopaminergic cultures. Society for Neuroscience 21:301.7.

Archer, J., (1975), Rodent sex differences in emotional and related behavior. Behav. Biol. 14:451.

Barfield, R.J., (1984), Reproductive hormones and aggressive behavior, in: Biological Perspectives on Aggression, eds. K.J. Flanelly, R.J Blanchard and D.C. Blanchard (Alan R. Liss, New York) p. 105.

Bates, S.E., Davidson, N.E., Valverius, E.M., Freter, C.E., Dickson, R.B., Tam, J.B., Kudlow, J.E., Lippman, M.E. and Salomon, D.S., (1988), Expression of transforming growth factorα and its messenger ribonucleic acid in human breast cancer: its regulation by estrogen and its possible functional significance. Mol. Endocrinol. 2:543.

Bernardi, M., Vergoni, A.V., Sandrini, M., Tagliavini, S. and Bertolini, A., (1989), Influence of ovariectomy, estradiol and progesterone on the behavior of mice in an experimental model of depression. Physiol. Behav. 45:1067.

Borsini, F. and Meli, A., (1988), Is the forced swimming test a suitable model for revealing antidepressant activity? Psychopharmacology 94:147.

Bradford, H.F., (1986), Chemical Neurobiology (W.H. Freeman and Company, New York).

Brain, P.F., Simon, V., Hasan, S., Martinez, M. and Castano, D., (1988), The potential of antiestrogens as centrally-acting antihostility agents: recent animal data. Intern J. Neurosci. 41:169.

Charney, D.S., Woods, S.W., Krystal, J.H. and Heninger, G.R., (1990), Serotonin function and human anxiety disorders. Ann. NY Acad. Sci. 600:558.

Clarke, R., Dickson, R.B. and Lippman, M.E., (1992), Hormonal aspects of breast cancer. Growth factors, drugs and stromal interactions. Crit. Rev. Oncol. Hematol. 12:1

Coffey, R.J., Sipes, N.J., Bascom, C.C., Graves-Deal, R., Pennington, C.Y., Weissman, B.E. and Moses, H.L., (1988), Growth modulation of mouse keratinocytes by transforming growth factors. Cancer Res. 48:1596.

Derynck, R., Goeddel, D.V., Ullrich, A., Gutterman, J.U., Williams, R.D., Bringman, T.S. and Berger, W.H. (1987), Synthesis of mRNAs for transforming growth factor receptor by human tumors. Cancer Res. 47:707.

Enna, S.J., Mann, E., Kendall, D. and Stancel, G.M., (1981), Effect of chronic antidepressant administration on brain neurotransmitter receptor binding, in: Antidepressants: Neurochemical, Behavioral and Clinical Perspectives, eds. S.J. Enna, M. Kuhar and J. Coyle (Raven Press, New York) p. 91.

Enna, S.J., Malick, J.B. and Richelson, E., (1981), Antidepressants: Neurochemical, Behavioral, and Clinical Perspectives (Raven Press, New York).

Fallon, J.H., Di Salvo, J., Loughlin, S.E. and et al., (1992), Localization of acidic fibroblast growth factor within the mouse brain using biochemical and immunocytochemical techniques. Growth Factors 6:139.

Fernstrom, J.D., (1988), Tryptophan availability and serotonin synthesis in brain, in: Amino Acid Availability and Brain Function in Health and Disease, Vol. 20th , ed. G. Huether (Springer-Verlag, Berlin) p. 138.

Fernstrom, J.D. and Wurtman, R.J., (1972), Brain serotonin content:physiological dependence on plasma tryptophan levels. Science 173:149.

Fuller, R.W. and Wong, D.T., (1990), Serotonin uptake and serotonin uptake inhibition, in: The Neuropharmacology of Serotonin, eds. P.M. Whitaker-Azmitia and S.J. Peroutka (The New York Academy of Sciences, New York) p. 68.

Garcia-Marquez, C. and Armario, A., (1987), Interaction between chronic stress and clomipramine treatment in rats. Effects on exploratory activity, behavioral despair, and pituitary-adrenal function. Psychopharmacology 93:77.

Goudsmit, E., Feenstra, M.G. and Swaab, D.F., (1990), Central monoamine metabolism in the male Brown-Norway rat in relation to aging and testosterone. Brain Res. Bull. 25:755.

Green, A.R. and Backus, L.I., (1990), Animal models of serotonin behavior, in: The Neuropharmacology of Serotonin, eds. P.M. Whitaker-Azmitia and S.J. Peroutka (The New York Academy of Sciences, New York) p. 237.

Griffin, J.E. and Ojeda, S.R., (1988), Textbook of Endocrine Physiology (Oxford University Press, New York).

Hasan, S.A., Brain, P.F. and Castano, D., (1988), Studies on effects of tamoxifen (ICI 46474) on antagonistic encounters between pairs of intact mice. Horm. Behav. 22:178.

Hasselbacher, G.K., Schwab, M.E., Pasi, A. and Humbel, R.E., (1985), Insulin-like growth factor II (IGF II) in human brain: regional distribution of IGF II and of higher molecular mass forms. Proc. Natl Acad. Sci. USA 82:2153.

Hilakivi, L.A., Lister, R.G., Durcan, M.J., Eskey, R.L., Mefford, I. and Linnoila, M., (1989), Behavioral, hormonal and neurochemical characteristics of aggressive alpha mice. Brain Res. 502:158.

Hilakivi-Clarke, L.A., Arora, P.K., Clarke, R., Lippman, M.E., and Dickson, R.B., (1992a), Differences in cancer appearance and hormonal, immune and behavioral parameters between female and male transgenic TGFα mice. AACR 83:1620.

Hilakivi-Clarke, L.A., Arora, P.K., Sabol, M.B.., Clarke, R., Dickson, R.B. and Lippman, M.E., (1992b), Alterations in behavior, steroid hormones and natural killer cell activity in male transgenic TGFα mice. Brain Res. 588:97.

Hilakivi-Clarke, L.A., Dickson, R.B., Taira, T., Boulay, V. and Clarke, R., (1992c), The role of transforming growth factor α in influencing brain monoamines and behavior: Studies in transgenic TGFα mice. Growth Factors Peptides and Receptors XII, Abstract 49.

Hilakivi-Clarke, L.A., Taira, T., Goldberg, R. and Clarke, R., (1992d), Serotonin and altered behavior in the models of depression and aggression in male transgenic TGFα mice. Society for Neuroscience 22:621.11.

Hilakivi-Clarke, L.A., (1993), TGFα alters sex differences in locomotor activity, depression and aggression: studies with transgenic mice. Physiol. Behav., submitted.

Hood, K.E., (1984), Aggression among female rats during the estrus cycle, in: Biological Perspectives on Aggression, eds. K.J. Flanelly, R.J. Blanchard and D.C. Blanchard (Alan R. Liss, New York) p. 181.

135

Jhappan, C., Stahle, C., Harkins, R.N., Fausto, N., Smith, G.H. and Merlino, G.T., (1990), TGFα overexpression in transgenic mice induces liver neoplasia and abnormal development of the mammary gland and pancreas. Cell 61:1137.

Kennett, G.A., Chaouloff, F., Marcou, M. and Curzon, G., (1986), Female rats are more vulnerable than males in an animal model of depression: the possible role of serotonin. Brain Res. 382:416.

Lazar, L.M. and Blum, M., (1992), Regional distribution and developmental expression of epidermal growth factor and transforming growth factor α mRNA in mouse brain by a quantitative nuclease protection assay. J. Neurosci. 12:1688.

Levi-Montalcini, R., (1987), The nerve growth factor 35 years later. Science 237:1154.

Maggi, A. and Perez, J. (1985), Role of female gonadal hormones in the CNS: clinical and experimental aspects. Life Sci. 37:893.

Marks, J.L., Porte, D. Jr. and Baskin, D.G. (1991), Localization of type I insulin-like growth factor receptor messenger RNA in the adult rat brain by in situ hybridization. Mol. Endocrinol. 5:1158.

Matussek, N., (1988), Catecholamines and mood: Neuroendocrine aspects, in: Neuroendocrinology of Mood, eds. D. Ganten and D. Pfaff (Springer-Verlag, Berlin) p. 141.

Meltzer, H.Y., (1990), Role of serotonin in depression, in: The Neuropharmacology of Serotonin, eds. P.M. Whitaker-Azmitia and S.J. Peroutka (The New York Academy of Sciences, New York) p. 486.

Miczek, K.A., (1987), The psychopharmacology of aggression, in: The Handbook of Psychopharmacology, eds. L.L. Iversen, S.D. Iversen and S.H. Snyder (Plenum Press, NY) p. 183.

Morand, C., Young, S.N. and Ervin, F.R., (1983), Clinical response of aggressive schizophrenics to oral tryptophan. Biol. Psychiatry 18:575.

Ojeda, S.R., Urbanski, H.F., Costa, M.E., Hill, D.F. and Moholt-Siebert, M., (1990), Involvement of transforming growth factor α in the release of luteinizing hormone-releasing hormone from the developing female hypothalamus. Proc. Natl Acad. Sci. USA 87:9698.

Pare, W.P. and Redei, V.A., (1991), Behavioral responses as a function of estrus cycle in WKY rats. Society for Neuroscience 21:197.13.

Porsolt, R.D., Bertin, A. and Jalfre, M., (1977), Behavioral despair in mice: A preliminary screening test for antidepressants. Arch. Int. Pharmacoldyn. 229:327.

Racagni, G., Ellison, G. and Nelson, L., (1982), In vivo studies on central noradrenergic synaptic mechanisms after acute and chronic antidepressant treatment: biochemical and behavioral comparison. J. Pharm. Exp. Ther. 223:227.

Raisman, R., Briley, M.S. and Langer, S.Z., (1980), Specific tricyclic antidepressant binding sites in rat brain characterized by high-affinity 3H-imipramine binding. Eur. J. Pharm. 61:373.

Reisert, I. and Pilgrim, C., (1992), Sexual differentiation of monoaminergic neurons - genetic or epigenetic? Trends Neurosci. 14:468.

Sandgren, E.P., Luetteke, N.C. Palmiter, R.D., Brinster, R.L. and Lee, D.C., (1990), Overexpression of TGFα in transgenic mice: Induction of epithelial hyperplasia, pancreatic metaplasia, and carcinoma of the breast. Cell 61:1121.

Scholtens, J., Roozen, M., Mirmiran, M. and van de Poll, N.E., (1990), Role of noradrenaline in behavioral changes after defeat in male and female rats. Behav. Brain Res. 36:199.

Seroogy, K.B., Han, V.K.M. and Lee, D.C., (1991), Regional expression of transforming growth factor α in the rat central nervous system. Neurosci. Lett. 125:241.

Shelton, D.L. and Reichardt, L.F., (1986), Studies on the expression of the β nerve growth factor (NGF) gene in the central nervous system: level and regional distribution of NGF mRNA suggest that NGF functions as a trophic factor for several distinct populations of neurons. Proc. Natl Acad. Sci. USA 83:2714.

van de Poll, N.E., Bowden, N.J., van Oyen, H.G., de Jonge, F.H. and Swanson, H.H., (1985), Gonadal hormone influences upon aggressive behavior in male and female rats, in: Psychopharmacology of Sexual Disorders, ed. M. Segal (Libbey, London) p. 63.

Weiss, J.M., Goodman, P.A., Lositi, B.G., Corrigan, S., Charry, M. and Bailey, W.H. (1981), Behavioral depression produced by an uncontrollable stressor: Relationship to norepinephrine, dopamine, and serotonin levels in various regions of rat brain. Brain Res. Rev. 3:167.

Wilcox, J.N. and Derynck, R., (1988), Localization of cells synthesizing transforming growth factor-alpha mRNA in the mouse brain. J. Neurosci. 8:1091.

GROWTH FACTORS AND BRAIN INJURY

David J. Berlove and Seth P. Finklestein

CNS Growth Factor Research Laboratory
Department of Neurology
Massachusetts General Hospital
Boston, MA 02114

INTRODUCTION

Growth factors are generally defined as polypeptides that, at very low concentrations, act through specific receptor-mediated mechanisms to initiate and sustain complex cellular processes. These processes include cell proliferation in the case of cells capable of mitosis (such as brain glia and endothelial cells), and, in the case of post-mitotic cells such as neurons, include cell survival, outgrowth, and biochemical differentiation. An increasing number of polypeptide growth factors have been identified over the past few decades, most now grouped into several gene "superfamilies." The identification of many of these peptides in the mammalian central nervous system (CNS) suggests that these factors play important roles in brain development, and also in the response of the mature brain to injury. Indeed, several studies have now documented increased expression of polypeptide growth factors following brain injury. Using the neurotrophin and fibroblast growth factor (FGF) superfamilies as specific examples, this review will explore two major themes: (1) that the increased "endogenous" expression of growth factors after brain injury may contribute to the cascade of cellular changes that underlie structural reorganization and functional recovery, and (2) that the "exogenous" addition of growth factors may serve to augment such recovery.

NEUROTROPHINS

Nerve growth factor (NGF), the original neurotrophin, was observed initially as a diffusible substance secreted from mouse sarcoma that promotes hyperinnervation of sympathetic and sensory ganglia in the chick embryo[1]. NGF was later found to be present in a variety of sources, with particular concentrations in snake venom and mouse salivary gland[2]. NGF was found to be essential for normal neuronal development; antibody-neutralization of endogenous NGF in the neonatal rodent causes massive loss of sympathetic ganglion cells (immunosympathectomy)[3,4]. In the adult animal, while NGF is not essential for the survival of dorsal root ganglion (DRG) sensory neurons, it does enhance axonal regeneration[5].

NGF is present in many regions of the CNS, but particularly high levels are found in brain regions containing cholinergic neurons (including septum, basal nucleus, diagonal band) and in regions innervated by cholinergic neurons (including hippocampus, cortex and olfactory bulb)[6]. NGF mRNA is present principally in the latter regions, but not in cholinergic cell bodies; taken together with studies of ^{125}I-labeled NGF transport[7], this suggests that NGF is synthesized by target neurons and retrogradely transported to cholinergic cell bodies.

NGF appears to function as an important survival factor for central cholinergic neurons in the adult brain. Cholinergic neurons from the septum project via the fimbria-fornix to the hippocampal formation; transection of this pathway leads to a loss of choline-acetyltransferase (ChAT) immunoreactive septal cells[8,9]. Evidence that cell loss in septum is a consequence of trophic factor deprivation has been obtained by studies of NGF administration following fimbria-fornix lesions. Intraventricular NGF infusion in both rodents and primates reduces the apparent loss of septal cholinergic neurons following fimbria-fornix transection[10,11]; in the primate, neuronal loss was reduced from 55% to 20%[12]. In addition to an increase in the actual number of surviving neurons, a qualitative change is observed in the surviving neurons. In vehicle-treated monkeys, many of the remaining cells are pale and shrunken, while in NGF-treated animals the cells are generally larger and more intensely labeled. Rescue of septal neurons may even occur when NGF is administered weeks after the lesion[13], although this finding is controversial[14]. The importance of NGF for cholinergic cells has also been demonstrated by infusion into intact animals; exogenous NGF induces a dose-dependent increase in ChAT activity in both hippocampus and septum[15].

Several studies implicate a role for NGF in the wound-healing response after CNS injury. Following cortical cavity lesions, NGF levels in the cavity fluid increase rapidly, peaking 16 hours after injury and remaining elevated for three to six days[16]. In another model, a small unilateral electrolytic lesion is placed in the dentate gyrus of the hippocampus to produce seizures. Following the lesion, there is a rapid, transient bilateral rise in NGF mRNA in the granule cells of dentate gyrus[17].

In a model of transient forebrain ischemia in the rat, NGF mRNA levels are increased transiently in the dentate gyrus, a region relatively resistant to ischemia, at four hours after ischemia, at a time before neural death occurs in other vulnerable regions (e.g., CA1 hippocampus)[18]. NGF protein levels are also increased in the hippocampus at a later time (5 days) after ischemia, well after neuronal death has occurred[19]. These results suggest that NGF expression is biphasic in the ischemic brain, and may play at least two roles: (1) an early role in protecting some brain neurons, and (2) a later role in stimulating axonal sprouting and new synapse formation. This former role is supported by studies showing that the exogenous administration of NGF can protect especially vulnerable neurons in CA1 hippocampus after ischemia[20]. These results are intriguing, given that non-cholinergic intrinsic hippocampal neurons are not normally considered to be targets of NGF action.

Atrophy of cholinergic neurons has been observed in the aged rodent forebrain where it is associated with cognitive impairment including loss of spatial memory[21,22]. A reduction in the density of cholinergic axons has also been found in aged human hippocampus where it is closely paralleled by a loss of NGF-receptor immunoreactivity[23]. Long-term NGF infusion in rats can partly reverse age-dependent cholinergic cell body atrophy, and improve spatial memory retention[24]. This is of particular significance, as degeneration of forebrain cholinergic neurons and cognitive impairment are highly characteristic of Alzheimer-type dementia[25].

In summary, NGF appears particularly promising as a neuroprotective agent to increase neuronal survival in brain injury, ischemia, and aging. However, it is unclear to what degree these effects apply primarily to central cholinergic neurons or may extend to other neuronal types in brain as well.

A second member of the neurotrophin family is brain-derived neurotrophic factor (BDNF). BDNF is a small, highly basic protein that was originally purified from pig brain, where it is present in concentrations of only a few μg/gr[26]. The primary structure of BDNF is highly homologous to that of NGF, but the distribution and biological activity of BDNF are unique. BDNF mRNA is localized principally in the CNS; it is widely distributed throughout the brain at levels higher than NGF mRNA[27,28]. Like NGF, BDNF mRNA is most heavily expressed in hippocampus and cortex, but it is also expressed in other CNS regions, including spinal cord and superior colliculus[27,29]. Unlike NGF, BDNF levels are relatively low in the developing nervous system and reach their highest levels in the adult[30]. Like NGF, BDNF appears to function as a retrogradely transported survival factor for cholinergic neurons of the basal forebrain which contain the peptide, but not the mRNA[29,31].

The biological activity of BDNF is distinctive; *in vitro* studies indicate no effect on survival or neurite outgrowth from sympathetic or parasympathetic ciliary ganglion neurons[32]. Peripheral sensory neurons (neural placode-derived) for which NGF is not trophic are highly responsive to BDNF, as are NGF-sensitive dorsal root ganglion (DRG) sensory neurons. BDNF is also trophic for a variety of CNS neurons. In culture, BDNF supports the survival and axonal elongation of retinal ganglion cells[33] and the survival of mesencephalic dopaminergic neurons[34]. BDNF increases the survival of rat septal cholinergic neurons and levels of their cholinergic enzymes; its effects on these neurons are additive with NGF[35]. BDNF neuroprotection against neuronal death has also been demonstrated *in vitro*. Cultured mesencephalic dopaminergic neurons are vulnerable to the neurotoxin MPP+. When these cells are exposed to 1 μM MPP+ for 48 hours, more than 75% of them are lost. But when the cells are pretreated with BDNF, less than 32% of them die; this protective effect is not seen with either NGF or bFGF[34]. The potential significance of this finding is considerable, since selective loss of these neurons occurs in Parkinson's disease.

Like NGF, BDNF has been reported to rescue septal cholinergic neurons after fimbria-fornix transection *in vivo*[36], but may not enhance cholinergic innervation of hippocampus[37]. The effects of brain injury on endogenous BDNF expression vary with region and type of injury, but in many paradigms a rapid increase in BDNF expression has been observed. Unilateral hippocampal needle injury or saline injection into hippocampus leads to a rapid (one hour) but transient ipsilateral increase in BDNF mRNA expression[38]. Injection of kainic acid leads to a more widespread, intense, and sustained increase of BDNF mRNA levels in numerous ipsilateral (and, to a lesser extent contralateral) regions, including hippocampus, cortex, amygada, claustrum, and hypothalamus[38]. Unilateral intrahippocampal injection of the excitotoxin quinolinic acid, which produces seizures, also alters BDNF gene expression. An increase in BDNF mRNA levels is found in the ipsilateral hippocampus in regions not directly in contact with the injection, suggesting transynaptic activation of this gene[39].

Brain ischemia also stimulates BDNF expression in hippocampus. Both global ischemia and hypoglycemia in the rat increase BDNF expression in dentate gyrus by 300-500% at two hours after the insult[18]. These changes appear to be mediated by non-NMDA glutamate receptors[18,40]. By contrast to brain ischemia, BDNF mRNA levels are decreased in post-mortem hippocampi of human patients with Alzheimer's disease, while NGF mRNA levels are unchanged[41], suggesting a possible link between decreased BDNF expression and the progressive neuronal loss that is characteristic of this disease.

A third member of the neurotrophin family, NT-3, was discovered by screening DNA for sequences homologous to NGF and BDNF[42-45]. NT-3 is a small basic protein that shares important structural regions with NGF and BDNF; the three neurotrophins are approximately 50% homologous. The distribution of NT-3, however, varies considerably from that of its fellow neurotrophins. NT-3 mRNA is present in all adult tissues surveyed at levels comparable to or higher than in brain[43]. Within the adult rat brain, NT-3 mRNA is relatively scarce, with the hippocampus and cerebellum showing the greatest expression[42,44,45]. NT-3 in adult rat hippocampus is primarily confined to the medial portions of CA1 and CA2, and the granular cells of dentate gyrus[45]. In contrast to BDNF, NT-3 expression is highest in the immature CNS. All three neurotrophin mRNAs are widely distributed in E12 rat embryos, but NT-3 is by far the most abundant[30]. While NGF mRNA expression varies within individual regions, and displays no consistent pattern of change during development, NT-3 and BDNF mRNA levels generally display a reciprocal relationship in the developing brain. NT-3 mRNA expression steadily decreases in the maturing brain as BDNF increases; by adulthood levels of the two neurotrophins are similar[30]. The precise timing of these changes varies considerably with the region of CNS examined.

The biological activity of NT-3 overlaps with that of the other neurotrophins, but is distinct. NT-3 has either no or very weak neurotrophic effects on parasympathetic ciliary ganglion cells[42-45]. Like NGF, NT-3 promotes survival and neurite outgrowth from sympathetic neurons and dorsal root ganglia[42,45]. Like BDNF, NT-3 is trophic for NGF-insensitive neural placode-derived sensory neurons[42,45]. Although both NGF and BDNF promote the differentiation of basal forebrain cholinergic neurons, NT-3 has no such effect[46].

The role of NT-3 has not yet been well explored in brain injury. Following unilateral hippocampal injection of the excitotoxin quinolinic acid, there is a transient induction of NT-3 mRNA in the contralateral hippocampus, followed by a later increase in NT-3 mRNA in regions of the ipsilateral hippocampus (CA1, CA4) that undergo delayed neuronal degeneration[39]. This may represent an attempted compensatory response of cells in an effort to survive. Injection of NT-3 into the lesioned rat spinal cord has been reported to increase sprouting of injured corticospinal axons[47]. Expression of NT-3 mRNA after cerebral ischemia differs greatly from that of other neurotrophins. While NGF and BDNF mRNA levels rise sharply in hippocampus after global cerebral ischemia in rats, NT-3 mRNA levels fall gradually, returning slowly to baseline at 24 hours[18]. Hypoglycemia does not alter NT-3 mRNA levels[18]. Currently, it is difficult to predict the potential therapeutic applications of NT-3, given our limited understanding of its role in brain injury and ischemia. It appears less likely than NGF or BDNF to promote central cholinergic activity. On the other hand, the abundance of NT-3 in the developing brain suggests that this factor may play an important role in the formation of neural connections.

FIBROBLAST GROWTH FACTORS

The fibroblast growth factors (FGFs) currently comprise a family of seven homologous polypeptides[48-50]. Both acidic FGF (aFGF) and basic FGF (bFGF) are found in relatively high concentrations in the mammalian brain. These 18 kiloDalton, 154 amino acid polypeptides bind tightly to heparin, reflecting their ability to bind to heparan sulfate proteoglycans in the extracellular space[51]. However, both the aFGF and bFGF genes lack signal peptide sequences[52], making it uncertain how and under what circumstances these factors may be released from cells in which they are synthesized.

Acidic and basic FGF have multipotential trophic effects on cultured CNS cells. Both of these polypeptides support the survival and/or outgrowth of neurons from embryonic rat cortex, hippocampus, striatum, septum, thalamus, hypothalamus, retina and spinal cord *in vitro*[53-57]. Basic FGF also supports the survival and outgrowth of cultured embryonic rat brainstem and postnatal mouse cerebellar neurons as well as the proliferation of embryonic rat neuroblasts *in vitro*[58-60]. In addition to these neuronotrophic and neurite-promoting properties, acidic and basic FGF also promote the proliferation of cultured brain capillary endothelial cells[61], as well as the proliferation of brain glial cells, including astroglia and oligodendroglia[62,63].

The cellular localization of FGFs within the mature mammalian brain remains controversial. Some immunohistochemical studies suggest that bFGF has a widespread neuronal localization within the mature rodent brain[64-66], while other studies suggest that bFGF is localized primarily within astrocytes and certain select neuronal populations (including neurons in CA2 hippocampus, cingulate cortex, septohippocampal nucleus, and cerebellum)[67,68]. *In situ* hybridization studies show prominent bFGF mRNA expression in neurons of CA2 hippocampus, cingulate cortex, indusium griseum, and fasciola cinereum[69]. Immunohistochemical studies suggest that aFGF is found primarily in ependymal cells, tanycytes, and scattered glial cells in the mature rat brain[70], although *in situ* hybridization studies show aFGF mRNA in neurons of hippocampus, cortex, cerebellum, substantia nigra, and locus ceruleus[71]. The FGF receptor *flg*, which binds both aFGF and bFGF, is expressed in neurons throughout the rat CNS, especially those in hippocampus, hypothalamus, brainstem, cerebellum, and spinal cord[72].

Several recent studies have addressed changes in FGF expression after focal brain injury in rats. Levels of aFGF in tissue and extracellular fluid surrounding focal brain wounds increase rapidly and peak within a day after injury[73]. Levels of bFGF and bFGF mRNA also appear to increase within the first day, and to remain elevated for at least a week after injury[74,75]. Studies in our laboratory using heparin-affinity high performance liquid chromatography (HPLC) coupled to specific bioassay and immunoassay techniques show a 7-fold increase in bFGF levels (and only a 1.5-fold increase in aFGF levels) in tissue surrounding focal mechanical brain wounds at one week after injury[76]. Immunohistochemical studies at this same time point show that bFGF immunoreactivity is localized largely in reactive astroglia surrounding lesions[66]. In similar studies of ischemic brain injury (stroke), we found a 2-3 fold increase in FGF levels in tissue surrounding focal infarcts occurring over the first three weeks after ischemia; at its peak, this increase was due largely to increased levels of bFGF[77]. Recent *in situ* hybridization studies by other investigators show increased bFGF mRNA expression in tissue surrounding focal mechanical brain wounds at 4-14 days after injury[74]. Recent immunocytochemical studies show that bFGF immunoreactivity is localized in invading macrophages and/or microglia at early time points (first few days) and confirm that bFGF immunoreactivity becomes localized to reactive astroglia at later time points (1 week) after injury[67,74]. Overall, these studies show increased FGF expression after brain injury. The patterns of aFGF and bFGF expression appear to be different, with aFGF levels peaking early and increased bFGF levels lasting for a week or more after injury. Secondly, in the case of bFGF, reactive glial cells appear to have the capacity to "turn on" the synthesis of this factor after injury. Finally, because of their multipotential trophic effects on CNS cells, FGFs are likely to play an important role in the cascade of cellular responses that define wound healing in the CNS, including glial and vascular proliferation, and neural sprouting and synaptogenesis[66,76].

Insights into the *in vivo* effects of FGFs in injured brain can be gained by studies of the exogenous addition of these factors in models of CNS injury. Indeed, several studies show neuronotrophic effects of FGFs on a variety of neural types in the injured brain. Application of low doses of aFGF or bFGF to the proximal stump of the transected optic nerve in rats results in the increased survival of retinal ganglion cells, presumably via retrograde transport of growth factor[78]. Intraventricular administration of bFGF in rats with fimbria-fornix transection results in rescue of septal cholinergic neurons projecting to hippocampus, as well as preservation of choline acetyltransferase levels (an index of cholinergic innervation) in hippocampus[79,80]. Implantation of bFGF-soaked gelfoam implants into the striatum of MPTP-lesioned mice results in enhanced sprouting of remaining dopaminergic fibers from injured substantia nigra, as well as increased restoration of dopamine levels and dopamine turnover in striatum[81]. Intracisternal administration of bFGF following focal cortical ischemic infarcts results in sparing of "secondary" neuronal degeneration in thalamic nuclei projecting to ischemic cortex[82]. In addition to these neuronotrophic effects, exogenously-administered bFGF has also been reported to promote glial and vascular proliferation ("angiogenesis") in rat brain[80,83].

In summary, using members of the neurotrophin and FGF growth factor families as prime examples, we have reviewed evidence of increased growth factor expression after brain injury and ischemia. Each of these factors are likely to affect different neuronal populations in brain; FGFs, in particular, may also stimulate glial and blood vessel proliferation after injury. Growth factor expression appears to proceed in two general phases after injury: an early phase, occurring before neuronal death, that may serve to protect vulnerable brain neurons; and a later phase, occurring after neural death, that contributes to wound healing and functional reorganization. Moreover, evidence has been presented that the exogenous application of growth factors in some experimental models can rescue vulnerable neurons and enhance adaptive responses to injury. Further study of these phenomena are likely to lead to new therapeutic approaches to the clinical problems of brain injury and stroke.

ACKNOWLEDGMENTS

Supported by NS-10828, AG-08207, and a Grant-in-Aid from the American Heart Association.

REFERENCES

1.   R. Levi-Montalcini and V. Hamburger. Selective growth-stimulating effects of mouse sarcoma on the sensory and sympathetic nervous system of the chick embryo. J. Exp. Zool. 116:321 (1951).
2.   R. Levi-Montalcini. The nerve growth factor 35 years later. Science 237:1154 (1987).
3.   R. Levi-Montalcini. Growth control of nerve cells by protein factor and its antiserum. Science 143:105 (1964).
4.   S. Cohen. Purification of a nerve-growth promoting protein from the mouse salivary gland and its neuro-cytoxic antiserum. Proc. Natl. Acad. Sci. (USA) 46:302 (1960).
5.   R.M. Lindsay. Nerve growth factors (NGF,BDNF) enhance axonal regeneration but are not required for survival of adult sensory neurons. J. Neurosci. 8:2394 (1988).
6.   S. Korsching, G. Auburger, R. Heumann, J. Scott and H. Thoenen. Levels of nerve growth factor and its mRNA in the central nervous system of the rat correlate with cholinergic innervation. EMBO J. 4:1389 (1985).
7.   M. Seiler and M.E. Schwab. Specific retrograde transport of nerve growth factor (NGF) from neocortex to nucleus basalis in the rat. Brain Research 300:33 (1984).
8.   F.H. Gage, K. Wictorin, W. Fischer, L.R. Williams, S. Varon and A. Bjorklund. Retrograde cell changes in medial septum and diagonal band following fimbria-fornix transection: Quantitative temporal analysis. Neurosci. 19:241 (1986).
9.   F.H. Gage, D.M. Armstrong, L.R. Williams and S. Varon. Morphological responses of axotomized septal neurons to nerve growth factor. J. Comp. Neurol. 269:147 (1988).
10.   F. Hefti. Nerve growth factor promotes survival of septal cholinergic neurons after fimbrial transections. J. Neurosci. 6:2155 (1986).
11.   L.R. Williams, S. Varon, G.M. Peterson, K. Wictorin, W. Fischer, A. Bjorklund and F.H. Gage. Continuous infusion of nerve growth factor prevents forebrain neuronal death after fimbria-fornix transection. Proc. Natl. Acad. Sci. (USA) 83:9231 (1986).

12. M.H. Tuszynski, H.S. U, D.G. Amaral and F.H. Gage. Nerve growth factor infusion in the primate brain reduces lesion-induced cholinergic neuronal degeneration. J. Neurosci. 10:3604 (1990).

13. T. Hagg, M. Manthorpe, H.L. Vahlsing and S. Varon. Delayed treatment with nerve growth factor reverses the apparent loss of cholinergic neurons after acute brain damage. Exp. Neurol. 101:303 (1988).

14. Montero and F. Hefti. Rescue of lesioned septal cholinergic neurons by nerve growth factor: Specificity and requirement for chronic treatment. J. Neurosci. 8:2986 (1988).

15. M. Fusco, B. Oderfeld-Nowak, G. Vantini, N. Schiavo, M. Gradowska, M. Zaremba and A. Leon. Nerve growth factor affects uninjured, adult rat septo-hippocampal neurons. Neurosci. 33:47 (1989).

16. R. Ishikawa, K. Nishikori and S. Furukawa. Appearance of nerve growth factor and acidic fibroblast growth factor with different time courses in the cavity-lesioned cortex of the rat brain. Neurosci. Lett. 127:70 (1991).

17. N. Rocamora, J.M. Palacios and G. Mengod. Limbic seizures induce a differential regulation of the expression of nerve growth factor, brain-derived neurotrophic factor and neurotrophin-3 in the rat hippocampus. Mol. Brain Res. 13:27 (1992).

18. O. Lindvall, P. Ernfors, J. Bengzon, Z. Kokai, M. Lis, B. Siesjo and H. Persson. Differential regulation of mRNAs for nerve growth factor, brain-derived growth factor, and neurotrophin 3 in the adult rat brain following cerebral ischemia and hypoglycemic coma. Proc. Natl. Acad. Sci. (USA) 89:648-652 (1992).

19. H. Lorez, F. Keller, G. Ruess and U. Otten. Nerve growth factor increases in adult rat brain after hypoxic injury. Neurosci. Lett. 98:339 (1989).

20. T. Shigeno, T. Mima, K. Takakura, D.I. Graham, G. Kato, Y. Hashimoto, and S. Furukawa. Amelioration of delayed neuronal death in the hippocampus by nerve growth factor. J. Neurosci. 11:2914 (1991).

21. R.T. Bartus, R.L. Dean, B. Beer and A.S. Lippa. The cholinergic hypothesis of geriatric memory dysfunction. Science 217:408 (1982).

22. J.C. Hornberger, S.J. Buell, D.G. Flood, T.H. McNeill and P.D. Coleman. Stability of numbers but not size of mouse forebrain cholinergic neurons to 53 months. Neurobiol. Aging 6:269 (1985).

23. J. Kerwin, C. Morris, A. Oakley, R. Perry and E. Perry. Distribution of nerve growth factor receptor immunoreactivity in the human hippocampus. Neurosci. Lett. 121:178 (1991).

24. W. Fischer, K. Wictorin, A. Bjorklund, L.R. Williams, S. Varon and F.H. Gage. Amelioration of cholinergic neuron atrophy and spatial memory impairment in aged rats by nerve growth factor. Nature 329:65 (1987).

25. B. Tomlinson and J. Corsellis. In "Greenfield's Neuropathology," J. Adams, J. Corsellis and L. Duchen (eds), J. Wiley & Sons, New York (1984).

26. Y.A. Barde, D. Edgar and H. Thoenen. Purification of a new neurotrophic factor from mammalian brain. EMBO J. 1:549 (1982).

27. J. Leibrock, F. Lottspeich, A. Hohn, M. Hofer, B Hengerer, P. Masiakowski, H. Thoenen and Y.A. Barde. Molecular cloning and expression of brain derived neurotrophic factor. Nature 341:149 (1989).

28. M. Hofer, S.R. Pagliusi, A. Hohn, J. Leibrock and Y.A. Barde. Regional distribution of brain-derived neurotrophic factor mRNA in the adult mouse brain. EMBO J. 9:2459 (1990).

29. C. Wetmore, Y. Cao, R. Pettersson and L. Olson. Brain-derived neurotrophic factor: Sub cellular compartmentalization and inter neuronal transfer as visualized with anti-peptide antibodies Proc. Natl. Acad. Sci. (USA) 88:9843 (1991).

30. P.C. Maisonpierre, L. Belluscio, B. Friedman, R.F. Alderson, S.J. Wiegand, M.E. Furth, R.M. Lindsay, and G.D. Yancopoulos. NT-3, BDNF, and NGF in the developing rat nervous system: Parallel as well as reciprocal patterns of expression. Neuron 5:501 (1990).

31. H.S. Phillips, J.M. Hains, G.R. Laramee, A. Rosenthal, and J.W. Winslow. Widespread expression of BDNF but not NT3 by target areas of basal forebrain cholinergic neurons. Science 250:290 (1990).

32. R.M. Lindsay, H. Thoenen and Y.A. Barde. Placode and neural crest-derived sensory neurons are responsive at early developmental stages to brain-derived neurotrophic factor. Dev. Biol. 112:319 (1985).

33. S. Thanos, M. Bahr, Y. A. Barde and J. Vanselow. Survival and axonal elongation of adult rat retinal ganglion cells: In vitro effects of lesioned sciatic nerve and brain-derived neurotrophic factor (BDNF). Eur. J. Neurosci. 1:19 (1989).

34. C. Hyman, M. Hofer, Y.A. Barde, M. Juhasz, G.D. Yancopoulos, S. Squinto and R.M. Lindsay. BDNF is a neurotrophic factor for dopaminergic neurons of the substantia nigra. Nature 350:230 (1991).

35. R.F. Alderson, A.L. Alterman, Y.A. Barde and R.M. Lindsay. Brain-derived neurotrophic factor increases survival and differentiated functions of rat septal cholinergic neurons in culture. Neuron 5:297 (1990).

36. J.K. Morse, R.F. Alderson, Y. You, N. Cai, C.A. Altar, S.J. Wiegand, and R.M. Lindsay. Brain-derived neurotrophic factor (BDNF) increases the survival of basal forebrain cholinergic neurons following a fimbria-fornix transection. Soc. Neurosci. Abstr. 18:1295 (1992).

37. P.A. Lapchak, D.M. Araujo and F. Hefti. Chronic neurotrophin administration to fimbriectomized rats: Effects on presynaptic hippocampal cholinergic function and weight gain. Soc. Neurosci. Abstr. 18:1295 (1992).

38. M. Ballarin, P. Enfors, N. Lindfors and H. Persson. Hippocampal damage and kainic acid injection induce a rapid increase in mRNA for BDNF and NGF in the rat brain. Exp. Neurol. 114:35 (1991).

39. N. Rocamora, L. Massieu, E. Boddeke, G. Mengod and J.M. Palacios. Neurotrophin-3 mRNA expression is stimulated in degenerating pyramidal cell layers of CA1 and CA4 hippocampal fields of the rat. Soc. Neurosci. Abstr. 18:1294 (1992).

40. F. Zafra, B. Hengerer, J. Leibrock, H. Thoenen and D. Lindholm. Activity dependent regulation of BDNF and NGF mRNAs in the rat hippocampus is mediated by non-NMDA glutamate receptors. EMBO J. 9:3545 (1990).

41. H.S. Phillips, J.M. Hains, M. Armanini, G.R. Laramee, S.A. Johnson and J.W. Winslow. BDNF mRNA is decreased in the hippocampus of individuals with Alzheimer's disease. Neuron 7:695 (1991).

42. A. Rosenthal, D. Goeddel, T. Nguyen, M. Lewis, A. Shih, G. Laramee, K. Nikolics and J.M. Winslow. Primary structure and biological activity of a novel human neurotropic factor. Neuron 4:767 (1990).

43. P.C. Maisonpierre, L. Belluscio, S. Squinto, N. Ip, M.E. Furth, R.M. Lindsay and G.D. Yancopoulos. Neurotrophin-3: A neurotrophic factor related to NGF and BDNF. Science 247:1446 (1990).

44. A. Hohn, J. Leibrock, K. Bailey, and Y.A. Barde. Identification and characterization of a novel member of the nerve growth factor/brain-derived growth factor family. Nature 344:339 (1990).

45. P. Ernfors, C. Ibanez, T. Ebendal, L. Olson and H. Persson. Molecular cloning and neurotrophic activities of a protein with structural similarities to nerve growth factor: Developmental and topographical expression in the brain. Proc. Natl. Acad. Sci. (USA) 87:5454 (1990).

46. B. Knusel, J.W. Winslow, A. Rosenthal, L.E. Burton, D.P. Seid, K. Nikolics and F. Hefti. Promotion of central cholinergic and dopaminergic neuron differentiation by brain-derived neurotrophic factor but not neurotrophin. Proc. Natl. Acad. Sci. (USA) 88:961 (1991).

47. M.E. Schwab, R. Kolbeck, Y.A. Barde and L. Schnell. Neurotrophin-3 increases the regenerative sprouting of the lesioned rat corticospinal tract. Soc. Neurosci. Abstr. 18:1296 (1992).

48. A. Baird, M. Klagsbrun (eds). The Fibroblast Growth Factor Family. Annals New York Acad Sci., New York, Vol. 638 (1991).

49. M. Klagsbrun. The fibroblast growth factor family: Structural and biological properties. Prog. Growth Factor Res. 1:207 (1989).

50. D. Gospodarowicz. In "Neuroimmune Networks: Physiology and Diseases." Alan R. Liss, Inc., 163 (1989).

51. M. Klagsbrun and A. Baird. A dual receptor system is required for basic fibroblast growth factor activity. Cell 67:229 (1991).

52. J.A. Abraham, A. Mergia, J.L. Whang, A. Tumolo, J. Friedman, K.A. Hjerrild, D. Gospodarowicz and J.C. Fiddes. Nucleotide sequence of a bovine clone encoding the angiogenic protein, basic fibroblast growth factor. Science 233:545 (1986).

53. R.S. Morrison, A. Sharma, J. DeVellis and R.A. Bradshaw. Basic fibroblast growth factor supports the survival of cerebral cortical neurons in primary culture. Proc. Natl. Acad. Sci. (USA) 83:7537(1986).

54. P.A. Walicke. Basic and acidic fibroblast growth factors have trophic effects on neurons from multiple CNS regions. J. Neurosci. 8:2618 (1988).

55. S.A. Lipton, J.A. Wagner, R.D. Madison and P.A. D'Amore. Acidic fibroblast growth factor enhances regeneration of processes of postnatal mammalian retinal ganglion cells in culture. Proc. Natl. Acad. Sci. (USA) 85:2388 (1988).

56. K. Unsicker, H. Reichert-Priesbsch, R. Schmidt, B. Pettman, G. Labourdette and M. Sensenbrenner. Astroglial and fibroblast growth factor have neurotrophic functions for cultured peripheral and central nervous system neurons. Proc. Natl. Acad. Sci. (USA) 84:5459 (1987).

57. I. Torres-Aleman, F. Naftolin and R.J. Robbins. Trophic effects of basic fibroblast growth factor on fetal rat hypothalamic cells: Interactions with insulin-like growth factor I. Dev. Brain Res. 52:253 (1990).

58. M.E. Hatten, M. Lynch, R.E. Rydel, J. Sanchez, J. Joseph-Silverstein, D. Moscatelli and D.B. Rifkin. *In vitro* neurite extension by granule neurons is dependent upon astroglial-derived fibroblast growth factor. Dev. Biol. 125:280 (1988).

59. G. Ferrari, M.-C. Minozzi, G. Toffano, A. Leon, and S.D. Skaper. Basic fibroblast growth factor promotes the survival and development of mesencephalic neurons in culture. Dev. Biol. 133:140 (1989).

60. C. Gensburger, G. Labourdette and M. Sensenbrenner. Brain basic fibroblast growth factor stimulates the proliferation of rat neuronal precursor cells *in vitro*. FEBS Lett. 217:1 (1987).

61. D. Gospodarowicz, S. Massoglia, J. Cheng and D.K. Fujii. Effect of fibroblast growth factor and lipoproteins on the proliferation of endothelial cells derived from bovine adrenal cortex, brain cortex, and corpus luteum capillaries. J. Cell. Physiol. 127:121 (1986).

62. B. Pettman, M. Weibel, M. Sensenbrenner and G. Labourdette. Purification of two astroglial growth factors from bovine brain. FEBS Lett. 189:102-108 (1985).

63. P.A. Eccleston and D.H. Silberberg. Fibroblast growth factor is a mitogen for oligodendrocytes *in vitro*. Dev. Brain Res. 21:315 (1985).

64. B. Pettman, G. Labourdette, M. Weibel and M. Sensenbrenner. The brain fibroblast growth factor (FGF) is localized in neurons. Neurosci. Lett. 68:175 (1986).

65. T. Janet, M. Miehe, B. Pettman, G. Labourdette and M. Sensenbrenner. Ultrastructural localization of fibroblast growth factor in neurons of rat brain. Neurosci. Lett. 80:153 (1987).

66. S.P. Finklestein, P.J. Apostolides, C.G. Caday, J. Prosser, M.F. Philips and M. Klagsbrun. Increased basic fibroblast growth factor (bFGF) immunoreactivity at the site of focal brain wounds. Brain Research 460:253 (1988).

67. F. Gomez-Pinilla, J.W.-K. Lee and C.W. Cotman. Basic FGF in adult rat brain: Cellular distribution and response to entorhinal lesion and fimbria-fornix transection. J. Neurosci. 12:345 (1992).

68. W.R. Woodward, R. Nishi, C.K. Meshul, T.E. Williams, M. Coulombe and F.P. Eckenstein. Nuclear and cytoplasmic localization of basic fibroblast growth factor in astrocytes and CA2 hippocampal neurons. J. Neurosci. 12:142 (1992).

69. N. Emoto, A.-M. Gonzalez, P.A. Walicke, E. Wada, D.M. Simmons, S. Shimasaki, and A. Baird. Basic fibroblast growth factor (FGF) in the central nervous system: Identification of specific loci of basic FGF expression in the rat brain. Growth Factors 2:21 (1989).

70. I. Tooyama, H. Akiyama, P.L. McGeer, Y. Hara, O. Yasuhara and H. Kimura. Acidic fibroblast growth factor-like immunoreactivity in brain of Alzheimer patients. Neurosci. Lett. 121:155 (1991).

71. B.J. Wilcox and J.R. Unnerstall. Expression of acidic fibroblast growth factor mRNA in the developing and adult rat brain. Neuron 6:397 (1991).

72. A. Wanaka, E.M. Johnson and J. Milbrandt. Localization of FGF receptor mRNA in the adult rat central nervous system by *in situ* hybridization. Neuron 5:267 (1990).

73. M. Nieto-Sampedro, R. Lim, D.J. Hicklin and C.W. Cotman. Early release of glia maturation factor and acidic fibroblast growth factor after rat brain injury. Neurosci. Lett. 86:361 (1988).

74. S.A. Frautschy, P.A. Walicke and A. Baird. Localization of basic fibroblast growth factor and its mRNA after CNS injury. Brain Research 553:291-299 (1991).

75. A. Logan, S.A. Frautschy and A. Baird. In "The Fibroblast Growth Factor Family." A. Baird and M. Klagsbrun (eds), New York Acad. Sci., New York, 638:474 (1992).

76. S.P. Finklestein, P.J. Fanning, C.G. Caday CG, P.P. Powell, J. Foster, E.M. Clifford and M. Klagsbrun. Increased levels of basic fibroblast growth factor (bFGF) following focal brain injury. Rest. Neurol. Neurosci. 1:387 (1990).

77. S.P. Finklestein, C.G. Caday, M. Kano, D.J. Berlove, C.Y. Hsu, M. Moskowitz and M. Klagsbrun. Growth factor expression after stroke. Stroke 21:III-122 (1990).

78. J. Sievers, B. Hausmann, K. Unsicker and M. Berry. Fibroblast growth factors promote the survival of adult rat retinal ganglion cells after transection of the optic nerve. Neurosci. Lett. 76:157 (1987).

79. K.J. Anderson, D. Dam, S. Lee and C.W. Cotman. Basic fibroblast growth factor prevents death of lesioned cholinergic neurons *in vivo*. Nature 332:360 (1988).

80. C. Barotte, F. Enclancher, A. Ebel, G. Labourdette, M. Sensenbrenner, and B. Will. Effects of basic fibroblast growth factor (bFGF) on choline acetyltransferase activity and astroglial reaction in adult rats after partial fimbria transection. Neurosci. Lett. 101:197 (1989).

81. D. Otto and K. Unsicker. Basic FGF reverses chemical and morphological deficits in the nigrostriatal system of MPTP-treated mice. J. Neurosci. 10:1912 (1990).

82. K. Yamada, A. Kinoshita, E. Kohmura, T. Sakaguchi, J. Taguchi, K. Kataoka and T. Hayakawa. Basic fibroblast growth factor prevents thalamic degeneration after cortical infarction. J. Cereb. Blood Flow Metab. 11:472 (1991).

83. M. Puumala, R.E. Anderson and F.B. Meyer. Intraventricular infusion of HBGF-2 promotes cerebral angiogenesis in the Wistar rat. Brain Research 534:283 (1990).

THE BASIC FIBROBLAST GROWTH FACTOR ISOFORMS: ENDOGENOUS AND EXOGENOUS BEHAVIOR

Natalina Quarto, Gérard Bouche, Béatrix Bugler, Catherine Chailleux,

Hervé Prats, Anne-Catherine Prats, Ana-Maria Roman, Isabelle Truchet,

and François Amalric

Laboratoire de Biologie Moléculaire Eucaryote, C.N.R.S.

118, Route de Narbonne 31062 Toulouse Cédex, France

INTRODUCTION

The fibroblast growth factors (FGFs) constitute a family of at least seven structurally related polypeptides sharing 35-55% sequence homology. The acidic FGF (FGF-1) and basic FGF (FGF-2) are the prototypes. Other members of this family include int-2 (FGF-3), K-fgf (FGF4), FGF-5, FGF-6 and KGF (FGF-7). They are potent modulators of cell function proliferation, motility, differentiation and survival. Furthermore, they play an important role in normal physiological processes such as embryonic development, angiogenesis, nervous cell system differentiation and wound-repair (Rifkin and Moscatelli, 1989; Burgess and Maciag, 1989). The FGFs exert their biological activity through a complex interaction with high and low affinity receptors. To date, four human genes have been identified each encoding a distinct high affinity receptor (Kd of 10^{-11}). Each of these genes encodes multiple proteins derived from alternative mRNA splicing (Houssaint et al., 1990; Johnson et al., 1990; Partanen et al., 1991). Despite their structural similarities, these high affinity receptors may differ in their ability to bind various members of FGF family. The heparin sulfate proteoglycans (HSPGs) are the low affinity receptors (Kd of 10^{-9}) which are essential for bFGF activity and high affinity binding (Moscatelli, 1987; Yayon et al., 1991; Rapraeger et al., 1991). Recently, syndecan, an integral membrane HSPG, has been identified as a low affinity binding site for bFGF (Kiefer et al., 1990). Basic FGF has several unique properties that differentiate it from other growth factors. One of these is the lack of a signal peptide normally required for vectorial transfer into the endoplasmic reticulum and for secretion (Abraham et al., 1986a). A second property of bFGF, shared with FGF-3, is the use of CUG codons for the initiation of translation of High Molecular Weight (HMW) isoforms, in addition to the canonic AUG codon used to generate the Low Molecular Weight bFGF (LMWbFGF) form (Florkiewicz and Sommer, 1989; Prats et al., 1989). In this paper, we will review some developments in the biology of bFGF(FGF-2), which have recently been achieved by ourselves and other authors.

Growth Factors, Peptides, and Receptors, Edited
by T.W. Moody, Plenum Press, New York, 1993

ENDOGENOUS bFGF

In vitro, numerous cell types, including fibroblasts, smooth muscle cells, granulosa cells, endothelial cells, and carcinoma cells, synthesize four bFGF isoforms of 24,23,22 and 18 kDa (Moscatelli, 1986b; Baird et al., 1986). Adult bovine aortic endothelial (ABAE) cells are the most likely of all cell types to synthesize all isoforms of bFGF, and furthermore are responsive to exogenous bFGF. ABAE cells, cultured at low density, require exogenous added bFGF to support their initial growth, while high density ABAE cell cultures are able to promote their growth without requirement of exogenous bFGF. As illustrated in Fig. 1A the growth of ABAE cells, which have been seeded at low density and in absence of bFGF, is preceded by a lag-period.

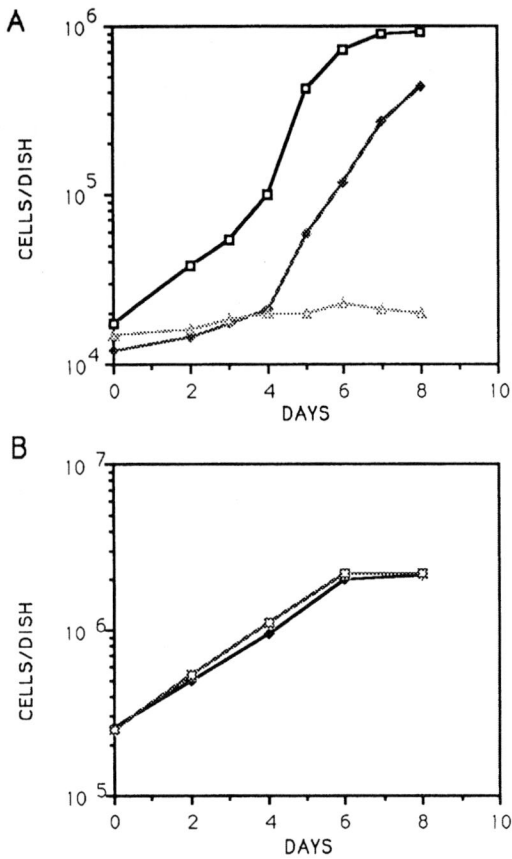

Figure 1(A). Growth curve of ABAE cells seeded at low density (2×10^4/35mm dish) in presence of 1 ng/ml bFGF (open squares) or absence, without medium change (filled squares), and with medium change (triangles) (B). Growth curve of ABAE cells seeded at high density (2×10^5/35mm dish) in presence of 1 ng/ml bFGF (open squares) or absence (filled squares).

In fact, cells start to divide after four days, and continue to grow only without changing the medium. When the medium is changed every two days, sparse cell cultures stop growing. In contrast, ABAE cells seeded at high density support their initial growth, even in absence of exogenous added bFGF. Indeed, the growth rate is similar to that observed in ABAE cells grown in presence of bFGF (Fig. 1B). These experimental observations indicate two interesting aspects: first, in ABAE cells expressing bFGF, this factor might act in an autocrine fashion; second, bFGF is released from the cells in a very inefficient manner as a consequence of the lack of a classical signal peptide.

So far, the intimate molecular mechanism by which bFGF is exported out of the cell remains unexplained, but several observations indicate that this growth factor acts extracellularly: 1) The number of high affinity receptors for a cell appears to be inversely proportional to the amount of bFGF which is expressed, suggesting a down-regulation of the cell receptors (Moscatelli, 1987; Moscatelli and Quarto, 1989). 2) Neutralizing antibodies to bFGF alter several properties of bFGF-producing cells, including morphology, growth in soft agar, plasminogen activator synthesis and cell migration (Sasada et al., 1988; Sato and Rifkin, 1988). 3) Significant amounts of bFGF are found extracellularly associated with heparin sulfate proteoglycans of the extracellular matrix (Moscatelli, 1987; Vlodavsky et al., 1987; Bashkin et al., 1989; Saksela and Rifkin, 1990).

Furthermore, Mignatti et al. (1991,1992) recently demonstrated, by studying the migration of a single isolated cell expressing bFGF that bFGF is released by the living cell via a pathway independent of the endoplasmic reticulum-Golgi complex. Thus, bFGF stimulates the cell migration via a "true" autocrine mechanism.

SUBCELLULAR LOCALIZATION OF ENDOGENOUS bFGF

Analysis of human bFGF cDNA by in vitro transcription/translation, transient COS cell expression, and constitutive expression in different cell lines, has demonstrated that the multiple bFGF forms are co-translated from a single mRNA transcript (Florkiewicz and Sommer, 1989; Prats et al., 1989). In addition, these reports suggested that the translation of the high molecular weight forms (HMW), but not the 18kDa form, initiates from CUG (leucine) codons as opposed to the classical AUG (methionine) codons.

Moreover, we have recently identified five different cis-acting elements within the bFGF mRNA leader, each having a specific effect on global or alternative translation of bFGF isoforms (Prats et al., 1992). Alternative translation could serve as post-transcriptional mechanism for regulating bFGF gene expression and sub-cellular localization.

The observation by Renko et al. (1990) that HMW forms of bFGF are found predominantly in the nucleus, while the 18kDa form which is the AUG initiated translation product, is found primarily in the cytoplasm, suggested that this alternative initiation of translation at CUG or AUG codons determines the different subcellular localization of bFGF isoforms. In light of this report, we decided to investigate whether the additional amino-terminal extention found in HMWbFGF could result in nuclear targeting (NLS). To test this hypothesis, a series of cDNA constructs encoding different chimeric (NLS) bFGF-CAT proteins was engineered and transiently expressed in COS cells by transfection techniques (Fig. 2).

Next, the subcellular distribution of the different chimeric proteins was monitored by indirect immunofluorescence. The chimeric NH_2bFGF-CAT protein containing the entire amino-terminal extension of HMWbFGF was localized strictly in the nucleus. In contrast, the chimeric protein containing the 18kDa FGF fused to CAT was strictly cytoplasmic as the wild type CAT protein. Cytoplasmic and nuclear immunofluorescence were observed in COS cells transfected with a plasmid bearing the entire bFGF reading frame fused to the reporter gene CAT (Bugler et al., 1991). Similar results have been obtained also by using different reporter genes, such as β-galactosidase or pyruvate-kinase (Quarto et al., 1991). Thus, this first set of experiments provided us with strong evidence that the NH_2-terminal extention of HMWbFGF forms accounts for their nuclear localization. So far, this 56 amino acid peptide contains nuclear localization sequences. Additional experiments carried out with chimeric proteins containing only the first 25 amino acids of the NH_2-terminal extention or the last 26 amino acid fused to CAT or β-galactosidase suggested that the nuclear localization of HMWbFGF isoforms is governed by multiple NLSs. In fact, by immunofluorescence and cell fractionation analysis, these two chimeric proteins were still detectable in the nucleus even though less accumulation was observed (N. Quarto, unpublished results). Furthermore, directed mutagenesis experiments showed that arginines at position 47 and 50 are required to provide complete translocation of HMWbFGF forms to the nucleus (B. Bugler, unpublished results). To date, no consensus nuclear localization sequence has emerged. NLS are usually a short stretch of positively charged amino acids that can be found either at the NH_2 or COOH terminus of a protein (Dingwall and Laskey, 1991). Within the amino-terminal extention of HMWbFGF isoforms there are several short sequences with homologies to the NLS of high mobility group (HMG) proteins as well as histones (Boulikas, 1987).

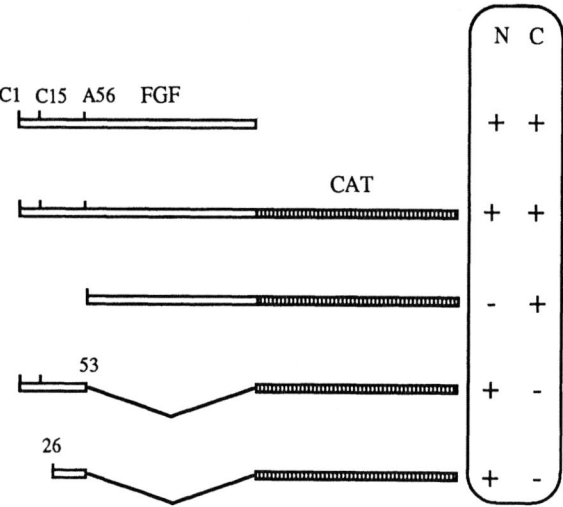

Figure 2. Subcellular localization of bFGF isoforms and bFGF-CAT chimeric proteins. Shown are the different constructs used to transfect cells.

A peculiarity of this NH_2-terminal extension is the presence of a gly-arg motif. It is of interest that this gly-arg motif is present also within the amino-terminal extention of int-2 (FGF-3) a CUG translated product, which also localizes in the nucleus (Acland, 1990). Furthermore, the gly-arg motif in the guinea pig and in human bFGF contains some methylated arginine residues (Sommer et al., 1989; Burgess et al., 1991). The occurrence of methylarginine in the sequence gly-arg has been previously described, and several of the proteins that contain these residues are found associated with nucleic acids (Boffa et al., 1987; Christensen et al., 1987; Lapeyre et al., 1986). It is possible that the gly-arg repeats ensure rapid sequestration and accumulation of HMWbFGF isoforms into the nucleus.

Puzzling observations have been made in the study concerning the subcellular localization of the LMWbFGF isoform (18 kDa). Experiments carried out by several groups using different cell lines and techniques such as cell-fractionation and indirect-immunofluorescence, have yielded observations suggesting that the subcellular localization of the endogenous LMWbFGF is primarily cytoplasmic, but a nuclear distribution of LMWbFGF has also been reported (Renko et al., 1990; Dell'Era et al. 1991; Neufeld et al., 1991; Florkiewicz et al., 1991). However, the nuclear "residency" of this isoform reflects a dynamic situation, because several findings suggest that in most cell types a shuttling of endogenous LMWbFGF between the cytoplasm and the nucleus occurs during the cell-cycle (Kardami et al., 1991; Hill and Logan, 1992).

Thus, the LMWbFGF is continuously shuttled between cytoplasm and nucleus, but does not accumulate in the nucleus. Indeed, the cytoplasmic localization might imply the presence of specific retention signals, as well as a "piggy-back" mechanism which could be responsible for the nuclear transport, leading to an apparent distribution of LMWbFGF between cytoplasmic and nuclear compartments. Further study concerning the cellular traffic of the endogenous bFGF isoforms is needed to understand the mechanism of action of these molecules.

FUNCTIONAL ROLE OF HWMbFGFs

The production of alternative forms of bFGF through the utilization of non-AUG initiation codons is a novel mechanism for the generation of growth factor diversity, and has previously ben described for the cellular proteins c-myc and int-2 (Hann et al., 1988; Acland et al., 1990).

To study the properties of the HMWbFGF isoforms, we established permanent cell lines that express selectively only the HMWbFGFs and analyzed their phenotype. This was accompanied by transfection with bFGF cDNA mutants encoding only for HMWbFGF or LMWbFGF. Results obtained from this study strongly suggest that the expression of HMWbFGFs, but not that of LMWbFGF in NIH 3T3 cells, which normally express very low levels of endogenous bFGF, resulted in a decrease in the rate of cell growth, and in a bizarre cell morphology. In fact, NIH 3T3 transfectants expressing only HMWbFGFs were growth impaired when compared with the parental cells or with NIH 3T3 cells expressing only the LMWbFGF or all bFGF isoforms. Moreover, these slow growing cells had a tendency to form multinucleated giant cells (Quarto et al., 1991). In contrast, an unbalanced expression of HMWbFGFs apparently conferred, to ABAE cells an immortalized phenotype (Couderc et al., 1991). These data strongly suggest a specific functional role for the HMWbFGF isoforms. Although the molecular mechanism for the differential effect of HMWbFGFs observed on the cells that we have studied is unknown, the nuclear localization of these isoforms could suggest a partial explanation. It is possible that HMWbFGFs affect transcription by binding specific DNA sequences and this interaction, in part, accounts for the altered phenotype of the cells. However, it is also reasonable to speculate that these nuclear bFGF isoforms interact with other molecules involved in cell cycle regulation and thereby could decrease the rate of cell division. This is currently to being investigated.

EXOGENOUS bFGF

Exogenous bFGF exerts a mitogenic response on a wide variety of cells through a dual receptor system consisting of high affinity cell-surface receptors that possess intrinsic tyrosine kinase activity and of low affinity receptors which are heparin sulfate proteoglycans (HSPGs).

Addition of bFGF (18 KDa form) to quiescent ABAE or Swiss 3T3 cells induces G1-S transition as shown by a ^{3}H-thymidine incorporation assay (Bouche et al., 1987). In order to follow the cellular uptake and the fate of internalized bFGF, iodinated 18 kDa bFGF was added the medium of sparse and confluent ABAE cell cultures. Cell-fractionation analysis revealed the presence of internalized bFGF in the cytoplasm and in the nucleus of exponentially growing cells, but not in the nucleus of confluent cells. Further, indirect immunofluorescence, using affinity purified anti-bFGF antibodies, detected a nucleolar localization of bFGF in exponentially growing cells. Table 1 shows that the amount of ^{125}I-bFGF localized in the nucleus represented 7-10% compared to that present in the cytoplasm (Bouche et al., 1987). In the cytoplasm, bFGF remained in its native 18kDa form during the first hour and afterward was converted to 16kDa form, which was slowly degraded. While in the nucleus, the 18kDa form appeared more stable (half-life of 3 hr). In cells treated with chloroquine the internalized bFGF was not degraded, suggesting that the degradation occurs through a lysosomal pathway (Amalric et al., 1991).

Table 1. Intracellular Distribution of [125]IbFGF in Exponential Growing ABAE and Swiss 3T3 Cells

Cells	Nucleus			Cytoplasm	
	pg[a]	Molecules[b]	Percent[c]	pg	Molecules
ABAE	55 ± 10	1800	8.1	610 ± 63	20400
Swiss 3T3	54 ± 12	1750	6.6	820 ± 123	26500

[a]Values (pg per 10^6 cells) calculated from specific activity of radiolabeled bFGF are the mean ± standard error of the mean from three experiments.
[b]The number of bFGF molecules per subcellular fraction corresponding to one cell was calculated using Avogadro's number and the specific activity of the [125]I bFGF.
[c]Percentage of bFGF in the nucleus.

Recently, additional evidence, supporting the correlation between nuclear translocation of bFGF and the cell-cycle has been reported. Previously we reported on the nuclear translocation of internalized bFGF during the cell cycle (Baldin et al., 1990). When quiescent G1 arrested ABAE or 3T3 cells were stimulated with exogenous bFGF, the growth factor entered the cell and accumulated into the cytoplasm continuously with a maximum in G2 between 8 and 10 hours after the stimulation. Then, after mitosis, the amount of bFGF decreased markedly. A nuclear translocation of bFGF was observed only at G1-S transition (Fig. 3A). Indeed, the fate of the internalized bFGF was quite different in the nucleus versus in the cytoplasm. A maximal nuclear localization was achieved at the beginning of the S phase, followed by a sharp decrease. The amount of nuclear bFGF was almost undetectable at the beginning of the G2 phase (Fig. 3B). These results strongly supported the idea that the intracellular distribution of bFGF is cell cycle dependent.

Similar observations have been recently reported by other investigators with endogenous bFGF. Hill and Logan (1992) have shown a cell cycle dependent localization of bFGF to the cytoplasm and nucleus in fetal growth plate chondrocytes. In these cells during late G1 a brief translocation of bFGF from the cytoplasm to the nucleus occurs.

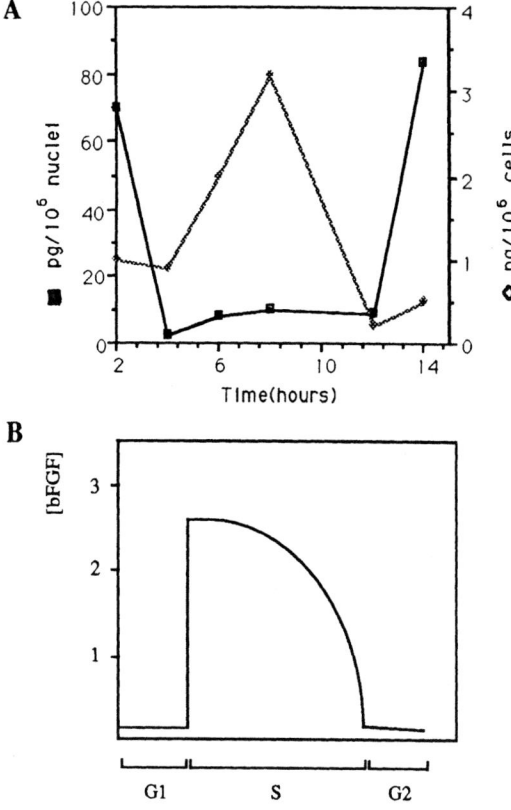

Figure 3. bFGF uptake along the cell cycle. G1 arrested ABAE cells were first stimulated with unlabeled bFGF (5 ng/ml). At different times in the cell cycle, cells were pulse-labeled with iodinated bFGF for 2 hours and harvested. After cellular fractionation, cytoplasmic and nuclear bFGF uptake was quantified. A: Filled circles, nuclei; open circles, cytoplasm. B: Quantification of exogenous bFGF recovered in nuclei during the cell cycle.

In addition, Kardami et al. (1991) observed an intense nuclear anti-bFGF staining in a fraction of interphase myocytes of proliferating cells, and this staining persisted even after the dissolution of the nuclear envelope in prophase myocytes. Thus, the pattern of subcellular localization of bFGF suggests a role in regulating the cell cycle and possibly, an involvement in the mechanism of mitotic division.

NUCLEAR TARGETS OF bFGF

The identification of specific bFGF nuclear targets represents one of the main goals to be achieved. Our previous experiments correlated the cell cycle dependent nuclear uptake of bFGF to an increased transcription of genes coding for ribosomal RNA (Bouche et al., 1987). Run-on experiments carried out on nuclei purified from quiescent sparse (G0) ABAE cells stimulated with bFGF have shown that transcription of rDNA, which is 10% in the nuclei of resting cells compared to growing cells, increased rapidly after the addition of exogenous bFGF to the cells, achieving the maximal rate after 2 hours. These experiments were paralled by in vitro experiments in which bFGF was directly added in the run-on assays. Again, an increase of rDNA transcription by 70% was observed only in nuclei purified from quiescent cells. Furthermore, addition of bFGF to the nuclei resulted in an increased phosphorylation of a subset of proteins, including the substrates of CK II.

A similar result was obtained by addition of exogenous CK II to the assay. Thus, the protein kinase II which regulates the activity of several factors involved in rDNA transcription, as well as transactivators of RNA polymerase II, might represent a potential nuclear target of bFGF. These results prompted us to investigate whether bFGF could act on rDNA transcription, either by binding to DNA in regulatory

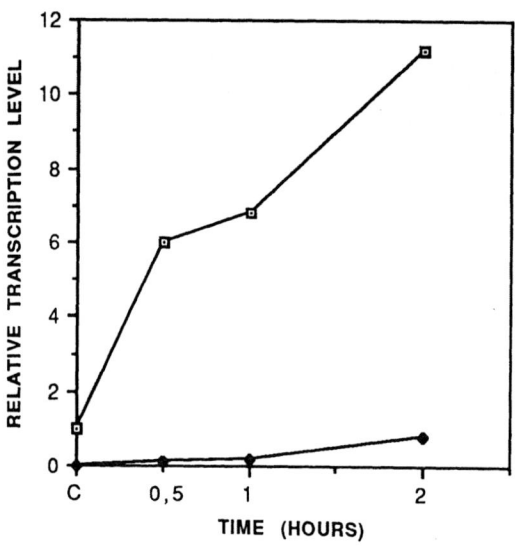

Figure 4. rDNA transcription in ABAE cells nuclei. Nuclei (3 x 10⁵) were prepared from quiescent sparse cells (C) or from cells stimulated at different times with bFGF (5 ng/ml) Run-on assays were carried out as previously described (Bouche et al., 1987). The relative transcriptional level was obtained after normalization of the densitometric scanning of autoradiograms. rDNA (filled squares); KC (open squares).

sequences or by being part of a trans-acting multimeric complex. In order to distinguish between these two possibilities we have tested for a direct interaction of bFGF and DNA. In vivo cross-linking experiments using trans-DDP, a nucleic acids-protein cross-linker, were performed on Swiss 3T3 growing cells after stimulation with ^{125}I-bFGF. Using this experimental approach, we isolated a 362 bp DNA fragment upstream from the spacer promoter of rRNA genes which binds bFGF specifically (Roman et al., submitted). At present, the functional meaning of this bFGF-DNA interaction is not yet known. It is tempting to speculate that by its binding to upstream sequence of the spacer promoter, bFGF could act directly a trans-activator of rDNA transcription, or indirectly through another component.

CONCLUDING REMARKS

Several features distinguish bFGF from other growth factors. Basic fibroblast growth factor is endowed with a peculiar structure that directs the localization and the specificity of its signalling. The presence and utilization of CUGs plus an AUG codon to co-synthesize multiple bFGF isoforms functionally diversifies the single copy gene, and probably regulates, post-transcriptionally, its expression. The different bFGF isoforms localize to different subcellular compartments: the HMW forms are nuclear, while the LMW form is primarily cytoplasmic. In addition, the selective expression of the nuclear isoforms confers a unique phenotype to the cells. As a consequence, these observations suggest that the different isoforms might have different functions dependent upon localization. Therefore, a detailed analysis and identification of specific cellular targets of HMW and LMW isoforms would be a first step towards understanding and distinguishing functionally their individual physiological significance.

The presence of nuclear localization signals (NLSs) within the HMWbFGFs and further, the lack of a classic signal peptide in the LMWbFGF imply that bFGFs may act, mainly, as an intracellular factors through an "intracrine" mechanism. Thus, the specificity of growth factor signallings is derived not only from their binding to cell-surface receptors, but also through interaction with cytoplasmic receptors or other components. However, several findings suggest that bFGF, despite the lack of a signal peptide, acts also in an autocrine and/or paracrine fashion. How bFGF is released from the cell and whether all isoforms or only the LMW form is released is not yet known.

The study of the action of exogenous bFGF on responsive cells as ABAE and 3T3 cells has provided several fascinating and attractive ideas suggesting that bFGF might have also other functions beside its growth factor activity. For example, the close relationship existing between the appearance of bFGF in the nucleus and the start of DNA replication argues in favor of a direct involvement of bFGF in DNA-replication. Moreover, the stimulation of rDNA transcription exerted by bFGF and its direct interaction with specific rDNA sequences strongly suggest that bFGF regulate directly or indirectly the transcription of some genes. Based upon all of these observations we believe that is of crucial importance to distinguish the behavior of endogenous versus exogenous bFGF and furthermore, get insight into their pathways in order to design future experiments suited to a better understanding of the molecular mechanism of these potent regulators of the cell.

ACKNOWLEDGEMENTS

This work was supported by grants from the Centre National de la Recherche Scientifique, Association pour le Dévelopment de la Recherche sur le Cancer and Foundation pour la Recherche Médicale. N. Quarto was supported by a fellowship from la Ligue National contre le Cancer.

REFERENCES

Abraham, J.A., Mergia, A., Whang, J.L., Tumolo, A., Friedman, J. Gospodarowicz, D., and Fiddes, J.C., 1986a, Nucleotide sequence of a bovine clone encoding the angiogenic protein, basic fibroblast growth factor, Science (Wash. DC) 233:545.

Acland, P., Dixon, M., Peters, G., and Dickson, C., 1990, Subcellular fate of int-2 oncoprotein is determined by choice of initiation codon, Nature 343:662.

Amlric, F., Baldin, V., Bosc-Bierne, I., Bugler, B., Couderc, B., Guyader, M., Patry, V., Prats, H., Roman, A.M., and Bouche, G., 1991, Nuclear translocation of basic fibroblast growth factor: "The Fibroblast Growth Factor Family" Ann. Acad. N.Y. Sci vol. 638.

Baird, A., Esch, F., Mormede, P., Ueno, N., Ling, N, Bohlen, P., Ying, S.Y., Wehremberg, W.B., and Guillemin, R., 1986, Molecular characterization of fibroblast growth factor: distribution and biological activities in various tissues, Recent Prog. Horm. Res. 42:143.

Baldin, W., Roman, A.M., Bosc-Bierne, I., Amalric, F., and Bouche, G., 1990, Translocation of bFGF to the nucleus is G1 phase cell cycle, EMBO J. 9:1511.

Bashkin, P., Doctrow, S., Klagsbrun, M., Svahn, C.M., Folkman, J., and Vlodavsky, I., 1989, Basic fibroblast growth factor binds to subendothelial extracellular matrix and is released by heparitinase and heparin-like molecules, Biochemistry 28:1737.

Boffa, L.C., Karn, J., Vidali, G., and Allfrey, V.G., 1977, Distribution of a NG, NG-dimethylarginine in nuclear protein fractions, Biochem. Biophys. Res. Commun. 74:969.

Bouche, G., Gas, N., Prats, H., Baldin, V., Tauber, V., Teissie, J., and Amalric, F., 1987, Basic fibroblast growth factor enters the nucleolus and stimulates the transcription of ribosomal genes in ABAE cells undergoing G0-G1 transition, Proc. Natl. Acad. Sci. USA 84:6770.

Boulikas, T., 1987, Nuclear envelope and chromatin structure, Int. Rev. Cytol. Suppl. 17:493.

Bugler, B., Amalric, G. and Prats, H., 1991, Alternative initiation of translation determines cytoplasmic or nuclear localization of bFGF, Mol. Cell. Biol. 11:543.

Burgess, W.H., and Maciag, T., 1989, The heparin-binding fibroblast growth factor family of proteins, Annu. Rev. Biochem. 58:575.

Burgess, W.H., Bizik, J., Mehlman, T., Quarto, N., and Rifkin, D.B., 1991, Direct evidence for methylation of arginine residues in high molecular weight forms of basic fibroblast growth factor, Cell Regul. 2:87.

Couderc, B., Prats, H., Bayard, F., and Amalric, F., 1991, Potential oncogenic effects of basic fibroblast growth factor requires cooperation between CUG and AUG initiated forms, Cell Regul. 2:709.

Dell'Era, P., Presta, M., and Ragnotti, G. 1991, Nuclear localization of endogenous basic fibroblast growth factor in cultured endothelial cells, Exp. Cell Res. 192:505.

Dingwall, C., and Laskey, R.A., 1991, Nuclear targeting sequences-a consensus?, TIBS 16:478.

Florkiewicz, R.Z., Baird, A., and Gonzalez, A.M., 1991, Multiple forms of bFGF: differential nuclear and cell surface localization, Growth Factors 4:265.

Florkiewicz, R.Z., and Sommer, A., 1989, The human bFGF gene encodes four polypeptides: three initiate translation from non-ATG codons, Proc. Natl. Acad. Sci. USA. 86:3978.

Hann, S.R., King, M.W., Bentley, D.L., Anderson, C.W., and Einsenman, R.N., 1988, A non-AUG translation initiation in c-myc exon 1 generates an N-terminally distinct protein whose synthesis is disrupted in Burkitt's lymphomas, Cell 52:185.

Hill, D.J., and Logan, A., 1992, Cell-cycle dependent localization of immunoreactive basic fibroblast growth factor to cytoplasm and nucleus of isolated bovine fetal growth plate chondrocytes, Growth Factors 7:215.

Houssaint, E., Blanquet, P.R., Champion-Arnaud, P., Gesnel, M.C., Torriglia, A., Courtois, Y., and Breathnach, R., 1990, Related fibroblast growth factor receptor genes exist in human genome, Proc. Natl. Acad. Sci. USA 87:8180.

Johnson, D.E., Lee, P.L., Lu, J., and Williams, L.T., 1990, Diverse forms of receptor for acidic and basic fibroblast growth factors, Mol. Cell Biol. 10:4728.

Kardami, E., Liu, L., and Doble, B.W., 1991, Basic fibroblast growth factor in cultured cardiac myocytes: "The Fibroblast Growth Factor Family", Ann. Acad. N.Y. Sci. vol. 638.

Kiefer, M.C., Stephens, J.C., Crawford, K., Okino, K., and Barr, P.J., 1990, Ligand-affinity cloning and structure of a cell surface heparin sulfate proteoglycan that binds basic fibroblast growth factor, Proc. Natl. Acad. Sci. USA 87:6985.

Lapeyre, B., Amalric, F., Ghaffari, S.H., Venkatarama Rao, S.V., Dumbar, R.S., and Olson, M.O.J., 1986, Protein and cDNA sequence of a glycin-rich, dimethylarginine containing region located near the carboxyl-terminal end of nucleolin, J. Biol. Chem. 261:9167.

Mignatti, P., Morimoto, T., and Rifkin, D.B., 1991, Basic fibroblast growth factor released by single, isolated cells stimulates their migration in an autocrine manner, Proc. Natl. Acad. Sci. USA 88:11007.

Mignatti, P., Morimoto, T., and Rifkin, D.B., 1992, Basic fibroblast growth factor, a protein devoid of secretory signal sequence, is released by cells via a pathway independent of the endoplasmic reticulum-golgi complex, J. Cell. Physiol. 151:81.

Moscatelli, D., 1987, High and low affinity binding sites for basic fibroblast growth factor on cultured cells, J. Cell. Physiol. 131:123.

Moscatelli, D., and Quarto, N., 1989, Transformation of NIH 3T3 cells with basic fibroblast growth factor or K-fgf oncogene causes down regulation of the fibroblast growth factor receptor: reversal of morphological transformation and restoration of receptor number by suramin, J. Cell Biol. 109:2519.

Partanen, J., Makela, T.P., Eerola, E., Korhonen, J., Hirvonen, H., Claesson-Welsh, L., and Alitalo, K., 1991, FGFR-4, a novel acidic fibroblast growth factor receptor with a distinct expression pattern, EMBO J. 10:1347.

Prats, H., Kaghad, H., Prats, A.C., Klagsbrun, M., Lilias, J.M., Liazun, P., Chalon, P., Tauber, P., Amalric, F., Smith, J.A., and Caput, D., 1989, High molecular mass forms of basic fibroblast growth factor are initiated by alternative CUG codons, Proc. Natl. Acad. Sci. USA 86:1836.

Quarto, N., Finger, P.F., and Rifkin, D.B., 1991, The amino-terminal extention of high molecular weight bFGF is a nuclear targeting signal, J. Cell. Physiol. 147:311.

Quarto, N., Talarico, D., Florkiewicz, R., and Rifkin, D.B., 1991, Selective expression of high molecular weight basic fibroblast growth factor confers an unique phenotype to NIH 3T3 cells, Cell Regul. 2:699.

Rapraeger, A.C., Krufka, A., and Olwin, B.B., 1991, Requirement of heparin sulfate for bFGF-mediated fibroblast growth and myoblast differentiation, Science (Wash. DC) 252:1705.

Renko, M., Quarto, N., Morimoto, T., and Rifkin, D.B., 1990, Nuclear and cytoplasmic localization of basic fibroblast growth factor species, J. Cell Physiol. 144:108.

Rifkin, D.B., Moscatelli, D., 1989, Recent developments, in the cell biology of basic fibroblast growth factor, J. Cell Biol. 109:1.

Saksela, O., and Rifkin, D.B., 1990, Release of basic fibroblast growth factor-heparin sulfate complexes from endothelial cells by plasminogen activator mediated proteolytic activity, J. Cell Biol. 110:767.

Sasada, R., Kurokawa, T., Iwana, M., and Igarashi, K., 1988, Transformation of mouse BALB/c 3T3 cells with human basic fibroblast growth factor cDNA, Mol. Cell. Biol. 8:588.

Sato, Y., and Rifkin, D.B., 1988, Autocrine activities of basic fibroblast growth factor: regulation of endothelial cell movement, plasminogen activator synthesis, and DNA synthesis, J. Cell Biol. 107:1129.

Sommer, A.D., Moscatelli, D., and Rifkin, D.B., 1989, An amino terminally extended and post-translationally modified form of a 25kD basic fibroblast growth, Biochem. Biophys. Res. Commun. 144:543.

Yayon, A., Klagsbrun, M., Esko, J.D., Leder, P., and Ornizt, D.M., 1991, Cell surface, heparin-like molecules are required for binding of basic fibroblast growth factor to its high affinity receptor, Cell 64:841.

Part III
Peptide Receptors

STRUCTURAL ANALYSIS OF LIGAND BINDING CHARACTERISTICS FOR THE BOMBESIN/GASTRIN-RELEASING PEPTIDE RECEPTOR

David H. Coy[1] and Robert T. Jensen[2]

[1]Peptide Research Laboratories
Tulane University Medical Center
New Orleans, LA 70112
[2]Digestive Diseases Branch
National Institutes of Health
Bethesda, MD 20892

INTRODUCTION

Bombesin (Bn)[1] and its mammalian counterpart, gastrin-releasing peptide (GRP)[2] have amino acid sequences which are closely related in their C-terminal regions (see Figure 1). They share a wide range of CNS and peripheral biological activities which have been recently reviewed[3]. Notable among them are increased grooming[4], decreased food intake[5], and altered hormone secretion[6] after central injection and stimulation of gastrin[7] and gastric acid release[8] and pancreatic amylase secretion[9] after peripheral injection.

Bn/GRP immunoreactivity is widely distributed throughout the GI tract[10], the brain[11], and lungs[12]. Specific binding sites for the peptides have been characterized on pancreatic acinar cells[13] and in the rat brain[14] where the distribution is markedly similar to substance P. Gastrin-releasing peptide was isolated[2] from porcine non-antral gastric tissue and was found to be a 27-residue peptide, the C-terminal seven amino acids of which are identical to bombesin. The known biological activities of GRP are thus far identical to bombesin.

Minamino et al. also isolated two peptides, named neuromedins B (NMB) and C, from pig brain and spinal fluid[15,16]. Neuromedin B contains a Leu replacement for Gln in

Bn	<D-Q-R-L-G-N-Q-W-A-V-G-H-L-M-NH$_2$
GRP(14-27)	-M-Y-P-R · · H · · · · · · · ·
Litorin	<D · · · · · · F · ·
NMB	· · L · · T · · F · ·
Phyllolitorin	<D · L · · T · S · F · ·

Figure 1. Comparison of the amino acid sequences of several Bn-like peptides.

Growth Factors, Peptides, and Receptors, Edited
by T.W. Moody, Plenum Press, New York, 1993

position 7, Thr for Val-10 and Phe for Leu-13 (see Figure 1) and thus has a distinct homology to litorin. Differences in the amino acid sequences of peptides within the same family have now led to the discovery of numerous instances of receptor subtypes with distinct ligand binding specificities - for instance, CCK/gastrin, VIP/secretin/glucagon, and the opiate peptides come to mind, and a similar situation has now been found to exist for Bn/NMB with the discovery[17] of NMB specific receptors in rat esophageal cells. NMB receptors have been localized throughout the brain, spinal column and nerve endings in the GI tract and it appears that important new physiological neuromodulatory functions for this peptide, some of which were perhaps previously attributed to Bn/GRP, will be found.

Bn/GRP also have mitogenic activity on several types of tissues and cells, including bronchial epethelial cells[18], SCLC cells[19,20], and murine Swiss 3T3 cells[21,22] and GRP appears to be an important autocrine growth factor for SCLC thus providing the rationale for this grant. Bn production from carcinogen-induced rat hepatocellular tumors has also been reported[23]. Bn receptors have also been found on PC3 and PMU23 human prostate tumor cells and Bn stimulated the growth of these cells in culture[24]. GRP receptors have also just been reported to be present in some human breast cancer cell lines[25] but not in normal breast tissue which suggests that Bn antagonists should be tested for inhibition of breast tumor growth also. Of importance in terms of a therapeutic application for Bn agonist analogues are recent reports of Bn-induced receptor down regulation in pancreatic islet cells[26] and Swiss 3T3 cells[27].

The most important feature of the Bn/GRP/NMB peptides from a therapeutic viewpoint is their secretion by and growth promoting effects on certain strains of human small cell lung carcinoma (HSCLC) and other types of tumor. They are presently perhaps the only recognized cellular growth factors which are within easy synthetic range of current chemical peptide synthesis techniques. This is reflected in the large number of synthetic analogues which have been reported and the rapid development of many potent receptor antagonists, in this and other laboratories, which have obvious therapeutic potential in blocking the autocrine functions of the endogenous peptides.

BOMBESIN/GRP STRUCTURE-ACTIVITY RELATIONSHIPS

Competitive Receptor Antagonists of Bn/GRP

The first progress in the search for competitive receptor antagonists of Bn/GRP was made in 1984 when Jensen et al.[29] reported that analogues in the "spantide" series of substance P antagonists (for instance, D-Arg-Pro-Lys-Pro-Gln-Gln-D-Trp-Phe-D-Trp-Leu-LeuNH$_2$) were also able to bind to bombesin receptors on pancreatic acini and function as weak bombesin antagonists with IC$_{50}$'s in the mM potency range. Some of these analogues were subsequently found to be capable of preventing Bn-stimulated growth of murine Swiss 3T3 cells at similar dose levels[30], however, their lack of specificity and low potency has severely limited their usefulness. We discovered a new series of antagonists based on the replacement of His[12] in bombesin with a D-Phe residue alone[31] or, preferably, in combination with a D-Phe residue in position 6. However, these also displayed relatively low binding affinity and were of little practical value. Our totally new design strategy of peptide bond rather than amino acid side-chain alteration yielded[32] the first really potent bombesin antagonist, [Leu^{14},ψ13-14]-bombesin, in which the CONH peptide bond between positions 13 and 14 was replaced with a CH$_2$NH group (often called a reduced peptide bond or ψCH$_2$NH group). This peptide exhibited an IC$_{50}$ and K$_i$ of 35, 60, and 20 nM, respectively, for inhibition of Bn-stimulated amylase release from and binding to guinea pig pancreatic acini cells and inhibition of growth of 3T3 cells, respectively. It has been used extensively and successfully as a general and specific Bn antagonist by many investigators. Replacement of the 26-27 CONH group with an ether ψCH_2O linkage also resulted in a potent antagonist in Ac-GRP(20-27)[33].

An entirely different method of producing bombesin antagonists was devised by Heimbrook et al.[34] who removed the C-terminal Met residue from GRP(20-27) to produce a weak competitive antagonist. Several series of acylated GRP(20-26) analogues with various C-terminal modifications such as esters and alkylamides were very potent antagonists[35,36].

The minimal chain length for preservation of full binding affinity for the Bn receptor encompasses Bn residues 6-14 (Asn-Gln-Trp-Ala-Val-Gly-His-Leu-Met-NH$_2$)[34]. Presently, these fragments have been converted into two principal families of very potent Bn/GRP receptor antagonist analogues[35-39] containing the two main types of C-terminal modifications. These can be represented by [D-Phe[6],Cpa[14],ψCH_2NH(13-14)]Bn(6-14), in which the C-terminal peptide bond is replaced by the reduced peptide bond, and [D-Phe[6]]Bn(6-13)OMe, in which the C-terminal amino acid is deleted and replaced by a number different groups, in this case a methyl ester. Many of these short chain analogues have K_i and IC_{50} values in the 1 nM vicinity in several assay systems. We have conducted some research into the pharmacokinetics of Bn antagonists in the rat and found[40] that the ester analogues were particularly long acting. Lipophilic modifications at the N-terminus were also very effective in further prolonging activity so that D-2,3,4,5,6-pentafluoro-Phe-Asn-Trp-Ala-Val-D-Ala-His-Leu-OMe displayed inhibitory activity up to 6 h after a single bolus i.v. injection[41] compared to 1 h or less for other typical compounds.

Two types of Bn receptor have now been cloned and characterized, one being relatively specific for Bn/GRP ligands[42,43] and the other for NMB[44]. Of great interest and some practical value has been the observation that all of the potent Bn receptor antagonists examined thus far have little or no affinity for the latter recepor[45].

Conformational Aspects of Bombesin/GRP Agonist and Antagonist Binding

The Role of the C-terminal Dipeptide Unit. A great deal of structural information has now been gathered which points to the position 13 and 14 residues as being critical for both the receptor binding and activation processes and which pinpoints certain groups in the 13-14 peptide bond and amide C-terminus as being primarily responsible for each event. Figure 2 gives a summary of some structural modifications which lead to either agonist or antagonist activity. When it was originally found that replacement of both the 9-10 and 13-14 peptide bonds (see structure C, Figure 2) CO groups in Bn with a CH_2 group produced competitive antagonists, we proposed[32] that intramolecular hydrogen bonding through these CO groups could be playing an important role in preserving a folded conformation of Bn capable of activating its receptor and that the destruction of these H-bonding points might allow a C-terminal conformational shift to occur which, whilst allowing binding, did not allow receptors to be activated. It now appears that the position 14 CO in the amide C-terminus group could be responsible for receptor triggering since its complete removal in the desMet[14] analogues also creates receptor antagonists. In the reduced peptide bond series, we proposed that free rotation around the 13-14 CH_2NH group results in incorrect positioning of the position 14 carboxamide group which does, however, manage to still

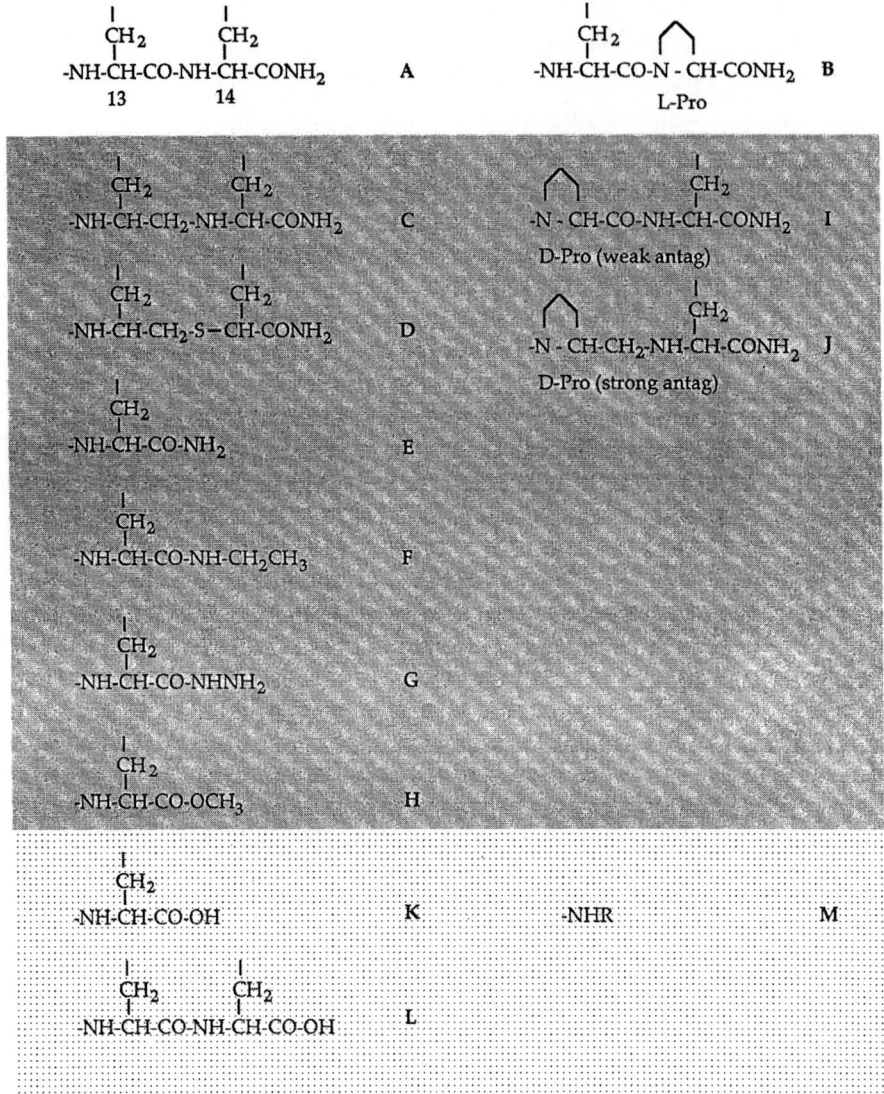

Figure 2. Effect of some C-terminal modifications on the biological properties of Bn/GRP analogues. No shading indicates normal agonist activity, dark shading antagonist activity, and light shading little or no biological activity.

retain an important role in peptide binding. Lack of involvement of the 13-14 peptide bond NH group in H-bonding is suggested by the high antagonist activity preserved with a CH_2S replacement group[46] (structure D, Figure 2) and also by the full agonist activity of a Pro^{14} analogue (structure B, Figure 2) in which the NH proton is also absent.

An important role in binding *via* possible H-bonding with the position 13 CO group in the $desMet^{14}$ analogues is indicated by the very low binding affinity of the $desMet^{14}$ free acid (structure K, Figure 2) and the increased affinities of analogues with groups that push electrons into the CO group (alkyl groups, hydrazides, esters etc.). Thus, it appears that either the position 13 or the position 14 CO group can take over receptor binding duties in the absence of the other given sufficient rotational flexibility at the C-terminus.

The presence of a conformationally restricting D-Pro residue in position 13 converts [D-Phe[6],Leu[14]]-Bn(6-14) into a new relatively weak antagonist (IC$_{50}$ on rat pancreatic acinar cells ~200 nM) (structure I, Figure 2). The same modification in a 13-14 CH$_2$NH peptide bond analogue, however, yielded some extremely potent antagonists with IC$_{50}$'s on 3T3 cells in the low nM region[47] (structure J, Figure 2). Thus, once again removal of the position 13 CO group seems to allow the chain to adopt a more favorable binding conformation in the presence of the Pro[13] alteration.

Cyclic Bn/GRP Analogues. On the basis of early analogue studies[48], it was proposed that Bn possibly bound to its receptor in a folded conformation involving a β-turn around the 10-13 residues. This, together with the additional reasoning presented above, prompted us to prepare[49] a large series of analogues in which the N- and C-termini of Bn(6-14) analogues were joined either by a disulfide bridge or a head-to-tail peptide bond. Some of these are shown in Figure 3 and it was found that significant receptor affinity and biological activity could be preserved using a wide range of cyclic analogue design strategies. These results eliminated the possibility of a functional role for the position 14 side-chain and demonstrated conclusively that the ends of the molecule were in close proximity in the receptor binding conformation. Furthermore, the enhanced potency of the D-Ala[11] cyclic peptides strongly suggested the presence of a type II' β-bend which is known to be stabilized by D-amino acids in a number of peptide hormones including GnRH, somatostatin, and opiate peptides. The smallest cyclic chain length to retain reasonable

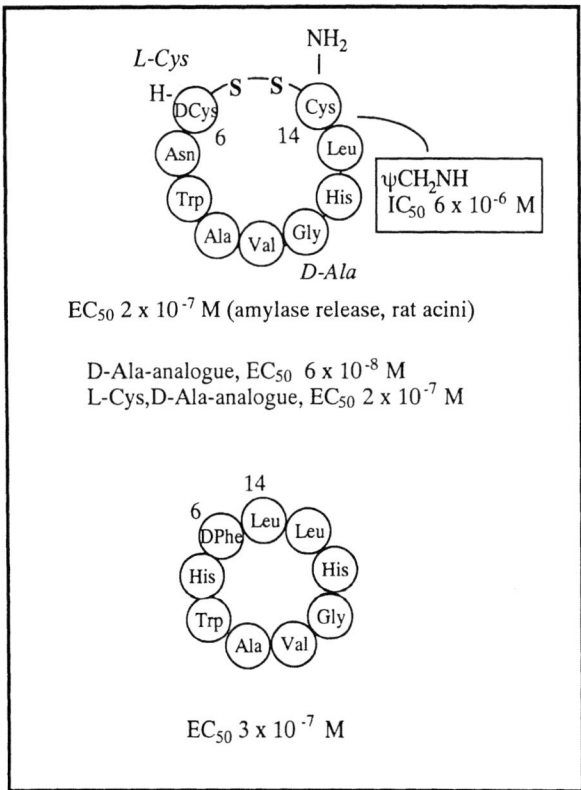

Figure 3. Main structural types of cyclic Bn/GRP analogues and biological potencies of key peptides.

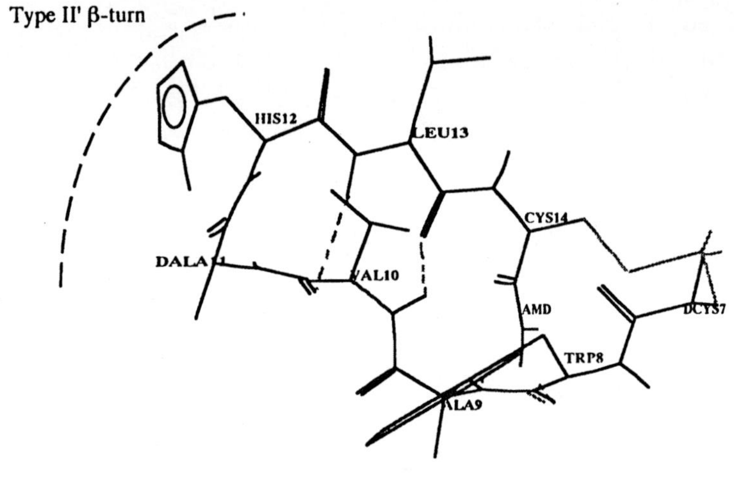

Type II' β-turn

[D-Cys^7,D-Ala^{11},Cys^{14}] Bn(7-14)   -33 Kcal

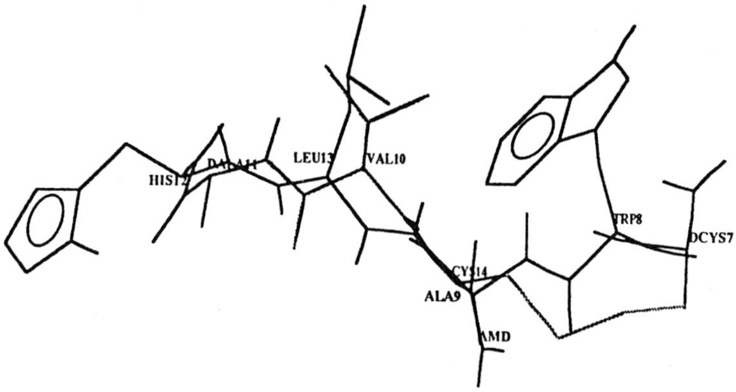

- side view

Figure 4. Computer-generated molecular model (top and side-view) of cyclic Bn octapeptide based on known somatostatin octapeptide conformation. Intramolecular H-bonds are shown by the dotted lines (left).

affinity is D-Cys-Trp-Ala-Val-D-Ala-His-Leu-Cys-NH$_2$ which has an IC$_{50}$ of 1×10^{-7} M and this analogue has been used in computer assisted molecular modeling studies employing the Tripos Assoc. SYBYL software set. The known[50] conformational aspects of the somatostatin octapeptide analogues which are folded *via* a type II' β-bend to form two antiparallel, H-bond stabilized (Leu^{13}NH - D-Ala^{11}CO; Leu^{13}CO - Val^{10}NH) β-sheets with close N and C-terminal proximity were used as a template for this octapeptide and the results are shown in Figure 4. The molecule was energy minimized (Kollman force-field method) with the interchain H-bond restraints (shown by dotted lines) in place. The striking feature of both octapeptide somatostatins and this proposed Bn structure is the formation of a largely hydrophobic surface in which hydrophobic side chains (Trp^8, Val^9 and Leu^{13}) project above the plain of the molecule (see Figure 4, side-view). These particular side chains are known to be critical for high biological potency of either agonist or antagonist structures and suggest that this surface is critical for receptor binding. The C-terminal amide (proposed earlier to be responsible for receptor activation) tends to project below the plane of the molecule on the more hydrophilic surface.

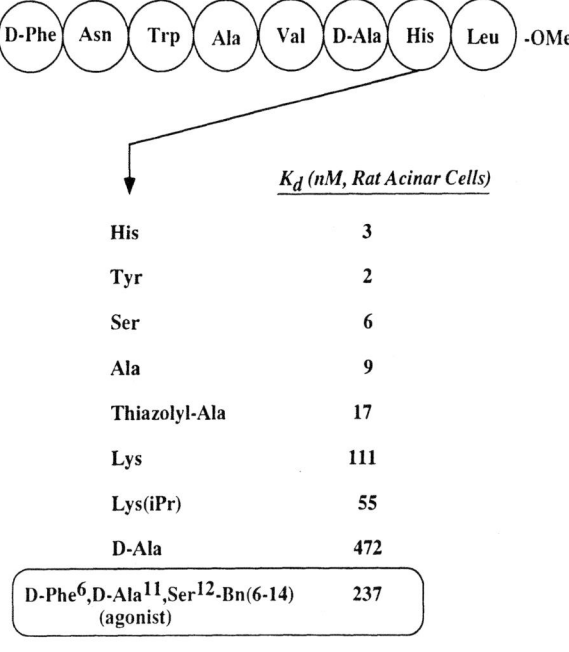

	K_d (nM, Rat Acinar Cells)
His	3
Tyr	2
Ser	6
Ala	9
Thiazolyl-Ala	17
Lys	111
Lys(iPr)	55
D-Ala	472
D-Phe[6],D-Ala[11],Ser[12]-Bn(6-14) (agonist)	237

Figure 5. Effect of various amino acid substitutions in place of His[12] in Bn antagonist and antagonist structures.

The side-chain of His[12] projects from the β-bend region of the molecule but is not really an important part of the proposed critical regions of the chain. This is partly born out by some recent studies aimed at elucidating the role of the position 12 imidazole ring in Bn/GRP which are displayed in Figure 5. His was replaced by a number of different amino acids using [D-Phe6]Bn(6-13) methyl ester, a potent Bn antagonist, as the base structure. Its replacement by Tyr, Ser, or Ala had little effect on affinity and even the substitution of amino acids with quite large side-chains of various degrees of basicity resulted in retention of some affinity. As expected from our model, the stereochemistry of this position was far more important since a D-Ala substitution resulted in more than a 100-fold loss of potency. Interestingly, the agonist structure, [D-Phe6,D-Ala11]Bn(6-14), was far less able to tolerate alterations to this position than the antagonist since Ser[12] resulted in about a 100-fold loss of affinity compared to a 2-fold loss in the latter. We feel that the difference in the SAR of the agonist and antagonist might reflect the proposed involvement of the hydrophilic surface, of which the imidazole and C-terminal amide group form a part, in agonist activation of the receptor.

CONCLUSIONS

The large number of extremely potent and receptor-specific Bn/GRP antagonists which are now available have already contributed much towards the elucidation of the many physiological roles of these important peptides. Antagonist use in the investigation of Bn/GRP physiology and pathophysiology will soon be extended to a clinical setting where inhibitory effects on Bn/GRP-secreting tumors might eventually result in new approaches to cancer therapy.

At a basic level, the elucidation of the receptor conformations of these peptides at an increasingly detailed level and with more and more conformational restraint of the backbone and side-chains should enable high resolution NMR data to be obtained and result in the design of more potent peptides and of peptide mimetic structures with improved pharmacological properties. None of the present antagonists have much affinity for the NMB receptor and we now know that NMB receptor ligands must have entirely different receptor binding conformations despite their extremely close structural homology (Figure 1). Indeed, the presence of D-Ala in position 11 of Bn/GRP or NMB greatly reduces affinity of the latter. The cyclized analogues just described have even less affinity thus emphasizing fundamental differences in receptor binding/activation mechanisms possibly involving peptide folding in a different region of the chain. The design of selective NMB agonists and of NMB competitive receptor antagonists is presently being given top priority since little is known concerning the biological roles of this peptide. Additionally, NMB also could presumably function as a tumor growth factor which could be blocked by a suitable compound.

ACKNOWLEDGEMENT

This work was supported in part by USPHS NIH grant CA-40153.

REFERENCES

1. A. Anastasia, V. Erspamer, and H. Bucci, Isolation and structure of bombesin and alytensin, two analogous active peptides from the skin of the European amphibians Bombina and Alytes, *Experientia* 27: 166 (1971).

2. T.J. McDonald, H. Jornvall, G. Nilsson, M. Vagne, M. Ghatei, S.R. Bloom, and V. Mutt, Characterization of a gastrin-releasing peptide from porcine non-antral gastric tissue, *Biochem. Biophys. Res. Commun.* 90: 227 (1979).

3. M.E. Sunday, L.M. Kaplan, E. Moroyama, W.W. Chin, and E.R. Spindel, Gastrin-releasing peptide (mammalian bombesin) gene expression in health and disease, *Lab. Invest.* 59: 5 (1988).

4. D.E. Gmerek and A. Cowan, Studies on bombesin-induced grooming in rats, *Life. Sci.* 31: 2229 (1982).

5. J. Gibbs, Effect of bombesin on feeding behavior, Life Sci. 37: 147-153 (1985).

6. M. Gunion and Y. Tache, Bombesin microinfusion into the paraventricular nucleus suppresses gastric acid secretion in the rat, *Brain Res.* 422: 118 (1987).

7. G.D. Fave, A. Kohn, L. de Magistori, M. Manuso, and C. Sparvoli, Effect of bombesin-stimulated gastrin on gastric acid secretion in man, *Life Sci.* 27: 993 (1980);

8. I.M. Modlin, C.B.H. Lamers, and J.H. Walsh, Stimulation of canine pancreatic polypeptide, gastrin, and gastric acid secretion by ranatensin, litorin, bombesin and substance P, *Reg. Peptides* 1: 279 (1981).

9. P. Deschodt-Lanckman, P. Robberecht, P. De Neef, M. Lammers, and J. Christophe, In vitro action of bombesin and bombesin-like peptides on amylase secretion, calcium efflux and adenyl cyclase activity in rat pancreas, *J. Clin. Invest.* 58: 891 (1976).

10. J.M. Polak, R. Hobbs, S.R. Bloom, E. Solcia, and A.G.E. Pearse, Distribution of a bombesin-like peptide in human gastrointestinal tract, *Lancet* 2: 1109 (1976).

11. M.R. Brown, R. Allen, J. Villarreal, J. Rivier, and W. Vale, Bombesin-like activity radioimmunologic assessment in biological tissues, *Life Sci.* 23: 2721 (1978).

12. J. Wharton, J. Polak, S. Bloom, M. Ghatei, E. Solcia, M. Brown, and A.G.E. Pearse, Bombesin immunoreactivity in the lung, *Nature* 273: 769 (1978).

13. R.T. Jensen, T. Moody, C. Pert, J.E. Rivier, and J.D. Gardner, Interaction of bombesin and litorin with specific membrane receptors on pancreatic acinar cells, *Proc. Natl. Acad. Sci. U.S.A.* 75: 6139 (1978).

14. T.W. Moody, C.B. Pert, J. Rivier, and M.R. Brown, Bombesin: specific binding to rat brain membranes, *Proc. Natl. Acad. Sci. USA* 75: 5372 (1978).

15. N. Minamino, K. Kangawa, and H. Matsuo, Neuromedin B: a novel bombesin-like peptide identified in porcine spinal cord, *Biochem. Biophys. Res. Commun.* 114: 541 (1983).

16. N. Minamino, K. Kangawa, and H. Matsuo, Neuromedin C: a bombesin-like peptide identified in porcine spinal cord, *Biochem. Biophys. Res. Commun.* 119: 14 (1984).

17. T. von Schrenk, P. Heinz-Erian, T. Moran, S.A. Mantey, J.D. Gardner, and R.T. Jensen, Potent bombesin receptor antagonists distinguish receptor subtypes, *Amer. J. Physiol.* 256: G747 (1989).

18. J.C. Willey, J.F. Lechner, and C.C. Harris, Bombesin and C-terminal tetradecapeptide of gastrin-releasing peptide are growth factors for normal human bronchial epithelial cells, *Exp. Cell Res.* 153: 245 (1984).

19. D.N. Carney, F. Cuttita, T.W. Moody, and J.D. Minna, Selective stimulation of small cell lung carcinoma clonal growth by bombesin and gastrin-releasing peptide, *Cancer. Res.* 47: 821 (1987).

20. F. Cuttita, D.N. Carney, J. Mulshine, T.W. Moody, J. Fedorko, A. Fischler, and J.D. Minna, Bombesin-like peptides can function as autocrine growth factors in human small-cell lung cancer, *Nature* 316: 823 (1985).

21. E. Rozengurt and J. Sinnet-Smith, Bombesin stimulation of DNA synthesis and cell division in cultures of Swiss 3T3 cells, *Proc. Natl. Acad. Sci. USA* 80: 2936 (1983).

22. I. Zachary and E. Rozengurt, High affinity receptors for peptides of the bombesin family in Swiss 3T3 cells, *Proc. Natl. Acad. Sci. USA* 82: 7616 (1985).

23. P.O. Seglen, H. Skomedal, G. Saeter, P.E. Schwartz, and J.M. Nesland, Neuroendocrine dysdifferentiation and bombesin production in carcinogen-induced hepatocellular rat tumours, *Carcinogenesis* 10: 21 (1989).

24. M. Bologna, C. Festuccia, P. Muzi, L. Biordi, and M. Ciomei, Bombesin stimulates growth of human prostatic cancer cells in vitro, *Cancer* 63: 1714 (1989).

25. S. Giacchetti, C. Gauville, P. de Cremaux, L. Bertin, P. Berthon, J-P. Abita, F. Cuttitta, and F. Calvo, Characterization, in some human breast cancer cell lines, of gastrin-releasing peptide-like receptors which are absent in normal breast epithelial cells, *Int. J. Cancer* 46: 293 (1990).

26. S.L. Swope and A. Schonbrunn, Desensitization of islet cells to bombesin involves both down-modulation and inhibition of receptor function, *Mol. Pharmacol.* 37: 758 (1990).

27. J.B.A. Millar and E. Rozengurt, Chronic desensitization to bombesin by progressive down-regulation of bombesin receptors in Swiss 3T3 cells, *J. Biol. Chem.* 265: 12052 (1990).

29. R.T. Jensen, S.W. Jones, K. Folkers, and J.D. Gardner, A synthetic peptide that is a bombesin receptor antagonist, *Nature* 308: 61 (1984).

30. P. Heinz-Erian, D.H. Coy, M. Tamura, S.W. Jones, J.D. Gardner, and R.T. Jensen, [D-Phe-12]bombesin analogues: a new class of bombesin receptor antagonists, *Amer. J. Physiol.* 252: G439-G442 (1987).

31. Z.A. Saeed, S.C. Huang, D.H. Coy, N-Y. Jiang, P. Heinz-Erian, S. Mantey, J.D. Gardner, and R.T. Jensen, Effect of substitutions in position 12 of bombesin on biologic activity in pancreatic acini, *Peptides* 10: 597-604 (1989).

32. D.H Coy, P. Heinz-Erian, N-Y. Jiang, Y. Sasaki, J. Taylor, J-P. Moreau, W.T. Wolfrey, J.D. Gardner, and R.T. Jensen, Probing peptide backbone function in bombesin: a reduced peptide bond analogue with potent and specific receptor antagonist activity, *J. Biol. Chem.* 263: 5056-5060 (1988).

33. W.S. Saari, D.C. Heimbrook, A. Friedman, T.W. Fisher, and A. Oliff, A gastrin-releasing peptide antagonist containing a $\psi(CH_2O)$ amide bond surrogate, *Biochem. Biophys. Res. Commun.* 165: 114 (1989).

34. D.C. Heimbrook, M.E. Boyer, V.M. Garsky, N.L. Balishin, D.M. Kiefer, A. Oliff, and M.W. Riemen, Elucidation of a novel gastrin releasing peptide antagonist by minimal ligand analysis, *UCLA Symposium On Molecular And Cellular Biology, New Series,* 86: 295 (1989).

35. D.C. Heimbrook, W.S. Saari, N.L. Balishin, A. Friedman, K.S. Moore, M.W. Riemen, D.M. Kiefer, N.S. Rotberg, J.W. Wallen, and A. Oliff, Carboxyl-terminal modification of a gastrin-releasing peptide derivative generates potent antagonists, *J. Biol. Chem.*.264: 11258-11262 (1989).

36. R. Camble, R. Cotton, A.S. Dutta, A. Garner, C.F. Hayward, V.E. Moore, and P.B. Scholes, N-Isobutyryl-His-Trp-Ala-Val-D-Ala-His-Leu-NHMe a potent in vivo antagonist analogue of bombesin/gastrin-releasing peptide derived from the C-terminal sequence lacking the final Met residue, *Life Sci.* 45: 1521 (1989).

37. D.H. Coy, J.E. Taylor, N.Y. Jiang, S.H. Kim, L.H. Wang, S.C. Huang, J-P. Moreau, J.D. Gardner, and R.T. Jensen, Short-chain pseudopeptide bombesin receptor antagonists with enhanced binding affinities for pancreatic acini and Swiss 3T3 cells display strong anti-mitotic activity, *J. Biol. Chem.* 264: 14691-14697 (1989).

38. L.H. Wang, D.H. Coy, J.E. Taylor, N-Y. Jiang, J-P. Moreau, S.C. Huang, S.A. Mantey, H. Frucht, and R.T. Jensen, Des-methionine alkylamide bombesin analogues: a new class of bombesin receptor antagonists with potent antisecretory activity in pancreatic acini and antimitotic activity in Swiss 3T3 cells, *Biochemistry* 29: 616-622 (1990).

39. L-H. Wang, D.H. Coy, J.E. Taylor, N-Y. Jiang, J-P. Moreau, S.H. Huang, H. Frucht, B.M. Haffar, and R.T. Jensen, Des-Met carboxyl-terminally modified analogues of bombesin function as potent bombesin receptor antagonists, partial agonists or agonists, *J. Biol. Chem.* 265: 15695-1570 (1990).

40. N. Alptekin, R.V. Yagci, A. Ertan, N-Y. Jiang, J.C. Rice, M. Sbeiti, W.J. Rossowski, and D.H. Coy, Comparison of prolonged in vivo inhibitory activity of several potent bombesin (BN) antagonists on BN-stimulated amylase release in the rat, *Peptides* 12: 749-75 (1991).

41. D.H. Coy, Z. Mungan, W.J. Rossowski, B.L. Cheng, J-T. Lin, J.E. Mrozinski, and R.T. Jensen, Development of a potent bombesin receptor antagonist with prolonged in vivo inhibitory activity on bombesin-stimulated amylase and protein release in the rat, *Peptides,* in press (1992).

42. E.R. Spindel, E. Giladi, P. Brehm, R.H. Goodman, and T.P. Segerson, Cloning and functional characterization of a complementary DNA encoding the murine fibroblast bombesin/gastrin-releasing peptide receptor, *Molec. Endocrinol.* 4: 1956 (1990).

43. J.F. Battey, J.M. Way, M.H. Corjay, H. Shapira, K. Kusano, R. Harkins, J.M. Wu, T. Slattery, E. Mann, and R.I. Feldman, Molecular cloning of the bombesin/gastrin-releasing peptide receptor from Swiss 3T3 cells, *Proc. Natl. Acad. Sci. USA* 88: 395 (1991).

44. E. Wada, J. Way, H. Shapira, K. Kusano, A.M. Lebacq-Verhayden, D.H. Coy, R.T. Jensen, and J. Battey, cDNA cloning, characterization and brain regio-specific expression of a neuromedin B-preferring bombesin receptor, *Neuron* 6: 421-430 (1991).

45. T. Von Schrenck, L.H. Wang, D.H. Coy, M.L. Villanueva, S. Mantey, and R.T. Jensen, Potent bombesin receptor antagonists distinguish receptor subtypes, *Amer. J. Physiol.* 259: G468-G473 (1990).

46. J.V. Edwards, B.O. Fanger, E.A. Cashman, S.R. Eaton, and L.R. McLean, Amide bond substitutions and conformational constraints applied to bombesin antagonists, in: "Proceeding of the 12th American Peptide Symposium", J.A. Smith and J.E. Rivier, eds., ESCOM, Leiden (1992).

47. F.C. Kull and J.J. Laban, International Patent PCT/GB91/01289 (1992).

48. J.E. Rivier and M.R. Brown, Bombesin, bombesin analogues and related peptides: effects on thermoregulation, *Biochemistry* 17: 1766-1771 (1978).

49. D.H. Coy, N-Y. Jiang, S.H. Kim, J-P. Moreau, J-T. Lin, H. Frucht, J-M. Qian, L-W. Wang, and R.T. Jensen, Covalently-cyclized agonist and antagonist analogues on bombesin and related peptides, *J. Biol. Chem.* 25: 16441-16447 (1991).

50. G. Van Binst and D. Tourwe, Backbone modifications in somatostatin analogues: relation between conformation and activity, *Peptide Res.* 5: 8-12 (1992).

IDENTIFICATION OF ZETA (ζ) OPIOID RECEPTOR BINDING POLYPEPTIDES IN RAT CEREBELLUM

Ian S. Zagon and Patricia J. McLaughlin

Department of Neuroscience and Anatomy
The Pennsylvania State University
College of Medicine
Hershey, PA 17033

INTRODUCTION

Growth factors are known to be crucial in dictating the course of cellular proliferation and differentiation[1-4]. Developing cells and tissues may be subjected to a host of elements that can stimulate or inhibit growth processes, ultimately serving to determine the number and type(s) of cells composing the mature organ. Nervous system development is a carefully orchestrated process that consists of a cascade of events (i.e., cell proliferation, migration, differentiation) which lead to the harmonious functioning unit vital to the organism[5]. One such event that is critical to achieving a successfully working nervous system is determination of the number of neuronal and glial cells. In humans, literally billions of cells are replicated and differentiated in a very short time, with most of the process of cellular replication occurring during prenatal life and in the early postnatal period. Indeed, by 18 months of age, the human nervous system has largely formed the neurons and many of the glia constituting the adult. Two forces are integral to determining the proper cell number. Cells must be allowed or even stimulated to proliferate, and repressed when the appropriate number has been achieved. Needless to say, if this process of cell replication malfunctions, short- and long-term damage may occur. Thus, dysfunction of these processes could lead to abnormally fewer cells which would compromise neural integration and signalling, or it could result in too many cells that may lead to neural neoplasia. While a number of growth factors are known to excite and stimulate cell proliferation, our knowledge of factors that curtail cell proliferation is extremely limited.

OPIOID GROWTH FACTOR

Endogenous opioids and opioid receptors, collectively termed endogenous opioid systems, are expressed during early life and participate in the development of the nervous system[6-8]. Endogenous opioid peptides have been implicated in a number of crucial events in neuroembryology, including cell proliferation, differentiation, and survival[7-11]. At least two different types of opioid interactions may occur during growth of the nervous system. First, opioid peptides that serve as neurotransmitters may, directly or indirectly, also have a trophic role during the ontogeny of these systems[6]. Secondly, and the focus of this discussion, opioid peptides can serve as growth factors related to cell proliferation and differentiation[7]. Early work utilizing opioid antagonists such as naltrexone and naloxone showed that continuous blockade of opioid receptors stimulated somatic and neurobiological growth in a stereospecific manner. These investigations suggested that opioid(s) played an important role in regulating developmental events and did so in a tonic manner. Following a series of extensive investigations, an opioid growth factor (OGF), [Met[5]]-enkephalin, has been identified in developing rat brain[12], mouse neural tumor cells[13], and a wide variety of eukaryotic cells and tissues[14], as well as prokaryotes[15]. This naturally occurring pentapeptide is derived from proenkephalin A and serves as an inhibitory growth factor that is especially targeted to cell proliferation, but also appears to be important in cellular differentiation and survival. Localization of OGF by light and electron microscopy reveals an association with replicating cells and developing macroneurons[16,17], and the results are consistent with a temporal and spatial expression that is related to development. In order to determine the source of opioid growth factor, in situ hybridization studies with an oligonucleotide probe for preproenkephalin (PPE) was utilized with sections of neonatal retina[18]. PPE mRNA was localized to the proliferating neuroblasts, as well as ganglion cells, suggesting that the opioid growth factor is produced in an autocrine (i.e., neuroblasts) and paracrine (i.e., ganglion cells) manner.

ZETA (ς) OPIOID RECEPTOR: HISTORY

Opioid peptides are known to interact with opioid receptors in order to elicit a response. Mediation of opioid activity by an opioid receptor is concluded when the influence of opioids is blocked by an opioid antagonist and when opioid interfacing is reliant on stereospecificity, with the (-) isomer being active and the (+) isomer inactive[19]. [Met[5]]-enkephalin, the OGF, is known to interact with a number of opioid receptor types, especially mu (μ) and delta (δ) opioid receptors when this peptide serves as a neurotransmitter. Whether these same receptors subsumed a role in growth modulation when OGF was active, was unclear. However, earlier work with ligands selective for many of the opioid receptor types showed that none of these synthetic and natural peptides known to interact with previously reported receptors had any effect on growth. Yet even when small concentrations of [Met[5]]-enkephalin had an extraordinary influence on developmental processes. These data inferred that during embryogenesis, [Met[5]]-enkephalin may act to control developmental events by way of an opioid receptor, but one that differs from those mediating neurotransmission. With this in mind, the strategy was

172

selected to use receptor binding analysis with radiolabeled OGF (i.e., [Met^5]-enkephalin) and homogenates of cerebellum from 6-day old rats[20]. Binding was found to be specific and saturable and the data were consistent with a single binding site. Scatchard analysis yielded a binding affinity (K_d) of 2.2 nM and a binding capacity (B_{max}) of 22.3 fmol/mg protein. Binding was linear with protein concentration, dependent on time, temperature and pH. Optimal binding required protease inhibitors, and pretreatment of the homogenates with trypsin markedly reduced binding, suggesting that the binding site was proteinaceous in character. The [Met^5]-enkephalin binding site was an integral membrane protein located in the nuclear fraction. Competition experiments indicated that [Met^5]-enkephalin was the most potent displacer of [^{3}H][Met^5]-enkephalin, and that binding was stereospecific. Binding to [^{3}H][Met^5]-enkephalin was not detected in the nuclear fraction of adult rat cerebellum. Thus, the functional (i.e., growth), pharmacological, biochemical, and structural/spatial characteristics, as well as the subcellular distribution (i.e., present in the nuclear fraction and not the membrane fraction as do other opioid receptors) and temporal expression of the [Met^5]-enkephalin binding site suggest the presence of a unique opioid receptor. This opioid receptor has been named from the Greek word 'zoe' (life): zeta (ς), the sixth letter of the Greek alphabet.

Recently, the ontogeny of the zeta receptor in the whole rat brain and cerebellum has been evaluated[21]. Using [^{3}H][Met^5]-enkephalin, specific and saturable binding was recorded at the earliest time sampled: prenatal day 15 (E15). In the whole brain, binding capacity was two-fold greater at E15 than at E18 and E20. The quantity of zeta receptor appeared to increase in the first postnatal week, reaching a maximum on postnatal day 8. Binding decreased the remainder of the 2nd week and between postnatal days 15 and 25 binding was no longer recorded. In the cerebellum, binding capacity increased from E20 to the 2nd postnatal week, reaching a maximum on postnatal days 8-10. The binding capacity of the zeta receptor decreased precipitously on postnatal day 11, being 5.4-fold lower than on postnatal day 10. Between postnatal days 21 and 30, no binding was observed. The binding affinities of the whole brain and cerebellum were 2.3 and 2.7 nM, respectively, and no differences between ages could be detected. Continuous opioid receptor blockade from birth to postnatal day 6 increased body weight, the binding capacity of the zeta receptor in the whole brain and cerebellum (but not the binding affinity), and the number of layers of germinal cells populating the developing cerebellum. These results defined the temporal expression of the zeta receptor in the rat brain, as well as some regulatory properties, and supports the concept that the zeta opioid receptor is primarily related to the proliferation of cells in the nervous system.

Finally, the spatial distribution of the zeta receptor during neuro-ontogeny has been examined. Studies using radiolabeled [Met^5]-enkephalin and in vitro autoradiographic observations of the developing rat retina have shown that that the neonatal, but not adult retina, has binding sites for OGF[11]. Thus, the temporal and spatial course of OGF and zeta receptor is consistent with a role in development.

IDENTIFICATION AND CHARACTERIZATION OF ZETA RECEPTOR BINDNG POLYPEPTIDES

Despite the importance of the zeta receptor in governing developmental events, little is known about its structure. To address this problem, we have combined techniques and information in protein chemistry, bio-chemistry, and pharmacology, and have utilized a method known as ligand blotting to identify zeta receptor binding subunits[22]. The basis of this technique is that proteins processed in sodium dodecyl sulfate (SDS)-polyacrylamide gel

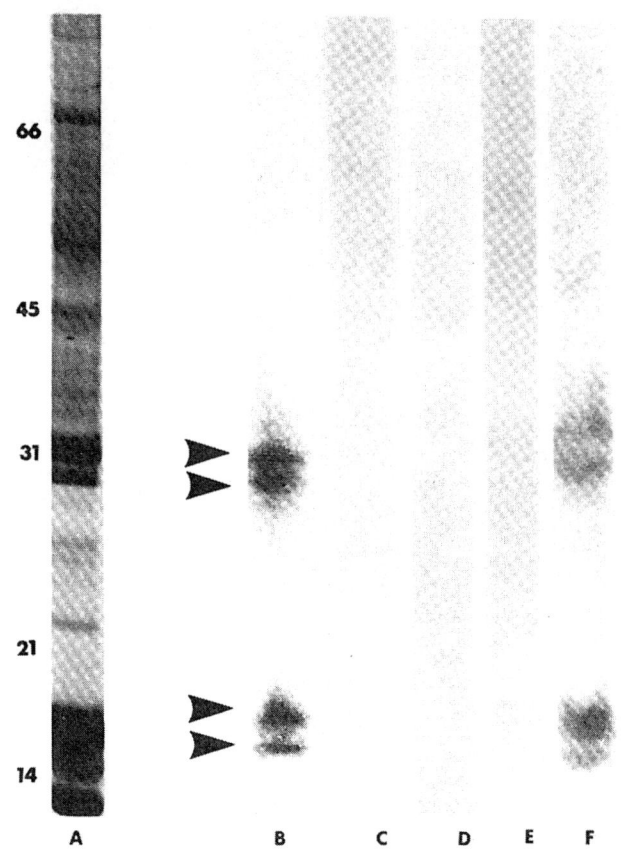

Figure 1. Cerebellar homogenates from 6-day-old (A-C,E,F) and adult (D) rats separated on SDS-PAGE (80 μg protein lane). Tissues were prepared according to earlier reports and a P0 fraction obtained by centrifugation at 39,000 x g in a solution of Tris buffer and protease inhibitors. A. Coomassie Blue stained gel showing proteins. B-F. SDS-PAGE gels electroblotted to nitrocellulose and incubated with 1.5 nM ^{125}I-[Met^5]-enkephalin (B,D) plus $10^{-6}M$ [Met^5]-enkephalin (C), $10^{-6}M$ (-) naloxone (E), or $10^{-6}M$ (+) naloxone (F). Note that binding subunits of 32, 30, 17, and 16 kD are revealed in B and F, binding was blocked by an excess of cold peptides (C) or by the (-) isomer of an opioid antagonist (E), and was stereospecific (compare E and F). In contrast to the 6-day old cerebellum (B,F), no binding was recorded in adult cerebellum (D). Molecular weights (x 10^{-3}) markers are indicated. Arrowheads show 32, 30, 17, and 16 kD polypeptides. (Reproduced from Zagon et al.[22] with permission of the publisher).

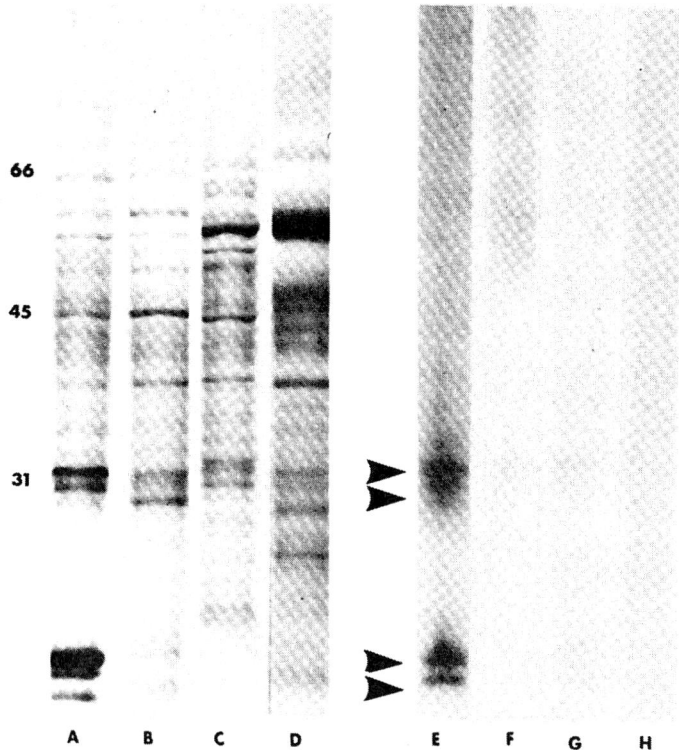

Figure 2. Subcellular fractionation of 6-day-old rat cerebellum through a sucrose density gradient, prepared for SDS-PAGE, and stained with Coomassie Blue (A-D) or electroblotted and incubated in ^{125}I-[Met^5]-enkephalin (E-H). Lanes A, E = nuclear fraction (P1) (2,200 x g); lanes B, F = membrane fraction (P2) (39,000 x g); lanes C, G = microsomal fraction (P3) (100,000 x g pellet); and lanes D, H = soluble fraction (S3) (100,000 x g supernatant). Binding of ^{125}I-[Met^5]-enkephalin to polypeptides of 32, 30, 17, and 16 kD can be observed in the autoradiograms of the P1 (E). Some faint bands of radioactivity corresponding in molecular weight to those in P1 could sometimes be detected in preparations of P3 (G). Molecular weight (x 10^{-3}) markers are indicated. Each lane contained 80 μg protein. (Reproduced from Zagon et al.[22] with permission of the publisher).

electrophoresis (PAGE) are electrotransferred to nitrocellulose paper and incubated with radiolabeled [Met^5]-enkpehalin under appropriate buffer conditions. SDS-PAGE and electrotransfer are rather harsh conditions, but in many cases the proteins can renature and display characteristics of a biological nature. Ligand blotting of one-dimensional gels of 6-day cerebellum from rats revealed four reactive bands: 32, 30, 17, and 16 kD on autoradiograms (Fig. 1). The preparations of homogenates were crude nuclear membrane fractions obtained by centrifugation and termed P0; an array of protease inhibitors was included to maintain the integrity of the proteins and prevent proteolysis. When cold [Met^5]-enkephalin was added to the incubation mixture (Fig. 1C) or when the (-) naloxone or naltrexone were included with the radiolabeled [Met^5]-enkephalin (Fig. 1E) no reaction product was visible (Fig. 1E). However, the binding subunits were apparent when 10^{-6}M (+) naloxone was

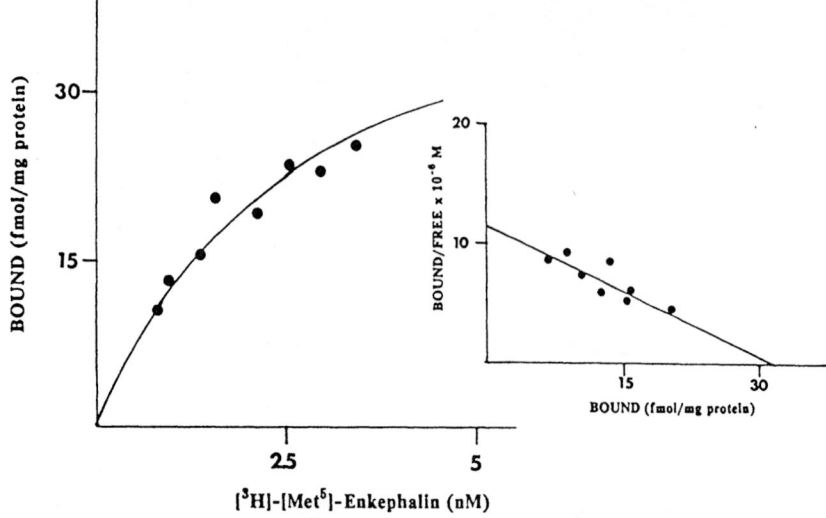

Figure 3. Representative saturation binding isotherm and Scatchard plot (inset) of the specific binding of [³H]-[Met⁵]-enkephalin to the nuclear fraction (P1) of 6-day-old rat cerebellum. (Reproduced from Zagon et al.[22] with permission of publisher).

added to the radiolabeled ligand solution (Fig. 1F). Preliminary experiments indicated that incubation times of at least 5.5 hr produced an optimal reaction, with shorter times resulting in diminished binding. No binding was visible when homogenates of adult rat cerebellum were utilized (Fig. 1D).

In order to localize the cellular disposition of the zeta receptor binding subunits, subcellular fractionation studies of 6-day old rat cerebellum in concert with ligand blotting were performed. The four binding subunits were present in autoradiograms of the nuclear fraction (P1) (Fig. 2). Little or no binding to [125]I-[Met⁵]-enkephalin could be detected in the other fractions consisting of membranes (P2), microsomes (P3), or cytosol (S3) (Fig. 2). Receptor binding assays of the four fractions showed that specific and saturable binding was only recorded in the P1 fraction (Fig. 3), with a binding affinity of 2.1 nM and a binding capacity of 27.0 fmol/mg protein recorded. Occasionally, specific but not saturable binding was detected in the P3 fraction, and 32, 30, 17, and/or 16 kD bands could sometimes be faintly observed in autoradiograms.

In order to establish even further information about the receptor subunits, two-dimensional gel electrophoresis using isoelectric focusing and SDS-PAGE along with ligand blotting of the electrotransferred proteins with [125]I-[Met⁵]-enkephalin was performed under acidic conditions. No binding to polypeptides was discovered under this protocol. However, ligand blotting of electroblotted proteins prepared with nonequilibrium pH gel electrophoresis (NEPHGE) and SDS-PAGE under basic conditions showed the four binding polypeptides (Fig. 4). The 32, 30, 17, and 16 kD binding subunits migrated to a pH of 7.8, 8.0, 8.4, and 8.5, respectively.

The relationship between the four binding polypeptides was explored by utilizing two-dimensional tryptic peptide mapping analysis and ^{125}I-Bolton Hunter labeled polypeptides

NEPHGE

7.0 ⟶ 9.0

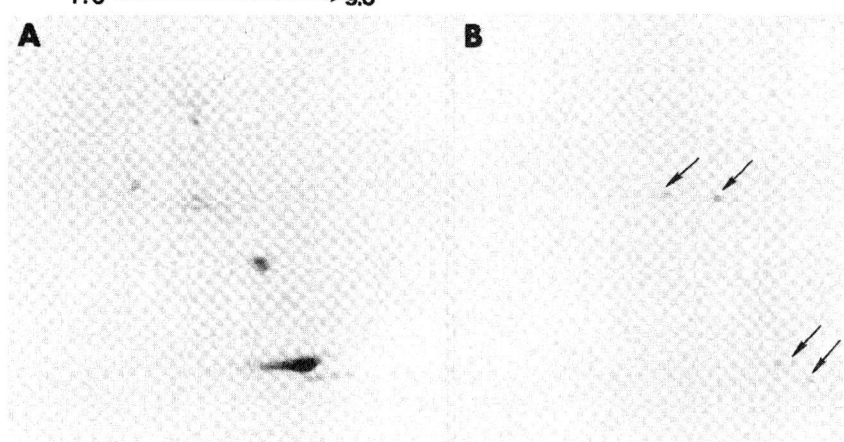

Figure 4. Preparations of P1 from the 6-day old rat cerebellum evaluated by two-dimensional nonequilibrium pH gel electrophoresis (NEPHGE)-SDS-PAGE and stained with Coomassie Blue (A) or electrotransferred to nitrocellulose and ligand blotted with [125]I-[Met[5]]-enkephalin (B). The position of 32, 30, 17, and 16 kD subunits of the zeta receptor are denoted by arrows. Molecular weight (x 10^{-3}) markers are indicated. (Reproduced from Zagon et al.[22] with permission of the publisher).

(Fig. 5). The 32 and 30 kD polypeptides were nearly identical, having a spot homology of >95%. Only a partial homology (~60% spot overlap) was detected between the 17 and 16 kD binding subunits, and each of these polypeptides had a spot homology of ~40% with the 32/30 subunits.

DISCUSSION

The zeta opioid receptor is related to the growth of the nervous system, mediating the action of opioid growth factor, [Met[5]]-enkephalin. The present results enhance our understanding of the structure of the zeta receptor. Four binding subunits of molecular weight 32, 30, 17, and 16 kD are demonstrated, all having a basic isoelectric point. Subcellular fractionation studies revealed that these binding polypeptides are located in the nuclear fraction, confirming earlier reports utilizing receptor binding analysis[20]. Peptide mapping of the binding subunits suggests that the 32 and 30 kD polypeptides are closely related, suggesting that the 30 kD subunit may be either a proteolytic fragment or alternately spliced form of the 32 kD polypeptide. Moreover, the data indicate that the 32 and 30 kD subunits do not have substantial spot overlap with the 17 and 16 kD subunits. Finally, the data show that the 17 and 16 kD polypeptides differ from each other.

The technique employed to identify the structural basis of the zeta opioid receptor, termed ligand blotting, is dependent on the renaturation of the polypeptides related to

Figure 5. Comparison of the binding subunits of the zeta receptor by two-dimensional α-chymotryptic mapping analysis of polypeptides radioiodinated within the gel slice with ^{125}I-Bolton Hunter reagent. The chymotryptic maps of the 32 and 30 kD subunits are identical, but differ from the 17 and 16 kD polypeptides. The 17 and 16 kD subunits exhibited only partial homology. (Reproduced from Zagon et al.[22] with permission of the publisher).

the receptor. Evidence that the renatured binding subunits were associated with the zeta receptor came from a number of factors. First, the binding affinity of the zeta receptor in the developing cerebellum is in the range of 2-3 nM[20,21], and this peptide concentration is known to have a profound influence on DNA synthesis[12,13]. The concentration of radiolabeled OGF utilized in the ligand blotting assays was 1.5 nM, indicating that the interaction of the renatured subunits with the ligand was physiologically relevant. Second, the ligand blotting reaction was consistent with two hallmarks of opioid receptors: displacement by opioid antagonists and stereospecificity[19]. For example, (-) naloxone but not (+) naloxone blocked the binding of [125]I-[Met[5]]-enkephalin. Third, the specificity of the relationship of the ligand to the binding subunits was demonstrated by the blocking of reaction with the addition of cold [Met[5]]-enkephalin. Fourth, in keeping with earlier results showing the temporal relationship of zeta receptor expression[20,21], binding polypeptides were recorded in the developing but not adult cerebellum. Thus, the biological, pharmacological, and biochemical characteristics of the binding subunits appear to be consonant with the detection of the binding polypeptides associated with the zeta opioid receptor.

ACKNOWLEDGEMENTS

The work summarized in this chapter was supported by NIH grant NS-20500.

REFERENCES

1. Dalsgaard, C.-J., A. Hultgardh-Nilsson, A., A. Haegerstgrand, and J. Nilsson, Neuropeptides as growth factors. Possible roles in human diseases, *Regulatory Peptides* 25:1 (1989).
2. Hollenberg, M.D., Growth factors, their receptors and development, *Amer. J. Med. Genetics* 34:35 (1989).
3. Keski-Oja, J. and H.L. Moses, Growth inhibitory polypeptides in the regulation of cell proliferation, *Med. Biol.* 65:13 (1987).
4. Mercola, M. and C.D. Stiles, Growth factor superfamilies and mammalian embryogenesis, *Development* 102:451 (1988).
5. Jacobson, M. "Developmental Neurobiology", 3rd edition, Plenum Press, New York (1991).
6. Loughlin, S.E. and F.M. Leslie, Opioid receptors and the developing nervous system, *in:* "Receptors in the Developing Nervous System," I.S. Zagon and P.J. McLaughlin, eds., Chapman and Hall, London, in press.
7. Zagon, I.S. and P.J. McLaughlin, Opioid growth factor in the developing nervous system, *in:* "Receptors in the Developing Nervous System," I.S. Zagon and P.J. McLaughlin, eds., Chapman and Hall, London, in press.
8. McDowell, J. and I. Kitchen, Development of opioid systems: peptides, receptors and pharmacology, *Brain Res. Rev.* 12:397 (1987).
9. Meriney, S.D., M.J. Ford, D. Oliva, and G. Pilar, Endogenous opioids modulate neuronal survival in the developing avian ciliary ganglion, *J. Neurosci.* 11:3705 (1991).
10. Hauser, K.F., P.J. McLaughlin, and I.S. Zagon, Endogenous opioid systems and the regulation of dendritic growth and spine formation, *J. Comp. Neurol.* 281:13 (1989).
11. Isayama, T., P.J. McLaughlin, and I.S. Zagon, Endogenous opioids regulate cell proliferation in the retina of developing rat, *Brain Res.* 544:79 (1991).
12. Zagon, I.S. and P.J. McLaughlin, Identification of opioid peptides regulating proliferation of neurons and glia in the developing nervous system, *Brain Res.* 542:318 (1991).
13. Zagon, I.S. and P.J. McLaughlin, Endogenous opioid systems regulate growth of neural tumor cells in culture, *Brain Res.* 490:14 (1989).
14. Zagon, I.S., R.E. Rhodes, and P.J. McLaughlin, Localization of enkephalin immunoreactivity in diverse tissues and cells of the developing and adult rat, *Cell Tissue Res.* 246:561 (1986).
15. Zagon, I.S. and P.J. McLaughlin, An opioid growth factor regulates the replication of microorganisms, *Life Sci.* 50:1179 (1992).
16. Zagon, I.S., R.E. Rhodes, and P.J. McLaughlin, Distribution of enkephalin immunoreactivity in germinative cells of developing rat cerebellum, *Science* 227:1049 (1985).
17. Zagon, I.S. and P.J. McLaughlin, Ultrastructural localization of enkephalin-like immunoreactivity in developing rat cerebellum, *Neuroscience* 34:479 (1990).
18. Isayama, T. and I.S. Zagon, Localization of preproenkephalin A mRNA in the neonatal rat retina, *Brain Res. Bull.* 27:805 (1991).

19. Pert, C.B. and S.H. Snyder, Opiate receptor: demonstration in nervous tissue, *Science* 179:1011 (1973).

20. Zagon, I.S., D.M. Gibo, and P.J. McLaughlin, Zeta (ς), a growth opioid receptor in developing rat cerebellum: identification and characterization, *Brain Res.* 551:28 (1991).

21. Zagon, I.S., D.M. Gibo, and P.J. McLaughlin, Ontogeny of zeta (ς), the opioid growth factor receptor, in the rat brain, *Brain Res.*, in press.

22. Zagon, I.S., S.R. Goodman, and P.J. McLaughlin, Zeta (ς), the opioid growth factor receptor: identification and characterization of binding subunits, *Brain Res.*, in press.

IDENTIFICATION AND CHARACTERIZATION OF SOMATOSTATIN (SRIF), GASTRIN RELEASING PEPTIDE (GRP), AND NEUROMEDIN B (NMB) RECEPTORS ON ESTABLISHED TUMORS AND TUMOR CELL LINES

John E. Taylor

Biomeasure Inc.
9-15 Ave. E
Hopkinton, MA 01748

INTRODUCTION

Since its initial isolation in 1973 and characterization as an inhibitor of pituitary growth hormone (GH) secretion, the observed biological actions of the tetradecapeptide, somatostatin (SRIF-14) have rapidly proliferated.to include a number of non-pituitary activities such as inhibition gastrointestinal endocrine and exocrine secretion, gastrointestinal motility and blood flow, and central nervous system activity (Reichlin, 1983a, 1983b). Several recent lines of evidence also suggest that in addition to having an indirect, systemic effect on growth related process mediated by GH, SRIF-14 may have direct effects on cell proliferation in numerous types of neoplastic tissues (De Feudis and Moreau, 1986; Moreau and De Feudis, 1987; Lamberts et al., 1987; Lamberts, 1988; Parmer, 1989; Lamberts et al., 1991). For example, we initially reported that the *in vivo* growth of the rat prostate tumor, R-3327, and the *in vitro* proliferation and *in vivo* growth of the NCI-H69 human small cell lung carcinoma cell line (SCLC), were inhibited by the potent octapeptide analogue of SRIF-14, BIM-23014 (3-(2-naphthyl)-D-Ala-cyclo[Cys-Try-D-Trp-Lys-Val-Cys]-Thr-NH$_2$) (Murphy et al., 1987; Taylor et al., 1988a, 1988b). Further evaluation of this phenomenom showed that the *in vivo* growth of several other tumor lines and solid tumors (prostate, breast, hepatic, chondrosarcoma, neuroblastoma) was retarded by chronic BIM-23014 administration (Bogden et al., 1988; Bogden at al., 1990a, 1990b). In some cases, the termination of SRIF analogue treatment, resulted in tumor regrowth suggesting that the antitumor activity of SRIF peptides is oncostatic rather than oncolytic in nature.

In addition, a number of studies have also demonstrated that gastrin-releasing peptide (GRP), and possibly neuromedin B (NMB), which are mammalian members of the bombesin-family of peptides, can function as autocrine or paracrine growth factors for SCLC proliferation (Moody et al., 1983; Cuttitta et al., 1985; Carney et al, 1991, 1992; Cardona et

Growth Factors, Peptides, and Receptors, Edited
by T.W. Moody, Plenum Press, New York, 1993

al., 1991; Corjay et al., 1991; Woll, 1991a, 1991b), and that GRP antagonists have antiproliferative activity against various tumors *in vitro* and *in vivo* (Staley et al., 1991; Mahmoud et al., 1991; Thomas et al, 1991; Bogden et al., 1992; Davis et al., 1992).

It is evident that a knowledge of the presence of peptide growth factors (e.g., GRP/NMB), anti-growth factors (e.g., SRIF), and their respective receptors in various other tumors, in addition to SCLC, may lead to a more rational approach to the development and clinical application of peptides for the diagnosis and treatment cancer. In this regard, we have undertaken a comprehensive study to examine a number of established tumors and tumor cell lines for the presence of SRIF and GRP/NMB receptors employing $[^{125}I\text{-}Tyr^{11}]$SRIF-14 and $[^{125}I\text{-}Tyr^4]$bombesin as the receptor probes, respectively.

RESULTS

Several types (lung, breast, prostate, pancreatic, hepatic, skin, glial) of established tumors or tumor cell lines have been examined in this study, and are described in Table 1. In addition, two fibroblast cell lines, which respond to various neuropeptides have been included for comparison.

$[^{125}I\text{-}Tyr^{11}]$SRIF-14 Receptor Binding

With respect to $[^{125}I\text{-}Tyr^{11}]$SRIF-14 binding, the rat acinar pancreatic cell line, AR42J, exhibited the highest concentration of SRIF receptor sites (Table 2). The SCLC cell lines, NCI-H69 and NCI-H345, and the R-3327 androgen-independent prostate tumor, were

Table 1. Tumor Characteristics

Code	Description	Type
SCLC NCI-H69	human small cell lung carcinoma (classic),	*in vitro* cell line
SCLC NCI-H345	human small cell lung carcinoma (classic),	*in vitro* cell line
SCLC NCI-N417	human small cell lung carcinoma (variant),	*in vitro* cell line
NSCLC A549:	human non-small cell lung carcinoma,	*in vitro* cell line
AR42J	rat pancreatic (acinar) carcinoma	*in vitro* cell line
MCF-7	human estrogen-dependent breast carcinoma	*in vitro* cell line
C6	rat glioma	*in vitro* cell line
B16	mouse melanoma	*in vitro* cell line
3T3	mouse fibroblast	*in vitro* cell line
WI-38	human fetal fibroblast	*in vitro* cell line
R-3327	rat androgen-sensitive prostate carcinoma	solid tumor
2PR-121D(1)/S	rat androgen-sensitive prostate carcinoma	solid tumor
2PR-121D(1)/R	rat androgen-resistant prostate carcinoma	solid tumor
M 5123	rat hepatoma	solid tumor

Table 2. Somatostatin Receptor Concentration: [^{125}I-Tyr^{11}]SRIF-14 Receptor Binding

Tumor	Bmax (fmol/mg protein)
AR42J Pancreas	231 ± 20
SCLC NCI-H69	129 ± 8
SCLC NCI-H345	83 ± 20
R-3327 Prostate	54 ± 20
SCLC NCI-H209	15 ± 4
2PR-121(1)D/R Prostate	12 ± 1.2
2PR-121(1)D/S Prostate	4.4 ± 1.1
B16 Melanoma	barely detectable
SCLC NCI-N417	not detectable
NSCLC A549	non detectable
MCF-7 Breast	not detectable
C6 Glioma	not detectable

Membranes for the SRIF receptor binding assay were obtained by homogenizing the cells or tumor samples (Polytron, setting 6, 15 sec) in ice-cold 50 mM Tris-HCl and centrifuging twice at 39,000 x g (10 min), with an intermediate resuspension in fresh buffer. The final pellets were resuspended in 10 mM Tris-HCl for assay. Aliquots of the crude membrane preparation were incubated for 25 min at 30 ° C with [^{125}I]SRIF(Tyr^{11})-14 (2000 Ci/mmol, Amersham Corp.) in 50 mM HEPES (pH 7.4) containing bovine serum albumin (10 mg/ml; fraction V, Sigma Chemical Co.), MgCl$_2$ (5 mM), Trasylol (200 KIU/ml), bacitracin (0.02 mg/ml), and phenylmethylsulphonyl fluoride (0.02 mg/ml). The final assay volume was 0.3 ml. The incubations were terminated by rapid filtration through GF/C filters (pre-soaked in 0.3% polyethylenimine). Each tube and filter were then washed three times with 5 ml aliquots of ice-cold buffer. Specific binding was defined as the total [^{125}I]SRIF(Tyr^{11})-14 bound minus that bound in the presence of 200 nM SRIF-14.

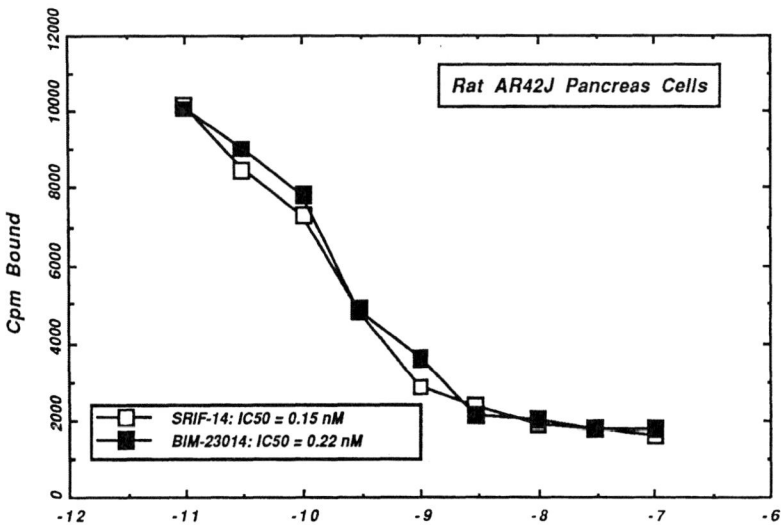

Figure 1. *In Vitro* Inhibition of [^{125}I-Tyr^{11}]SRIF-14 Binding

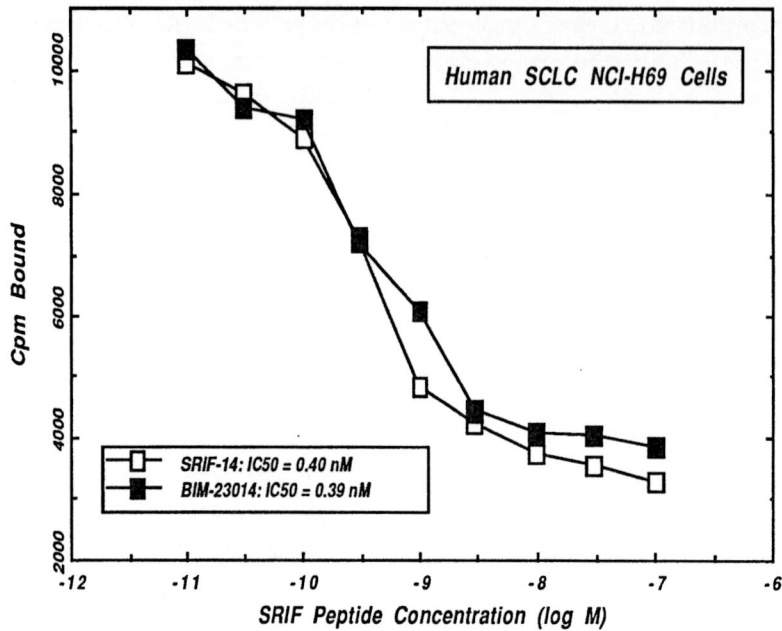

Figure 2. *In Vitro* Inhibition of [^{125}I-Tyr^{11}]SRIF-14 Binding

also highly enriched and have been shown to be responsive to the antiproliferative actions of SRIF peptides (Murphy et al., 1987; Siegel et al., 1988; Bogden at al., 1990a, 1990b). Several representative binding curves are shown in Figures 1 and 2. Interestingly, the androgen-independent subline of the rat 2PR-121D(1) prostate tumor was approximately three times more enriched in SRIF receptors than the dependent subline, indicating that androgen resistant prostate tumors may be more amenable to the antitumor activity of SRIF peptides.

The growth of some tumors (e.g., SCLC NCI-N417, B16 melanoma) can be effectively retarded by the SRIF analogue BIM-23014 (Bogden at al., 1990b, 1991, but exhibit no or very little [^{125}I-Tyr^{11}]SRIF-14 binding. We were also unable to demonstrate detectable binding to MCF-7 breast tumor cells, although other investigators have observed a more direct correlation between binding to breast carcinomas and tumor inhibition (Nelson et al., 1989; Setyono-Han et al., 1987; Papotti et al., 1989; Weber et al., 1989).

[^{125}I-Tyr^4]Bombesin

[^{125}I-Tyr^4]Bombesin binding was also detected in the majority of the cell lines and tumors examined, with the AR42J pancreas and fibroblast cell lines being the most highly enriched (Table 3). Significant binding was also observed in the C6 glioma and androgen-resistant 2PR-121D(1) prostate tumor. Low receptor levels were detected in the SCLC line, NCI-H345, MCF-7 breast, androgen-sensitive 2PR-121D(1) prostate, and M 5123 hepatoma. Further characterization of the GRP/NMB selectivity of [^{125}I-Tyr^4]bombesin binding indicated that the C6 glioma was of the NMB subtype, SCLC NCI-H345 cells exhibited mixed NMB/GRP binding, and the remainder of the tumors examined were highly selective for the GRP receptor (Table 4). Representative competition curves for the NMB type C6

Table 3. Bombesin Receptor Concentration: [^{125}I-Tyr^4]Bombesin Receptor Binding

Tumor	Bmax (fmol/mg protein)
AR42J Pancreas	8915 ± 227
3T3 Fibroblasts	913 ± 96
WI-38 Fibroblasts	574 ± 218
C6 Glioma	352 ± 75
2PR-121D(1)/R	262 ± 16
SCLC NCI-H345	42 ± 12
MCF-7 Breast	26 ± 8
2PR-121D(1)/S	10 ± 0.5
M 5123 Hepatoma	3.3 ± 0.7
SCLC NCI-H69	not detectable
SCLC NCI-N417	not detectable
NSCLC A549	not detectable

Membranes for the GRP receptor binding assay were obtained by homogenizing cultured cells or tumor samples (Polytron, setting 6, 15 sec) in ice-cold 50 mM Tris-HCl (Buffer A) and centrifuging (twice at 39,000 x g (10 min), with an intermediate resuspension in fresh buffer. The final pellets were resuspended in the 50 mM Tris-HCl, containing 0.1 mg/ml bacitracin, and 0.1% BSA (Buffer B), and held on ice for the receptor binding assay. For assay, aliquots (0.4 ml) were incubated with 0.05 ml [^{125}I-Tyr^4]bombesin (~2200 Ci/mmol, New England Nuclear) and Buffer B, with and without 0.05 ml of unlabeled competing peptides. After a 30 min incubation (4 °C), the bound [^{125}I]-Tyr^4]bombesin was separated from the free by rapid filtration through GF/B filters which had been previously soaked in 0.1% polyethyleneimine. The filters were then washed three times with 5 ml aliquots of ice-cold Buffer A. Specific binding was defined as the total [^{125}I-Tyr^4]bombesin bound minus that bound in the presence of 1 μM unlabeled bombesin.

Table 4. Inhibition of [^{125}I-Tyr^4]Bombesin Receptor Binding: GRP/NMB Receptor Selectivity

Tumor	GRP Ki (nM)[1]	NMB Ki (nM)[1]	NMB/GRP Ratio	Selectivity
AR42J Pancreas	2.1	17	8.1	GRP
3T3 Fibroblasts	5.3	48	9.1	GRP
WI-38 Fibroblasts	1.5	308	205	GRP
C6 Glioma	23	5.0	0.22	NMB
2PR-121D(1)/R	2.5	34	14	GRP
2PR-121D(1)/S	4.0	64	16	GRP
SCLC NCI-H345	2.6	3.6	1.3	GRP/NMB
MCF-7 Breast	2.5	216	86	GRP
M 5123 Hepatoma	14	249	18	GRP

[1]Ki values are the means of 2-4 determinations.

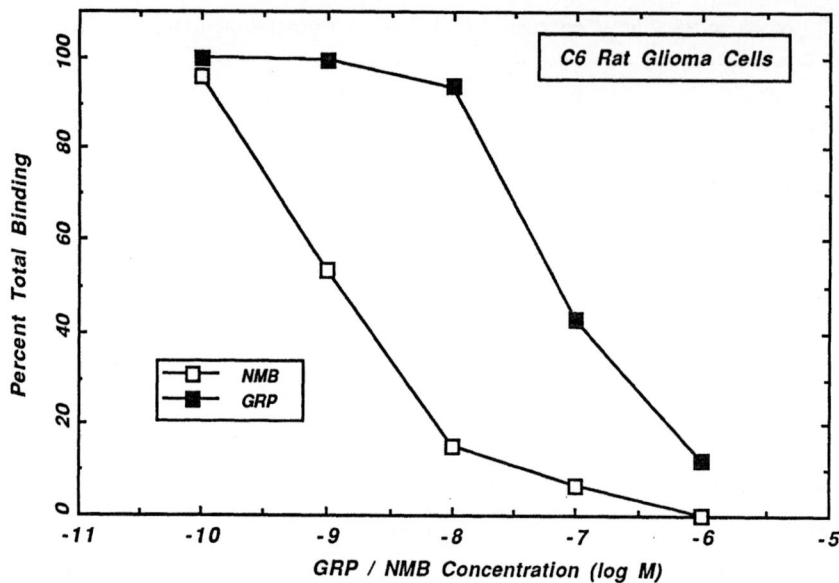

Figure 3. *In Vitro* Inhibition of [^{125}I-Tyr^4]Bombesin Binding: GRP / NMB Selectivity

glioma and GRP type M 5123 hepatoma are shown in Figures 3 and 4. Comparison of the receptor data with *in vivo* tumor proliferation experiments (Bogden et al., 1992) with have shown that GRP-specific antagonists variably inhibit the growth of the GRP receptor-positive tumors

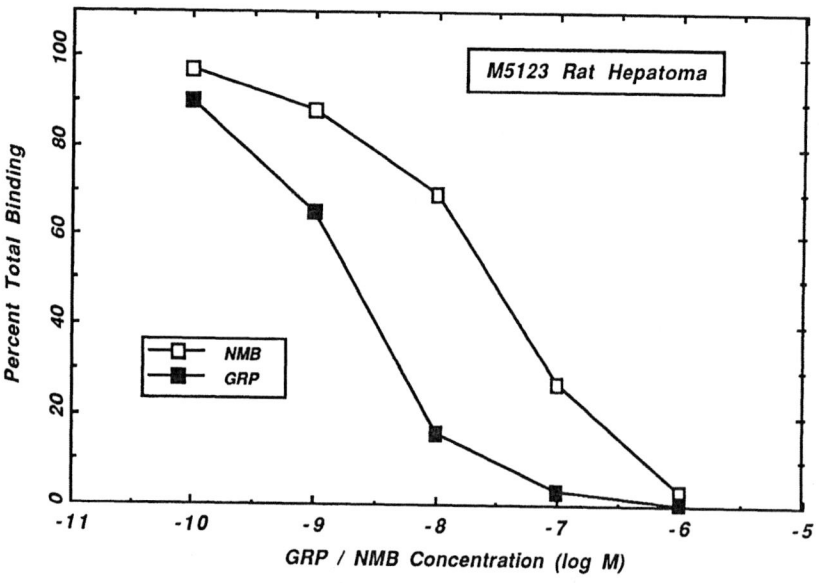

Figure 4. *In Vitro* Inhibition of [^{125}I-Tyr^4]Bombesin Binding: GRP / NMB Selectivity

CONCLUSION

The presence of SRIF receptors correlates with our previous observations that the SRIF octapeptide, BIM-23014 inhibits both the *in vitro* cell proliferation (SCLC NCI-H69, AR42J) and the *in vivo* growth of established tumors (SCLC NCI-H69, SCLC NCI-H345, AR42J, R-3327 prostate) (Bogden et al, 1988, 1990a,1990b; Taylor et al., 1988a, 1988b). However, the observed *in vivo* antitumor activity against receptor-negative tumors is difficult to explain. The presence of SRIF receptor subtypes (Yamada et al., 1992) sensitive to growth inhibition by certain SRIF analogues, but weakly detected by [^{125}I-Tyr^{11}]SRIF-14 could provide a partial explanation. For the tumors or tumor cell lines examined, pharmacological analysis of [^{125}I-Tyr^4]bombesin binding showed that most were selective for GRP receptor subtype. However, the SCLC NCI-H345 line, exhibited both subtypes, and the C6 glioma cell line was characterized as the NMB subtype. The presence of GRP/NMB receptors on various tumors and the sensitivity to GRP-antagonist growth inhibition (Staley et al., 1991; Mahmoud et al., 1991; Thomas et al., 1991; Bogden et al., 1992) suggests that these peptides may also have a regulatory role in the growth of certain types of cancer. In summary, these observations indicate that SRIF or GRP/NMB receptor density may be an excellent marker for the diagnosis of receptor-positive tumors and may be predictive in many cases of the antitumor activity by SRIF peptides or GRP/NMB antagonists.

REFERENCES

Bogden, A.,E, Taylor, J.E,. Moreau J-P, Coy, D.H., Moreau, S., and LePage, D.J., 1988, *In vivo* responsiveness of human and animal tumors to somatostatin (SRIF) analogue BIM-23014C (DC13-116), *Proc. Am. Assoc. Cancer Res.* Abst. #56.

Bogden, A.E. Taylor, J.E., Moreau, J.-P., and Coy, D.H., 1990a,. Treatment of R-3327 prostate tumors with a somatostatin analogue (Somatuline) as adjuvant therapy following surgical castration, *Cancer Res.* 50:2646.

Bogden, A.E., Taylor, J.E., Coy, D.H., Keyes, S.R., Prevost, G., Kim, S.H., and Moreau, J.-P., 1992, *In vivo* screening of agonist and antagonist analogues of bombesin/gastrin releasing peptide versus human lung tumor xenografts, *Proc. Am. Assoc. Cancer Res.*, Abst. #415

Bogden, A.E., Taylor, J.E., Keyes, S.R., Kim, S.H., and Moreau, J.-P., 1991, Inhibition of intradermal melanomas by topical application of Somatuline, a somatostatin analogue (BIM-23014C), *Proc. Am. Assoc. Cancer Res.*, Abst. #334.

Bogden, A.E., Taylor, J.E., Moreau, J.-P., Coy, D.H., and LePage, D.J., 1990b, Response of human lung tumor xenografts to treatment with a somatostatin analogue (Somatuline), *Cancer Res.* 50:4360.

Cardona, C., Rabbitts, P.H., Spindel, E.R., Ghatei, M.A., Bleehen, N.M., Bloom, S.R., Reeve, J.G., 1991, Production of neuromedin B and neuromedin B gene expression in human lung tumor cell lines, *Cancer Res.* 51:5205.

Carney, D.N., 1991, Lung cancer biology, *Eur. J. Cancer* 27:366.

Carney, D.N., 1992, Biology of small-cell lung cancer, *Lancet* 339:843.

Corjay, M.H., Dobrzanski, D.J., Way, J.M., Viallet, J., Shapira, H., Worland, P., Sausville, E.A., and Battey, J.M., 1991, Two distinct bombesin receptor subtypes are expressed and functional in human lung carcinoma cells, *J. Biol. Chem.* 266:18771.

Cuttitta, F., Carney, D.N., Mulshine, J., Moody, T.W., Fedorko, J., Fischler, A., and Minna, J.D., 1985, Bombesin-like peptides can function as autocrine growth factors in human small-cell lung cancer, *Nature* 316:823.

Davis, T.P., Crowell, S., Taylor, J.E., Coy, D.H., Staley, J., and Moody, T.W., 1992, Metabolic stability and tumor inhibition of bombesin/GRP antagonists, *Peptides* 13:401.

DeFeudis, F.V., and Moreau, J.-P., 1986, Studies on somatostatin analogues might lead to new therapies for certain types of cancer, *TiPS* 7:384.

Lamberts, S.W.J., Koper, J.W., and Reubi, J.-C., 1987, Potential role of somatostatin analogues in the treatment of cancer, *Eur. J. Clin. Invest.* 17:281.

Lamberts, S.W.J., Krenning, E.P., and Reubi, J.-C., 1991, The role of somatostatin and its analogs in the diagnosis and treatment of tumors, *Endocrine Rev.* 12:450.

Mahmoud, S. Staley, J. Taylor, J.E., Bodgen, A.E., Moreau, J.-P., Coy, D.H., Avis, I., Cuttitta, F., Mulshine, J.L., and Moody, T.W., 1991, (psi[13,14])Bombesin analogs inhibit the growth of small cell lung cancer *in vitro* and *in vivo*, *Cancer Res.* 51:1798.

Moody, T.W., Bertness, V., and Carney, D.N., 1983, Bombesin-like peptides and receptors in human tumor cell lines, *Peptides* 4:683.

Moreau, J.P., and DeFeudis, F.V., 1987, Pharmacological studies of somatostatin and somatostatin analogues: therapeutic advances and perspectives, *Life Sci.* 40:419.

Murphy, W.A., Lance, V.A., Moreau, S., Moreau, J.-P., and Coy, D.H., 1987, Inhibition of rat prostate tumor growth by an octapeptide analog of somatostatin, *Life Sci.* 40: 2515.

Nelson, J., Cremin, M., and Murphy, R.F., 1989, Synthesis of somatostatin by breast cancer cells and their inhibition by exogenous somatostatin and sandostatin, *Br. J. Cancer* 59:739.

Papotti, M., Macri, L., Bussolati, G., and Reubi J.-C., 1989, Correlative study of neuro-endocrine differentiation and presence of somatostatin receptors in breast carcinomas, *Int. J. Cancer* 43:365.

Parmer, H., Bodgen, A., Mollard, M., de Rouge, B., Phillips, R.H., and Lightman, S.L., 1989, Somatostatin and somatostatin analogues in oncology, *Cancer Treat. Rev.* 16: 95

Reichlin, S., 1983a, Somatostatin, *N. Eng. J. Med.* 309:1495.

Reichlin, S., 1983b, Somatostatin, *N. Eng. J. Med.* 309:1556.

Setyono-Han, B., Henkelman, M.S., Foekens. J.A., and Klijn, J.G.M., 1987, Direct inhibitory effects of somatostatin (analogues) on the growth of human breast cancer cells, *Cancer Res.* 47:1566.

Siegel, R.A., Tolcsvai, L., and Rudin, M., 1988, Partial inhibition of the growth of transplanted Dunning rat prostate tumors with the long-acting somatostatin analogue Sandostatin (SMS 201-995), *Cancer Res.* 48:4651.

Staley, J., Coy, D.H., Taylor, J.E., Moreau, J.-P., and Moody, T.W., 1991, [Des-Met[14]]bombesin analogues function as small cell lung cancer bombesin receptor antagonists, *Peptides* 12:145.

Taylor, J.E., Bogden, A.E., Moreau, J.-P., and Coy, D.H., 1988b, In vitro and in vivo inhibition of human small cell lung carcinoma (NCI-H69) growth by a somatostatin analogue. *Biochem. Biophys. Res. Comm.* 153:81.

Taylor, J.E., Coy, D.H., and Moreau, J.-P., 1988a, High affinity binding of [^{125}I-Tyr[11]]somatostatin-14 to human small cell lung carcinoma (NCI-H69), *Life Sci.* 43: 421.

Thomas, F., Keyes, S., Taylor, J.E., Moreau, J.-P., and Bogden, A.E., 1991, *In vivo* inhibition of human small cell lung carcinoma (SCLC) NCI-H345 with BIM-26189 a bombesin (BN) antagonist, *6th Eur. Conf. Clinical Oncology & Cancer Nursing* Abst. # 1112.

Weber, C., Merriam, L., Koschitzky, T., Karp, F., Benson, M., Forde, K., and LoGerfo, P., 1989, Inhibition of growth of human breast carcinomas *in vivo* by somatostatin analog SMS 201-995: Treatment of nude mouse xenografts, *Surgery* 106:416.

Woll, P.J., 1991a Neuropeptide growth factors and cancer, *Br. J. Cancer* 63:469.

Woll, P.J., 1991b, Growth factors and lung cancer, *Thorax* 46:924.

Yamada, Y., Post, S.R., Wang, K, Tager, H.S., Bell, G.I., and Seing, S., 1992, Cloning and functional characterization of a family of human and mouse somatostatin receptors expressed in brain, gastrointestinal tract, and kidney, *Proc. Natl. Acad. Sci. USA* 89: 251.

NEUROPEPTIDE RECEPTOR SUB-TYPES: NEUROPEPTIDE Y AND CALCITONIN-GENE RELATED PEPTIDES AS NEW EXAMPLES

Rémi Quirion[1,2], Denise van Rossum[1,2] and Yvan Dumont[1,2]

[1]Dept. of Psychiatry and Pharmacology and Therapeutics,
Faculty of Medicine, McGill University, Montreal, Canada
[2]Neuroscience Division
Douglas Hospital Research Centre
6875 Blvd. LaSalle
Verdun, Québec, Canada H4H 1R3

INTRODUCTION

Receptor multiplicity for a given neurotransmitter and/or modulator is now the rule rather than the exception. Over the past decade, this has been clearly demonstrated with the cloning of multiple classes of receptors for various transmitters and hormones. For example, five muscarinic receptors have been cloned[1], as well as multiple α sub-units of the nicotinic receptor[2], and an ever increasing number of serotonin receptors (up to a dozen at this time)[3].

Similarly, multiple classes of neuropeptide receptors were pharmacologically characterized before their more recent cloning. This is exemplified by the neurokinin/substance P receptor family which were the first three peptide receptors to be cloned in the mid - late eighties[4]. Since then, various other peptide receptors have been cloned including those of neurotensin[5], vasopressin[6], somatostatin[7], bombesin[8] and atrial natriuretic peptides[9]. Interestingly, opiate receptors still have to be cloned in spite of major efforts over the past decade.

Among the various other peptide families, detailed information about neuropeptide Y (NPY) and calcitonin-gene related peptide (CGRP) receptors have only recently begun to accumulate with the characterization of suitable pharmacological assays and the development of receptor sub-type selective agonists. The recent cloning of the Y_1 receptor sub-type[10,11] should also allow for a better characterization of the functional role of this receptor class. In this chapter, we briefly review the major characteristics of NPY and CGRP receptor sub-types focusing on pharmacological aspects.

NEUROPEPTIDE Y RECEPTOR SUB-TYPES

NPY is one of the most abundant brain peptides isolated thus far (for a recent review[12]). This 36 amino acid polypeptide shares important sequence homology with peptide YY (PYY) and pancreatic polypeptides (PP). Therefore, all three peptides are members of the PP family[13].

NPY and homologues possess multiple biological effects including a most potent stimulation of food intake, the inhibition of sexual behavior, the facilitation of learning and memory and the modulation of various cardiovascular parameters (for review[12]). The unique profile of effects induced by this peptide family certainly relates to their

Table 1. Characteristics of NPY receptor sub-types

	Y_1	Y_2	Y_3
Potency of Homologues	PYY$\geq$NPY	PYY$\geq$NPY	NPY$>>>$PYY
Preferential Agonists	[Leu^{31},Pro^{34}]PYY [Pro^{34}]PYY	C-terminal fragments (PYY$_{13-36}$)	none
Selective Antagonist	Not available	Not available	Not available
Prototypic Tissues	Rat fronto-parietal cortex Rat saphenous vein	Rat vas deferens Rat hippo-campus	Bovine chromaffin cells Rat colon
Transduction Mechanisms	Multiple depending upon tissues	Inhibition of cAMP	?
Molecular Information	Cloned Rhodopsin-like family 384 a.a. (human)	Not available (but see[55])	Possibly cloned Rhodopsin-like family

activation of highly specific receptors present in various tissues of the organism including the central nervous system (CNS).

Wahlestedt *et al*[14] were first in 1986 to propose the existence of two classes of NPY receptors (Y_1 and Y_2) on the basis of *in vitro* bioassay data. They observed that while the complete NPY/PYY sequence is required to activate the Y_1 receptor, long C-terminal fragments (NPY$_{13-36}$ and others) as well as the native peptides, are able to bind, with high affinity to another class of sites classified as Y_2 (Table 1). Since then, various groups have confirmed and extended these findings (for review[12]). The recent development of [Leu^{31},Pro^{34}]NPY and [Pro^{34}]NPY as selective Y_1 receptor agonists was especially useful to characterize further the Y_1 and Y_2 receptor sub-types[15]. For example, it now appears that the potent stimulating effect of NPY on food intake is likely mediated by a Y_1 receptor sub-type (albeit atypical[16,17]) while effects on learning demonstrate a Y_2-ligand selectivity profile[18]. Cardiovascular actions of NPY homologues are likely related to the activation of both receptor sub-types depending upon tissues and physiological states (for review[12]). It appears that the Y_1 and Y_2 receptors are members of the G-protein coupled receptor family; the Y_1 sites being coupled to phosphoinositide turnover[19,20] (and others, see[11]) while the Y_2 sub-type is linked to the inhibition of adenylate cyclase[21-23].

Very recently, the Y_1 receptor has been cloned both from human[11] and rat[10] tissues. Both are highly homologous and members of the seven-transmembrane rhodopsin receptor super-family. Interestingly, the human version, composed of 384 amino acids, can either inhibit cAMP production or elevate intracellular Ca^{++} depending upon the expression system used[11]. Thus, a given G-protein receptor sub-type can be coupled to various transduction mechanisms depending upon tissues and/or cell types. The cloning of the Y_2 sub-type has yet to be reported.

Grundemar, Wahlestedt and colleagues[24,25] have recently suggested the possible existence of yet another class (Y_3) of NPY receptors, mediating certain cardiovascular and respiratory effects of NPY[24], and being present in bovine chromaffin cells[25]. The uniqueness of the Y_3 sub-type relates to its exquisite affinity for NPY as well as previously thought to be selective Y_1 ([Leu^{31},Pro^{34}]NPY) and Y_2 (NPY$_{13-36}$) agonists, and its very poor affinity for PYY. Thus, in contrast to Y_1 and Y_2 receptors, the Y_3 sub-type can clearly distinguish between NPY and PYY[24,25]. Recently, Rimland *et al*[26] reported the putative cloning of this receptor class using homology screening of "orphan" G-protein rhodopsin type receptors. However, doubts have been expressed on the genuine nature of this cloned protein[27] and it appears that the race is still on to clone the putative Y_3 receptor sub-type. In any case, the structure-activity profile of this newly characterized class of NPY receptors is unique and suggests that PYY C-terminal fragments and [Leu^{31},Pro^{34}]PYY should be used as selective Y_2 and Y_1 agonists, respectively; NPY-related molecules also having affinity for the Y_3 sites (Table 1).

BRAIN NPY RECEPTOR SUB-TYPES

Over the past few years, we investigated the discrete localization of NPY binding sites in the mammalian brain. We were first to report on the discrete localization of these sites in the rat brain, high densities being particularly seen in the hippocampal formation and various cortical areas[28]. Subsequent studies[29,30] suggested possible differences in the comparative localization of NPY and PYY binding sites. For example, greater amounts of [^{125}I]PYY than [^{125}I]Bolton-Hunter NPY sites were detected in various brainstem nuclei[30]. More recently, using [^{125}I]PYY as ligand and [Leu^{31},Pro^{34}]NPY and NPY$_{13-36}$ competitors, we[31,32] and others[33] reported on the differential distribution of Y_1 and Y_2 sites in the rat brain (the Y_3 sub-type not being investigated since the radioligand [^{125}I]PYY has very low affinity for this receptor[24,25]). As shown in Figure 1, Y_1 receptor binding sites appear to be most concentrated in cortical areas while the Y_2 sub-type is more broadly distributed, and especially concentrated in the hippocampus, olfactory bulb and hypothalamus (see[32] for details). The distinct anatomical distribution most likely bears significance in regard to the comparative functional relevance of Y_1 and Y_2 receptor sub-types in the CNS.

The discrete brain distribution of the Y_3 sub-type remains to be established but these sites are likely present in the nucleus tractus solitarius and/or the area postrema where they could mediate cardiovascular and respiratory effects of NPY and its fragments[24]. Hopefully, the cloning of the various NPY receptor sub-types will soon allow for the development of highly specific antibodies which could be used as tools for detailed ultrastructural studies of the localization of a given receptor sub-type. Moreover, *in situ* hybridization is currently underway and should permit the comparison of sites of receptor synthesis vs protein location in various neuronal pathways. This could provide clues as to the functional relevance of NPY receptor sub-types in the CNS. The existence of other putative classes of NPY/PYY receptor have also been proposed (sigma, PP, etc.) but data are still too preliminary to allow for their distinct classification (for review[12]).

CGRP RECEPTOR SUB-TYPES

CGRP is a 37 amino acid polypeptide produced from the alternate tissue-specific splicing of the calcitonin gene mRNA[34,35]. α- and β forms of CGRP are present in a variety of tissues mostly of neural origin[36]. In fact, CGRP is often co-localized with substance P in sensory nerves, with acetylcholine in motoneurons and with various other transmitters in the brain while calcitonin is mostly found in non-neural tissues (for review[37]).

CGRP-like peptides have been shown to induce a variety of peripheral and central effects including a most potent vasodilation, cardiac acceleration, modulation of nicotinic receptor turnover at the neuromuscular junction, increase in body temperature and decrease in food intake following direct injection into the CNS (for review[37]). Early on, it was clearly established that the biological effects of CGRP were mediated by highly specific receptors, distinct in most tissues from those of calcitonin (for review[37]). Over the past few years, we further characterized CGRP receptors using a series of fragments and analogues. This allowed us to be the first to propose a distinct classification for CGRP receptor sub-types[38-42].

The existence of at least two classes of CGRP receptors, CGRP$_1$ and CGRP$_2$, was suggested by the differential activities of CGRP-related molecules in using *in vitro*

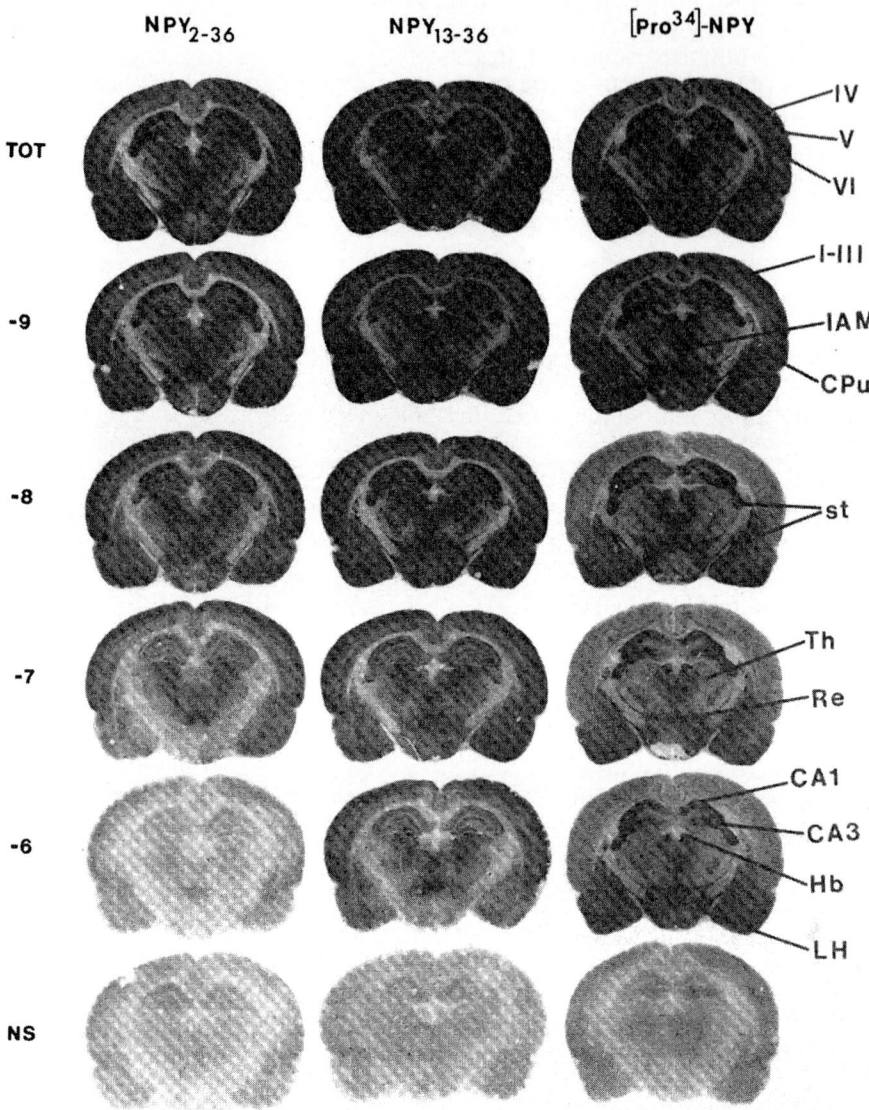

Figure 1. Photomicrographs of the autoradiographic distribution of [^{125}I]PYY binding at the level of the rat hippocampus. Adjacent coronal sections were incubated with 25 pM [^{125}I]PYY (total) or in the presence of either 1 μM NPY (non specific binding, NS) or increasing concentrations (M) of competitors such as NPY$_{2-36}$, NPY$_{13-36}$ and [Pro^{34}]NPY. Note that cortical labeling is highly sensitive to [Pro^{34}]NPY (Y$_1$-enriched) while hippocampal binding is competed by the two fragments (Y$_2$-enriched). CA$_1$ and CA$_3$, fields of the Ammon's horn of the hippocampus; CPu, caudate-putamen; Hb, habenular nucleus; IAM, interanteromedial thalamic nucleus; LH, lateral hypothalamus; Re, reuniens thalamic nuclei; st, stria terminalis; Th, thalamus; I-IV, cortical laminae. From Dumont *et al* with permission[32].

Table 2. Characteristics of CGRP receptor sub-types

	CGRP$_1$	CGRP$_2$	CGRP/sCT
Potency of homologues	αCGRP$\geq\beta$CGRP $>>$Amylin$>$sCT	αCGRP$\geq$ βCGRP$>>$ Amylin$>$sCT	αCGRP$\geq\beta$CGRP $\geq$sCT
Preferential agonist	Not available	[Cys(ACM)^{2,7}] hCGRPα	Not available
Antagonist	CGRP$_{12-37}$ CGRP$_{8-37}$ (potent antagonists)	CGRP$_{8-37}$ (weak antagonist)	?
Prototypic tissues	Atria	Vas deferens	Nucleus accumbens
Prototypic behavior	Hypophagia	Hyperthermia	?
Transduction mechanisms	G-protein coupled Activation of adenylate cyclase	G-protein coupled likely cyclase activation	
Molecular information	Not available	Not available	Likely not available (but see[52])

sCT, salmon calcitonin
See[37] for more details about transduction mechanisms associated with CGRP receptor activation.

bioassays and *in vivo* behavioral effects. We observed that fragments lacking the N-terminal disulfide bridge (between amino acid residues 2 and 7) such as CGRP$_{8-37}$ and CGRP$_{12-37}$ behave as potent competitive antagonists in certain, but not all, *in vitro* and *in vivo* assays[38,39] (Table 2). For example, CGRP$_{12-37}$ is able, at μM concentrations, to block the inotropic and chronotropic effects of CGRP in the guinea pig atria but not the action of this peptide in the electrically-stimulated rat vas deferens[38]. Similarly, but with greater potencies, CGRP$_{8-37}$ demonstrated potent antagonistic properties in atrial preparations while being significantly weaker in the vasa deferentia[39,41,42]. In the CNS, intracerebroventricular injections (icv) of CGRP$_{8-37}$ antagonized the effect of CGRP on food intake but failed to alter the hyperthermia observed following an icv injection of the parent molecule[39,42]. Other groups have recently reported similar findings in a variety of preparations[43-45]. This strongly suggests the existence of CGRP receptor sub-types which are differentially sensitive to the antagonistic properties of C-terminal CGRP fragments. We classified these receptors as the CGRP$_1$ receptor sub-type.

On the other hand, a linear analogue, [Cys(ACM)^{2,7}]hCGRPα behaved as a more potent **agonist** in vas deferens than atrial preparations[38]. This analogue has thus been used as a prototypic agonist of a CGRP$_2$-receptor sub-type; this class being relatively resistant to the antagonistic properties of CGRP$_{8-37}$[41,42] (Table 2). Recent data has extended our findings as Stangl *et al*[46] observed the presence of [Cys(ACM)^{2,7}]hCGRPα-sensitive receptors mediating cAMP production in the rat liver while Giuliani and colleagues observed the likely existence of CGRP$_{8-37}$-resistant, likely CGRP$_2$, receptors in the guinea pig urinary bladder[48].

Additionally, amylin (a peptide sharing important homologies with CGRP)-sensitive receptors are present in various peripheral tissues. It is a matter of debate as to whether this peptide acts on its own receptor sub-type or via CGRP-related receptors[47-49].

BRAIN CGRP RECEPTOR SUB-TYPES

Certain behavioral effects of CGRP are blocked by $CGRP_{8-37}$ while others are fully resistant[42,50] (Table 2). We thus attempted to provide evidence for the existence of distinct $CGRP_1$ and $CGRP_2$ receptor sub-types in the CNS. As shown in Figure 2,

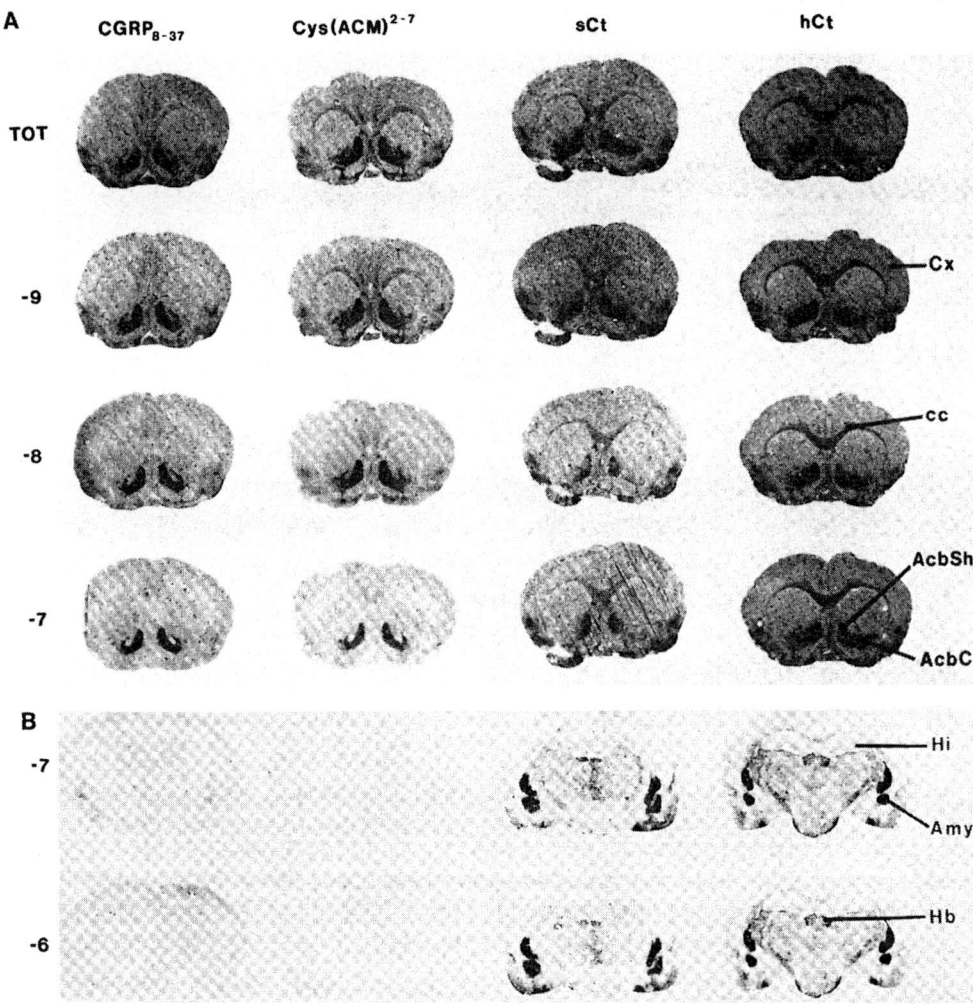

Figure 2. Photomicrographs of the autoradiographic distribution of $[^{125}I]hCGRP\alpha$ binding at the level nucleus accumbens (A) and the hippocampal formation (B) in the rat brain. Adjacent coronal sections were incubated with 25 pM $[^{125}I]hCGRP\alpha$ (Total) or in the presence of increasing concentrations of $CGRP_{8-37}$, $[Cys(ACM)^{2-7}]hCGRP\alpha$, salmon calcitonin (sCT) and human calcitonin (hCT). Note that while binding in the accumbens is sensitive to sCT (but not hCT) it is not the case in most other areas. AcbSh, shell of the nucleus accumbens; Amy, amygdaloid body; CC, corpus callosum; CPu, caudate-putamen; Cx, cortex; Hb, habenular nuclei; Hi, hippocampus; Par, parietal cortex; Pir, periform cortex. From Quirion *et al* with permission[42].

competition experiments using increasing concentrations of either $CGRP_{8-37}$ or $[Cys(ACM)^{2,7}]hCGRP\alpha$ failed to clearly reveal the differential distribution of these two receptor classes in the rat CNS[42]. We are now attempting an alternative approach based on the use of rather selective analogues as radiolabelled probes ($[^{125}I]CGRP_{8-37}$, $[^{125}I]CGRP_{12-37}$, etc.) a very often preferable strategy leading to more definitive results. Hopefully, this approach combined with detailed autoradiographic studies will reveal the presence of $CGRP_1$ and $CGRP_2$ receptors in the CNS, as already strongly suggested by behavioral data[39,42,50].

Interestingly, however, $[^{125}I]hCGRP\alpha$ binding in certain areas of the rat brain is apparently sensitive to salmon (sCT), but not human (hCT) calcitonin[40,51]. As shown in Fig. 2A, $[^{125}I]hCGRP\alpha$ labelling in the nucleus accumbens is competed by increasing concentrations of sCT while this is not seen in most other areas (Fig. 2B). This extended results reported earlier by Sexton et al[51] suggests the existence of a genuine class of CGRP/sCT receptor in certain areas of the mammalian brain. It is certainly most puzzling that only the salmon homologue of CT is able to compete for CGRP labelling in certain areas of the rat brain; the human counterpart being totally inactive. Could this suggest the existence of sCT-like peptides in the mammalian brain? While this hypothesis has generated some interest in the past, we believe that it may be worthwhile to seriously reconsider it in view of these recent observations. The possible existence of distinct amylin receptors in the CNS has yet to be fully explored.

As for molecular information, CGRP receptors have yet to be cloned. However, Lin et al[52] recently reported on the cloning of the calcitonin receptor. Interestingly, this G-protein coupled receptor bears rather little similarity with previously reported seven transmembrane rhodopsin-type receptors. Only the parathyroid hormone[53] and the secretin[54] receptors shared sequence homologies with the calcitonin receptors and these authors have thus suggested the existence of an additional super-family of G-protein-coupled receptors distinct from the better known rhodopsin family. It will be of interest to find out if the putative $CGRP_1$ and $CGRP_2$ receptor sub-types belong to this newly characterized super family.

CONCLUSION

It is now evident that NPY and CGRP receptor sub-types are present and widely distributed in the organism. The future development of highly receptor sub-type selective agonists and antagonists, and the cloning of these various receptors, should allow for a better characterization of their respective functional relevance in the organism.

ACKNOWLEDGEMENTS

This research is supported by the Medical Research Council of Canada. R.Q. is a "Chercheur-Boursier" of the "Fonds de la Recherche en Santé du Québec" while D.v.R. holds a fellowship from the FCAR.

REFERENCES

1. T.I. Bonner. The molecular basis of muscarinic receptor diversity. Trends Neurosci. 12:148-154(1989).
2. J. Patrick. Molecular Biology of multiple nicotinic receptors in the CNS. Prog. Brain Res. In press.
3. S.J. Peroutka and A.J. Sleight. In "The Role of Serotonin In Psychiatric Disorders" S-L. Brown and H.M. van Praag (eds), pp. 8-26, Brunner/Mazel, New York (1991).
4. S. Nakanishi. Mammalian tachykinin receptors. Annu. Rev. Neurosci. 14:123-136(1991).
5. K. Tanaka, M. Masu and S. Nakanishi. Structure and functional expression of the cloned rat neurotensin receptor. Neuron 4:847-854(1990).
6. S.J. Lolait, A.M. O'Carroll, O. Wesley McBride, M. Kong, A. Morel and M.J. Brownstein. Cloning and characterization of a vasopressin V2 receptor and possible link to nephrogenic diabetes insopiders. Nature 357:336-339(1992).

7. F.W. Klunen, C. Bruns and H. Lübbert. Expression cloning of a rat brain somatostatin receptor cDNA. Proc. Natl. Acad. Sci. USA 89:4618-4611(1992).

8. E.R. Spindel, E. Giladi, T.P. Brehm, R.H. Goodman and T.P. Segerson. Cloning and functional characterization of a cDNA encoding the murine fibroblast bombesin/GRP receptor. Mol. Endocrinol. 4:1956(1990).

9. M. Chinkers, D.L. Garbers, M.S. Chang, D.G. Lowe, H. Chin, D.V. Goeddel and S. Schulz. A membrane form of guanylate cyclase is an atrial natriuretic peptide receptor. Nature 338:78-84(1989).

10. Larhammar, D., A.G. Blomqvist, E. Jazin, H. Yoo, D.J. Reis and C. Wahlestedt. Cloning and functional expression of a human neuropeptide Y/peptide YY receptor of the Y1 type. J. Biol. Chem. 267:10935-10939(1992).

11. H. Herzog, Y.J. Hort, H.J. Ball, G. Hayes, J. Shine and L.A. Selbie. Cloned human neuropeptide Y receptor couples to two different second messenger systems. Prog. Natl. Acad. Sci. USA, 89:5794-5799(1992).

12. Y. Dumont, J-C. Martel, A. Fournier, S. St-Pierre and R. Quirion. Neuropeptide Y and Neuropeptide Y receptor sub-types in brain and peripheral tissues. Prog. Neurobiol. 38:125-167(1992).

13. K. Tatemoto, M. Carlquist and V. Mutt. Neuropeptide Y: a novel brain peptide with structural similarities to peptide YY and pancreatic polypeptide. Nature 296:659-660(1982).

14. C. Wahlestedt, N. Yanaihara and R. Häkanson. Evidence for different pre- and post-junctional receptors for neuropeptide Y and related peptide. Regul. Pept. 13:307-318(1986).

15. J. Fuhlendorff, U. Gether, L. Aakerlund, N. Langeland-Johansen, H. Thogersen, S.G. Melberg, U.B. Olsen, O. Thastrup and T.W. Schwartz. [Leu^{31},Pro^{34}]neuropeptide Y: a specific Y$_1$ receptor agonist. Proc. Natl. Acad. Sci. USA 87:182-186(1990).

16. F.B. Jolicoeur, J.N. Michaud, D. Menard, A. Fournier and S. St-Pierre. *In vivo* structure activity study supports the existence of heterogeneous neuropeptide Y receptors. Brain Res. Bull. 26:309-311(1991).

17. S.F. Leibowitz and J.T. Alexander. Analysis of neuropeptide Y-induced feeding dissociation of Y$_1$ and Y$_2$ receptor effects on natural meal patterns. Peptides 12:1251-1260(1991).

18. J.F. Flood, and J.E. Morley. Dissociation of the effects of neuropeptide Y on feeding and memory evidence for pre- and postsynaptic mediation. Peptides 10:963-966(1989).

19. J. Hinson, C. Rauh and J. Coupet. Neuropeptide Y stimulates inositol phospholipid hydrolysis in the rat brain miniprisms. Brain Res. 446:379-382(1988).

20. P.S. Widdowson and A.E. Halaris. Effects of neuropeptide Y on alpha 1- and beta-adrenoceptor-stimulated second messenger systems in rat frontal cortex. Peptides 11:661-665(1990).

21. A. Härfstrand, B. Fredholm and K. Fuxe. Inhibitory effects of neuropeptide Y on cyclic AMP accumulation in slices of the nucleus tractus solitarius region of the rat. Neurosci. Lett. 76:185-190(1987).

22. S. Petrenko, M.C. Olianas, P. Onali and G.L. Gessa. Neuropeptide Y inhibits forskolin-stimulated adenylate cyclase activity in rat hippocampus. Eur. J. Pharmacol. 136:425-428(1987).

23. A. Westlind-Danielsson, A. Undén, J. Abens, S. Andell and T. Bartfai. Neuropeptide Y receptors and the inhibition of adenylate cyclase in the human frontal and temporal cortex. Neurosci. Lett. 74:237-242(1987).

24. L. Grundemar, C. Wahlestedt and D.J. Reis. Neuropeptide Y acts as an atypical receptor to evoke cardiovascular depression and to inhibit glutamate responsiveness in the brainstem. J. Pharmacol. Exp. Ther. 258:633-638(1991).

25. C. Wahlestedt, S. Regunathan and D.J. Reis. Identification of cultured cells selectively expressing Y1-, Y1-, or Y3-type receptors for neuropeptide Y/peptide YY. Life Sci. 50:PL7-PL12(1992).

26. J. Rimland, W. Sin, P. Sweetnam, K. Saijoh, E.J. Nestler and D.S. Duman. Sequence and expression of a neuropeptide Y receptor cDNA. Mol. Pharmacol. 40:869-875(1991).

27. A. Blomqvist, E. Jain, D. Larhammar, F. Yee, H. Yoo, D.J. Reis and C. Wahlestedt. XIIth Washington Spring Symposium, Washington, DC., Abstract 87(1992). This volume.

28. J-C. Martel, S. St-Pierre and R. Quirion. Neuropeptide Y receptors in rat brain: autoradiographic localization. Peptides 7:55-60(1986).

29. D.R. Lynch, M.W. Walker, R.J. Miller and S.H. Snyder. Neuropeptide Y receptor binding sites in rat brain: differential localization with [^{125}I]peptide YY and [^{125}I]neuropeptide Y imply receptor heterogeneity. J. Neurosci. 9:2607-2619(1989).

30. J-C. Martel, A. Fournier, S. St-Pierre and R. Quirion. Quantitative autoradiographic distribution of [^{125}I]Bolton-Hunter neuropeptide Y receptor binding sites in rat brain. Comparison with [^{125}I]peptide YY receptor sites. Neuroscience 36:255-283(1990).

31. Y. Dumont, A. Fournier, S. St-Pierre, T.W. Schwartz and R. Quirion. Differential distribution of neuropeptide Y1 and Y2 receptors in the rat brain. Eur. J. Pharmacol. 191:501-503(1990).

32. Y. Dumont, A. Fournier, S. St-Pierre and R. Quirion. Comparative characterization and autoradiographic distribution of neuropeptide Y receptor subtypes in the rat brain. J. Neuroscience(1992). In press.

33. S.A. Aicher, M. Springston, S.B. Berger, D.J. Reis and C. Wahlestedt. Receptor-selective analogs demonstrate NPY/PYY receptor heterogeneity in rat brain. Neurosci. Lett. 130:32-36(1991).

34. S.G. Amara, V. Jonas, M.G. Rosenfeld, E.S. Ong and R.M. Evans. Alternative RNA processing in calcitonin gene expression generates mRNAs encoding different polypeptide products. Nature 296:240-244(1982).

35. M.G. Rosenfeld, J.J. Mermod, S.G. Amara, L.W. Swanson, P.E. Sawchenko, J. Rivier, W.W. Vale and R.M. Evans. Production of a novel neuropeptide encoded by the calcitonin gene via tissue-specific RNA processing. Nature 304:129-135(1983).

36. S.G. Amara, J.L. Arriza, L.W. Swanson, R.M. Evans and M.G. Rosenfeld. Expression in brain of a messenger RNA encoding a novel neuropeptide homologous to calcitonin gene-related peptide. Science 119:1094-1097(1985).

37. Y. Tache, P. Holzer, M.G. Rosenfeld (eds) Calcitonin Gene Related Peptide, Ann. NY Acad. Sci. 657:561pp(1992).

38. T. Dennis, A. Fournier, S. St-Pierre and R. Quirion. Structure-activity profile of calcitonin gene-related peptide in peripheral and brain tissues. Evidence for receptor multiplicity. J. Pharmacol. Exp. Ther. 251:718-725(1989).

39. T. Dennis, A. Fournier, A. Cadieux, F. Pomerleau, F.B. Jolicoeur, S. St-Pierre and R. Quirion. hCGRP$_{8-37}$ a calcitonin gene-related peptide antagonist revealing CGRP receptor heterogeneity in brain and periphery. J. Pharmacol. Exp. Ther. 254:123-128(1990).

40. T. Dennis, A. Fournier, S. Guard, S. St-Pierre and R. Quirion. Calcitonin gene-related peptide (hCGRPα) binding sites in nucleus accumbens. Atypical structural requirements and marked phylogenic differences. Brain Res. 539:59-66(1991).

41. M. Mimeault, A. Fournier, Y. Dumont, S. St-Pierre and R. Quirion. Comparative affinities and antagonistic potencies of various human calcitonin gene-related peptide fragments on CGRP receptors in brain and periphery. J. Pharmacol. Expt. Ther. 258:1084-1090(1991).

42. R. Quirion, D. van Rossum, Y. Dumont, S. St-Pierre and A. Fournier. Characterization of CGRP$_1$ and CGRP$_2$ receptor sub-types. Ann. N.Y. Acad., Sci. 657:88-105(1992).

43. T. Chiba, A. Yamaguchi, T,. Yamatani, A. Nakamura, T. Morishita, T. Inui, M. Fukase, T. Noda and T. Fujita. Calcitonin gene-related peptide receptor antagonist human CGRP (8-37). Am. J. Physiol. 256:E331-E335(1989).

44. S.M. Gardiner, A.M. Compton, P.A. Kemp, T. Bennett, C. Bose, R. Foulkes and B. Hughes. Antagonistic effect of human α-CGRP [8-37] on the *in vivo* regional hemodynamic actions of human α-CGRP. Biochem. Biophys. Res. Commun. 171:938-943(1990).

45. C.A. Maggi, T. Chiba and S. Giuliani. Human α-calcitonin gene-related peptide (8-37) as antagonist of exogenous and endogenous calcitonin gene-related peptide. Eur. J. Pharmacol. 192:85-88(1991).

46. D. Stangl, R. Muff, C. Schmolck and J.A. Fischer. Photoaffinity labeling of rat calcitonin gene-related peptide receptors and adenylate cyclase activation: Identification of receptor subtypes. Endocrinology (1992). In press.

47. S. Giuliani, S.J. Wimalawansa and C.A. Maggi. Involvement of multiple receptors in the biological effects of calcitonin gene-related peptide and amylin in rat and guinea pig preparations. Brit. J. Pharmacol. (1992). In press.

48. S.J. Wimalawansa. Isolation purification and characterization of calcitonin gene-related peptide receptors. Ann. NY Acad. Sci. 657:70-87(1992).

VASOACTIVE INTESTINAL PEPTIDE AND ASTROCYTE MITOGENESIS: *IN VITRO* AND *IN VIVO* STUDIES

Joanna M. Hill and Douglas E. Brenneman

Section on Developmental and Molecular Pharmacology
Laboratory of Developmental Neurobiology
NICHD, NIH
Bethesda, MD 20892

INTRODUCTION

In the central nervous system, in addition to their recognized action on synaptic transmission, neurotransmitters and neuromodulators may also regulate the numbers and functions of various cell types, including astroglia. Communication between neurons and glial cells is of fundamental importance in the regulation of nervous system development with the flow of interacting substances moving in both directions. Neurons provide substances that influence glial function[1,2,3] and glia produce substances important for the survival and growth of neurons[4,5,6,7,8,9].

Astroglia play a significant role during neurodevelopment. These cells have been shown to be involved in axonal guidance, neuronal migration, neuritic outgrowth, and trophic factor production[10]. Thus, an understanding of the regulation of glial cellular activity and number is essential to our knowledge of mechanisms which influence the maturation and development of the whole nervous system. Astroglial proliferation can be stimulated by the following substances: interleukin-1[11]; plasminogen activator[12]; fibroblast growth factor[13]; thrombin[14]; platelet-derived growth factor[15]; and epidermal growth factor[16]. In addition, previous studies have suggested that vasoactive intestinal peptide (VIP) is a neuronal signal molecule that influences glial function[8,9].

VIP is a 28 amino acid neuropeptide which is involved in diverse regulatory functions including vasodilation, gastric secretion, glycogenolysis and growth[17]. In the central nervous system (CNS), VIP exhibits neurotransmitter and neuromodulator functions, however, it has also been shown to have growth and survival-promoting actions[8,9,18,19,20,21]. Furthermore, the treatment of rat neonates from birth to two weeks of age with an antagonist to VIP has been shown to result in the retardation in the appearance of several complex motor behaviors[22]. These data suggest that VIP may have growth factor functions in the developing nervous system.

Growth Factors, Peptides, and Receptors, Edited
by T.W. Moody, Plenum Press, New York, 1993

Binding studies with CNS membranes have revealed both high and low affinity VIP receptors[23,24,25,26] and *in vitro* autoradiography has shown that VIP receptors are distributed throughout the adult CNS and are localized to distinct cytoarchitectural sites[26,27,28,29,30]. Based on their sensitivity to guanine nucleotides, VIP receptors can be differentiated into two subtypes in the brain[30]. These two receptor subtypes differ in their affinity for VIP and also have different distribution patterns in brain[30]. Furthermore, radioligand binding studies have indicated that cerebral cortical astrocytes grown in cell culture exhibited two classes of binding sites. A high affinity site (Kd: 70 pM) and a low affinity site (Kd: 0.8 μM) can be distinguished with each showing distinct pharmacological and functional properties[31].

We have examined both the stimulatory effect of VIP on astroglial mitosis *in vitro*, and traced the *in vivo* appearance and pattern of distribution of VIP receptors during the development of the nervous system. These studies suggest that, in addition to other potential growth-regulatory functions, VIP has a critical role in the mitogenesis of astrocytes during pre- and post-natal neural development.

IN VITRO STUDIES

The effect of VIP on the mitosis of astrocytes was tested on spinal cord cultures under conditions identical to those employed to study the neuron survival-promoting effects of VIP[18,32,33]. Two methods were used to evaluate the effect of a range of concentrations of VIP on gliogenesis: 1) ^{3}H-thymidine incorporation in dissociated spinal cord cultures was measured; and 2) cultures were stained with glial fibrillary acidic protein (GFAP) and the number of GFAP-positive cells counted[9].

[3]H-thymidine incorporation DNA was done in standard low-density cultures composed of both neurons and glia. Cultures were treated with 1 μM tetrodotoxin to prevent activity-induced release of endogenous VIP by neurons. As shown in Figure 1, a dose-dependent increase in the amount of ^{3}H-thymidine incorporation was observed within 5 days of VIP treatment. The effective concentration range was 10^{-11} to 10^{-7} M VIP. To substantiate that VIP was an astroglial mitogen, the number of astrocytes were counted as well as the number of silver grain-positive cells after ^{3}H-thymidine autoradiography. A concentration-dependent increase in the number of GFAP-positive cells after VIP application is shown in Figure 1. Again the small but potent response to VIP is similar to that observed for DNA synthesis.

To determine if astroglia were among the cell types that exhibited an increase in thymidine uptake after VIP treatment, cultures were stained immunocytochemically for GFAP and developed for thymidine autoradiography as shown in Figure 2. These analyses indicated that some, but not all the cells displaying silver grains were GFAP-positive. In addition, not all GFAP-positive cells possessed silver grains. Cell counts of these cultures are shown in Figure 3.

These studies revealed that the number of GFAP-positive cells that were labeled with the ^{3}H-thymidine was increased significantly by treatment of 0.1 nM VIP as compared to that of control cultures. In contrast, the number of cells that were labeled with ^{3}H-thymidine but were not GFAP-positive showed no increase after VIP treatment. These non-responsive cells were

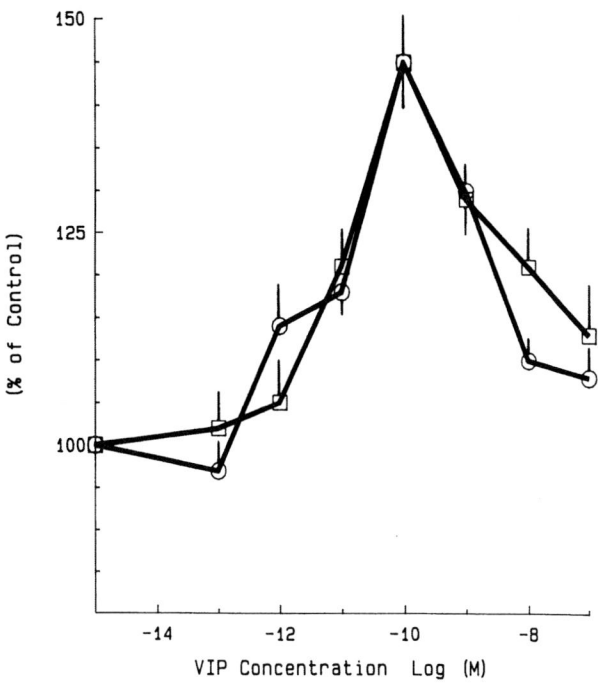

Figure 1. Effect of VIP on ^{3}H-thymidine incorporation into DNA (squares) and on astroglial cell number (circles): concentration dependency. Treatment of the standard cultures (neurons plus non-neuronal cells) with peptide was started 9 days after plating[9]. A complete change of medium was made prior to the addition of the peptide. Cultures received one application of peptide and the duration of the treatment was 5 days. Cell counts were made from 100 regularly spaced microscopic fields (0.5 mm^2 each). For thymidine incorporation, each value is the mean of 10-12 determinations from four experiments. For the cell counts, each value is the mean of 5-7 dishes from two experiments. The error bars represent the standard error. Statistical comparisons were made by analysis of variance with the Newman-Kuels multiple comparison of means test. For ^{3}H thymidine incorporation, significant increases (p<0.002) from controls were detected from 10^{-11} to 10^{-7} M VIP. Similarly, for cell counts, significant increases (p<0.02) from controls were observed at 10^{-12} to 10^{-7} M VIP.

probably meningeal fibroblasts. Therefore the effects of VIP on thymidine uptake showed cellular specificity. Neurons in these cultures were post-mitotic.

The *in vitro* studies outlined here demonstrate that low concentrations of VIP can increase GFAP-positive cells, but that other cell types, for example fibroblasts, are not affected. This suggests a physiological role for VIP in regulating the number of support cells in the developing nervous system.

IN VIVO STUDIES

As a measure of the extent of the involvement of VIP in the growth and differentiation of the central nervous system, VIP receptor distribution was traced in the rat brain during development from embryonic day 14 through to the adult. Guanine nucleotides inhibit VIP binding by increasing the dissociation of VIP from its receptor[23,34,35]. However, *in vitro* autoradiography in the presence of GTP, reveals that only a subset of VIP binding sites are GTP-sensitive; that is, VIP is displaced from these receptors,

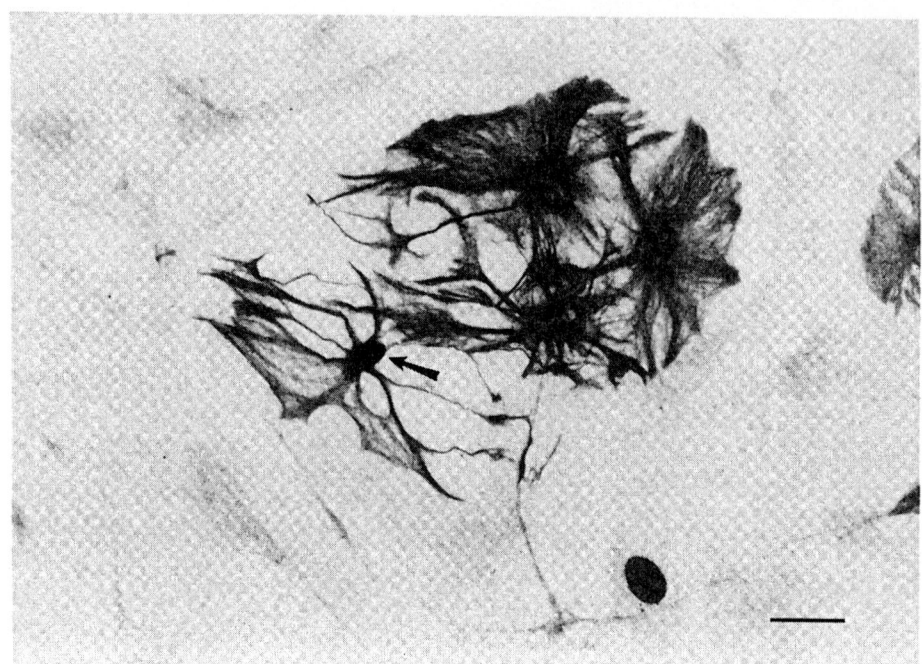

Figure 2. ^{3}H-thymidine autoradiogram of standard spinal cord cultures stained immunohistochemically with antiserum to GFAP. Cultures were analyzed at 2 weeks. Label was added 24 h before harvest. The arrow identifies a grain-positive astrocyte. The calibration bar is 50 μm.

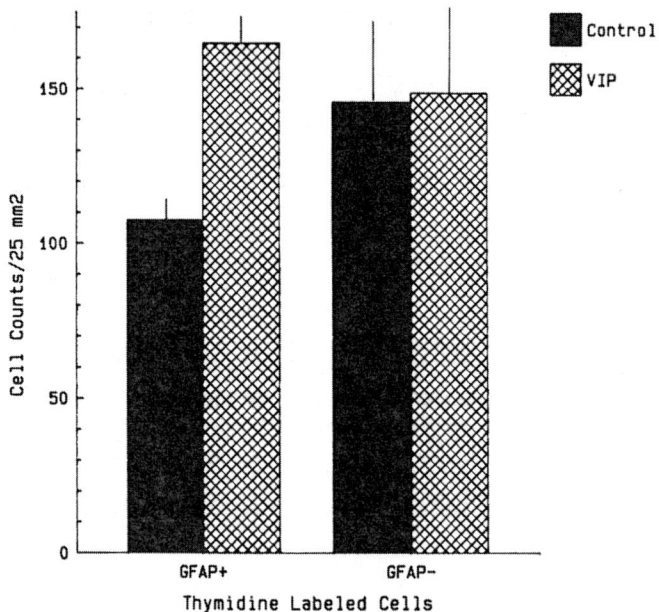

Figure 3. Cell counts of standard spinal cord cultures after ^{3}H-thymidine autoradiography and GFAP immunocytochemistry: comparison of VIP stimulation between GFAP-positive and non-GFAP cells. Counts were obtained from 100 fields with a total area of 0.25 mm[2] each. Values are the mean of 6 determinations. The number of fibroblasts relative to the number of GFAP-positive cells varied significantly between preparations, but there was no evidence of an increase in these cells after VIP treatment in any of the experiments.

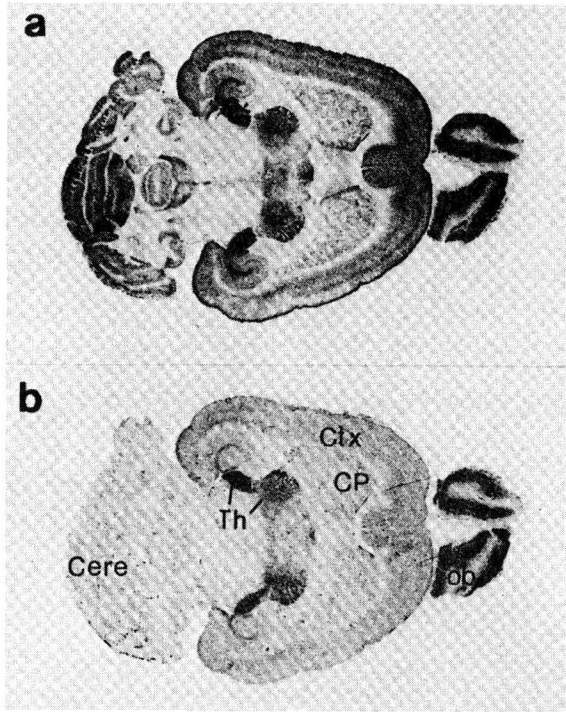

Figure 4. Film autoradiographs of ^{125}I-VIP binding in horizontal rat brain sections. Tissue slices were incubated as described previously[30] . Briefly, fresh frozen brain tissue was cryostat cut at 20 µm and dehydrated at 0°C overnight. A 30 min preincubation in 10 mM HEPES with 130 mM NaCl, 4.7 mM KCl, 5 mM MgCl$_2$, 5 mM MnCl$_2$, 1 mM ethylenediaminetetraacetate and 1% bovine serum albumin with pH adjusted 7.4 with NaOH[27], was followed by a 1 h incubation in the same buffer at room temperature with 1 mg/ml bacitracin and 50 pM ^{125}I-labelled VIP (Amersham) with and without 1 µM VIP (Peninsula) or 10 µM of a stable guanosine 5'-triphosphate analogue, guanylyl-imidodiphosphate (GMP-PNP, Boehringer Mannheim). Sections were then washed for one min in each of 3 washes of cold PBS and dried under a stream of air. Sections were placed in a cassette with Amersham Hyperfilm-^{3}H for 4 days. The film was developed in Kodak D-19 at 20°C for 4 min. a. Total ^{125}I-VIP binding. b. ^{125}I-VIP binding in the presence of 10 µM GMP-PNP revealing GTP-insensitive binding sites. Cere, cerebellum; CP, caudate/putamen; Ctx, cortex; ob, olfactory bulb; Th, thalamus.

and that a second subset is insensitive to the inhibiting effects of GTP[30]. As shown in Figure 4 (a - total binding, b - binding in the presence of 10 µM of the non-hydrolyzable GTP-analogue, guanylyl-imidodiphosphate [GMP-PNP]), in most brain regions, including the cortex (Ctx) and caudate/putamen (CP), GTP reduced VIP binding between 40 and 60%. However, in some regions, such as the cerebellum (Cere), supraoptic nucleus, locus coeruleus, interpeduncular nucleus, facial nucleus, olfactory tubercle and periventricular hypothalamic nucleus, 80% or more of VIP binding was inhibited. In other brain regions, including the olfactory bulbs (ob), and ventral thalamic nuclei and medial geniculate (Th), GTP had little effect on VIP binding[30]. Since GTP sensitivity is characteristic of receptors linked to adenylate cyclase second messenger mechanisms[36], the GTP-sensitive VIP receptor in the brain is probably the adenylate cyclase-linked VIP receptor which stimulates cAMP pathways. The GTP-insensitive site is probably linked to an alternate second messenger system.

In our examination of VIP receptor binding throughout development, we observed that from the earliest age studied (E14), VIP receptors, both GTP-sensitive and GTP-insensitive, were widespread and abundant in the CNS and expressed changing patterns of distribution. The pattern of VIP binding was seen to occur in three phases and was characterized by the transient appearance of GTP-insensitive and GTP-sensitive VIP receptors. During the first phase, E14-16, VIP binding was relatively uniform and dense throughout the brain stem and spinal cord; but in brain, VIP receptors were present only in discrete sites including the posterior commissure, thalamus and cortex. The uniformly dense VIP binding was GTP-sensitive; however, intense VIP binding in the midline regions of the floor and roof plates of the brain stem and spinal cord were GTP-insensitive. Throughout the second phase, E19-P14, while VIP binding was higher in the germinal regions of the periventricular layers of the lateral, third and fourth ventricles and other brain regions undergoing gliogenesis, it tended to be uniformly dense throughout the brain with little regional differentiation. The uniformly dense VIP binding was GTP-sensitive and the periventricular regions expressed primarily GTP-insensitive receptors. During third phase, beginning at P21, the adult pattern began to emerge with both GTP-sensitive and GTP-insensitive VIP binding sites unevenly distributed and related to specific cytoarchitectural sites.

Although VIP binding was dense throughout the spinal cord, beginning at E14 the densest binding, 90% of which was GTP-insensitive, occurred along midline structures including the roof and floor plates. These are exclusive sites of macroglial formation[37]. Gliogenesis begins in the floor and roof plates on E13 and continues, with a high level of fasciculation of the glial fibers[38], in midline structures until E19. The dense GTP-insensitive VIP binding at these midline structures occurring at the time of gliogenesis and glial fasciculation strongly indicates an important role for VIP in these developmental events. Furthermore, as will be shown below, throughout the development of the brain, the transient appearance of dense GTP-insensitive VIP binding is correlated both temporally and spatially with gliogenesis and/or glial fasciculation.

In the brain stem during prenatal development the most intense binding, which was GTP-insensitive, also occurred in midline structures, including the posterior commissure. As in the spinal cord, these are also sites of gliogenesis and radial glial fasciculation. In the thalamus of the E16 embryo VIP binding appeared to be maximal in the intermediate medial region which, at this age, is occupied by numerous fasciculated radial glial fibers[38].

In the forebrain, as in the brain stem and spinal cord, dense GTP-insensitive VIP binding occurred in the periventricular neuroepithelial regions of during periods of glial proliferation and fasciculation. In the cortex (Figure 5) of the E19 through to P7 brain, the most dense VIP binding, which was GTP-insensitive, occurred in the germinative ventricular zone (cx) and in the molecular layer (1). At this time, the intense mitotic activity occurring in this layer is associated with the production of glial cells[39,40]. Neuronal production in the germinative zone has all but ceased. Meanwhile, the molecular layer is also a major site of gliogenesis during the first postnatal

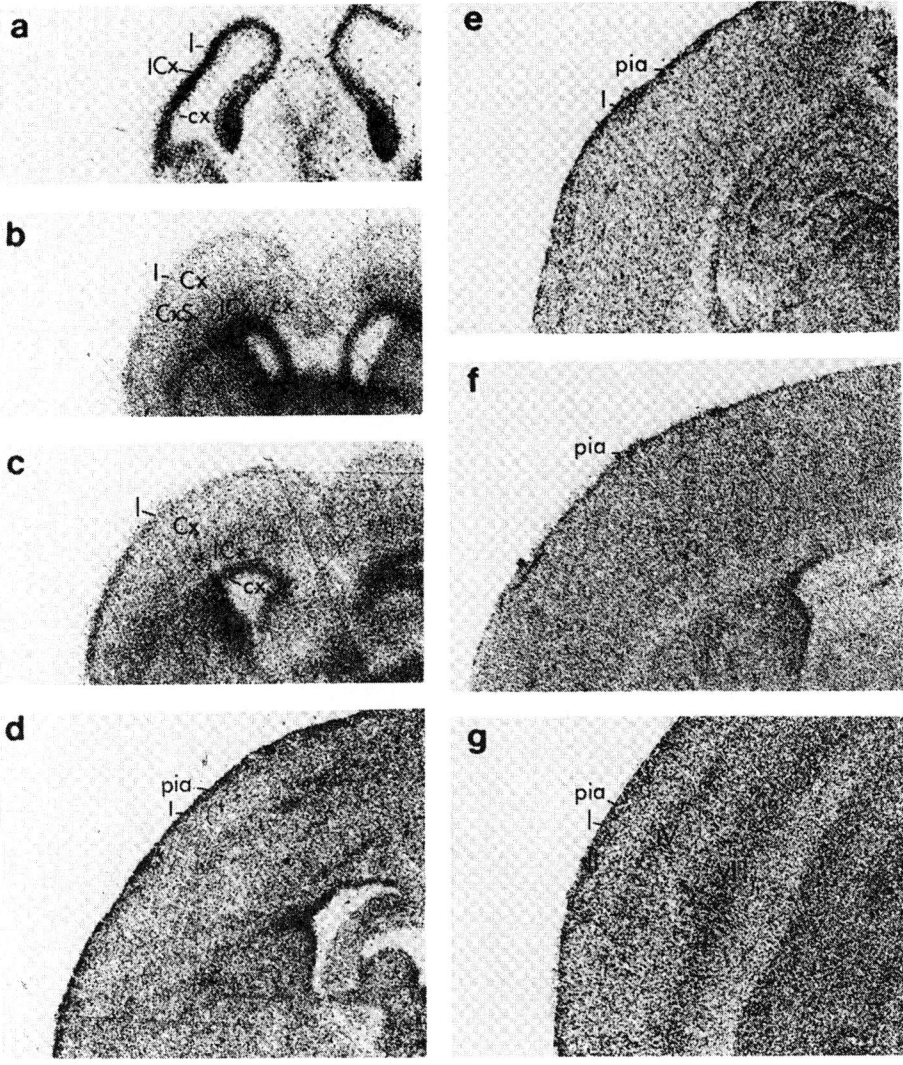

Figure 5. Autoradiographs of ^{125}I-VIP binding in coronal sections of cortex throughout development of E16 (a) to adult (g). Tissue slices were incubated under conditions described in Figure 4. a. E16 embryo. b. E19 embryo. c. P0 rat pup. d. P7 rat pup. e. P14 rat pup. f. P21 rat pup. g. adult (P60) rat. I-VI, cortical layers I through VI; Cx, cortical plate; cx, cortical neuroepithelium, CxS, cortical subplate zone; ICx, intermediate cortical layer.

week[40]. With the disappearance of the neuroepithelial layer (cx) at P7, the VIP binding also disappeared.

In the E16 hippocampus (Figure 6), a high density of VIP receptors was present throughout the cortical plate (Hi) and the dentate gyrus (DG), areas where radial glia are increasing to guide the prospective migrating neurons[41,42]. From E19 to P7, the highest density of VIP receptors, most of which were GTP-insensitive, were localized in the germinative neuroepithelium (hi) and in the molecular layer of the dentate gyrus (mDG). This last region has been described as a major site of perinatal

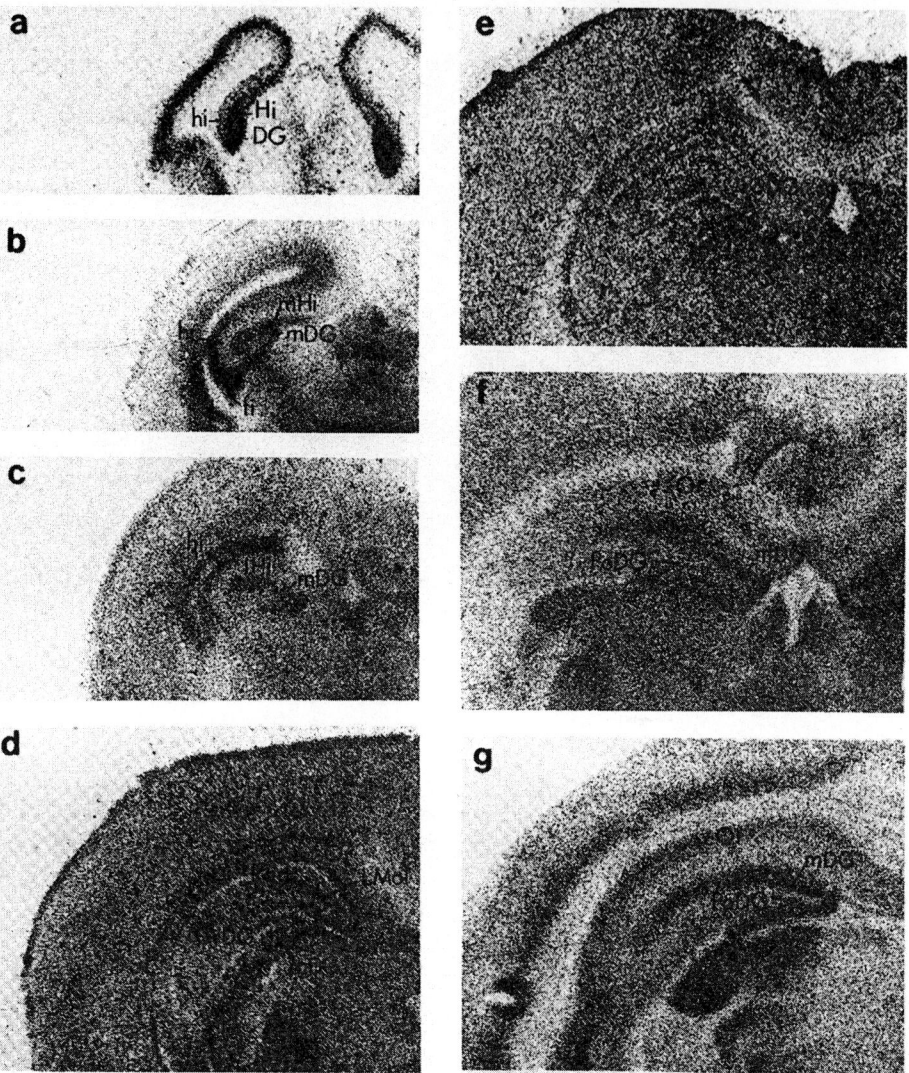

Figure 6. Autoradiographs of ^{125}I-VIP binding in coronal sections of hippocampus throughout development from E16 (a) to adult (g). Tissue slices were incubated under conditions described in Figure 4. a. E16 embryo. b. E19 embryo. c. P0 rat pup. d. P7 rat pup. e. P14 rat pup. f. P21 rat. g. adult (P60) rat. DG, dentate gyrus; fi, fimbria; GrDG, granule layer of dentate gyrus; Hi, cortical plate of hippocampus; hi, hippocampal neuroepithelium; IHi, intermediate layer of hippocampal cortex; LMol, lacunosum moleculare; mDG, molecular layer of dentate gyrus; mHi, marginal zone of the hippocampal cortex; Or, stratum oriens; PoDG, polymorph zone of dentate gyrus; Py, pyramidal cell layer; Rad, stratum radiatum.

astrocytogenesis[42,43]. We can assume that since neuronal production in the germinative zone had all but ceased, the postnatal germinative neuroepithelium was committed to gliogenesis.

In the cerebellum (not shown), between E16 and P7, VIP binding was the highest in the area external to the Purkinje layer, including the external granular layer. At this developmental period, this is a site of intense production and subsequent fasciculation of Bergman radial glia[38]. The perinatal astrocytogenesis in the deep nuclei was also paralleled by an increase of VIP binding.

Therefore, we show that during the development of the brain and spinal cord there is a transient population of GTP-insensitive VIP binding sites which are numerous in the generative epithelial zones and other regions of the brain during the time when these cells are undergoing gliogenesis and glial fasciculation.

In addition, a uniformly dense blanket of GTP-sensitive VIP receptors was seen throughout the spinal cord and brain stem of the E14-E16 embryo and beginning late in gestation and continuing until at least P14, the period of the "brain growth spurt" of the rat[44], this uniformly dense pattern of GTP-sensitive VIP binding was seen throughout the brain. In both spinal cord and brain, these periods are characterized by multitude of developmental events including neuronal differentiation, axonal growth, synaptogenesis, the appearance of neurotransmitter phenotypes, neuronal death, gliogenesis and angiogenesis. At present, it is not possible to identify which of these event(s) may be regulated by VIP. Nevertheless, *in vitro* studies have demonstrated the presence of VIP receptors on axonal growth cones[45] and that VIP can promote neurogenesis and neuronal growth directly[20,21] or neuronal survival indirectly through the release of trophic factors by astrocytes[8,9].

SUMMARY

To investigate the role of VIP on glial function during development, the present study has sought to integrate observations made on astroglia mitosis in CNS-derived cultures with studies of VIP binding sites in the developing brain. The *in vitro* experiments clearly demonstrated the potent influence of VIP in specifically increasing the mitogenic activity of astroglia but not meningeal fibroblasts[9]. This effect of VIP occurred at low concentrations (EC50: 50 pM), consistent with observations of previous radioligand binding studies which indicated the existence of a high affinity receptor (Kd: 70 pM) on cultured astroglia from cerebral cortex[31]. A goal of the present paper was to attempt to place these observations made on *in vitro* systems into the context of the developing brain. Based on the growing knowledge of the anatomical localization of gliogenesis during ontogeny and the description of two GTP-distinguishable binding sites for VIP during development, several correlations are suggested. During development of the CNS, there is a transient population of GTP-insensitive VIP binding sites which are numerous and localized to the generative epithelial zones and other regions of the brain during the time when these cells are undergoing gliogenesis and glial fasciculation. It is our hypothesis that these GTP-insensitive sites are non-adenylate cyclase-mediated receptors that correspond to the high affinity VIP receptor observed on glial cells *in vitro*. These data further suggest that the high affinity, GTP-insensitive sites are the receptors which mediated the mitogenic and trophic properties of VIP on astroglia. Several lines of evidence have suggested that VIP, in addition to stimulating cAMP pathways, may function through a phospholipid transduction system[46,47,48,49,50,51,52]. The GTP-sensitive sites, which are presumed to be linked to cAMP generation, mediate such functions as glycogenolysis or neurogenesis, as previously suggested[20,21,53]. Together, these data support the idea that VIP and developing neurons can influence the number and function of astroglia and that these effects can have widespread influence as indicated by the presence of VIP receptors in the developing brain.

REFERENCES

1. S.B. Masters, T.K. Harden, and J.H. Brown, Relationships between phosphoinositide and calcium responses to muscarinic agonists in astrocytoma cells, *Mol. Pharmacol.* 26:149 (1984).
2. J.J. DeGeorge, P. Morell, K.D. McCarthy, and E.G. Lapetina, Adrenergic and cholinergic stimulation of arachidonate and phosphatidate metabolism in cultured astroglial cells, *Neurochem. Res.* 11:1061 (1986).
3. B. Pearce, J. Albrecht, C. Morrow, and S. Murphy, Astrocyte glutamate receptor activation promotes inositol phospholipid turnover and calcium flux, *Neurosci. Lett.* 72:335 (1986).
4. G.A. Banker, Trophic interactions between astroglial cells and hippocampal neurons in culture, *Science* 209:809 (1980).
5. H.W. Muller, S. Beckh, and W. Seifert, Neurotrophic factor for central neurons, *Proc. Natl. Acad. Sci. USA* 81:1248 (1984).
6. K.L. Eagleson, T.R. Raju, and M.R. Bennett, Motoneurone survival is induced by immature astrocytes from developing avian spinal cord, *Dev. Brain Res.* 17:95 (1985).
7. D.E. Brenneman, S. Fitzgerald, and P.G. Nelson, Interaction between trophic action and electrical activity in spinal cord cultures, *Dev. Brain Res.* 15:211 (1984).
8. D.E. Brenneman, E.A. Neale, G.A. Foster, S. d'Autremont, G.L. Westbrook, Nonneuronal cells mediate neurotrophic action of vasoactive intestinal peptide, *J. Cell Biol.* 104:1603 (1987).
9. D.E. Brenneman, T. Nicol, D. Warren, and L.M. Bowers, Vasoactive intestinal peptide: a neurotrophic releasing agent and an astroglial mitogen, *J. Neurosci. Res.* 25:386 (1990).
10. J.S. Manthorpe, J.S. Rudge, and S. Varon, Astroglia cell contributions to neuronal survival and neurite growth, *in:* "Astrocytes," S. Federoff and A. Vernadakin, eds., Academic Press, New York (1986).
11. D. Guilian and L.B. Lachman, Interleukin-1 stimulation of astroglial proliferation after brain injury, *Science* 228:497 (1985).
12. G. Moonen, M-P Grau-Wagemans, I. Selak, P.P. Lefebvre, B. Rogister, J.D. Vassalli, and D. Belin, Plasminogen activator is a mitogen for astrocytes in developing cerebellum, *Dev. Brain Res.* 20:41 (1985).
13. R.S. Morrison and J. de Vellis, Growth of purified astrocytes in a chemically defined medium, *Proc. Natl, Acad. Sci. USA* 78:7205 (1981).
14. F. Perraud, F. Besnard, M. Sensenbrenner, and G. Labourdette, Thrombin is a potent mitogen for rat astroblasts but not for oligodendroblasts and neuroblasts in primary culture, *Int. J. Dev. Neurosci.* 5:181 (1987).
15. C.H. Heldin, B. Westermark, and A. Wasteson, Platelet-derived growth factor: purification and partial characterization, *Proc. Natl. Acad. Sci USA* 76:3722 (1979).
16. B. Westermark, Density dependent proliferation of human glial cells stimulated by epidermal growth factor, *Biochem. Biophys. Res. Commun.* 69:304 (1976).
17. I. Gozes and D.E. Brenneman, VIP: molecular biology and neurobiological function, *Mol. Neurobiol.* 3:201 (1989).
18. D.E. Brenneman, L.E. Eiden, and R.E. Siegel, Neurotrophic action of VIP on spinal cord cultures, *Peptides* 6 (suppl.2):35 (1985).
19. D.E. Brenneman and L.E. Eiden, Vasoactive intestinal peptide and electrical activity influence neuronal survival, *Proc. Natl. Acad. Sci .USA* 83:1159 (1986).
20. D.W. Pincus, E.M. DiCicco-Bloom, and I.B. Black, Vasoactive intestinal peptide regulation of neuroblast mitosis and survival: role of cAMP, *Brain Res.* 514:355 (1990).
21. D.W. Pincus, E.M. DiCicco-Bloom, and I.B. Black, Vasoactive intestinal peptide regulates mitosis, differentiation and survival of cultured sympathetic neuroblasts, *Nature (Lond.)* 343:564 (1990).
22. J.M. Hill, I. Gozes, J.L. Hill, M. Fridkin, and D.E. Brenneman, Vasoactive intestinal peptide antagonist retards the development of neonatal behaviors in the rat, *Peptides* 12:87 (1991).
23. P. Robberecht, P. DeNeef, M. Lammens, M. Deschodt-Lanckman, and J. Christophe, Specific binding of vasoactive intestinal peptide to brain membranes from the guinea pig, *Eur. J.Biochem.* 90:147 (1978).

24. J. Fahrenkrug, Receptors for vasoactive intestinal polypeptide on isolated synaptosomes from rat cerebral cortex. Heterogeneity of binding and desensitization of receptors, *J. Neurochem.* 39:1242 (1982).

25. P. Staun-Olsen, B. Ottesen, P.D. Bartels, M.H. Nielsen, S. Gammeltoft, and J. Fahrenkrug, Receptors for vasoactive intestinal polypeptide on isolated synaptosomes from rat cerebral cortex. Heterogeneity of binding and desensitization of receptors, *J. Neurochem.* 39:1242 (1982).

26. J. Besson, Distribution and pharmacology of vasoactive intestinal peptide receptors in the brain and pituitary, *Ann. N.Y. Acad. Sci.* 527:204 (1988).

27. M.M. Shaffer and T.W. Moody, Autoradiographic visualization of CNS receptors for vasoactive intestinal peptide receptor, *Peptides* 7:283 (1986).

28. J-L. Martin, M.M. Dietl, P.R. Hof, J.M. Palacios, and P.J. Magistretti, Autoradiographic mapping of [mono(^{125}I)iodo-Tyr^{10}, MetO^{17}] vasoactive intestinal peptide binding sites in the rat brain, *Neurosci.* 23:539 (1987).

29. C.J. Wiedermann, K. Sertl, B. Zipser, J.M. Hill, and C.B. Pert, Vasoactive intestinal peptide receptors in rat spleen and brain: A shared communication network, *Peptides 9* (Suppl):21 (1988).

30. J.M. Hill, A. Harris, and D.I. Hilton-Clarke, Regional distribution of guanine nucleotide-sensitive and guanine nucleotide-insensitive vasoactive intestinal peptide receptors in rat brain, *Neurosci.* 48:925 (1992).

31. I. Gozes, S.K. McCune, L. Jacobson, D. Warren, T.W. Moody, M. Fridkin, and D.E.Brenneman, An antagonist to vasoactive intestinal peptide affects cellular functions in the central nervous system, *J. Pharmacol. Exp. Ther.* 257:959 (1991).

32. D.E. Brenneman, E.A. Neale, W.H. Habig, L.M. Bowers, and P.G. Nelson, Developmental and neurochemical specificity of neuronal deficits produced by electrical impulse blockade in dissociated spinal cord cultures, *Dev. Brain Res.* 9:13 (1983).

33. D.E. Brenneman, S. Fitzgerald, and M. J. Litzinger, Neuronal survival during electrical blockade is increased by 8 bromo cyclic adenosine 3',5' monophosphate, *J. Pharmacol, Exp. Ther.* 233:402 (1985).

34. JA. Couvineau, S. Gammeltoft, and M. Laburthe, Molecular characteristics and peptide specificity of vasoactive intestinal peptide receptors from rat cerebral cortex, *J. Neurochem.* 47:1469 (1986).

35. A. Couvineau, C. Rouyer-Fessard, T. Voisin, and M. Laburthe, Functional and immunological evidence for stable association of solubilized vasoactive intestinal peptide receptor and stimulatory guanine-nucleotide-binding protein from rat liver, *Eur. J. Biochem.* 187:605 (1990).

36. T.W. Moody, D.P. Taylor, and C.B. Pert, Effects of guanine nucleotides on CNS neuropeptide receptors, *J. Supramole. Struct.* 15:153 (1981).

37. J. Altman and S.A. Bayer, The development of the rat spinal cord, *Adv. Anat. Embryol. Cell Biol.* 85:1 (1984).

38. M.A. Edwards, M. Yamamoto, and V.S. Caviness Jr., Organization of radial glia and related cells in the developing murine CNS. An analysis based upon a new monoclonal antibody marker, *Neurosci.* 36:121 (1990).

39. D. Schmechel and P. Rakic, Arrested proliferation of radial glial cells during midgestation in rhesus monkey, *Nature* 177:303-305 (1979).

40. P. Gressens, C. Richelme, H.J. Kadhim, J.F. Gadisseus, and P. Evrard, The germinative zone produces the most cortical astrocytes after neuronal migration in the developing mammalian brain, *Biol. Neonate* 61:4 (1992).

41. M. Rickmann, D.G. Amaral, and W.M. Cowan, Organization of radial glial cells during the development of the rat dentate gyrus, *J. Comp. Neurol.* 264:449 (1987).

42. J.A. del Rio, B. Heimrich, E. Soriano, H. Schwegler, and M. Frotscher, Proliferation and differentiation of glial fibrillary acidic protein-immunoreactive glial cells in organotypic slice cultures of rat hippocampus, *Neurosci.* 43:335 (1991).

43. S.A. Bayer, Development of the hippocampal region in the rat. II. Morphogenesis during embryonic and early postnatal life, *J. Comp. Neurol.* 190:115 (1980).

44. J. Dobbing, The later development of the brain and its vulnerability, *in*: "Scientific Foundation of Paediatrics," J.A. Davis and J Dobbing, eds., Saunders, Philadelphia (1974).

45. R.O. Lockerbie, K. Herve, G. Blare, J.P. Tassin, and J. Glowinski, Isolated neuronal growth cones from developing rat forebrain possess adenylate cyclase activity that can be augmented by various receptor agonists, *Devel. Brain Res.* 38:19 (1988).

46. S. Audigier, C. Barberis, and S. Jard, Vasoactive intestinal polypeptide increases inositol phospholipid breakdown in the rat superior cervical ganglion, *Brain Res.* 376:363 (1986).

47. T. Durroux, C. Barberis, and S. Jard, Vasoactive intestinal polypeptide and carbachol act synergistically to induce the hydrolysis of inositol containing phospholipids in the rat superior cervical ganglion, *Neurosci. Lett.* 75:211 (1987).

48. C.L. Weill, VIP activates protein kinase C in nuclei isolated from rat hippocampus, *Soc. Neurosci. Abstr.* 15:1364 (1989).

49. J.T. Russell, A. Fatatis, P.G. Nelson, and D.E. Brenneman, Vasoactive intestinal polypeptide (VIP) causes intracellular calcium oscillations in astrocytes, *Soc. Neurosci. Abstr.* 16:994 (1990).

50. D.V. Agoston, L.E. Eiden, D.E. Brenneman, and I. Gozes, Spontaneous electrical activity regulates vasoactive intestinal peptide expression in dissociated spinal cord cell cultures, *Mol. Brain Res.* 10:235 (1991).

51. Z. Olah, S. Komoly, D. Warren, D.E. Brenneman, and D.V. Agoston, Protein kinase C-related signal transduction pathway is triggered by subnanomolar concentration of VIP in astrocytes, *Soc. Neurosc. Abstr.* 17:603 (1991).

52. I. Tatsuno, T. Yada, S. Vigh, H. Hidaka, and A. Arimura, Pituitary adenylate cyclase activating polypeptide and vasoactive intestinal peptide increase cytosolic free calcium concentration in cultured rat hippocampal neurons, *Endo.* 131:73 (1992).

53. P.J. Magistretti, M. Manthorpe, F.E. Bloom, and S. Varon, Functional receptors for vasoactive intestinal polypeptide in cultured astroglia from neonatal rat brain, *Regul. Pept.* 6:71 (1983).

SYNTHESIS AND BIOLOGICAL ACTIVITY OF

CHOLECYSTOKININ ANTAGONISTS

Martha Knight[1], Kazuyuki Takahashi[1], Terrence R. Burke, Jr.[1],
J. Desiree Pineda[1], Vitaly A. Fishbeyn[2], Robert T. Jensen[2] and
Terry W. Moody[3]

[1]Peptide Technologies Corporation
125 Michigan Ave. N.E.
Washington, D.C. 20017

[2]Digestive Disease Branch
National Institute of Metabolic, Digestive
and Kidney Diseases
National Institutes of Health
Bethesda, MD 20892

[3]Biochemistry and Molecular Biology
George Washington University Medical Center
2300 Eye St. N.W.
Washington, D.C. 20037

INTRODUCTION

Cholecystokinin (CCK) is a gastrointestinal hormone[1] that stimulates satiety[2] and pancreatic enzyme secretions in digestion. The active portion of the hormone is the carboxyl terminal octapeptide sequence. CCK is present as a neuropeptide in the CNS[3] with the octapeptide being the active released species from neurons. CCK-8 is present in vagal afferents and in the brain, it is co-localized with dopamine[4]. Thus, besides a role in mediation of gastrointestinal functions, a CNS role for CCK in satiety and in modulation of dopamine release or function is investigated.

CCK interacts with specific receptors with high affinity. The gastrointestinal type exists on pancreatic acinar cells[5] and the neural type is present on brain membranes and more recently has been characterized on a small cell lung carcinoma (SCLC) cell line[6]. The receptor type specific for gastrin which has the CCK carboxyl tetrapeptide sequence has been characterized in brain as well. To date two molecular species of receptors for CCK have been cloned, a CCK_A[7] receptor subtype and a $CCK_{B/gastrin}$ subtype[8]. The receptors are chiefly distinguished by the degree of specificity of the receptors for the unsulfated form of CCK-8 (or CCK 26-33). There are also differences in the affinities of the CCK-8 carboxyl fragments, CCK-5 and

CCK-4. The CCK_A receptors bind these species much less potently than CCK-8 whereas the $CCK_{B/gastrin}$ receptors bind the other peptides with similar affinity[5]. The amino terminal fragment peptides missing the carboxyl terminal amino acids (Phe[33], Asp[32]-Phe[33] and Met[31]-Asp[32]-Phe[33]) have lower affinities for the receptors but, importantly, were discovered to be specific CCK receptor inhibitors and thus comprise a peptide class of CCK antagonists[9].

In the present study analogues of the active octapeptide which are agonists and analogues of the amino-terminal fragment peptides, CCK(26-32) and (26-31) which are known to be antagonists at the guinea pig receptor[9-11] were synthesized with particular amino acid substitutions to confer increased stability and potency. Solid-phase methods were used for their synthesis and unique methods of countercurrent chromatography[12] and reversed-phase chromatography were utilized as steps in their purification. The biological activities of the peptides were determined in the pancreatic acinar cell secretion assay. The results of these studies indicated further structural modifications to increase potency. More CCK antagonist peptides were then synthesized and tested. The effects of these peptides were also studied on $CCK_{B/gastrin}$ receptors of small cell lung carcinoma cells.

STRUCTURE-ACTIVITY RELATIONSHIPS OF CHOLECYSTOKININ

The methionines in positions 28 and 31 of the naturally occuring peptide were substituted by isosteric norleucines which do not form oxidation products. In one analogue N-methyl leucine was substituted for Met[31] to prevent enzymatic breakdown at the Trp[30]-Met[31] peptide bond. Gly[29] was substituted in most of the peptides by D-alanine to confer stabilization of a ß bend in the octapeptide, possibly induced by a D-amino acid conformation[13]. Previously L- and D-Ala substituted analogues were studied for binding to guinea pig brain cortical receptors and the D-Ala analogue was found to interact preferentially[14]. The D-Ala[29] substitution also prevents the initial proteolytic cleavage step in the sequential enzymatic degradation that takes place in some in vitro tissue preparations[15-17]. Another substitution was made at position 29 using N-methyl alanine to prevent proteolysis. L- and D-Trp bearing an indole side group were substituted for Gly to observe the large hydrophobic side group's effect on the biological activity. All of the analogues have either an acetyl or a ß-naphthoyl amino-protecting group to prevent aminopeptidase degradation thus increasing in vitro and in vivo stability. The role of the carboxyl group in Asp[26] was studied by substituting Asn in one peptide.

Pancreatic acinar cell assay

Dispersed acini were prepared by collagenase digestion of guinea pig pancreas with previously described modifications[18] of the method of Peikin[19]. For agonist activity, amylase release from acini incubated in the presence of various concentrations of peptide analogue was measured. For antagonism determination, the effect of various concentrations of the peptide on amylase release stimulated by 0.1 nM CCK(26-33), a half-maximally effective concentration (EC_{50}) was measured. Acini of one pancreas was suspended in 125 ml of incubation solution consisting of 25.5 mM Hepes, pH 7.4, 98 mM NaCl, 6 mM KCl, 2.5 mM NaH_2PO_4, 5 mM Na pyruvate, 5 mM Na fumarate, 5 mM Na glutamate, 2 mM glutamine, 11.5 mM glucose, 0.5 mM $CaCl_2$, 1 mM $MgCl_2$, 1% essential vitamin mixture and equilibrated with 100% oxygen. Dispersed cell samples of 250 ul, containing peptides were incubated for 30 min at 30°C. Amylase activity was determined with the Phadebas reagent[18,20]. Amylase release was calculated as the percentage of amylase activity in the acini at the beginning of the incubation that was released into the extracellular medium during the incubation.

Activity of CCK octapeptide agonists in pancreatic acini

The activities of the agonist analogues in stimulation of pancreatic acinar cell secretion were determined as described above. The most potent compounds are D-Ala[29] and D-Trp[29] CCK (26-33) which were 1/100 as potent as CCK(26-33) with EC_{50} values of 8.6 and 7.6 nM, respectively. The next most potent compounds were the Nle substituted compounds Nle[28,31] D-Ala[29] and Asn[26] CCK(26-33) with potencies of 17.6 and 20.0 nM respectively. The remaining agonist peptides, a disulfated peptide, Ser[32]SO₃H CCK, L-Ala[29] CCK, L-Trp[29] CCK and N-MeAla[29],N-MeLeu[31] CCK had much lower potencies, EC_{50} values of 3 μM and higher. Interestingly, Asn[26] CCK, despite a reduced potency, did not display an inhibition at supramaximal concentrations characteristic of CCK as previously reported for CCK-4 and CCK-JMU-180[18,21]. The ß-carboxyl group of Asp may play a role in this effect.

The more potent agonist analogues have D-amino acid substitutions at Gly[29] thus confirming that the interaction at the CCK_A receptor is due to an orientation of the molecule stabilized by a D-conformation at this position. This is in relative agreement with the results of similar substitutions in other reported analogues[22]. The selectivity for the D-amino acid substitution is also characteristic of the brain $CCK_{B/gastrin}$ receptors[13,14]. The Nle substitution for Met does not decrease activity and enhances chemical stability. Although the disulfated analogue was not very active in the pancreatic in vitro assay, it inhibited feeding behavior and was the most potent of these compounds (unpublished results, Cohen and Knight).

CCK Fragment Peptide Inhibitory Analogues

Analogues of CCK(27-32) amide and CCK(27-31) amide with the above described modifications were examined for their ability to inhibit 0.1 nM CCK-stimulated amylase release from the pancreatic acinar cells. In one analogue a large amino-protecting group, ß-naphthoyl, was added in some compounds during the solid-phase synthesis to observe the effect on the potency. The effect of alkylation of the carboxyl amide on biological activity was studied. The same modification in LRH produced one of the first superpotent LRH analogues[23].

In Tables 1 and 2 are listed the concentrations of the sulfated and unsulfated peptides that produce half-maximal inhibition of CCK-stimulated amylase release (IC_{50}). Among the CCK(27-31) fragment analogues (Table 1), all the sulfated peptides are active in the concentration range of 5 μM, similar to CCK(26-31) NH₂, #3, the unmodified fragment.

The IC_{50} values for the sulfated fragment analogues ranged from 1.3 μM to 19.9 μM. [Nle[28,31]D-Ala[29]]CCK(27-31) NH₂, #2, without the Asp[26] is slightly more potent than the CCK(26-31) amide[10]. Each unsulfated analogue is 1/10 as potent as the corresponding sulfated analogue and all are more potent than the unsulfated CCK(26-31) amide fragment peptide. The large ß-naphthoyl group, #4, Table 1, did not increase activity. This was the least potent of the sulfated des Asp[32]-Phe[33] peptide fragment analogues. The analogue that is the most potent inhibitor of this group is peptide #1, [D-Ala[29]Nle[28,31]]CCK(27-31) bis-dimethylamide. The alkylated amide with an IC_{50} of 0.21 μM, is approximately 20-fold more potent than CCK(26-31) amide.

Most of the des Phe fragment peptides (Table 2) were weaker inhibitors than the unmodified peptide, #5, with IC_{50} values above 17 μM, except for the alkylated

Table 1. Activity of CCK(26-31) Fragment Analogues in Inhibition of 0.1 nM CCK(26-33)-Stimulated Amylase Release

No.	Peptide	I.C.$_{50}$, uM
1	[N-Ac,Nle^{28,31},D-Ala^{29}]CCK(27-31)[N(CH$_3$)$_2$]$_2$ Ac-Tyr-SO$_3$H-Nle-D-Ala-Trp-Nle-N[(CH$_3$)$_2$]$_2$	0.21 ± 0.01
2	[N-Ac,Nle^{28,31},D-Ala^{29}]CCK(27-31) NH$_2$ Ac-Tyr(SO$_3$H)Nle-D-Ala-Trp-Nle-NH$_2$	1.30 ± 0.5
1US*	[N-Ac,Nle^{28,31},D-Ala^{29}]CCK(27-31)[N(CH$_3$)$_2$]$_2$ US Ac-Tyr-Nle-D-Ala-Trp-Nle-N(CH$_3$)$_2$	2.10 ± 0.8
3	[N-Ac]CCK(26-31) NH$_2$ Ac-Asp-TyrSO$_3$H-Met-Gly-Trp-Met-NH$_2$	5.0[ref. 10]
4	[N-ß-Naph,Nle^{28,31},D-Ala]CCK(27-31) NH$_2$ N-ß-Naphthoyl-Tyr(SO$_3$H)-Nle-D-Ala-Trp-Nle-NH$_2$	19.9 ± 10
2US	[N-Ac,Nle^{28,31},D-Ala^{29}]CCK(27-31) NH$_2$ US Ac-Tyr-Nle-D-Ala-Trp-Nle-NH$_2$	23.0
3US	[N-Ac]CCK(26-31) NH$_2$ US Ac-Asp-Tyr-Met-Gly-Trp-Met-NH$_2$	50.0[ref. 10]

*US = unsulfated, des SO$_3$H

Table 2. Ability of Various CCK 26-32 Analogues to Inhibit 0.1 nM CCK(26-33)-Stimulated Amylase Release from Pancreatic Acini

No.	Peptide	IC$_{50}$, uM
464	[N-Ac,Nle^{28,31},D-Ala^{29}]CCK(27-32)[NHC$_2$H$_5$]$_2$ Ac-Tyr(SO$_3$H)-Nle-D-Ala-Trp-Nle-Asp-[NH-CH$_2$CH$_3$]$_2$	0.1
5	[N-Ac]CCK(26-32) NH$_2$ Ac-Asp-Tyr(SO$_3$H)-Met-Gly-Trp-Met-Asp-NH$_2$	6.0[ref. 8-9]
6	[N-Ac,Nle^{28,31},D-Ala^{29}]CCK(27-32) NH$_2$ Ac-Tyr(SO$_3$H)-Nle-D-Ala-Trp-Nle-Asp-NH$_2$	17.5 ± 2.8
6US	[N-Ac,Nle^{28,31},D-Ala^{29}]CCK(27-32) NH$_2$ US Ac-Tyr-Nle-D-Ala-Trp-Nle-Asp-NH$_2$	20.0 ± 10
464US	[N-Ac,Nle^{28,31},D-Ala^{29}]CCK(27-32)[NC$_2$H$_5$]$_2$ Ac-Tyr-Nle-D-Ala-Trp-Nle-Asp-[NCH$_2$CH$_3$]$_2$	44.0 ± 4.8

analogue D-Ala^{29}Nle^{28,31}]CCK(27-32)[NH(CH$_2$CH$_3$)]$_2$ #464 which has an IC$_{50}$ of 0.1 µM. The effect of this inhibitor at varying concentrations on CCK-stimulated amylase release is shown in Fig. 1. This compound which is twice as potent as the CCK(26-31) alkylamide, is the most potent new CCK$_A$ receptor antagonist found in this study. The des Phe CCK fragment peptide bis-ethylamide proved to be more potent than dibutyryl cyclic GMP and its derivatives and proglumide[24], equipotent to glutaramide analogues such as RI409, and 1/100 times as potent as the benzodiazepine-type inhibitor, L-364,718[5,25].

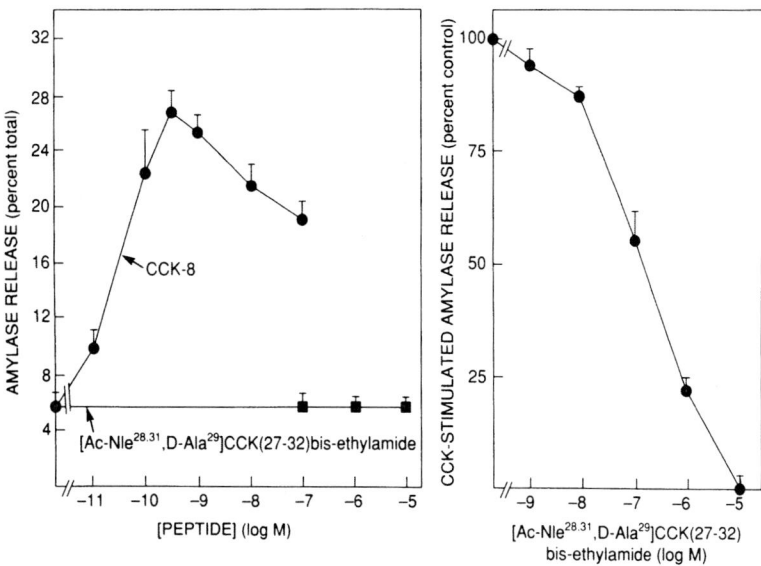

Figure 1. Left. Effect of varying concentrations of [Ac,Nle[28,31], D-Ala[29]] CCK(27-32) bis-ethylamide or CCK-8 on stimulation of amylase release from incubated pancreatic acinar cells as described in the text. Right. Percent inhibition of 0.1 nM CCK-8-stimulated enzyme release by increasing concentrations of [Ac,Nle[28,31], D-Ala[29]] CCK(27-32) bis-ethylamide.

Apparently, alkylation at the carboxyl end of the fragment peptides maintains a high affinity for the guinea pig pancreatic CCK_A receptor but the lack of the carboxyl residue Phe eliminates the intrinsic activity. A further series of stabilized CCK peptide inhibitors were synthesized with alkylamide groups of increasing size and aromaticity. The effect of a stronger acidic group ($SerSO_3$) in place of Asp^{32} was studied as well, #462 (Table 3). The inhibitory potencies of these peptides are presented below.

Table 3. Inhibition by CCK alkylamide peptides of Amylase Release (CCK_A Receptors)

No.	Peptide	IC_{50}, µM
462	$(Ac,Nle^{28},D-Ala^{29},Nle^{31},SerSO_3H^{32})CCK27-32NHCH_2CH_3$	1.5
468A	$(Ac,Nle^{28},D-Ala^{29},Nle^{31})CCK27-32NH(CH_2)_2\phi$	1.0
468AUS	" US	0.56
469	$(Ac,Nle^{28},D-Ala^{29},D-Trp^{30},Nle31)CCK27-32-[NHCH_2CH_3]_2$	>3
470US	$(Ac,Nle^{28},D-Ala^{29},D-Trp^{30},Nle31)CCK27-32-[NH(CH_2)_2\phi]_2$, US	2.9
471	$(Ac-D-Lys^{26},Nle^{28},D-Ala^{29},Nle^{31})CCK27-31NH_2$	>10

All of the peptides had IC_{50} values in the μM range. But none were more potent than the CCK 27-32 bis-ethylamide. Interestingly, the D-Trp[30] substituted alkylamides (469 and 470US) while not more potent than their corresponding L-Trp analogues, were not much less potent with the CCK_A receptors of pancreatic acinar cells. This substitution in a CCK peptide resulted in an agonist analogue that was significantly more potent on the CCK_A receptor than on $CCK_{B/gastrin}$ receptors[26]. Cyclizing the peptide from the position 26 Lys ϵ amino group to the carboxyl end, #471, reduced the potency.

ACTION OF ANTAGONISTS ON $CCK_{B/gastrin}$ RECEPTORS OF SMALL CELL LUNG CARCINOMA

Small cell lung cancer (SCLC) is a neuroendocrine tumor which synthesizes peptides and has peptide receptors. The first discovered secreted hormone was bombesin/gastrin releasing peptide, which functions as an autocrine growth factor[27]. Thus SCLC synthesizes and secretes gastrin releasing peptide which binds to cell surface receptors and stimulates growth. $CCK_{B/gastrin}$ receptors are also present on SCLC cells. In particular, $[^{125}I]CCK$-8 binds with high affinity (K_d = 2 nM) to a single class of sites (B_{max} = 1700/cell) using cell line NCI-H345[6]. Specific $[^{125}I]CCK$-8 binding is inhibited half maximally (IC_{50}) by 0.9 nM CCK-8 or 6.5 nM unsulfated CCK-8. Because CCK-8 unsulfated is not dramatically less potent than CCK-8, $CCK_{B/gastrin}$ receptors are present on NCI-H345. Further, specific $[^{125}I]CCK$-8 binding was inhibited with high affinity by the benzodiazepine-related $CCK_{B/gastrin}$ receptor antagonist, L-365,260 (IC_{50} = 0.2 nM) but not by L-364,718, the CCK_A specific receptor antagonist (IC_{50} = 500 nM)[28].

Agonist occupation of the $CCK_{B/gastrin}$ receptors on SCLC cells stimulates phosphatidylinositol turnover. When NCI-H345 was loaded with the fluorescent calcium indicator Fura-2 AM, 10 nM CCK-8 elevated the cytosolic Ca^{2+} from 150 to 180 nM[29]. The effects of CCK-8 were dose dependent and calcium was released from intracellular pools. The ability of CCK-8 antagonists to interact with $CCK_{B/gastrin}$ receptors in SCLC cells was thus investigated.

SCLC cytosolic Ca^{2+} level assay and CCK receptor binding

The human tumor cell line NCI-H345 is cultured in serum-supplemented medium (RPMI 1640 containing 10% heat inactivated fetal calf serum) in a humidified atmosphere of 5% CO_2 and 95% air at 37°C as described previously[29]. For the assay, cells are harvested by centrifugation at 1,000 x g for 10 min. Then the cells are washed and resuspended in assay buffer (SIT medium-RPMI 1640 containing 3 x 10^{-8} M Na_2SeO_3, 5 $\mu g/ml$ bovine insulin and 10 $\mu g/ml$ human transferrin).

For binding studies, SCLC cell line NCI-H345 cells are incubated at 37°C for one hr with 100,000 cpm of $[^{125}I]$-CCK-8 (2000 Ci/mmol) in the presence or absence of competitor; total assay volume was 250 μl. The cells were pelleted using a Beckman B microfuge for 1 min and the pellet with bound peptide was washed 2 x with 150 μl of assay buffer to remove remaining free peptide. Bound peptide was determined by counting the pellet in a LKB gamma counter.

For the cytosolic Ca^{2+} assays, the cells were harvested 24 hr after a medium change, washed, and resuspended in SIT media containing 20 mM HEPES, pH 7.4. SCLC cells (2.5 x 10^6/ml) were incubated with 5 μM Fura 2AM at 37°C for 30 min in a shaking water bath. Cells were washed by centrifuging at 150 x g for 2 min, and resuspended to the same concentration. Fluorescence of the stirred cells loaded with

Table 4. Potency of CCK peptides on SCLC cells

No.	Peptide	[^{125}I] CCK Binding IC$_{50}$, μM	Ca^{++} Stim.
	CCK26-33 (CCK-8)	0.001	+
	CCK-8 US	0.006	+
7	(Ac,Nle^{28},D-Ala^{29},Nle^{31},SerSO$_3$H^{32})CCK27-33NH$_2$	2	+
7US	" US	20	--
4	(ßNap,Nle^{28},D-Ala^{29})CCK27-31NH$_2$	>20	--
4US	" US	5	--
6	(Ac,Nle^{28},D-Ala^{29},Nle^{31})CCK27-32NH$_2$	>20	-
6US	" US	20	-
1	(Ac,Nle^{28},D-Ala^{29},Nle^{31})CCK27-31[N(CH$_3$)$_2$]$_2$	>20	-
1US	" US	10	-
461	(Ac,Nle^{28},Nle^{31})CCK27-32[NHCH$_2$CH$_3$]$_2$	>20	-
462	(Ac,Nle^{28},D-Ala^{29},Nle^{31},SerSO$_3$H^{32})CCK27-32NHCH$_2$CH$_3$	1	--
464	(Ac,Nle^{28},D-Ala^{29},Nle^{31})CCK27-32[NHCH$_2$CH$_3$]$_2$	20	-
464US	" US	>20	-
465	(Ac,Nle^{28},D-Ala^{29},Nle^{31})CCK27-32[NH(CH$_2$)$_3$CH$_3$]$_2$		-
465US	" US	>20	-
466	(Ac,Nle^{28},D-Ala^{29},Nle^{31})CCK27-32[NH(CH$_2$)$_5$CH$_3$]$_2$	10	--
466US	" US	>20	-
468	(Ac,Nle^{28},D-Ala^{29},Nle^{31})CCK27-32[NH(CH$_2$)$_2$$\phi$]$_2$	10	--
468US	" US	>20	-
468A	(Ac,Nle^{28},D-Ala^{29},Nle^{31})CCK27-32NH(CH$_2$)$_2$$\phi$	3	--
468AUS	" US	10	-
469	(Ac,Nle^{28},D-Ala^{29},D-Trp^{30},Nle31)CCK27-32-[NHCH$_2$CH$_3$]$_2$	>20	-
470	(Ac,Nle^{28},D-Ala^{29},D-Trp^{30},Nle31)CCK27-32-[NH(CH$_2$)$_2$$\phi$]$_2$	>20	-
470US	" US	>20	-
471	(Ac-D-Lys^{26},Nle^{28},D-Ala^{29},Nle^{31})CCK27-31NH$_2$ |_____|	>20	-
472	(Ac-D-Lys^{26},Nle^{28},D-Ala^{29},Nle^{31},Asp)CCK27-32NH$_2$ |_____|	>20	-

The ability of peptides to inhibit 10 nM CCK-stimulation of Ca^{++} levels (--);to stimulate cytosolic Ca^{++} (+); or have no effect (-) is indicated. CCK-8 = Asp-Tyr(SO$_3$)-Met-Gly-Trp-Met-Asp-Phe-NH$_2$.

Fura 2 was monitored using a spectrofluorometer at excitation and emission wavelengths of 340 nm and 510 nm, respectively. Changes in cytosolic Ca^{2+} was measured after addition of CCK antagonists or agonists.

Results

Inhibition of $[^{125}I]$-CCK binding to SCLC was measured for most of the peptides. The IC_{50} values are presented in the first column of Table 4. The effect of the compounds on inhibition of CCK-stimulation of Ca^{++} cytosolic levels is given in the second column. For more potent compounds an IC_{50} was determined. Compound #7 functioned as an agonist and elevated cytosolic Ca^{++}. Compounds #7US, 4, 4US, 462, 466, 468 and 468A were antagonists.

Compound #7, synthesized to observe the effect of a stronger acidic group in position 32[30], displays an equivalent agonist potency in both the SCLC and pancreas cells, EC_{50} of 2 μM and 3 μM, respectively (Table 4). The disulfated CCK (27-32) ethylamide, #462, was determined to be the most potent antagonist on the SCLC cells with an IC_{50} of 1 μM in the Ca^{++} level assay. Other more potent compounds were the CCK (27-32) phenylethylamide, #468A, and the unsulfated naphthoyl CCK (27-31) amide, #4. All the other peptides were less potent and none of the compounds had sub μM potency as some of the compounds had for CCK_A receptors.

The cyclic analogues were designed with the amide closure between an ϵ amino group of Lys in position 26 to the carboxyl residue of the des Asp-Phe and des Phe fragment peptides #471 and #472, respectively. Neither analogue had significant inhibitory activity in either receptor assay. This may be due to the necessity for the carboxyl end to be unconstrained. A CCK analogue cyclized from the amino end to position 29, where residues 30 to the end were linear, has been shown to retain high CCK brain receptor binding affinity but low affinity for CCK_A receptors[31]. Therefore, conformationally stabilized peptides cyclized in a similar fashion with the modifications that induce antagonism will have to be tried.

CCK RECEPTOR SELECTIVITY

Compounds with equivalent affinities in the μM range for CCK_A and $CCK_{B/gastrin}$ receptors are the disulfated CCK (27-32) ethylamide, #462 and the CCK 27-32 phenylethylamide, #468A. The D-Trp[30] CCK27-32 analogues, #469 and 470 were more potent on the CCK_A receptor in agreement with other reports of this substitution in the agonist octapeptide. The most potent inhibitor is CCK (27-32) bis-ethylamide, #464 which has sub μM potency for the CCK_A receptor and is 200 times less potent at the $CCK_{B/gastrin}$ receptor (see Table 5). The results of some of

Table 5. Potency of CCK antagonists on CCK_A receptors (pancreatic acinar cell secretion) and on $CCK_{B/gastrin}$ receptors (SCLC)

No.	Peptide		Secretion IC_{50}, μM	SCLC Binding IC_{50}, μM
462	$(Ac,Nle^{28},D-Ala^{29},Nle^{31},SerSO_3H^{32})CCK27-32NHCH_2CH_3$		1.5	1
4	$(\beta Nap,Nle^{28},D-Ala^{29})CCK27-31NH_2$		20	>20
4US	"	US	>20	5
6	$(Ac,Nle^{28},D-Ala^{29},Nle^{31})CCK27-32NH_2$		17.5	>20
1	$(Ac,Nle^{28},D-Ala^{29},Nle^{31})CCK27-31[N(CH_3)_2]_2$		0.21	>20
1US	"	US	2.1	10
464	$(Ac,Nle^{28},D-Ala^{29}Nle^{31})CCK27-32[NHCH_2CH_3]_2$		0.1	20
464US	"	US	20	>20

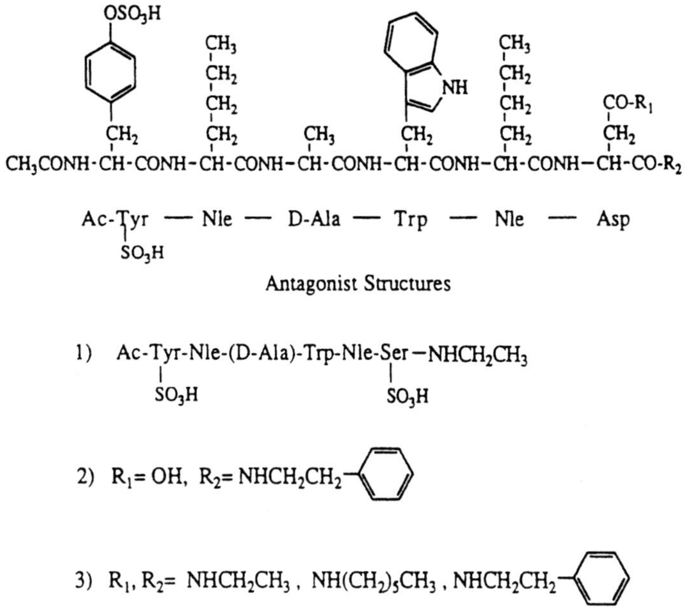
OSO3H
CH3
CH2
CH2
CH2
NH
CO-R1
CH3CONH-CH-CONH-CH-CONH-CH-CONH-CH-CONH-CH-CONH-CH-CO-R2
Ac-Tyr — Nle — D-Ala — Trp — Nle — Asp
SO3H
Antagonist Structures
1) Ac-Tyr-Nle-(D-Ala)-Trp-Nle-Ser-NHCH2CH3
SO3H
SO3H
2) R1= OH, R2= NHCH2CH2
3) R1, R2= NHCH2CH3, NH(CH2)5CH3, NHCH2CH2

Figure 2. Structure of the CCK (27-32) peptide that is the N-terminal CCK fragment known to be inhibitory. The analogues have the groups indicated. The CCK (27-31) peptide is the des Asp^{32} N-terminal fragment peptide whose dimethylamide proved to be an inhibitor (not shown).

the antagonists tested in both cellular assays are presented together in Table 5 for comparison of their activities with the receptor subtypes. The next most potent inhibitor is CCK (27-31) bis-dimethylamide which has 100 times less affinity for the $CCK_{B/gastrin}$ receptor. These alkylamide peptides are highly selective for the CCK_A receptor.

Although a variety of CCK receptor inhibitors have been developed from natural product structures[32,25] and rationally designed[33], study of structure-activity relationships of the amino-terminal part of CCK-8 may generate less gastrin-specific compounds. This is because the more potent inhibitors here do not have an acidic group near the carboxyl terminus and lack the amino acid residues in common with gastrin. Since there are many biological effects of CCK peptides, a particular antagonist may have a useful selectivity. Alkylamidation renders the peptides more hydrophobic and possibly more useful for in vivo administration where blood-brain barrier passage is desirable. Thus, if future structures achieve sub μM (not necessarily nM) potency, then they may display CCK antagonism upon injection in vivo. Behavioral effects as well as in vitro effects indicative of a particular CCK receptor interaction will have to be studied. The possibility of an antagonist peptide having an effect on dopamine function would be interesting to investigate. A potent $CCK_{B/gastrin}$ receptor antagonist peptide structure here is the di-sulfated CCK (27-32) fragment peptide which retains an acidic group at position 32. It may be that an acidic group group at the carboxyl end of CCK related peptides is essential for $CCK_{B/gastrin}$ antagonism.

A novel CCK antagonist structure has been discovered (see Fig. 2) where the Asp ß-carboxyl free acid group at the C-end, has been masked. The effect of this on CCK antagonism is important to study from the point of view of gastrin specificity as well as the peripheral and central CCK effects. Other analogues such as those of CCK (27-31) deserve further investigation as well, because of the similar lack of a carboxyl acidic group.

SYNTHESIS AND CHROMATOGRAPHY

The peptides were synthesized by solid-phase techniques[34] using either Boc chemistry which produced the di-alkylamides or Fmoc protocols for producing the mono-alkylamides. The antagonist peptides were started with either Boc-Bzl-Asp or Nle-p-Me-benzhydrylamine resin to which the Boc-amino acids were coupled in the desired sequence. The amino-terminus of the peptides were acetylated by acetylimidazole or were coupled with ß-Naphthoic acid. The alkylamide peptides were synthesized on the chloromethyl resin to which Boc-Bzl-Asp or Boc-Nle had been esterified. After the solid phase assembly, the protected peptide was ammonolyzed by 40% ethylamine, dimethylamine or other amine in DMF for three days. After drying, the peptide alkylamide was treated with HF to remove side chain protecting groups. The peptide and resin were washed with ethyl ether and dried; then the peptide was extracted from the resin in glacial acetic acid and lyophilized. The product was chromatographed in the horizontal flow-through coil planet centrifuge in chloroform, acetic acid and water or 0.01 N HCl[35]. An example is presented in Figure 3. The peptide was analyzed for homogeneity by HPLC. This proved to be a good method for the preparative purification of water insoluble peptides resulting in high recoveries. Countercurrent chromatography of CCK-related peptides has been described previously[36].

Sulfation was effected by reaction of 10^{-2}M peptide with 6-10 equivalents of pyridine sulfur trioxide complex (technical grade, Aldrich) in a 1:2 mixture of dimethylformamide and pyridine (silylation grade, Pierce Chemical Co.)[37]. After 36 hr, the solvent was evaporated off and 1 M sodium bicarbonate in a volume equal to 1/2 the volume of the original reaction was added to the residue. After stirring for 1 hr, the mixture was lyophilized and submitted to preparative HPLC as described in Figure 4. The peptide recovered by lyophilization was analyzed by TLC, HPLC, amino acid analysis, UV spectroscopic analysis to determine the presence of the sulfate group and by MS.

For the mono-alkylamidated peptides such as #468A the peptide was assembled via the Fmoc chemistry using Fmoc ß-Bzl-Asp hydroxybenzyl alcohol

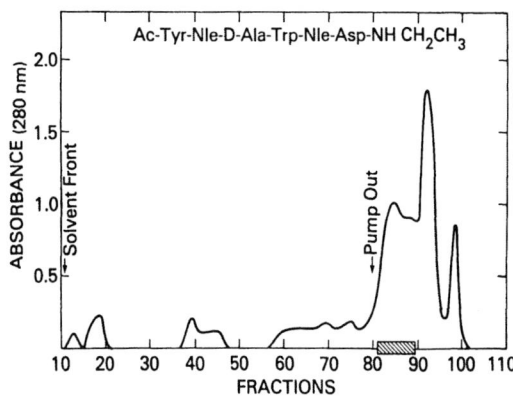

Figure 3. Purification of Ac-Tyr-Nle-D-Ala-Trp-Nle-Asp-[NH-CH₂CH₃]₂ after the solid phase synthesis. This is the countercurrent chromatography of 250 mg of the unsulfated peptide after ammonolysis and HF. The compound was chromatographed in chloroform, acetic acid and water (2:2:1, by volume) with the upper aqueous phase used as the mobile phase. The instrument was centrifuged at 400 rpm and mobile phase eluted at 24 ml/hr and fractions of 6 ml collected. After stopping rotation at fraction 80, contents were pumped out and fraction collection was continued. Fractions 81-89 contained 126 mg of pure peptide. The fractions 90-100 contained 31 mg of side product. The absorbance at 280 nm was determined manually.

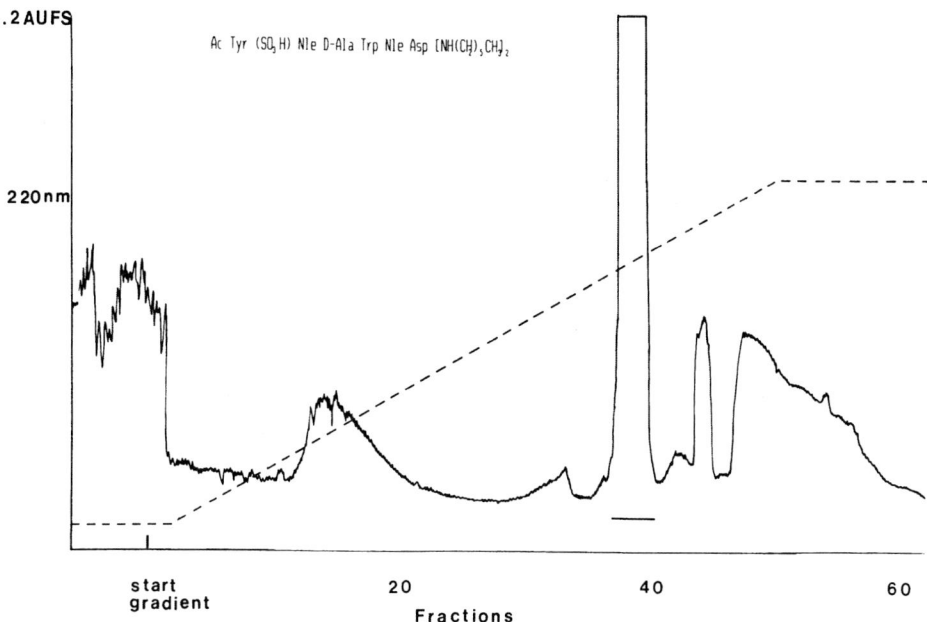

Figure 4. Preparative HPLC separation of Ac-Tyr(SO₃H)-Nle-D-Ala-Trp-Nle-Asp-bis-hexylamide # 466, from the sulfation reaction mixture on a 19 X 300 mm column of Delta Pak C₁₈, 15 μm 300 Å silica packing. Solvent A = 0.05M NH₄HCO₃; solvent B = acetonitrile at a flow of 12 ml/min with a gradient of 5% to 70% B in 50 min. One half of a sulfation reaction consisting of 36 mg peptide, 156 mg pyridine·sulfur trioxide complex and 2 mmoles Na HCO₃ was loaded. Fractions of 40 ml were collected and detection was by 215 nm absorbance. The dashed line is % acetonitrile. The purified peptide was contained in the peak at fractions 39-40. The amount of sulfated peptide recovered after lyophilization was 15 mg. The other sulfated peptides in loads of up to 100 mg were isolated by this method.

resin; then the peptide was removed from the resin by trifluoroacetic acid to give the Bzl-protected peptide which was coupled to phenylethylamine in the presence of the peptide bond forming agent pyrrolidone-BOP (benzotriazolyloxytris pyrrolidyl-phosphonium hexafluorophosphate[38]). The peptide was next reacted with HF to remove the Bzl group and then sulfated. The cyclic peptides were synthesized and cyclized prior to sulfation. Peptide #471 was prepared by the Fmoc chemistry. The resulting linear peptide was reacted with the coupling agent BOP-Cl in N-methyl pyrrolidone and the formation of the peptide bond between the ε-amino group of Lys and the carboxyl terminus was monitored by HPLC. The cyclic peptide was isolated, then sulfated as described above. Peptide #472 was prepared on the p-methyl-benzhydrylamine resin which gave the linear peptide amide which was cyclized between the ε-amino group of Lys and the ß-carboxyl of Asp.

ACKNOWLEDGMENTS

This research was supported by Small Business Innovation Research contract DK02262 and grants NS22319 and GM40833 from the National Institutes of Health.

REFERENCES

1. J.E. Jorpes and V. Mutt, "Secretin, Cholecystokinin, Pancreozymin and Gastrin", Springer, New York (1973).

2. J. Gibbs, R.C. Young and G.P. Smith, Cholecystokinin decreases food intake in rats, *J. Comp. Psychol.* 84:488 (1973).

3. G. Dockray, R.A. Gregory, J.B. Hutchinson, J.E. Harris and J. Renswick, Isolation, structure and biological activity of two cholecystokinin octapeptides from sheep brain, *Nature* 270:359 (1977).

4. T. Hokfelt, J.F. Rehfeld, L. Skirboll, B. Ivemark, M. Goldstein, and K. Markey, Evidence for coexistence of dopamine and CCK in meso-limbic neurones, *Nature* 285:476 (1980).

5. R.T. Jensen, S.A. Wank, W.H. Rowley, S. Sato and J.D. Gardner, Interaction of cholecystokinin with pancreatic acinar cells: a well studied model of the peripheral action of CCK, *Trends Pharmacol. Sci.* 10:418 (1989).

6. D. Yoder, and T.W. Moody, Cholecystokinin binds with high affinity to small cell lung cancer cells, *Peptides* 8:103 (1987).

7. R. Harkins, R.T.Jensen, H. Shapira, A. de Weerth and T. Slattery, Purification, molecular cloning, and functional expression of the cholecystokinin receptor from rat pancreas, *Proc. Nat'l. Acad. Sci. USA*, 89:3125 (1992).

8. A.S. Kopin, Y.M. Lee, E.W. McBride, L.J. Miller, M.Lu, H.Y. Lin, L.F. Kolakowski,Jr. and M. Beinborn, Expression, cloning and characterization of the canine parietal cell gastrin receptor, *Proc. Nat'l. Acad. Sci. USA* 89:3605 (1992).

9. M. Spanarkel, J. Martinez, C. Briet, R.T. Jensen and J.D. Gardner, Cholecystokinin 27-32 amide: a member of a new class of cholecystokinin receptor antagonists, *J. Biol. Chem.* 258:6746 (1983).

10. J.D. Gardner, M. Knight. V.E. Sutliff and R.T. Jensen, N-terminal fragments of CCK-26-33 as cholecystokinin receptor antagonists in guinea pig pancreatic acini, *Am. J. Physiol.* 248:G98 (1985).

11. J.D. Gardner, M. Knight, V.E. Sutliff, C.A. Tamminga and R.T. Jensen, Derivatives of cholecystokinin 26-32 as cholecystokinin receptor antagonists in guinea pig pancreatic acini, *Am. J. Physiol.* 246:G292 (1984).

12. M. Knight, A.M. Kask and C.A. Tamminga, Synthesis of amino-terminal fragments of cholecystokinin 26-33, *in:* "Peptides: Structure and Function", V. Hruby and D. Rich, eds., Pierce Chemical Co., Rockland IL, 759 (1984).

13. B.P. Roques, C. Durieux, G. Gacel, D. Pelaprat, M. Ruiz-Gayo, J. Belleney, E. Fellion, J.M. Zajac, M.C. Fornie Zaluskie, V. Dauge, I. Menant, P. Rossignol, B. Lux, D. Gerard, D. Begue, A. Sasaki and J.L. Morgat, Studies on the conformation, enzymatic degradation, pharmacological potency, binding properties in brain tissue of cholecyctokinin-8 and new related peptides, *in:* "Neuronal Cholecystokinin", J.J. Vanderhaegen and J.N. Crawley, eds. *Annal. N.Y. Acad. Sci.*, 448:61 (1985).

14. M. Knight, P. Barone, C.A. Tamminga, L. Steardo and T.N. Chase, Cholecystokinin octapeptide analogues stable to brain proteolysis, *Peptides* 6:631 (1985).

15. L. Steardo, M. Knight, C.A. Tamminga and T.N. Chase, Proteolysis of CCK-8 by synaptic membranes generates peptide fragments which interact with central receptors, *Neuroscience Lett.* 54:319 (1985).

16. J.R. McDermott, P.R. Dodd, J.A. Edwardson, J. A. Hardy and A.I. Smith, Pathway of inactivation of cholecystokinin octapeptide CCK-8 by synaptosomal fractions, *Neurochem. Internat.* 5:641 (1983).

17. L. Steardo, M. Knight, C.A. Tamminga, P. Barone, A.M. Kask and T.N. Chase, CCK 26-33 degrading activity in brain and nonneural tissue: a metalloendopeptidase, *J. Neurochem.* 45:784 (1985).

18. J.D. Gardner, and M.J. Jackson, Regulation of amylase release from dispersed pancreatic acinar cells *J. Physiol.* 270:439 (1977).

19. S.R. Peikin, A.J. Rottman, S. Batziri and J.D. Gardner, Kinetics of amylase release by dispersed acini prepared from guinea pig pancreas, *Am. J. Physiol.* 235:E743 (1978).

20. R.T. Jensen, G.F. Lemp and J.D. Gardner, Interaction of cholecystokinin with receptors on dispersed acini from guinea pig pancreas, *J. Bio. Chem.* 257:5554 (1982).

21. M.C. Galas, M.C. Lignon, M. Rodriquez, C. Mendre, P. Fulcrand, J. Lauer and J. Martinez, Structure-activity relationship studies on cholecystokinin analogues with partial agonist acitvity, *Am. J. Physiol.* 254:G176 (1988).

22. T.K. Sawyer, R.T. Jensen, T. Moran, P.J.K.D. Schreur, D.J. Staples, A.E. de Vaux and A. Hsi, Structure-activity relationships of cholecystokinin-8 analogs: comparison of pancreatic, pyloric sphincter and brain stem CCK receptor activities with in vivo anorexigenic effects, *in:*"Peptides: Chemistry and Biology" G.R. Marshall, ed., ESCOM, Leiden, 503 (1988).

23. M. Fugino, S. Shinagawa, I. Yamazaki, M. Kobayashi, M. Obayashi, T. Fukuda, R. Nakayama, W.F. White and R.H. Rippel, [Des-Gly-NH_2^{10},Pro-ethylamide^9]-LH-RH: a highly potent analog of luteinizing hormone releasing hormone, *Arch. Biochem. Biophys.* 154:488 (1973).

24. W.F. Hahne, R.T. Jensen, G.F. Lemp and J.D. Gardner, Proglumide and benzotript: members of a different class of cholecystokinin receptor antagonists, *Proc. Nat'l. Acad. Sci. USA* 78:6304 (1981).

25. B.E. Evans, M.G. Bock, K.E. Rittle, R.M. DiPondo, W.L. Whitter, D.F. Veber, P.S. Anderson and R.M. Friedinger, Design of potent, orally, effective, non peptidal antagonists of the peptide hormone cholecystokinin, *Proc. Nat'l. Acad. Sci. USA* 83:4918 (1986).

26. J. Martinez, M. Rodriquez, J. Bali and J. Lauer, A synthetic peptide derivative that is a cholecystokinin receptor antagonist, *J. Med. Chem.* 29:2201 (1986).

27. F. Cuttitta, D.N. Carney, J. Mulshine, T.W. Moody, J. Fedorko, A. Fischler, and J.D. Minna, Bombesin-like peptides can function as autocrine growth factors in human small cell lung cancer, *Nature* 316:823 (1985).

28. J. Staley, R.T. Jensen, and T.W. Moody, CCK antagonists interact with CCK-B receptors on human small cell lung carcinoma cells, *Peptides* 11:1033 (1990).

29. J. Staley, G. Fiskum, and T.W. Moody, Cholecystokinin elevates cytosolic calcium in small cell lung carcinoma, *Biochem. Biophys. Res. Commun.* 163:605 (1989).

30. B. Penke, F. Hajnal, J. Lonovics, G. Holzinger, T. Kadar, G. Telegdy and J. Rivier, Synthesis of potent heptapeptide analogues of cholecystokinin, *J. Med. Chem.* 27:845 (1984).

31. B. Charpentier, A. Dor, P. Roy, P. England, H. Pham, C. Durieux, and B.P. Roques, Synthesis and binding affinities of cyclic and related linear analogues of CCK selective for central receptors, *J. Med. Chem.* 32:1184 (1989).

32. R.S. Chang, V.J. Lotti, R.L. Monaghan, J. Birnbaum, E.O. Stapley, M.A..Goetz, G. Albers-Schonberg, A.B. Patchett, J.M. Liesch, et al., A potent nonpeptide cholecystokinin antagonist selective for peripheral tissues isolated from <u>Aspergillus alliaceus</u>, *Science* 230:177 (1985).

33. D.C. Horwell, J. Hughes, J.C. Hunter, M.C. Pritchard, R.S. Richardson, E. Roberts, and G.N. Woodruff, Rationally designed 'dipeptoid' analogues of CCK, α-methyltryptophan derivatives as highly selective orally active gastrin and CCK-B antagonists with potent anxiolytic properties, *J. Med.Chem.* 34:404 (1991).

34. J.M. Stewart and J.D. Young, "Solid Phase Peptide Synthesis", Pierce Chemical Co., Rockland, IL (1983).

35. M. Knight, J.D. Pineda and T.R. Burke, Jr., Solvent systems for the countercurrent chromatography of hydrophobic neuropeptide analogs and hydrophilic protein fragments *J. Liq. Chromatog.* 11:119 (1988).

36. M. Knight, Countercurrent chromatography for the purification of peptides, *in*: "Adv. Chromatog." vol. 31, C. Giddings, E. Grushka and P. Brown, eds., Marcel Dekker, New York, (1992).

37. M. Knight, C.A. Tamminga, Y. Ito, J.D. Gardner and T.N. Chase, Purification of a CCK fragment peptide on the horizontal flow-through coil planet centrifuge, *J. Chromatog.* 30:277 (1984).

38. J. Coste M.N. Dufove, D. Le-Nguyen and B. Castro, BOP and congeners: Present status and new developments, *in*: "Peptides: Chemistry, Structure and Biology," J.E. Rivier and G.R. Marshall, eds., ESCOM, Leiden, 885 (1990).

PEPTIDE-NEUROMEDIN B (NMB) RECEPTOR INTERACTIONS: COMPARISON WITH GASTRIN-RELEASING PEPTIDE (GRP) RECEPTOR INTERACTIONS

Robert T. Jensen,[1] Lu-Hua Wang,[1] Jaw-Town Lin,[1]
Richard V. Benya,[1] John E. Mrozinski, Jr.,[1] and David H. Coy[2]

[1]Digestive Diseases Branch
National Institutes of Health
Bethesda, MD 20892
[2]Peptide Research Laboratories
Tulane University Medical Center
New Orleans, LA 70112

INTRODUCTION

The amphibian peptide bombesin and the mammalian structurally-related peptides, gastrin-releasing peptide (GRP) and neuromedin B (NMB) elicit a wide range of biological responses (1-3). These include release of numerous gastrointestinal hormones (2-4), stimulation of pancreatic enzyme secretion (1,5), regulation of central nervous system functions such as maintenance of circadian rhythm and thermoregulation (2,6,7), stimulation of macrophages and phagocytes (8,9), regulation of TSH release (11), and functioning as a growth factor (11,12). Bombesin-related peptides stimulate growth of both normal tissues (11,13,14), neoplastic tissues such as prostatic, breast, hepatocellular carcinomas or small cell lung cancer cells, and have been proposed to have an autocrine growth function (11,15-18).

The natural occurring bombesin-related peptides structurally can be divided into three different groups based on chemical structure (1,3,19). Bombesin resembles other amphibian peptides such as alytesin and the mammalian peptide GRP in possessing a carboxy terminal Gly-His-Leu-Met-NH_2. The second class included the amphibian peptides ranatensin, litorin and the mammalian peptide neuromedin B in possessing a COOH terminal Gly-His-Phe-Met-NH_2. The third group comprise the amphibian peptide phyllolitorin and other structurally related amphibian peptides, have no mammalian counterpart at present and are characterized by having a carboxyl terminal Gly-Ser-Phe-Met-NH_2 or Gly-Ser-Leu-Met-NH_2. Recent cloning studies (20-23,57) confirm previous conclusions from binding and functional studies that at least two classes of receptors mediate the actions of the bombesin-related peptides (24-29). One class, the GRP receptor, has a high affinity for GRP and lower affinity for neuromedin B, whereas the reverse is true of the NMB receptor. Binding, functional or *in situ* hybridization studies

Growth Factors, Peptides, and Receptors, Edited
by T.W. Moody, Plenum Press, New York, 1993

provide evidence for GRP receptors in the central nervous system, pituitary cells, pancreatic acinar cells, and in numerous cell lines such as murine 3T3 cells, prostatic and breast adenocarcinoma cells and small cell lung cancer cells (2,12,15,18,23,25,31,57) and for NMB receptors in the central nervous system, gastrointestinal smooth muscle and various tumor cell lines such as rat glioblastoma C-6 cells or some small cell lung cancer cells (22,23,26,28,30,32,33,57).

The ability of bombesin-related peptides to interact with GRP receptors on murine 3T3 cells, pancreatic acinar cells, small cell lung cancer cells and on insulin secretory and pituitary cell lines has been extensively studied (5,12,30,31,33-37). These studies show that activation of GRP receptors by agonists increases phospholipase C, generates phosphoinositides, mobilizes cellular calcium, increases diacylglycerol and activates protein kinase C (5,12,30,36,37). Both binding studies and functional studies provide evidence that GRP receptors are regulated by guanaine nucleotides binding regulatory proteins (37,38). Extensive studies have shown that various tissues possessing GRP receptors internalize and degrade specific GRP receptor ligands (12,31,39,40). In contrast to GRP receptors, little is known about the cellular basis of action of agonists for NMB receptors either to alter cellular function or about ligand receptor interactions.

Recently we have shown that rat glioblastoma C-6 cells possess functional NMB receptors, activation of which activates phospholipase C, increases phosphoinositides and alters cytosolic calcium (32). We have used these cells to study the ability of ligands to interact with the NMB receptor, the ability to internalize or degrade NMB receptor specific ligands, the regulation of the NMB receptor by guanine nucleotide binding proteins, and the relationship between NMB receptor interaction and alterations in cellular function (32,41). In this study we review our recent results comparing the ligand-receptor interactions with C-6 cells and other cells possessing NMB receptors to those seen in pancreatic acini or 3T3 cells possessing GRP receptors as well as comparing the ability of agonists for each bombesin receptor subtype to alter cellular function.

Ligand-Receptor Interactions: Comparison of NMB and GRP Receptors

Kinetics. With NMB receptors either ^{125}I-Bolton Hunter labeled NMB or ^{125}I-[D-Tyr°]NMB have been found to be excellent ligands (26,32,43,44), whereas with GRP receptors, ^{125}I-[Tyr^4]Bn or ^{125}I-GRP are generally used (12,30,43,44). Without protease inhibitors present, with C-6 cells ^{125}I-[D-Tyr°]NMB bound in time- and temperature-dependent fashion with maximal binding occurring by 30 min at 37°C. Binding at 4°C was slower than at 22°C, whereas at 37°C binding reached a maximum rapidly (10 min), then decreased with time (32). With rat pancreatic acini or 3T3 cells with binding of ^{125}I-[Tyr^4]Bn similar association kinetics were seen. With either C-6 cells or various cells possessing GRP receptors (rat or guinea pig pancreatic acinar cells, 3T3 cells or AR42J cells) the dissociation rate and the completeness of dissociation of bound radiolabeled ligand depended on the association temperature. After a 30 min incubation at 37°C, only 40% of the ^{125}I-[D-Tyr°]NMB to C-6 cells or ^{125}I-[Tyr^4]Bn bound to 3T3 cells, AR42J cells or pancreatic acini dissociated. In contrast, after an incubation at 22°C for 60 min, >70% of the bound radioligands dissociated, suggesting that a temperature-dependent alteration of binding was occurring likely due to internalization.

Internalization and Ligand Degradation

HPLC analysis of the ligand after incubating ^{125}I-[D-Tyr°]NMB with C-6 cells or ^{125}I-[Tyr^4]Bn with 3T3 cells for 60 min at 37°C without protease inhibitors demonstrated >90% of both ligands were degraded. The addition of bacitracin (0.1%) during the incubation was effective at blocking degradation of either ligand, whereas leupeptin (500

Table 1. Internalization of radiolabeled ligands by GRP or NMB receptors

| | GRP Receptors | | |
| | 125I-[Tyr^4]Bn (percent internalized) | | |
Incubation Time	Guinea Pig Pancreatic Acini	3T3 Cells	AR42J Cells
10 min	22±3	64±4	30±3
60 min	23±1	83±5	36±2

| | NMB Receptors | |
| | 125I-[D-Tyr^0]NMB (percent internalized) | |
	C-6 Cells	NMB-R Transfected BALB 3T3 Cells
10 min	75±1	41±4
60 min	64±5	75±4

Guinea pig pancreatic acini, 3T3 cells, or AR42J cells were incubated with 50 pM ^{125}I-[Tyr^4]Bn for the indicated time and then acid strippability of bound ligand was determined using 0.2 M acetic acid and 0.5 M NaCl (pH 2.5). The internalized counts were the percentage of the saturably bound counts not removed by acid strippling. Rat glioblastoma C-6 cells or Balb 3T3 cells transfected with the NMB receptor from rat esophageal muscularis mucosa were incubated with 50 pM ^{125}I-[D-Tyr^0]NMB and treated in a similar manner.

μg/ml) was partially effective and amastatin or chymostatin ineffective. With 0.1% bacitracin present in the incubation solution during a time course of incubation for up to 120 min either at 37°C or at 22°C, binding did not decrease with time, suggesting the decreased binding with time seen without bacitracin present was due to ligand degradation.

Acid strippling of surface bound ligand using the method of Haigler et al., (45) demonstrated that with cells possessing NMB receptors (either C-6 cells or Balb 3T3 cells transfected with the NMB receptor cloned from rat esophageal muscularis mucosa) [NMB-R transfected cells] that by 10 min of incubation 25-45% of bound ligand was internalized and by 60 min, 64-75% (Table 1). With cells possessing GRP receptors (pancreatic acini, 3T3 cells or AR42J cells) internalization of bound ^{125}I-[Tyr^4]Bn also occurred but the degree varied in the three different cell systems (Table 1). After 60 min incubation at 37°C with guinea pig pancreatic acini, 3T3 cells or AR42J cells, 23%, 83% and 36% respectively, of the bound radioligand, was internalized (Table 1).

Comparison of Affinities for Naturally Occurring Bombesin-Related Peptides

Bombesin and the related peptides, neuromedin B, GRP, phyllolitorin, litorin and ranatensin each interacted with both GRP and NMB receptors; however, their affinities differed markedly (Table 2, Fig. 1). For GRP receptors on 3T3 cells or rat pancreatic acini, bombesin, litorin, or GRP all had high affinities, whereas neuromedin B and phyllolitorin were approximately 60-fold less potent (Table 2, Fig. 1, left panel). In contrast for NMB receptors on either C-6 cells or NMB-R transfected Balb cells, neuromedin B, ranatensin and litorin had high affinity, bombesin or phyllolitorin a 7- to 12-fold lower affinity and GRP a 100-fold lower affinity (Table 2, Fig. 1, right panel). These results demonstrated that these naturally occurring bombesin-related peptides can be divided into three different groups by their relative affinities for these two bombesin receptor subtypes. Litorin and ranatensin have an equal high affinity for each; bombesin a 5- to 10-fold and GRP a 100-fold higher affinity for GRP than NMB receptors; phyllolitorin a 10-fold and neuromedin B a 100-fold higher affinity for NMB receptors (Table 2).

Table 2. Affinity of various natural occurring Bombesin-related agonists for GRP and NMB receptors

	Affinity (nM)					
	GRP Receptors			Neuromedin B Receptors		
	Rat Pancreatic Acini		3T3 Cells	Glioblastoma C-6 Cells		NMB-R Transfected Balb 3T3 Cells
AGONIST	EC_{50}	K_d	K_d	EC_{50}	K_d	K_d
Bombesin	0.2±0.1	4±1	1.0±0.1	9±1	20±7	34±2
GRP	0.4±0.1	15±3	1.6±1.0	148±2	500±27	440±70
Phyllolitorin	1.5±0.3	237±46	55±12	ND	15±4	47±3
Litorin	0.4±0.1	6±1	0.4±0.1	0.6±0.1	3±1	6±1
Ranatensin	0.2±0.1	14±7	0.6±0.1	13±2	8±1	13±2
Neuromedin B	5±1	351±28	60±4	4±1	2±1	4±1

The EC_{50} was the concentration causing half-maximal amylase release in rat pancreatic acini or increase in [^{3}H]IP in C-6 cells. The K_d values were calculated by the method of Cheng and Prusoff from either inhibition of 50 pM ^{125}I-[Tyr^4]Bn to pancreatic acini or 3T3 cells and of 50 pM ^{125}I-[D-Tyr^0]NMB to C-6 cells or NMB receptor transfected cells. Abbreviations: GRP=gastrin-releasing peptide; NMB-R-transfected Balb 3T3 cells=Balb 3T3 cells into which the NMB receptor from rat esophageal mucosa was transfected as described previously (22); ND=no data. Data are from references 32,41,42,46,53.

Comparison of the Affinities for Different Classes of Bombesin Receptor Antagonists

Four classes of bombesin receptor antagonists have been described (46) and in a recent study (27) some of these antagonists were reported to differ in their selectivity for the two different bombesin receptor subtypes. Potent representative members of each of these four classes of bombesin receptor antagonists were compared for their abilities to interact with NMB or GRP receptors (Table 3, Fig. 2). The ability of the substance P

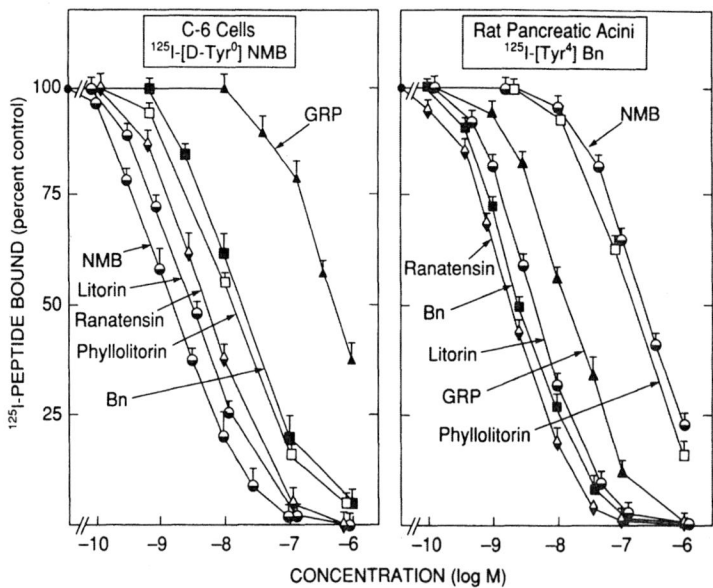

Figure 1. Ability of various natural occurring bombesin-related agonists to interact with NMB receptors on rat glioblastoma C-6 cells and GRP receptors on rat pancreatic acini. Pancreatic acini were incubated with 50 pM ^{125}I-[Tyr^4]Bn and C-6 cells with 50 pM ^{125}I-[D-Tyr^0]NMB. Results are expressed as the percentage of saturable binding seen without unlabeled peptide added. Data are means ± 1 SEM.

Table 3. Affinity of bombesin receptor antagonists for GRP and NMB receptors

Peptide	GRP Receptors			Neuromedin B Receptors		
	Rat Pancreatic Acini		3T3 Cells	C-6 Cells		NMB-R Transfected Balb 3T3 Cells
	IC_{50}	K_d	K_d	$IC_{50}[^3H]IP$	K_d	K_d
[D-Arg^1,D-Pro^2,D-Trp^{7,9},Leu^{11}]SP	agonist >1 µM	11,323±1780	8900±1600	ND	3980±6000	4105±780
[D-Phe^{12},Leu^{14}]Bn	>10,000	1300±20	10,400±200	11000±1000	3300±200	1870±130
[Leu^{14},ψ13-14]Bn	P. Agon. (11% max)	434±65	68±8	>10,000	18,160±1782	15,356±3355
D-[Phe^6,Cpa^{14},ψ13-14]Bn(6-14)	10±2	42±5	5±1	>10,000	2540±1000	2710±240
D-[Phe^9]Bn(6-13)propylamide	15±2	6±1	2±1	>10,000	2179±133	4601±586
D-[Phe^{6}Bn(6-13)methyl ester	3±1	10±2	1.1±0.2	>30,000	7466±413	7701±1068
N-Ac-GRP(20-26)methyl ester	15±2	17±11	2±1	>10,000	37,000±1700	65,641±4854

The IC_{50} was the concentration causing half-maximal inhibition of 0.3 nM stimulated amylase release in rat pancreatic acini or 3 nM stimulated [^{3}H]IP increase in C-6 cells. The K_d values were calculated by the method of Cheng and Prusoff from either inhibition of 50 pM ^{125}I-[Tyr^4]Bn to pancreatic acini or 3T3 cells and of 50 pM ^{125}I-[D-Tyr^0]NMB to C-6 cells or NMB receptor transfected cells. Abbreviations: GRP=gastrin-releasing peptide; NMB-R-transfected Balb 3T3 cells=Balb 3T3 cells into which the NMB receptor from rat esophageal mucosa was transfected as described previously (22); ND=no data; P. Agon. (% max)=partial agonist (percent maximal efficacy. Data are from references 26,27,32,41,42,46-51,53.

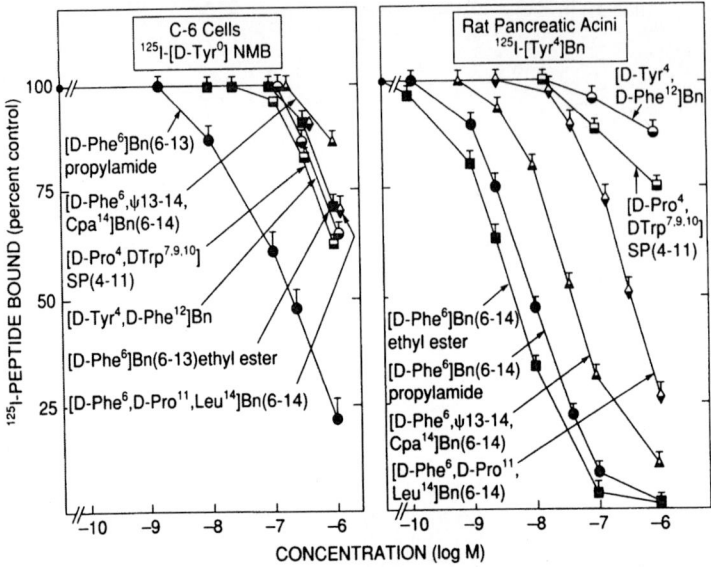

Figure 2. Ability of various classes of bombesin-receptor antagonists to interact with NMB receptors on rat glioblastoma C-6 cells or GRP receptors on rat pancreatic acini. Binding studies were performed as described in Figure 1 legend. Abbreviations: ψ13-14=peptide bond between Leu[13] and Cpa[14] reduced from -CONH- to -CH$_2$NH-; Cpa=chlorophenylalanine; SP=substance P; Bn=bombesin.

analogue, [D-Arg^1,D-Pro^2,D-Trp^{7,9},Leu^{11}]SP was studied, which is similar to a number of other substance P analogues in functioning both as substance P and bombesin receptor antagonists (46-48); [D-Phe^{12},Leu^{14}]Bn the reduced peptide bond analogues, [Leu^{14},ψ13-14]Bn and the more potent analogue [D-Phe^6,Cpa^{14},ψ13-14]Bn(6-14) (46,49); the des-Met^{14}Bn analogues [D-Phe^6]Bn(6-13)propylamide or methyl ester and N-Ac-GRP(20-26) methyl ester all reported to be potent bombesin receptor antagonists (46,50-52), were

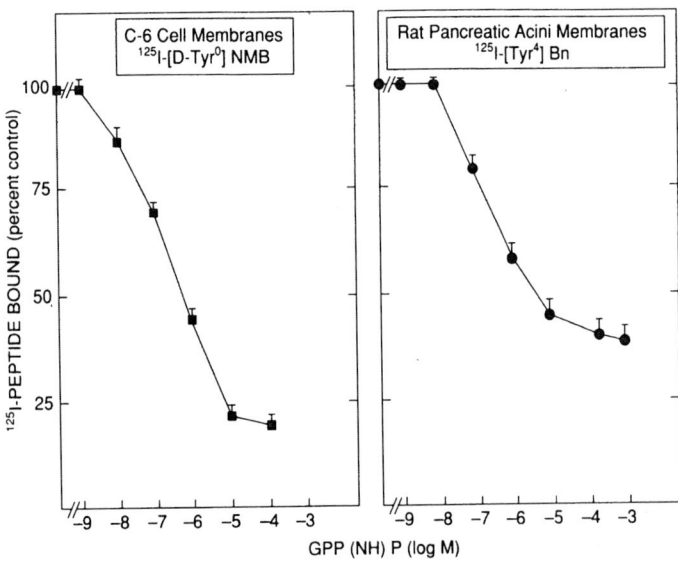

Figure 3. Effect of the guanine nucleotide, GPP(NH)P on binding of ^{125}I-[D-Tyr^o]NMB to glioblastoma C-6 cell membranes (left banel) or binding of ^{125}I-[Tyr^4]Bn to rat pancreatic acinar cell membranes (right panel). Results are expressed as the percentage of saturable binding in the absence of unlabeled peptide. Values are means $\pm$ 1 SEM .

compared (Table 3). The different classes of Bn receptor antagonists each interacted with both classes of bombesin receptor antagonists but differed markedly in potency (Fig. 2, Table 3). The different antagonists could be divided into two groups based on their relative affinities for each receptor. The des-Met[14]Bn or des-Met[27]GRP esters or the alkylamide analogue propylamide, had high affinity for GRP receptors (1-20 nM), in both rat pancreatic acini and 3T3 cells and low affinities for NMB receptors (>2 μM); therefore, N-Ac-GRP(20-26)methyl ester or [D-Phe[6]]Bn(6-13)propylamide or methyl ester had 750-, 1504-, and 3700-fold greater affinity for GRP than NMB receptors. The pseudopeptides, [D-Phe[6],Cpa[14],ψ13-14]Bn(6-14) and [Leu[14],ψ13-14]Bn had a 70- and 30-fold greater selectivity for GRP receptors (Table 3). In contrast, the substance P analogue, [D-Arg[1],D-Pro[2],D-Trp[7,9]Leu[11]]SP or [Phe[12],Leu[14]]Bn had low affinity for both receptor classes (>1 μM); however, they had a 3- and 14-fold higher affinity for NMB than GRP receptors. In contrast to reported in guinea pig pancreas or 3T3 cells (30,46,53), the SP analogue, [D-Arg[1],D-Pro[2],D-Trp[7,9]Leu[11]]SP functioned as a weak agonist (Table 3) and the pseudopeptide [Leu[14],ψ13-14]Bn functioned as a partial agonist in rat pancreatic acini (Table 3). Each of the other analogues functioned as an antagonist at GRP receptors in rat pancreatic acini (Table 3). However, because of their low affinities none of the classes of antagonists inhibited the action of neuromedin B at concentrations <10 μM (Table 3). These results demonstrate that potent selective antagonists exist for GRP receptors, but at present none exists for NMB receptors.

Effect of Guanine Nucleotide Protein Activation on Receptor-Ligand Interaction

To compare the effects of activation of a guanine nucleotide-binding protein on binding to either bombesin receptor subtype, the effect of increasing concentration of the nonhydrolyzable guanosine analogue, GPP(NH)P on binding of either ^{125}I-[D-Tyr$^{\circ}$]Bn to C-6 cell membranes (Fig. 3, left panel), or of binding of ^{125}I-[Tyr[4]]Bn to rat pancreatic acinar membranes (Fig. 3, right panel) or 3T3 cell membranes, was compared.

GPP(NH)P caused a dose-dependent decrease in binding to membranes possessing either bombesin receptor subtype (Fig. 3). GPP(NH)P caused half-maximal inhibition at 0.1-0.2 μM with each receptor subtype (Fig. 3). Analysis of the effect of a half-maximal concentration of GPP(NH)P on the dose-inhibition curve of bombesin for inhibiting binding of ^{125}I-[Tyr[4]]Bn to 3T3 or rat pancreatic acinar cell membranes or of neuromedin B for inhibiting binding of ^{125}I-[D-Tyr$^{\circ}$]NMB to C-6 cell membranes demonstrated that GPP(NH)P was decreasing binding by altering affinity and not changing the number of binding sites.

Alteration of Cellular Function

Activation of NMB receptors on C-6 cells activated phospholipase C, increasing cytosolic calcium and increasing the generation of phosphoinositides[32]. With 1 μM neuromedin B, a rapid increase in [^{3}H]IP$_1$, [^{3}H]IP$_2$, and [^{3}H]IP$_3$, was detected. Neuromedin B caused a half-maximal increase in [^{3}H]IP at 2 nM (Fig. 5) and GRP was greater than 50-fold less potent. Similarly, as reported previously[32], neuromedin B caused a rapid 3-fold increase in cytosolic calcium with a half-maximal effect at 0.2 nM (Fig. 5) and GRP was >50-fold less potent. Similar comparative studies were done on rat pancreatic acini and as reported previously similar to other cells possessing GRP receptors[5,12,34,36,37], bombesin caused a rapid activation of phospholipase C, with increases in cytosolic calcium and phosphoinositides (Fig. 4). For altering changes in both cytosolic calcium and phosphoinositides, bombesin was significantly more potent than neuromedin B.

In a previous study in 3T3 cells[54], activation of GRP receptors was reported not only to cause activation of phospholipase C, but also activation of adenylate cyclase with

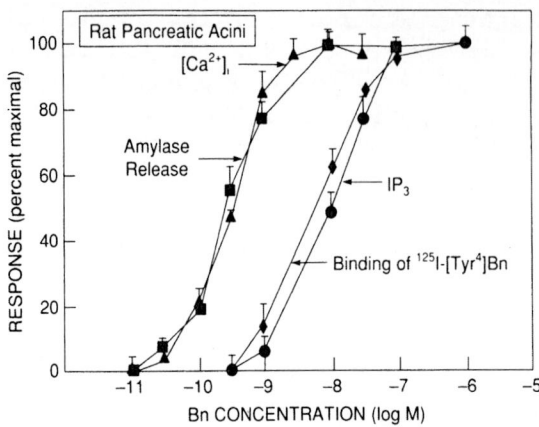

Figure 4. Comparison of the ability of bombesin to occupy GRP receptors on rat pancreatic acinar cells and alter cellular function. Results are expressed as a percentage of the maximal response seen with 1 μM bombesin. Values are means ± 1 SEM. Abbreviations: $[Ca^{2+}]_i$=increase in cytosolic calcium; IP_3=increase in $IP_3(1,4,5)$ determined using a radioreceptor assay.

small increases in cyclic AMP. In C-6 cells neuromedin B, when either added alone or in the presence of the phosphodiesterase inhibitor IBMX (100 μM), forskolin (25 μM) or cholera toxin (100 mg/ml), did not cause an increase in cyclic AMP. Furthermore, neuromedin B did not inhibit the increase in cyclic AMP caused by forskolin, IBMX, or cholera toxin, demonstrating that neuromedin B did not alter cellular function by also inhibiting adenylate cyclase.

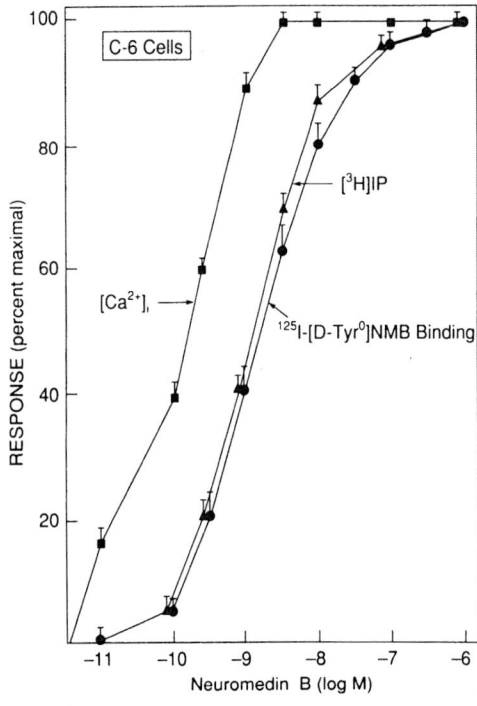

Figure 5. Comparison of the ability of NMB to occupy NMB receptors on glioblastoma C-6 cells and alter cellular function. Results are expressed as a percentage of the maximal response seen with 1 μM neuromedin B. Values are means ± 1 SEM. Abbreviations: $[Ca^{2+}]_i$=increase in cytosolic calcium; $[^3H]IP$=increase in total inositol phosphates.

Relationship of Receptor Occupation to Alterations in Cellular Function

In Fig. 5 for the NMB receptor on C-6 cells and Fig. 4 for the GRP receptor on rat pancreatic acini, the ability of an agonist to occupy the receptor and cause changes in cellular function are compared. For both GRP receptors and NMB receptors, the dose response-response curves for receptor occupation by either bombesin or neuromedin B respectively, are almost superimposible on their ability to increase phosphoinositides. In contrast, with both bombesin receptor subtypes, the dose-response curve for increases in cytosolic calcium is to the left of the receptor occupation or stimulation of increases on phosphoinositide dose-response curve (Fig. 4 and 5). For acinar cells similar to reported previously[30], the dose-response curve for stimulation of enzyme secretion is almost superimposible for that for stimulation of changes in cytosolic calicum (Fig. 4). These results demonstrate that, with both bombesin receptor subtypes, submaximal changes in phosphoinositides cause maximal changes in cytosolic calcium and in pancreatic acinar cells maximal changes in enzyme secretion, demonstrating in both cell systems marked amplification of the signal for changes in phosphoinositides for altering cellular calcium. The fact that the receptor occupation curves are superimposible on the dose-response curves for changes in phosphoinositides demonstrates that in contrast to other hormones such as with cholecystokinin[55,56], spare receptors do not exist for the bombesin receptor subtypes but rather, spare mediator.

CONCLUSIONS

These results indicate that the NMB and GRP receptors have similarities and differences in their abilities to interact with ligands and alter cellular function. The two bombesin receptor subtypes differ markedly in their affinities for natural occurring bombesin-related peptides such as GRP, NMB, phyllolitorin and bombesin. They also differ in their affinities for the four different classes of bombesin receptor antagonists recently described. Receptor antagonists such as the des-Met[14]Bn or des-Met[27]GRP derivatives have greater than 1000-fold selectivity for GRP receptors than for NMB receptors. Whereas potent antagonists are available for GRP receptors, at present no potent receptor antagonists exist for NMB receptors.

The two bombesin receptor subtypes are similar in that both rapidly internalize radiolabeled agonists and rapidly degrade these agonists. The interaction with radiolabeled agonists of both bombesin receptor subtypes is regulated by guanine nucleotide binding proteins with activation resulting in a decreased receptor affinity. Occupation of both bombesin receptor subtypes activates phospholipase C, increases cytosolic calcium and increases phosphoinositides. In contrast to recent studies of GRP receptors in 3T3 cells[54] which are coupled to both adenylate cyclase and changes in cytosolic calcium, in C-6 cells possessing NMB receptors, NMB receptor occupation neither increased cyclic AMP or inhibited the increase caused by IBMX, forskolin, or cholera toxin, demonstrating that agonist occupation of the NMB receptor does not alter the activity of adenylate cyclase. With both bombesin receptor subtypes, receptor occupation is closely coupled to increases in phosphoinositides and submaximal changes in phosphoinositides caused maximal changes in cytosolic calcium demonstrated marked amplification of the signal for changes in phosphoinositides.

Acknowledgments

This work was supported in part by USPHS NIH Grant CA-45153.

REFERENCES

1. V. Erspamar, and P. Melchiorri, Active polypeptides of the amphibian skin and their synthetic analogues, *Pure Appl. Chem.* 35:463-493 (1973).
2. Y. Tache, P. Melchiorri, and L. Negri, Bombesin-like peptides in health and disease, *Ann. N.Y. Acad. Sci.* 547:1-540 (1988).
3. E.R. Spindel, Mammalian Bombesin-like peptides, *Trends Neurosci.* 9:130-133 (1986).
4. M.A. Ghatei, R.T. Jung, J.C. Stevenson, C.J. Hillyard, T.C. Adrian, Y.C. Lee, N.D. Christofides, D.L. Sarson, K. Nashiter, I. MacIntyre, and S.R. Bloom, Bombesin action on gut hormones and calcium in man, *J. Clin. Endocrinol. Metab.* 54:980-985 (1982).
5. R.T. Jensen, D.H. Coy, Z.H. Saeed, P. Heinz-Erian, S. Mantey, and J.D. Gardner, Interaction of bombesin and related peptides with receptors on pancreatic acinar cells, *Ann. N.Y. Acad. Sci.* 547:138-149 (1988).
6. M.R. Brown, K. Carver, and L.H. Fisher, Bombesin: central nervous system actions to affect the autonomic nervous system, *Ann. N.Y. Acad. Sci.* 547:174-182 (1988).
7. H.E. Albers, S-Y. Liou, E.G. Stoper, and R.T. Zoeller, Interaction of colocalized neuropeptides: functional significance in the circadian timing system, *J. Neurosci.* 11:846-851 (1991).
8. M. Ruff, E. Schiffman, V. Terranova, and C.P. Pert, Neuropeptides are chemoattractants for human cells and monocytes: a possible mechanism for metastasis, *Clin. Immunol. Immunopathol.* 37:387-396 (1985).
9. G.F. Jin, Y.S. Guo, E.R. Smith, and C.W. Houston, The effect of bombesin-related peptides on the phagocytic function of mouse phagocytes *in vitro, Peptides* 11:393-396 (1990).
10. V. Rettori, C.C. Pazos-Moura, E.G. Moura, J. Polak, and S.M. McCann, Role of neuromedin B in control of release of thyrotropin in hypothyroid and hyperthyroid rats, *Proc. Natl. Acad. Sci.* 89:3035-3039 (1992).
11. A.M. Labacq-Verheyden, J. Trepel, E.A. Sausville, and J.F. Battey, Bombesin and gastrin-releasing peptide: neuropeptides, secretagogues, and growth factors, *in*: "Handbook of Experimental Pharmacology," M.N. Sporn, and A.B. Roberts, eds., Springer-Verlag, Berlin (1990).
12. E. Rozengurt, Bombesin-induction of cell proliferation in 3T3 cells, *Ann. N.Y. Acad. Sci.* 547:277-292 (1988).
13. J.C. Willey, J.F. Lechner, and C.C. Harris, Bombesin and C-terminal tetradecapeptide of gastrin-releasing peptide are growth factors for normal human bronchial epithelial cells, *Exp. Cell Res.* 153:245-248 (1984).
14. T. Endo, H. Fukue, M. Kanaya, M. Mizunuma, M. Fujii, H. Yamamoto, S. Tanaka, and M. Hashimoto, Bombesin and bradykinin increase inositol phosphates and cytosolic free Ca^{2+} and stimulate DNA synthesis in human endometrial stomal cells, *J. Endocrinol.* 131:313-318 (1991).
15. M. Bologna, C. Festuccia, P. Muzi, L. Biordi, and M. Ciomei, Bombesin stimulates growth of human prostatic cancer cells *in vitro, Cancer* 63:1714-1720 (1989).
16. F. Cuttitta, D.N. Carney, J. Mulshine, T.W. Moody, J. Fedorko, A. Fischler, and J.D. Minna, Bombesin-like peptides can function as autocrine growth factors in human small-cell lung cancer cells, *Nature Lond.* 316:823-826 (1985).
17. P.O. Seglen, H. Skomedal, G. Saeter, P.E. Schwartze, and J.M. Nesland, Neuroendocrine dysdifferentiation and bombesin production in carcinogen-induced hepatocellular rat tumours, *Carcinogenesis* 10:21-29 (1989).
18. J. Nelson, M. Donnelly, B. Walker, J. Gray, C. Shaw, and R.F. Murphy, Bombesin stimulates proliferation of human breast cancer cells in culture, *Br. J. Cancer* 63(6):933-936 (1991).
19. V. Erspamer, Discovery, isolation and characterization of bombesin-related peptides, *Ann. N.Y. Acad. Sci.* 547:3-9 (1988).

20. E.R. Spindel, E. Giladi, P. Brehm, R.H. Goodman, and T.P. Segerson, Cloning and functional characterization of a complementary DNA encoding the murine fibroblast bombesin-gastrin-releasing peptide receptor, *Mol. Endocrinol.* 4:1956-1963 (1990).

21. J.F. Battey, J.M. Way, M.H. Corjay, H. Shapira, K. Kusano, R. Harkins, J.M. Wu, T. Slattery, E. Mann, and R.I. Feldman, Molecular cloning of the bombesin/gastrin-releasing peptide receptor form Swiss 3T3 cells, *Proc. Natl. Acad. Sci. USA* 88:395-399 (1991).

22. E. Wada, J. Way, H. Shapira, K. Kusano, A.M. Labacq-Verhayden, D.H. Coy, R.T. Jensen, and J. Battey, cDNA cloning, characterization and brain region-specific expression of a neuromedin B-preferring bombesin receptor. *Neuron* 5:421-430 (1991).

23. M.H. Corgay, D.J. Dohrzanski, J.M. Way, J. Viallet, H. Shapira, P. Worland, E.A. Sausville, and J.F. Battey, Two distinct bombesin receptor subtypes are expressed and functional in human lung carcinoma cells, *J. Biol. Chem.* 266:18771-18779, (1991).

24. G. Falconieri Erspamer, G. Severini, V. Erspamer, P. Melchiorri, G. Delle Fave, and T. Nakajimi, Parallel bioassay of 27 bombesin-like peptides on 9 smooth muscle preparations, structure-activity relationships and bombesin receptor subtypes, *Regul. Pept.* 21:1-11 (1988).

25. C. Severi, R.T. Jensen, V. Erspamer, L. D'Arpino, A. Torsoli, and G. Delle Fave, Different subtypes of receptors mediate the action of bombesin-related peptides on gastric smooth muscle cells, *Am. J. Physiol.* 260:G683-G690 (1991).

26. T. von Schrenck, P. Heinz-Erian, T. Moran, S.A. Mantey, J.D. Gardner, and R.T. Jensen, Characterization of a neuromedin B-preferring receptor in esophagus muscle: evidence for subtypes of bombesin receptors, *Am. J. Physiol. (Gastrointest. Liver Physiol. 19)* 256:G747-G758 (1989).

27. T. von Schrenck, L-H. Wang, D.H. Coy, M.L. Villanueva, S. Mantey, and R.T. Jensen, Potent bombesin receptor antagonists distinguish receptor subtypes, *Am. J. Physiol.* 259:G468-G473 (1990).

28. E.E. Ladenheim, R.T. Jensen, S.A. Mantey, P.R. McHugh, and T.H. Moran, Receptor heterogeneity for bombesin-like peptides in the rat antral nervous system, *Brain Res.* 537:233-240 (1990).

29. M.C. Lee, R.T. Jensen, D.H. Coy, and T.W. Moody, Neuromedin B binds with high affinity to rat brain slices, *J. Mol. Cell Neurosci.* 1:161-167 (1991).

30. R.T. Jensen, T. Moody, C.P. Pert, J.E. Rivier, and J.D. Gardner, Interaction of bombesin and litorin with specific membrane receptors on pancreatic acinar cells, *Proc. Natl. Acad. Sci. USA* 75:6139-6143 (1978).

31. J.M. Westendorf, and A. Schonbrunn, Characterization of bombesin receptors in a rat pituitary cell line, *J. Biol. Chem.* 258:7527-7535 (1983).

32. L-H. Wang, J.F. Battey, E. Wada, J-T. Lin, S.A. Mantey, D.H. Coy, and R.T. Jensen, Activation of neuromedin B-preferring bombesin receptors on rat glioblastoma C-6 cells alters cellular calcium and phosphoinositides, *Biochem. J.* (in press).

33. T.W. Moody, J. Staley, F. Zia, D.H. Coy, and R.T. Jensen, Neuromedin B binds with high affinity, elevates cytosolic calcium and stimulates the growth of small cell lung cancer cell lines, *J. Pharmacol. Exp. Ther.* (in press).

34. S.L. Swope, and A. Schonbrunn, The biphasic stimulation of insulin secretion by bombesin involves both cytosolic free calcium and protein kinase C, *Biochem. J.* 253(1):193-202 (1988).

35. J.B. Fischer, and A. Schonbrunn, The bombesin receptor is coupled to a guanine nucleotide-binding protein which is insensitive to pertussis and cholera toxins, *J. Biol. Chem.* 263(6):2808-2816 (1988).

36. R. Heikkila, J.B. Trepel, F. Cuttitta, L.M. Neckers, and E.A. Sausville, Bombesin-related peptides induce calcium mobilization in a subset of human small cell lung cancer cell lines, *J. Biol. Chem.* 262(34):16456-16460 (1987).

37. T.W. Moody, A. Murphy, S. Mahmoud, and G. Fiskum, Bombesin-like peptides elevate cytosolic calcium in small cell lung cancer cells, *Biochem. Biophys. Res. Commun.* 147(1):189-195 (1987).

38. J. Sinnett-Smith, W. Lehmann, and E. Rozengurt, Bombesin receptor in membranes from Swiss 3T3 cells: binding characteristics, affinity labelling and modulation by guanine nucleotides, *Biochem. J.* 265(2):485-493 (1990).

39. K.D. Brown, S. Laurie, C.J. Littlewood, D.M. Blakely, and A.N. Corps, Characterization of the high affinity receptors on Swiss 3T3 cells which mediate the binding, internalization and degradation of the mitogenic peptide bombesin, *Biochem. J.* 252:227-235 (1988).

40. W-Y. Zhu, B. Goke, and J.A. Williams, Binding, internalization, and processing of bombesin by rat pancreatic acini, *Am. J. Physiol. (Gastrointest. Liver Physiol.* 261:G57-G64 (1991).

41. R.V. Benya, L-H. Wang, J-T. Lin, E. Wada, J.F. Battey, and R.T. Jensen, Neuromedin B receptor-peptide interaction: binding, regulation and cell activation, *Regul. Peptides* (abstract) (in press).

42. L-H. Wang, S. Mantey, J-T. Lin, and R.T. Jensen, Ligand-receptor interactions with bombesin receptor subtypes, *Gastroenterology* 102(#4):Part 2, A764 (1992) (abstract).

43. E.E. Ladenheim, T.M. Moran, and R.T. Jensen, Identification and characterization of neuromedin B receptors in the rat central nervous system, *Methods Neurosci.* (in press).

44. H. Shapira, E. Wada, R.T. Jensen, and J.F. Battey, Distinguishing bombesin receptors, *Methods Neurosci.* (in press).

45. H.T. Haigler, F.R. Maxfield, M.C. Willingham, and I. Pastan, Dansylcadaverine inhibits internalization of ^{125}I-epidermal growth factor in BALB 3T3 cells, *J. Biol. Chem.* 255:1239-1241 (1980).

46. R.T. Jensen, and D.H. Coy, Progress in the development of potent bombesin receptor antagonists, *TIPS* 12:13-18 (1991).

47. R.T. Jensen, P. Heinz-Erian, T. Moran, S.A. Mantey, S.W. Jones, and J.D. Gardner, Characterization of ability of various substance P antagonists to inhibit action of bombesin. *Am. J. Physiol. (Gastrointest. Liver Physiol.)* 17:G883-G890, (1988).

48. R.T. Jensen, S.W. Jones, K. Folkers, and J.D. Gardner, A stnthetic peptide that is a bombesin receptor antagonist, *Nature Lond.* 309:61-63 (1984).

49. D.H. Coy, L-H. Wang, N-Y. Jiang, and R.T. Jensen, Short chain bombesin pseudopeptides which are potent and more general bombesin receptor antagonists, *Eur. J. Pharmacol.* 190:31-38 (1990).

50. L-H. Wang, D.H. Coy, J.E. Taylor, N-Y. Jiang, S.H. Kim, J-P. Moreau, S.C. Huang, S. Mantey, H. Frucht, and R.T. Jensen, Desmethionine alkylamide bombesin analogues: a new class of bombesin receptor antagonists with potent antisecretory activity in pancreatic acini and antimitotic acitivity in Swiss 3T3 cells, *Biochemistry* 29:616-622 (1990).

51. L-H. Wang, D.H. Coy, J.E. Taylor, N-Y. Jiang, J-P. Moreau, S.C. Huang, H. Frucht, B.M. Haffar, and R.T. Jensen, Des-Met carboxyl-terminally modified analogues of bombesin function as potent bombesin receptor antagonists, partial agonists or agonists, *J. Biol. Chem.* 265:1569-1570 (1990).

52. D.C. Heimbrook, W.S. Saari, N.L. Balishin, A. Friedman, K.S. Moore, M.W. Riemen, D.M. Kiefer, N.S. Rotberg, J.W. Wallen, and A. Oliff, Carboxy-terminal modification of a gastrin-releasing peptide derivative generates potent antagonists, *J. Biol. Chem.* 264:11258-11262 (1989).

53. R.T. Jensen, J.E. Mrozinski, Jr., and D.H. Coy, Bombesin antagonists: different receptor classes and mechanisms of action, *in:* "Peptides in Oncology, Vol. II, Somatostatin And Bombesin Antagonists," K. Hoffken, ed., Springer-Verlag, Berlin (in press).

54. J.B.A. Millar, and E. Rozengurt, Bombesin enhancement of cAMP accumulation in Swiss 3T3 cells: evidence of a dual mechanism of aciton, *J. Cell Biol.* 137:212-214 (1988).

55. W.H. Rowley, S. Sato, S-C. Huang, D.M. Collado-Escobar, M.A. Beaven, L-H. Wang, J. Martinez, J.D. Gardner, and R.T. Jensen, Cholecystokinin-induced formation of inositol phosphates in pancreatic acini, *Am. J. Physiol.* 259:G655-G665 (1990).

56. R.T. Jensen, S.A. Wank, W.H. Rowley, S. Sato, and J.D. Gardner, Interactions of cholecystokinin with pancreatic acinar cells: a well studied model of a peripheral action of CCK, *TIPS* 10:418-423 (1989).

57. J. Battey, and E. Wada, Two distinct receptors for mammalian bombesin-like peptides, *Trends Neurosci.* 14:524-527 (1991).

Part IV

Growth Factor Receptors

NEUROTROPHINS IN THE ADULT BRAIN: EFFECTS ON HIPPOCAMPAL CHOLINERGIC FUNCTION FOLLOWING DEAFFERENTATION, AND REGULATION OF THEIR EXPRESSION BY PHARMACOLOGICAL AGENTS AND LESIONS

Paul A. Lapchak, Dalia M. Araujo, Timothy L. Denton, Millicent M. Dugich-Djordjevic, and Franz Hefti

Andrus Gerontology Center
Division of Neurogerontology
University of Southern California
Los Angeles, California
90089-0191

INTRODUCTION

This chapter describes recent progress in the understanding of the regulation of neurotrophin and neurotrophin receptor expression in the hippocampal formation following pharmacological manipulations or brain lesions. The chapter is subdivided into different sections which provide details about five different pharmacological treatment regimens and lesion paradigms found to influence neurotrophin mechanisms.

NEUROTROPHINS AND THEIR RECEPTORS

The protein family called neurotrophins, consists of nerve growth factor (NGF), brain-derived neurotrophic factor (BDNF) neurotrophin-3 (NT-3), neurotrophin-4 (NT-4), and neurotrophin-5 (NT-5) (see Thoenen, 1991 and Lapchak et al., 1992 for reviews). Neurotrophins are synthesized as precursor proteins from which an active protein of approximately 120 amino acids is cleaved at the N-terminal. Individual neurotrophins are highly conserved among mammalian species and share approximately 50% of the amino acids with other members of this group. The functional proteins are basic and occur as homodimers.

Recent evidence indicates that two proteins are required for the formation of receptors for neurotrophins, the low affinity NGF receptor protein (p75NGFR; Chao et al., 1986; Radeke et al., 1987) and products of *trk*-related proto-oncogenes (Bothwell, 1991, review). The *trk* gene products, exhibit protein kinase activity (Martin-Zanca et al., 1989; Barbacid et al., 1991; Kaplan et al., 1991a,b). Functional NGF receptors are formed by the *trk*A tyrosine protein kinase, leaving the functional role of p75NGFR unclear (Cordon-Cardo et al., 1991). BDNF binds to and stimulates *trk*B tyrosine protein kinase (Klein et al., 1991; Squinto et al., 1991). NT-3 is capable of interacting with *trk*A, *trk*B and *trk*C (Lamballe et al., 1991; Soppet et al., 1991; Squinto et al., 1991). NT-4 and NT-5 activate *trk*B and possibly *trk* A (Klein et al., 1992). The majority of available data which has accumulated has relied on cell lines in order to characterize *trk* receptors, whether or not the same specificity is observed in the adult rat nervous system remains to be examined. The description of biological actions of other neurotrophins and their responsive populations is still in its initial stages. This manuscript will serve to review recent progress from our laboratory concerning the functional significance and regulation of neurotrophins.

BDNF AND NGF TREATMENT IN RATS WITH LESIONS: EFFECTS ON HIPPOCAMPAL CHOLINERGIC FUNCTION AND WEIGHT GAIN

NGF is thought to be important in the normal function of basal forebrain cholinergic neurons in the adult brain (reviews, Hefti et al., 1992; Lapchak, 1992). The second neurotrophin to be discovered, BDNF, is present in many brain areas, and, in the hippocampal formation, its distribution seems to match the pattern of cholinergic terminal density (Phillips et al., 1990). The protein product of *trkB*, encodes a functional BDNF receptor, and this protein is abundantly expressed in brain including the hippocampal formation (Klein et al., 1990, 1991). These findings and the ability of BDNF to stimulate the developmental differentiation of septal cholinergic neurons in culture (Alderson et al., 1990; Knusel et al., 1991) indicated that BDNF may be involved in the adult function of cholinergic neurons of the septo-hippocampal pathway. We compared the effects of chronic treatment rhBDNF with those of rhNGF on presynaptic cholinergic function in the hippocampus following unilateral partial fimbrial transections. In addition, we determined whether rhBDNF or rhNGF treatment alters food intake of the animals.

Effects of Chronic rhNGF or rhBDNF Treatment on Presynaptic Hippocampal Cholinergic Function Following Partial Unilateral Fimbrial Transections

Female Wistar rats with partial unilateral fimbrial transections were administered rhBDNF or rhNGF (1.4 μg qid) via intraventricular cannulas for 21 days before the hippocampus was used for neurochemical analysis. Controls received intraventricular injections of cytochrome c, which was chosen because of its comparable biochemical properties to neurotrophins but its inability to stimulate their receptors. Partial fimbrial transections were performed as described previously by Lapchak and Hefti (1992). Cholinergic parameters were measured as described previously by Lapchak et al. (1991).

P2 synaptosomes were used in the present study to measure high affinity choline uptake (HACU), [^{3}H]acetylcholine (ACh) synthesis, and ChAT activity. Partial fimbrial transections reduced ChAT activity on the lesioned side by 40% when measured in a P2 synaptosomal preparation. Chronic rhBDNF treatment did not significantly alter measures of hippocampal ChAT activity, however, chronic rhNGF treatment attenuated the lesion-induced decrease of ChAT activity by 53-65%. In addition, in the synaptosomal P2 fraction of rhNGF-treated animals there was a slight but significant increase of ChAT activity on the contralateral unlesioned-side.

On the lesioned side of control animals, the quantity of [^{3}H]choline taken up by the HACU system of hippocampal synaptosomes was decreased by 53%. NGF treatment increased HACU by synaptosomes from the lesioned side by 118%, however, there was also a significant increase (approximately 25%) on the contralateral unlesioned-side. rhBDNF did not affect HACU on either the lesioned or unlesioned-side.

The amount of [^{3}H]ACh synthesized from [^{3}H]choline by hippocampal synaptosomes was measured following partial fimbrial transections. The lesion reduced the amount of [^{3}H]ACh formed by 63% compared to contralateral control values. Chronic NGF treatment produced an increase in the amount of [^{3}H]ACh formed by hippocampal synaptosomes, whereas rhBDNF was ineffective in altering this measure of presynaptic cholinergic function.

Table 1. Effect of chronic icv neurotrophin administration on hippocampal cholinergic markers following partial fimbrial transections

Treatment	CHOLINERGIC MARKER (% UNLESIONED CONTROL)		
	cc	rhNGF	rhBDNF
ChAT Activity	57.9%	75.1%	59.7%
HACU	46.8%	72.3%	52.2%
ACh Synthesis	36.3%	63.3%	34.4%

The last parameter measured was the average weight gain by fimbriectomized animals that received either cc, rhNGF or rhBDNF. Over the course of 21 days, lesioned control animals increased their body weight by an average of 47 grams. In contrast, both rhNGF- and rhBDNF-treated lesioned animals gained only 20 grams over the treatment period. In neurotrophin-treated animals there was a slow but steady increase of body weight until 12 days following the lesion and start of treatment. Thereafter, the body weight was maintained constant.

Our study was designed to determine whether the neurotrophin BDNF exerts similar trophic actions on adult septo-hippocampal neurons as has been established for NGF (Hefti et al., 1992; Lapchak, 1992). The results do not support the hypothesis that BDNF is a trophic factor for lesioned adult cholinergic neurons. Chronic BDNF treatment was ineffective at preventing lesion-induced decreases of a variety of cholinergic parameters including ChAT activity, HACU, and ACh synthesis, wherease these measures of presynaptic cholinergic function were elevated by NGF treatment. These findings, together with the limited protective effect on cholinergic cell bodies after axotomy (Knusel et al., 1992), indicate that NGF remains the agent of choice when attempting to protect cholinergic neurons from degeneration induced by injury or disease.

Despite the lack of effects on cholinergic function, BDNF administration reduced weight gain of the experimental animals. The demonstration of a similar effect of NGF confirms recent findings by Williams (1991) which demonstrated reduced food intake and weight gain in rats receiving chronic intraventricular treatment of NGF. While we did not measure food intake, the findings by Williams suggest that, as found in our study, the reduced weight gain of neurotrophin-treated animals reflects hypophagia. There was no direct correlation between the efficacy of neurotrophins to increase cholinergic function in the hippocampus and change in body weight since BDNF did not affect cholinergic function but it did significantly decrease body weight similar to that observed with chronic NGF treatment (Lapchak and Hefti, 1992).

The present findings suggest a physiological role for endogenous BDNF in the maintenance of appetite and food intake in the adult rat. The results do not support the hypothesis that BNDF plays a role in the function of adult cholinergic neurons comparable to that of NGF since, in contrast to NGF, BDNF did not attenuate lesion-induced deficits of hippocampal cholinergic function.

REGULATION OF HIPPOCAMPAL MUSCARINIC RECEPTOR FUNCTION BY CHRONIC NGF TREATMENT IN ADULT RATS WITH FIMBRIAL TRANSECTIONS

The section above included a description of the effects of chronic NGF treatment on presynaptic hippocampal cholinergic function following a selective lesion of the fimbria which is the route by which cholinergic afferents reach the hippocampal formation. The increases of presynaptic cholinergic function are associated with the attenuation or normalization of behavioral responses associated with memory consolidation and memory formation (Will and Hefti, 1985; Will et al., 1988, 1990). However, the effects of NGF treatment on postsynaptic cholinergic receptors and signalling mechanisms following partial fimbrial transections are not known. We therefore determined whether chronic NGF treatment regulates the function and maintenance of the cholinergic synapse, in particular at the postsynaptic ACh responsive cell. To test for such actions we assessed the effects of fimbrial transections on the distribution of muscarinic receptors in the hippocampal formation and the response of muscarinic receptors to the muscarinic agonist oxotremorine.

Effects of Chronic rhNGF Treatment on Hippocampal Muscarinic Receptor Density following Fimbrial Transections

Muscarinic receptor-selective ligands were used to determine the distribution of muscarinic receptors in lesioned animals treated with either the control protein cytochrome c or rhNGF. Twenty one days following cholinergic deafferentation of the hippocampal formation quantitative autoradiography was used to map the topographical distribution of M1 and M2 muscarinic receptor in the dorsal and ventral hippocampus. These experiments indicated that partial unilateral fimbrial transections did not significantly alter the density of muscarinic M1 receptors as labeled by $[^3H]$pirenzepine (PZ) or muscarinic M2 receptors as labeled by $[^3H]$AF-DX-384. Chronic intraventricular treatment with rhNGF, which was earlier shown to upregulate presynaptic cholinergic function in animals with partial fimbrial transections (Lapchak and Hefti, 1991; also see above), did not change either the density of muscarinic sites in animals receiving partial fimbrial transections or change the density of muscarinic sites on the unlesioned side of the hippocampus. In contrast, rhNGF treatment produced small, but significant increases in the density of sites labeled by $[^3H]$PZ and $[^3H]$AF-DX 384 in animals with full fimbrial transections. The increase of muscarinic receptors were evident in the hippocampal CA1 region, associated with the oriens, pyramidal and stratum radiatum layers.

Effects of Fimbrial Transections on Muscarinic Receptor-Mediated Second Messenger Production Following Chronic rhNGF Treatment

Although the density of muscarinic receptor populations in the hippocampal formation was largely unaffected by partial fimbrial transections at the time point studied (ie: 21 days post-lesion), there is some evidence in the literature showing that muscarinic M1 receptor coupling mechanisms are affected by similar lesions (Smith et al., 1989), suggesting that denervation-induced supersensitivity may result when the source of endogenous ACh is removed. Based on these earlier studies, we determined whether M1 muscarinic receptor coupling to phosphatidylinositol (PI) hydrolysis was altered by intraventricular rhNGF treatment of partially fimbriectomized animals by quantitating the production of inositol trisphosphate (IP_3). In cc-treated control animals there was an increase (61%) of oxotremorine-induced IP_3 production by hippocampal slices. This finding confirms that lesions of the septal cholinergic input result in functional hypersensitivity of hippocampal M1 receptors. The lesion-induced increase in IP_3 production was completely prevented by chronic intraventricular rhNGF treatment.

We furthermore determined whether muscarinic receptor-mediated production of cyclic nucleotides was altered under the same experimental conditions. M2 receptor-mediated cAMP formation was unaltered by partial fimbrial lesions. The tissue content of cAMP was similar on lesioned and unlesioned sides of cc-treated animals.

Table 2. Effect of chronic rhNGF administration to rats with partial fimbrial transections on the density of hippocampal muscarinic receptors

Treatment	Muscarinic receptor levels (% contralateral control)	
	cc	rhNGF
M1 receptors		
partial lesion	99%	101%
full lesion	106%	158%
M2 receptors		
partial lesion	100%	104%
full lesion	108%	134%

However, in rhNGF-treated animals there was a significant (35%) increase in the amount of cAMP formed by hippocampal slices. This effect was observed on both lesioned and unlesioned sides. The oxotremorine-induced formation of cGMP was not affected by partial fimbrial transections. In contrast to the effect of rhNGF treatment on cAMP formation, this treatment failed to alter the accumulation of cGMP by hippocampal slices.

The results of the present study indicate that chronic intraventricular NGF treatment of rats with fimbrial transections, results in functional changes of postsynaptic muscarinic receptors in the hippocampus. In animals with partial fimbrial transections, NGF treatment prevented the supersensitivity of muscarinic receptor-mediated PI hydrolysis and stimulated muscarinic receptor-induced cAMP formation. These findings suggest that the trophic effects of NGF treatment on cholinergic septo-hippocampal neurons translate into functional changes at the level of postsynaptic muscarinic receptors in the hippocampus. The findings of the present study extend the characterization of effects of chronic NGF treatment on forebrain cholinergic mechanisms to the level of postsynaptic cholinergic receptors. They show that trophic effects of NGF treatment on presynaptic cholinergic function correlate with functional changes

Table 3. Effect of chronic rhNGF treatment on oxotremorine-induced second messenger levels in hippocampal slices following fimbrial transections

Treatment	Second Messenger Levels (% untreated control)	
	cc	rhNGF
IP_3	161%	99%
cAMP	98%	135%
cGMP	102%	97%

at the level of postsynaptic cholinergic receptors (Lapchak et al., 1992b). These findings provide further evidence that behavioral effects mediated by intraventricular NGF treatment (Will and Hefti, 1985; Pallage et al., 1986; Will et al., 1988, 1990) reflect a functional upregulation of basal forebrain cholinergic systems. The results of the present study afford further support for the concept that NGF treatment should beneficially affect forebrain cholinergic neurons in Alzheimer's disease and attenuate the progression of those behavioral deficits directly reflecting cholinergic dysfunction (Hefti and Schneider, 1989; Lapchak, 1992). These results are important considering the ongoing clinical trials with NGF in Alzheimer's disease and the proposal to continue testing the effects of NGF in larger groups of Alzheimer disease patients (Phelps et al. 1989; Hefti and Schneider, 1991; Olson et al. 1992).

KAINIC ACID-INDUCED REGULATION OF HIPPOCAMPAL NGF, BDNF, NT-3 and trkB mRNA

Seizure-induced changes in the expression of growth factors may offer information toward a better understanding of the function of neurotrophins in the adult brain, since seizures can be considered a maximal form of neural activation and since they induce select neuronal populations to degenerate (Sloviter, 1987, 1989). Recent studies have shown that hippocampal and cortical NGF and BDNF mRNA levels are increased in response to limbic seizures induced by electrolytic lesions of the hilus (Gall and Isackson, 1989; Gall et al., 1991; Isackson et al., 1990), kindling epileptogenesis (Ernfors et al., 1991), and seizures following parenteral kainic acid (KA) adminstration (Dugich-Djordjevic et al., 1992a). Based on these findings it has been hypothesized that seizure-induced neuronal activity modulates the transcriptional response of multiple neurotrophic factor mRNAs in the CNS. However, our recent work suggests that seizure activity may not be sufficient to induce increases of BDNF mRNA (Dugich-Djordjevic et al., 1992b). Regulatory events changing neurotrophin functions may also include modulation of receptor levels. To investigate this possibility we examined the expression of trkB mRNA following kainic acid induced seizures in adult rats. Patterns of trkB mRNA expression were compared with those of BDNF and NT-3 mRNA.

Effect of kainic acid on neurotrophin and neurotrophin receptor mRNA levels in brain

Male Sprague-Dawley received either kainic acid (KA) (10-12 mg/kg, s.c.) or saline and were sacrificed 1h, 4h, 16h, 24h, 4d or 7d following seizure onset as described previously by Dugich-Djordjevic et al. (1992a,b). Animals were observed for the onset and duration of behavioral seizure activity.

In untreated control animals trkB mRNA was localized to the hippocampal granule and pyramidal cell layers and throughout cortical layer. Specific hybridization was also evident in thalamic and hypothalamic regions, however, the only areas discretely labeled were the ventromedial hypothalamic nuclei. Emulsion autoradiography for [^{35}S]trkB cRNA revealed that hybridization in control sections was localized to the neuronal layers of the hippocampus. Furthermore, more darkly stained cells were occasionally observed throughout the extent of the dentate molecular layer and the stratum radiatum of the CA1 region. Following 1 h of KA-induced seizure activity, [^{35}S]trkB cRNA hybridization increased nearly two-fold in the dentate granule cell layer. Four hours following the onset of seizure activity, hybridization in the dentate granule cell layer remained increased over control and a 30% increase was observed in the CA1 pyramidal region, while no change in hybridization density was noted in the CA3 region. Emulsion autoradiography confirmed that the increases occured over cell bodies in the dentate and the CA1 pyramidal layers. By 16 h following seizure actvity, [^{35}S]trkB cRNA hybridization in neuronal layers had returned to control levels and remained unchanged up to 7 d after seizures.

Since both, BDNF and NT-3 are ligands for the trkB receptor, we investigated the relationship of [^{35}S]trkB hybridization signals to those seen with [^{35}S]BDNF cRNA and [^{35}S]NT-3 cRNA on adjacent tissue sections. In control animals BDNF mRNA expression was evident in both pyramidal and granule cell layers of the hippocampus, in confirmation of our earlier studies (Dugich-Djordjevic et al., 1992a). NT-3 expression was confined to the dentate granule cell population and the CA2 region. Following systemic administration of kainic acid, BDNF mRNA expression was significantly elevated in the hippocampal subregions while NT-3 in the dentate granule cell population was decreased by 25-30%. Dentate and hilar region neuronal [^{35}S]BDNF cRNA hybridization was elevated up to 4 d following seizures. [^{35}S]NT-3 cRNA hybridization was decreased in the dentate granule cells at 1 h following seizure and restored to basal levels by 4 h following seizure onset.

Our study shows that seizures induced by systemic administration of kainic acid result in a rapid and pronounced increase in expression of trkB mRNA in hippocampal neuronal cell layers, which roughly parallels elevations of BDNF mRNA. Hippocampal seizures induced by systemic kainic acid administration or kindling lesions result in rapid and robust changes of the expression of genes coding for neurotrophins and their receptors. There are strong elevations of NGF, BDNF, and trkB mRNA expression, whereas the expression of NT-3 mRNA is slightly reduced. The differential regulation of the neurotrophins supports the conclusion that the induction of NGF, BDNF, and trkB mRNA represents a selective regulatory event rather than the consequence of a general and non-specific upregulation of gene expression. Concomitant upregulation of the expression of BDNF and its trkB receptor suggests a very effective regulatory system to activate BDNF mechanisms. It is tempting to speculate that such a mechanism, demonstrated in animals with seizures, is operational under physiological conditions. The earlier finding of a seizure-induced upregulation of NGF and BDNF expression led to the hypothesis that seizures should be considered an extreme form of neuronal stimulation and that also under physiological conditions, neurotrophin expression is regulated by neuronal activity (Gall and Isackson, 1989; Zafra et al., 1990,1991). Coordinated regulation of BDNF and its receptor in response to increased electric activity could enable a cell to very quickly adjust programs of intracellular metabolism and gene expression. Such metabolic activation could affect, by an autocrine mechanism, the stimulated cell or, by a paracrine mechanism, neighbouring neurons. Neurotrophin-mediated adjustments of intracellular metabolism and gene expression could serve to meet a higher metabolic demand in response to increased activity or, alternatively, be involved in mechanims of synaptic and neuronal plasticity. However, it has to be pointed out that neuronal activity does not seem to be the sole determinant of BDNF mRNA expression since, during postnatal development there is a clear dissociation of neuronal activitation and BDNF mRNA induction in the hippocampus (Dugich-Djordjevic et al., 1992b), and since removal of an excitatory input to the hippocampus in the adult brain increases rather than decreases BDNF mRNA expression (see below).

DIFFERENTIAL REGULATION OF BDNF AND NT-3 mRNA EXPRESSION IN THE RAT HIPPOCAMPUS BY GLUCOCORTICOIDS

Glucocorticoids have been shown to exert a profound effect on the integrity of neuronal populations in the normal adult rat hippocampus (Sapolsky, 1986; McNeill et al., 1991), including detrimental effects on hippocampal neurons and in the enhancement of age-related effects on selected hippocampal neuronal populations (Sapolsky and Pulsinelli, 1985; Sapolsky, 1985). During development, low levels of glucocorticoids maintain neuronal populations, while high levels can result in neuronal damage and loss (Sapolsky, 1986).

High serum levels of glucocorticoids can suppress cell proliferation during development while removal of endogenous gluccocorticoids reults in increased mitotic activity of both glial elements and neuronal populations in the cerebellum and hippocampus, regions which undergo extensive postnatal neurogenesis. A dose-dependent glucocorticoid-mediated attenuation of lesion-induced axonal sprouting responses in the dentate gyrus has been shown following ipsilateral destruction of the perforant pathway (Scheff et al., 1986). Thus, the possibility exists that glucocorticoids may interact with hippocampal neurotrophin producing cellular populations and thereby remodel hippocampal neurons. Based on these findings we investigated the effect of adrenalectomy alone (ADX) and adrenalectomy with corticosterone treatment (CORT) on the expression of BDNF and NT-3 mRNA in the rat hippocampus.

Effects of Corticosteroids on Hippocampal Neurotrophin Expression

Northern blot analysis revealed a significant increase in total hippocampal BDNF mRNA prevalence in 1 d ADX versus intact controls (INT). ADX animals which received daily injections of corticosterone (10 mg/rat) showed a 34% decrease in hippocampal BDNF mRNA prevalence as compared to intact control rats. BDNF mRNA expression in the dentate granule cell region was increased 31% over INT levels in the 1 d ADX group and decreased to 57% of INT in both CORT groups. In contrast, NT-3 mRNA expression in the dentate granule cells was decreased by 15-25% of INT in both ADX groups and increased over INT by 25% in the 3 d CORT group.

Following bilateral adrenalectomy, the BDNF mRNA levels in the CA3 and the dentate granule cell regions were increased to almost 150% of intact control. Hybridization of [^{35}S]BDNF cRNA in the CA1 pyramidal layer exhibited the largest increase compared to intact levels. Cortical levels and expression in thalamic and hypothalamic subregions also appeared to be increased. At 3 d following adrenalectomy, hybridization of [^{35}S]BDNF cRNA remained elevated in the pyramidal regions and had returned to intact levels in the dentate granule cell subregion. [^{35}S]NT-3 cRNA hybridization in the hippocampus granule cell region was decreased by nearly 50% in the dentate granule cell region in response to adrenalectomy. In a separate experiment, we compared sham operated ADX, 3 d ADX, sham operated CORT replaced and 3 d ADX CORT replaced groups. Although the SHAM ADX groups exhibited slightly greater (30%) BDNF mRNA levels than intact controls, the pattern of increases at 3 d ADX essentially replicated the present results. NT-3 levels in the dentate granule cell region were decreased by 15-25% in reponse to adrenalectomy, with the largest observed decrease in the 1 d ADX group. A complete loss of CA2 labeling was observed in some of the animals. Emulsion autoradiography revealed that the increases in BDNF mRNA expression observed were localized over cell bodies in the pyramidal and granule subregions.

Our results suggest that glucocorticoids regulate neurotrophin mRNA expression in the hippocampal formation and may act to physiologically inhibit BDNF mRNA expression in the adult hippocampus. Differential responsiveness of BDNF mRNA expression to glucocorticoids in the hippocampal pyramidal layers may reflect differences in the topographical distribution of Type I and Type II glucocorticoid receptor in brain. Furthermore, the reciprocal regulation of BDNF and NT-3 mRNA expression in the dentate granule cell region suggest that these neurotrophins are co-regulated via a glucocorticoid-dependent transduction mechanism and may be involved in the continued neuronal plasticity of this region in the adult rat.

REGULATION OF BDNF mRNA EXPRESSION IN THE HIPPOCAMPAL FORMATION FOLLOWING ENTORHINAL CORTEX LESIONS

We investigated the effects of ablation of excitatory input from the entorhinal cortex, which normally occupy the outer 2/3 of the hippocampal dentate molecular layer, on neurotrophin markers in the hippocampal formation. In this model the synaptic space formerly occupied by the inputs from the entorhinal cortex is replaced by sprouting of the cholinergic input from the inner 1/3 of the molecular layer. Male Fisher 344 rats were used for perforant pathway lesions which remove the entorhinal cortex input to the dentate gyrus. We used histochemical methods to confirm the accuracy of the lesion. Cytochrome oxidase, a mitochondrial enzyme, which has previously been used as a marker for the excitatory input to the hippocampus, (Borowsky and Collins, 1989) was shown to decrease in the outer 2/3 of the dentate molecular layer after 24 hours. Sprouting of cholinergic neurons into the denervated areas was shown using histochemical staining for acetylcholinesterase. The sprouting first appeared at four days post-lesion and was very marked at 14 days post-lesion. This is similar to the results obtained by Nadler et al. (1977). We then characterized the effects of entorhinal cortex lesions on hippocampal BDNF mRNA levels using both in situ and Northern blot analysis. Northern blot analysis showed a several-fold increase in hippocampal BDNF mRNA prevalance with a peak at two days post-lesion and a gradual decline to control levels by 14 days post-lesion. The observed time course of lesioned-induced BDNF mRNA increases is consistent with a role for BDNF in the cholinergic sprouting seen in the dentate gyrus. However, a somewhat surprising result was that the increase in BDNF mRNA was bilateral, that is, the unlesioned side of the brain showed a similar increase in BDNF message. This may be due to the presence of bilateral innervation of the hippocampal dentate from the entorninal cortex. This is in agreement with an earlier study of Nadler et al., (1977), which showed that there exists bilateral innervation of the dentate as evidenced by direct counts of the synaptic density. The level of innervation of the contralateral dentate is estimated to be decreased approximatley 30% by this lesion. In situ hybridization histochemistry confirmed that the increase in BDNF mRNA levels occured bilaterally, and furthermore showed that the response was localized to the dentate, as expected since these cells are directly affected by the deafferentation. Additionally, the in situ hybridization results showed that the response was localized to neuronal and not glial cells. The observation that the removal of the excitatory input to the hippocampus results in increases in BDNF mRNA levels suggests that BDNF mRNA levels are not a simple reflection of the activity of neurons expressing this factor, but that they are regulated by a variety of mechanisms (see also Dugich-Djordjevic et al., 1992a,b). In addition to the postulated role for BDNF in the reorganization and reinnervation of the hippocampus on the side of the lesion, the bilateral response of hippocampal BDNF mRNA following entorhinal cortex lesions suggests that BDNF may be involved in subtle realignments of circuitry which occur in the contralateral hippocampus as well.

CONCLUSIONS

The recently discovered neurotrophins, particularly BDNF, are widely expressed in the hippocampus and other areas of the adult brain, suggesting an important role of these growth factors in neuronal plasticity and in the response to brain injury. Our recent studies show that BDNF is less effective than NGF in protecting cholinergic neurons from degenerative events following lesions but that BDNF as well as NGF reduce weight gain of rats. NGF treatment increases presynaptic cholinergic function in rats with partial fimbrial transections and concomitantly prevents the development of postsynaptic muscarinic receptor hypersensitivity normally induced by the lesions. We found that BDNF and NT-3 synthesis, as assessed by measuring their mRNA levels, is affected by a variety of manipulations including seizures, glucocorticoids and lesions of hippocampal afferents. These manipulations produced concomitant but reciprocal changes in BDNF and NT-3 mRNA expression. Seizures induced by systemic administration of kainic acid were found to elevate levels of both, BDNF mRNA and the mRNA coding for its trkB receptor. Lesions of the perforant path, which remove an excitoratory input to the hippocampus, increase BDNF expression in this structure. These findings suggest multiple functional roles for neurotrophins and complex regulatory mechansims which warrant further investigation.

ACKNOWLEDGMENTS

This work was supported by was supported by PHS grants NS22933, AG10480, and State of California Alzheimer Research Program Contract 91-12965. PAL was supported by a long term postdoctoral fellowship from HFSPO (Japan) and a grant/fellowship from the French Foundation for Alzheimer Research (Los Angeles,CA). TLD was supported by an Allied-Signal/National Parkinson Foundation fellowship.

REFERENCES

Alderson, R.F., Alterman, A.L., Barde, Y.A., and Lindsay, R.M., 1990 Brain derived neurotrophic factor increases survival and differentiated functins of rat septal cholinergic neurons in culture. Neuron 5:297.

Barbacid, M., Lamballe, F., Pulido, D., and Klein, E., 1991, The trk family of tyrosine protein kinase receptors. BBA Rev. Cancer 1072:115.

Borowsky, I.W., and Collins, R.C., 1989, Histochemical changes in enzymes of energy metabolism in the dentate gyrus accompany deaffferentation and synaptic reorganization. Neurosci. 33: 253.

Bothwell, M., 1991, Keeping track of neurotrophin receptor. Cell 43:915.

Chao, M.V., Bothwell, M.A., Ross, H., Koprowski, A.A., Lanahan, C., Buck, C.R., and Sehgal, A., 1986, Gene transfer and molecular cloning of the human NGF receptor. Science 232:518.

Cordon-Cardo, C., Tapley, P., Jing, S., Nanduri, V., O'Rourke, E., Lamballe, F., Kovary, K., Klein, R., Jones, K.R., Reichardt, L.F., and Barbacid, M., 1991, The trk tyrosine protein kinase mediates the mitogenic properties of nerve growth factor and neurotrophin-3. Cell 66:173.

Dugich-Djordjevic, M.M., Lapchak, P.A., Tocco, G., Pasinetti, G., Baudry, M.and Hefti, F., 1992a, Regionally specific and rapid increases in BDNF mRNA in the adult rat brain following seizures induced by systemic administration of kainic acid. Neuroscience 47:303.

Dugich-Djordjevic, M., Toccco, G., Willoughby, D.A., Najm, I., Pasinetti, G., Thompson, R.F., Baudry, M., Lapchak, P.A., and Hefti, F., 1992b, BDNF mRNA expression in the developing rat brain following kainic acid-induced seizure activity. Neuron 8:1.

Ernfors, P., Bengzon, J., Kokaia, Z., Persson, H., and Lindvall, O., 1991, Increased levels of messenger RNAs for neurotrophic factors in the brain during kindling epileptogenesis. Neuron 7:165.

Gall, C.M., and Isackson, P.J., 1989, Limbic seizures increase neuronal production of messenger RNA for nerve growth factor. Science 245:758.

Gall, C.M., Murray, K., and Isackson, P.J., 1991, Kainic acid-induced seizures stimulate increased expression of nerve growth factor mRNA in rat hippocampus. Mol. Brain Res. 9:113.

Hefti, F., and Schneider, L.S., 1989, Rationale for the planned clinical trials with nerve growth factor in Alzheimer's disease. Psychiatric Development. 4:297.

Hefti, F., Lapchak, P.A., Knusel, B., Jenden, D.J., and Will, B., 1992, Correlation of molecular, cellular and behavioral effects of nerve growth factor administration to rats with partial lesions of the septo-hippocampal cholinergic pathway. J. Cognitive Neuroscience (in press).

Hempstead, B.L., Martin-Zanca, D., Kaplan, D.R., Parada, L.F., Chao, M.V., 1991, High-affinity NGF binding requires coexpression of the trk proto-oncogene and the low affinity NGF receptor. Nature 350:678.

Isackson, P.J., Huntsman, M.M., Murray, K.D. and C.M. Gall., 1990, BDNF mRNA expression is increased in adult rat forebrain after limbic seizures: temporal patterns of induction distinct from NGF. Neuron 6:937.

Kaplan, D.R, Hempstead, B.L., Martin-Zanca, D., Chao, M.V. and Parada, L.F., 1991a, The trk proto-oncogene product: a signal transducing receptor for nerve growth factor. Science 525:554.

Kaplan, D.R., Martin-Zanca, D., and Parada, L.F., 1991b, Tyrosine phosphorylation and tyrosine kinase activity of the trk proto-oncogene product induced by NGF. Nature 350:158.

Klein, R,, Nandurim V,, Jing, S., Lamballe, F., Tapley, P., Bryant, S., Cordon-Cardo, C., Jones, K.R., Reichardt, L.F., and Barbacid, M. , 1991, The trkB tyrosine protein kinase is a receptor for brain-derived neurotrophic factor and neurotrophin-3. Cell 66:395.

Klein, R., Lamballe, F., Bryant, S., and Barbacid, M., 1992, The trkB tyrosine protein kinase is a receptor for neurotrophin-4. Neuron. 8:947.

Knusel, B., Winslow, J.W., Rosenthal, A., Burton, L.E., Seid, D.P., Nikolics, K., and Hefti, F., 1991, Promotion of central cholinergic and dopaminergic neuron differentiation by brain-derived neurotrophic factor but not neurotrophin-3. Proc. Natl. Acad. Sci (USA) 88:961.

Knusel, B., Beck, K., Winslow, J.W., Rosenthal, A., Burton, L.E., Widmer, H.R., Nikolics, K., and Hefti, F., 1992, Brain-derived neurotrophic factor administration protects basal forebrain cholinergic but not nigral dopaminergic neurons from degenerative changes after axotomy in the adult rat brain. J. Neurosci. (in press).

Lamballe, F., Klein, R., and Barbacid, M., 1991, trkC, a new member of the trk family of tyrosine protein kinases is a receptor for neurotrophin-3. Cell 66:967.

Lapchak, P.A. and Hefti, F., 1991, Effect of recombinant human nerve growth factor of presynaptic function in rat hippocampal slices: measures of [^{3}H]acetylcholine synthesis, [^{3}H]acetylcholine release and choline acetyltransferase activity. Neuroscience 42:639.

Lapchak, P.A., Jenden, D.J., and Hefti, F., 1991, Compensatory elevation of acetylcholine synthesis in vivo by cholinergic neurons surviving partial lesions of the septohippocampal pathway. J. Neurosci. 11(9):2821.

Lapchak, P.A. and Hefti, F., 1992, BDNF and NGF treatment of lesioned rats: effect on cholinergic function and weight gain. NeuroReport 3(5):405.

Lapchak, P.A., Araujo, D.M., and Hefti, F., 1992a, Neurotrophins in the central nervous system. Reviews in the Neurosciences. 3(1):1.

Lapchak, P.A., Jenden, D.J., and Hefti, F., 1992b, Pharmacological stimulation reveals NGF-induced elevations of hippocampal cholinergic function measured in vivo in rats with partial fimbrial transections. Neuroscience (in press).

Lapchak, P.A., 1992, Therapeutic potential for nerve growth factor in Alzheimer's disease: insights from pharmacological studies using lesioned central cholinergic neurons. Reviews in the Neurosciences. 3(2):1.

Martin-Zanca, D., Oskam, R., Mitra, G., Copeland, T., and Barbacid, M, 1989, Molecular and biochemical characterization of the human trk proto-oncogene. Mol. Cell. Biol. 9:24.

McNeill, T.H., Masters, J.N. and Finch, C.E., 1991, Effect of chronic adrenalectomy on neuron loss and the distribution of sulfated glycoprotein-2 in the dentate gyrus of prepubertal rats. Exp. Neurol., 111:140.

Nadler, J.V., Cotman, C.W., and Lynch, G.S., 1977, Histochemical evidence of altered development of cholinergic fibers in the ratdentate gyrus following lesions. J.Comp.Neurol. 171: 561.

Olson, L., Nordberg, A., von Holst, H., Alafusoff, I., Amberla, C., Backman, L., Ebendal, T., Hartvig, P., Lilja, A., Lundqvist, H., Langstrom, B., Meyerson, B., Viitanen, M., Winblad, B., and Seiger, A., 1992, Nerve growth factor affects ^{11}C-nicotine binding, blood flow, EEG, and verbal episodic memory in an Alzheimer patient. J. Neural. Transm. (P-D Sect) 4:79.

Pallage, V., Toniolo, G., Will, B., and Hefti, F., 1986, Long-term effects of nerve growth factor and neural transplants on behavior or rats with medial septal lesions. Brain Res.386:197.

Phelps, C.H., Gage, F.H., Growdon, J.H., Hefti, F., Harbaugh, R., Johnson, MV, Khachaturian, Z.S., Mobley, W.C., Price, D.L., Raskind, M, Simpkins, J., Thal, L.J., and Woodcock, J., 1989, Potential use of nerve growth factor to treat Alzheimer's disease. Neurobiol. Aging 10:205.

Phillips, H.S., Hains, J.M., Laramee, G.R., Rosenthal, A., and Winslow, J.W., 1990, Widespread expression of BDNF but not NT3 by target areas of basal forebrain cholinergic neurons. Science 250:290.

Radeke, M.J., Misko, T.P., Hsu, C., Herzenberg, L.A., and Shooter, E.M., 1987, Gene transfer and molecular cloning of the rat nerve growth factor receptor: a new class of receptors. Nature 325:593.

Sapolsky, R.M., 1986, Glucocorticoid toxicity in the hippocampus: Reversal by supplementation with brain fuels. J. Neurosci., 6:2240.

Sapolsky, R.M. and Pulsinelli, W., 1985, Glucocorticoids potentiate ischemic injury to neurons: Therapeutic implications. Science, 229:1397.

Scheff, S.W., Hoff, S.F. and Anderson, K.J., 1986, Altered regulation of lesion-induced synaptogenesis by adrenalectomy and corticosterone in young adult rats, Exp. Neurol. 93:456.

Smith, C.J., Court, J.A., Keith, A.B., and Perry, E., 1989, Increases in muscarinic stimulated hydrolysis of inositol phospholipids in rat hippocampus following deafferentation are not parallelled by alterations in cholinergic receptor density. Brain Res. 485:317.

Soppet, D., Escadon, E., Maragos, J., Middlemas, D.S., Reid, S.W., Blair, J., Burton, L.E., Stanton, B.R., Kaplan, D.R., Hunter, T., Nikolics, K., and Parada, L.F., 1991, The neurotrophic factors BDNF and NT-3 are ligands for the trkB tyrosine kinase receptor. Cell 65:895.

Squinto, S.P., Stitt, T.N., Aldrich, T.H., Davis, S., Bianco, S.M., Radziejewski, C., Glass, D.J., Masiakowski, P., Furth, M.E., Valenzuela, D.M., DiStefano, P.S. and Yancopoulos, G.D., 1991, trkB encodes a functional receptor for brain-derived neurotrophic factor and neurotrophin-3 but not nerve growth factor. Cell 43:915.

Thoenen, H., 1991, The changing scene of neurotrophic factors. TINS 14(5):165.

Will, B. and Hefti, F., 1985, Behavioural and neurochemical effects of chronic intraventricular injections of nerve growthfactor in adult rats with fimbrial lesions. Behav. Brain Res. 17:17.

Will, B., Hefti, F., Pallage, V., and Toniolo, G., 1988, Nerve growth factor, Effect on CNS neurons and on behavioral recovery from brain damage, in Pharmacological approaches to the treatment of brain and spinal cord injury, pp. 339-360. Plenum Press, New York.

Will, B., Pallage, V., and Eclancher, F., 1990, Nerve growth factor and behavioral recovery after brain damage in rats, in Neurotrophic factors and Alzheimer's disease, Springer-Verlag, New York. pp. 117-130.

Williams, L.R., 1991, Hypophagia is induced by intraventricular administration of nerve growth factor. Exp. Neurol. 113:31.

Zafra, F., Hengerer, B., Leibrock, J., Thoenen, H., and Lindholm, D., 1990, Activity dependent regulation of BDNF and NGF mRNAs in the rat hippocampus is mediated by non-NMDA glutamate receptors. EMBO 9:3545.

Zafra, F., Castren, E., Thoenen, H., and Lindholm, D., 1991, Interplay between glutamate and γ-aminobutyric acid transmitter systems in the physiological regulation of brain-derived neurotrophic factor and nerve growth factor synthesis in hippocampal neurons. Proc. Natl. Acad. Sci (USA) 88:10037.

ANGIOSUPPRESSIVE AND ANTIPROLIFERATIVE ACTIONS OF SURAMIN: A GROWTH FACTOR ANTAGONIST

Shingo Takano[1], Stephen Gately[1], Herbert Engelhard[1], Ana Maria C. Tsanaclis[1], Janet L. Gross[2], William F. Herblin[2], Kristi Eidsvoog[2], Mary Neville[2], and Steven Brem[1,3]

[1]Division of Neurological Surgery, Northwestern Memorial Hospital and School of Medicine, Chicago, Illinois, 60611.

[2]Du Pont Merck Pharmaceutical Corporation, Wilmington, Delaware, 19880-0400

[3]To whom reprint requests should be addressed: 233 East Erie St, Suite 500, Chicago, Illinois, 60611-2906

INTRODUCTION

Suramin is a novel anticancer agent[1] that appears to be effective against advanced adrenocortical carcinoma[2,3], prostatic cancer[4,5], ovarian cancer[6], renal cell carcinoma[7] and certain refractory lymphomas[8]. The antiproliferative action is possibly related to the ability of suramin to block the binding of autocrine growth factors to their receptors[9-18], to inhibit a variety of cytoplasmic and intranuclear enzymes critical for cell maintenance and proliferation[19,20], and to disrupt cellular respiration and energy balance[21]. Cell migration and adhesion of B16 melanoma cells to the extracellular matrix is inhibited following suramin exposure[22, 23], suggesting a mechanism for suramin to inhibit tumor invasion.

Angiogenesis, the proliferation and migration of endothelial cells resulting in the formation of new blood vessels, is an essential event for tumor growth[24]. Suppression of angiogenesis has been proposed as a form of cancer therapy[25-28]. The inhibitory effect of suramin on tumor growth *in vivo* has been viewed as a direct action on the human cell[2], but it is also possible that suramin inhibits angiogenesis, thereby inhibiting the growth of tumors. Our laboratories are engaged in the study of angiosuppressive agents based on their selective actions on tumor and endothelial subpopulations *in vitro* and *in vivo*.

MECHANISM OF ANGIOSUPPRESSION

The angiosuppressive effects of suramin have received little attention previously. It is now recognized that angiogenesis takes place as a series of sequential steps[26]; 1) first, the local degradation of the basement membrane of the parent vessel by proteolytic enzymes; 2) the migration of endothelial cells toward an angiogenic stimulus; 3) DNA synthesis and mitosis modulated by growth factors; and 4) the formation of a lumen through which blood begins to flow. Suramin could inhibit one or more of these steps and consequently inhibit the formation of new vessels.

There is evidence that the angiosuppressive properties of suramin could be linked to the displacement of specific angiogenic growth factors, e.g. transforming growth factor-ß (TGF-ß), platelet derived growth factor (PDGF), vascular endothelial cell growth factor (VEGF), and basic fibroblast growth factor (bFGF) from their receptors[1,2,11,29]; the result of increasing circulating levels of the angiostatic glycosaminoglycans, heparan and dermatan sulphate[26,30]; the selective localization to the vascular endothelial cells following administration[31]; or the inhibition of endothelial cell migration[32].

ENDOTHELIAL CELL PROLIFERATION AND GROWTH

The antiproliferative effects of suramin, as measured by inhibition of cell growth *in vitro*, have been demonstrated in many human tumor cell lines, including prostate, sarcoma, glioblastoma, and rhabdomyosarcoma [1,18]. We have investigated the effect of suramin on the growth of bovine pulmonary artery endothelial cells after 144 hours of exposure to 10 - 1500 μg/ml of suramin. The effects of suramin on endothelial cell growth were biphasic: a low concentrations of 10 and 30 μg/ml, suramin slightly stimulated growth; it was inhibitory of cell growth at higher concentrations of 100, 250, 500, 750, 1000, and 1500 μg/ml (Figure 1). Cell viability was over 90% in all treatment conditions. The calculated IC_{50} of suramin for endothelial cells was 944 μg/ml, which is greater than the plasma concentration of 350μg/ml that is clinically toxic[34,35] and the concentration which is growth inhibitory against human cancer cell lines[2]. We measured BUdR labeling index of endothelial cells after 48 hours exposure to suramin. The BUdR labeling index was significantly and dose-dependently lower at suramin 100, 500, and 1000 μg/ml, compared to controls (Figure 2).

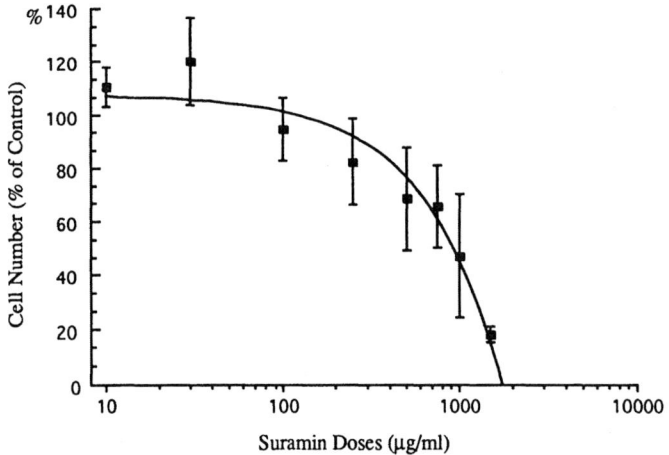

Figure 1. Effect of suramin on endothelial cell growth. Bovine pulmonary artery endothelial cells (5 x 10^3/well) were plated in 6-well plates. Twenty four hours later, the medium was replaced with medium containing suramin. Six days later, the number of cells was counted and compared to control. Each point represents the average cell number (% of control) from three different experiments from each of three wells. Bar represents the standard deviation. (From reference [33].)

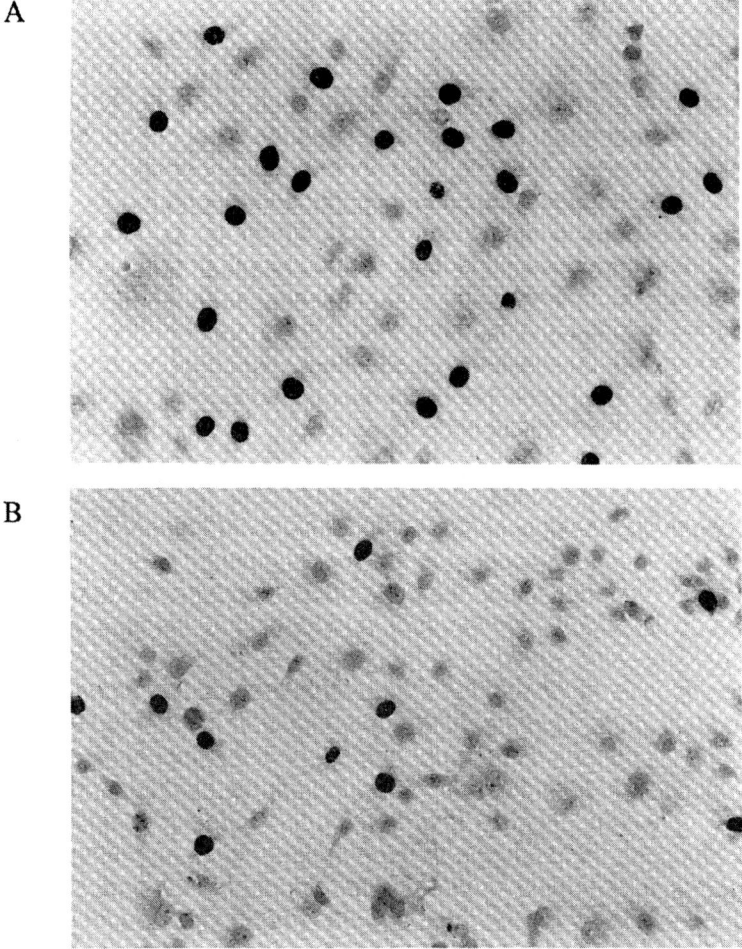

Figure. 2. BUdR labeling of cultured endothelial cells. Note the dark-staining nuclei, which represent cells in S-phase. Stained nuclei are clearly demonstrated in the control (A) but significantly reduced after treatment with 500 μg/ml of suramin (B).

We demonstrated a mild stimulatory effect of suramin at the dose of 10 and 30μg/ml on endothelial cells (Figure 1). Proliferative effects of suramin at low concentration have been detected on various human cancer cells[13,36-38]. The mechanism is unknown, but may reflect binding to or stimulation of growth factor receptors by suramin. Extension of the capillary network into the growing tumor is primarily dependent on the division of endothelial cells within capillary walls[39]. Endothelial cell proliferation is mediated by specific angiogenic, polypeptide, mitogenic factors, especially fibroblast growth factor[26,27], that bind to their receptors and are blocked by suramin.

Inhibition of DNA synthesis by suramin has been shown previously in human tumor cells as measured by ^{3}H-thymidine incorporation[17,37,40-42]. Cell cycle analysis by flow cytometry has revealed that suramin increases the percentage of tumor cells in the S phase of the cell cycle[40,43,44]. Forsbeck *et al*.[40] showed that suramin decreased DNA synthesis as measured by ^{3}H-thymidine incorporation of human histiocytic lymphoma cells in addition to increasing the S-phase cell number using flowcytometry. After six days exposure to suramin,

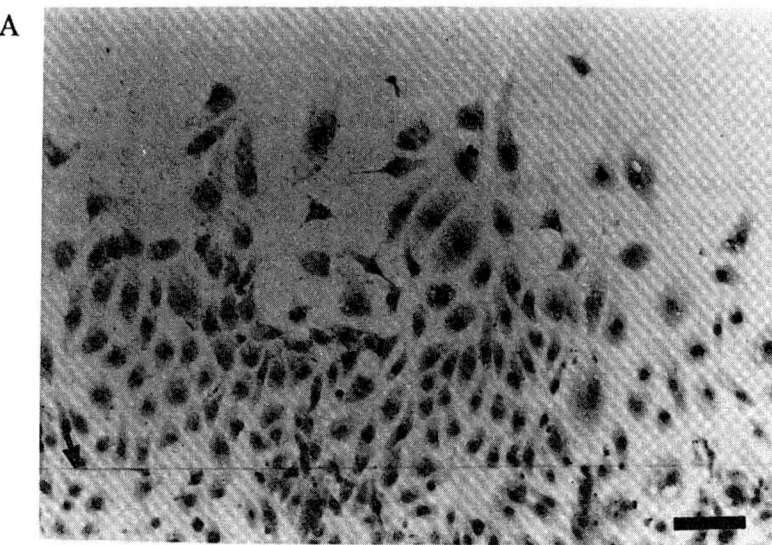

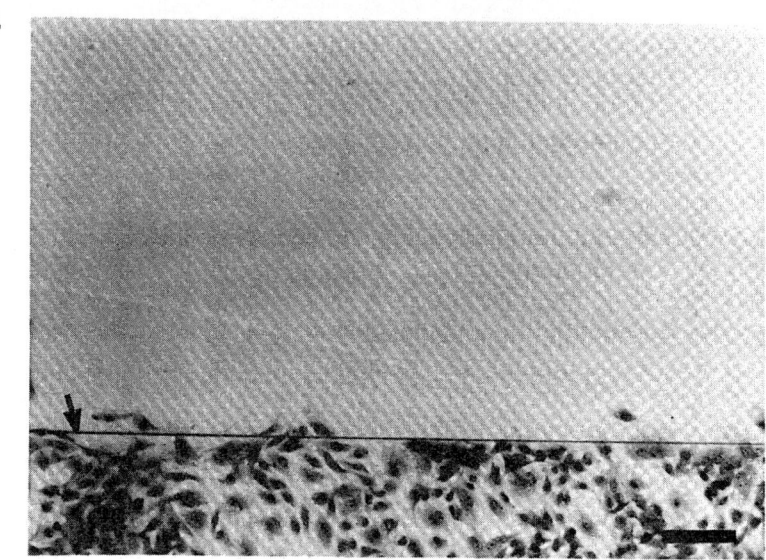

Figure. 3. Inhibition of endothelial cell migration. Confluent monolayers of endothelial cells were wounded with a razor blade. The cells were then incubated 16 hours in the media, fixed, stained and photographed. A: control, B: suramin 1000 µg/ml. The arrows point to the original edge of the wound. Bar=100 µm. (From reference[33].)

the percentage of endothelial cells in the S phase of the cell cycle increased slightly[33]. In addition, we demonstrated that suramin decreased, in a dose-dependent fashion, BUdR labeling index. These results suggest that suramin blocks the cell cycle during the S phase, as cells that are arrested in early S phase do not incorporate BUdR or actively synthesize DNA[45] and subsequently inhibits DNA synthesis in tumor as well as endothelial cells.

ENDOTHELIAL CELL MIGRATION

The migration of endothelial cells is one of the critical features of neovascularization and wound repair, two processes thought to involve basic fibroblast growth factor (bFGF)[32].

Sato *et al.* demonstrated that suramin reversibly inhibited bovine aortic endothelial cell movement; 0.5mM (715 µg/ml) suramin inhibited endothelial cell migration to 4% of control[32].

Figures 3A and B show the results of wounding endothelial cell monolayers and then incubating them for 16 hours in serum-free Dulbeco's modified Eagle's medium with and without suramin. The inhibition of endothelial cell migration with suramin occurred, in a dose-dependent manner, at 100, 500, 750, and 1000µg/ml. As the calculated IC_{50} of suramin for endothelial cells is 945µg/ml, our results not only confirm those of Sato *et al.*[32] but also indicate that suramin may perturb endothelial cell migration more than endothelial cell proliferation. Endothelial cell migration may, in turn, be regulated by protease production[46]. Preliminary data suggests that suramin inhibits the induction of urokinase-type plasminogen activator by bFGF (M. Neville *et al.*, unpublished results.).

CHICK CHORIOALLANTOIC MEMBRANE ASSAY

Chorioallantoic membrane (CAM) of the chick embryo is used as an important model for the study of angiogenesis[26,28,47]. Control disks containing 0.5% methylcellulose demonstrate normal vascularization on the CAM. Disks containing suramin produce an avascular zone on the CAM (Figure 4). This effect is dose-dependent: no avascular zones were obtained with 30 µg/disk, 25% with 250 µg, 50% with 500 µg, 62.5% with 1000 µg, and 100% with 1500 µg[33].

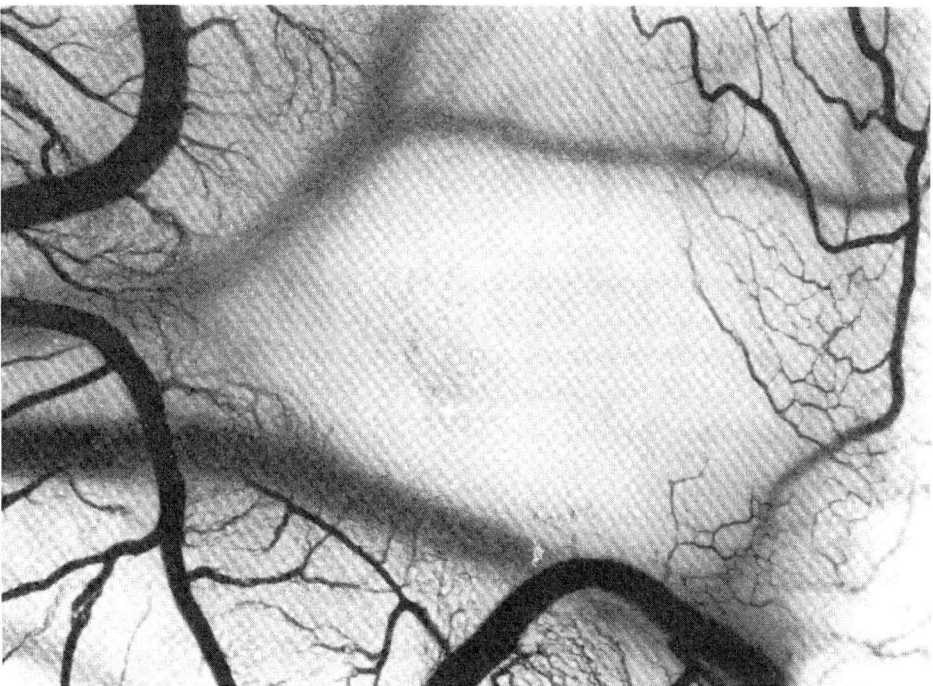

Figure 4. Inhibition of angiogenesis by suramin (1500µg/disk) in the 8-day old chorioallantoic membrane (CAM).

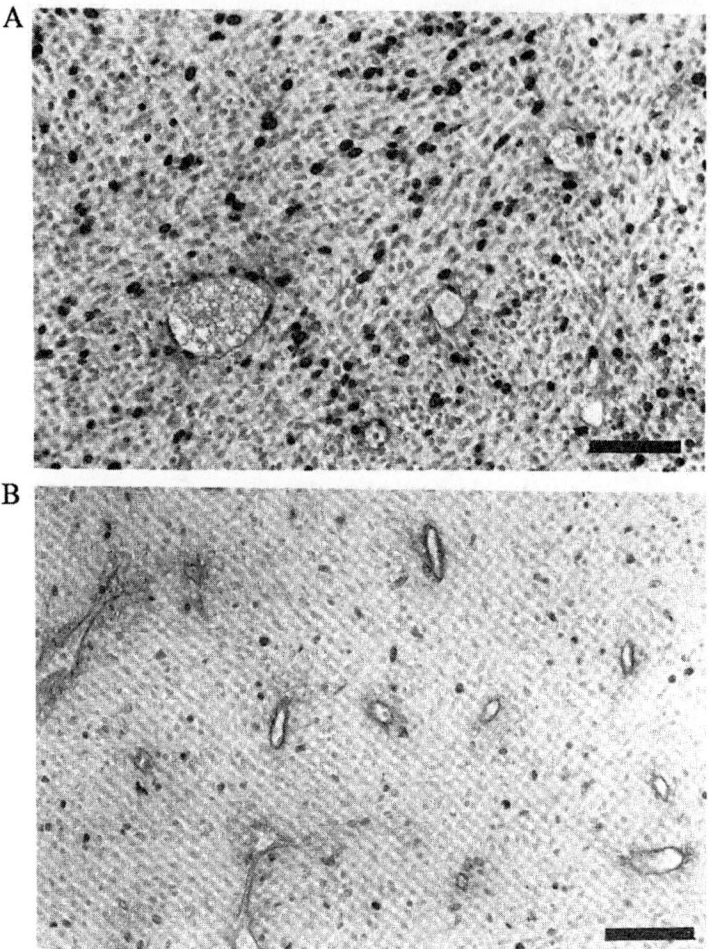

Figure. 5. Tissue sections of rat brain tumor double stained for BUdR and laminin, then counterstained with hematoxylin. A: control group - many of the tumor cell nuclei and endothelial cell nuclei are positive for BUdR. B: suramin 30 mg/kg. Note the significantly decreased labeling in both tumor and endothelial subpopulations. Bar=100 μm. (From reference[33].)

These results are compatible with a previous report in which 80% inhibition was obtained with 500 μg and 100% with 1000 μg/pellet[48]. In another study, suramin alone produced avascular zones in 30% of the CAMs; heparin alone or angiostatic steroids alone produced avascular zones in only 5%. By contrast, the co-administration of angiostatic steroids and suramin produced avascular zones in 70-100% of the embryos[49].

RAT BRAIN TUMOR MODEL

Because of the angiosuppressive effects of suramin *in vitro* and in the chick CAM assays, we examined the effect of suramin on angiogenesis and tumor growth in the brain, using a rat C6 glioma model. Intracerebral injection of C6 glioma cells in Wistar-Furth rats produces tumors having a prominent neovasculature and histology similar to that of human glioblastoma multiforme[50].

Twenty-four male Wistar-Furth rats were treated with intraperitoneal injection of suramin at doses of 10, 30, and 60mg/kg on alternative days. All animals were sacrificed 13 days following tumor implantation. In order to evaluate S-phase fraction of tumor and endothelial cells and the vascular density, paraffin embedded tumor sections were stained simultaneously for BUdR and laminin. Vascular density was determined for each tumors using immunostaining for laminin, a reliable vascular marker in brain neoplasms and normal brain[51,52], and defined as the highest number of vascular lumens per field (0.087mm^2) encountered at the margin of the tumor. The endothelial cell and the tumor cell BUdR labeling index were significantly lower in the treatment groups compared to the control group (Figure 5).

CONCLUSION

Suramin is shown to be angiosuppressive *in vitro* and *in vivo* in the CAMs. Studies of cultured cells demonstrate the inhibition of endothelial cell migration and proliferation even at the dose of 100 μg/ml. In addition, suramin inhibits endothelial cell proliferation *in situ* in the brain. Because the combination of angiostatic steroids and suramin increased lifespan of B16 melanoma-bearing mice compared to control and inhibited neovascularization in the CAM assay[49]; combination treatments of suramin and angiostatic steroids are suggested as anticancer agents[49]. The development of less toxic analogs of suramin[53] that potentiate the angiosuppressive effects, or combinations of this drug with other agents including angiostatic steroids[49], tumor necrosis factor[54], or interferons[55] show promise for future clinical application.

ACKNOWLEDEMENTS

The authors acknowledge the expert technical assistance of Marguerite Wotoczek, David Ivancic, and Holly Duncan. This work was supported, in part, by a grant from the Northwestern Memorial Foundation and the Du Pont Merck Pharmaceutical Corporation (to S.B.)

REFERENCES

1. R.V. La Rocca, C.A. Stein, and C.E. Myers, Suramin: Prototype of a new generation of antitumor compounds. *Cancer Cells.* 2: 106 (1990).
2. C.A. Stein, R.V. La Rocca, R. Thomas, N. McAtee, and C.E. Myers, Suramin: An anticancer drug with unique mechanism of action, *J Clin Oncol.* 7: 499 (1989).
3. R.V. La Rocca, C.A. Stein, R. Danesi, C. Jamis-Dow, G.H. Weiss, and C.E. Myers, Suramin in adrenal cancer: modulation of steroid hormone production, cytotoxicity in vitro and clinical antitumor effect. *J Clin Endocrinol Metab.* 71: 497 (1990).
4. R.V. La Rocca, M.R. Cooper, and M. Uhrich, The use of suramin in the treatment of prostate cancer refractory to conventional hormonal manipulation. *Urol Clin North Am.* 18: 123 (1991).
5. C.M. Myers, M. Cooper, C. Stein, La Rocca, R, M.M. Walther, G. Weiss, P. Choyke, N. Dawson, S. Steinberg, M.M. Uhrich, J. Cassidy, D.R. Kohler, J. Trepel, and D.R. Linehan, Suramin: a novel growth factor antagonist with activity in hormone-refractory metastatic prostate cancer. *J Clin Oncol.* 10: 881 (1992).
6. E. Reed, M.R. Cooper, R.V. La Rocca, F. Bostick-Bruton, and C.E. Myers, Suramin in advanced platinum-resistant ovarian cancer. *Eur J Cancer.* 28A: 864 (1992).
7. R.V. La Rocca, C.A. Stein, R. Danesi, M.R. Cooper, M. Uhrich, and C.E. Myers, A pilot study of suramin in the treatment of metastatic renal carcinoma. *Cancer.* 67: 1509-1513 (1991).
8. R.V. La Rocca, C.E. Myers, C.A. Stein, M.R. Cooper, and M. Uhrich, Effect of suramin in patients with refractory nodular lymphomas requiring systemic therapy. *Proc Am Soc Clin Oncol.* 9: 1041 (1990)

9. M. Hosang, Suramin binds to platelet-derived growth factor and inhibits its biological activity. *J Cell Biochem.* 29: 265 (1985).

10. C. Betsholtz, A. Johnson, C.H. Heldin, and B. Westermark, Efficient reversion of simian sarcoma virus-transformation and inhibition of growth factor-induced mitogenesis by suramin. *Proc Natl Acad Sci USA.* 83: 6440 (1986).

11. L.T. Williams, P. Tremble, M.F. Lavin, and M.E. Sunday, Platelet-derived growth factor receptors from a high affinity state in membrane preparations. Kinetics and affinity cross-linking studies. *J Biol Chem.* 259: 5287 (1984).

12. R.J. Coffey, E.B. Leof, G.D. Shipley, and H.L. Moses, Suramin inhibition of growth factor receptor binding and mitogenicity in AKR-28 cells. *J Cell Physiol.* 132: 143 (1987).

13. S. Olivier, P. Formento, J.L. Fischel, M.C. Etienne, and G. Milano, Epidermal growth factor receptor expression and suramin cytotoxicity *in vitro. Eur J. Cancer.* 26: 867 (1990).

14. D.A. Lappi, P.A. Maher, D. Martineau, and A. Baird, The basic fibroblast growth factor-saporin mitotoxin acts through the basic fibroblast growth factor receptor. *J Cell Physiol.* 147: 17 (1991).

15. A. Wellstein, R. Lupu, C. Zugmaier, S.L. Flamm, A.L. Cheville, P. Delli-Bovi, C. Basilico, M.E. Lippman, and F.G. Kern, Autocrine growth stimulation by secreted kaposi fibroblast growth factor but not by endogeneous basic fibroblast growth factor. *Cell Growth Diff.* 1: 63 (1990).

16. M. Pollak and M. Richard, Suramin blockade of insulin-like growth factor I-stimulated proliferation of human osteosarcoma cells. *J Nat Cancer Inst.* 82: 1349 (1990).

17. C.P. Minniti, M. Maggi, and L.J. Helman, Suramin inhibits the growth of human rhabdomyosarcoma by interrupting the insulin-like growth factor II autocrine growth loop. *Cancer Res.* 52: 1830 (1992)

18. G.B. Mills, N. Zhang, C. May, M. Hill, and A. Chung, Suramin prevents binding of interleukin-2 to its cell surface receptor: a possible mechanism for immunosuppression. *Cancer Res.* 50: 3036 (1990).

19. Z. Spiegelman, A. Dowers, S. Kennedy, D. DiSorbo, M. O'Brien, R. Barr, and R. McCaffrey, Antiproliferative effects of suramin on lymphoid cells. *Cancer Res.* 47: 4694 (1987).

20. K. Ono, H. Nakane, and M. Fukushima, Differential inhibition of various deoxyribonucleic and ribonucleic acid polymerases by suramin. *Eur J Biochem.* 172: 349 (1988).

21. R. Rago, J. Mitchen, A.L. Cheng, T. Oberley, and G. Wilding, Disruption of cellular energy by suramin in intact human prostatic carcinoma cells, a likely antiproliferative mechanism. *Cancer Res.* 51: 6629 (1991).

22. M. Nakajima, A. De Chavigny, C.E. Johnson, J. Hamada, and C.A. Stein, Suramin. A potent inhibitor of melanoma heparanase and invasion. *J Biol Chem.* 266: 9661 (1991).

23. V.S. Zabrenetzky, E.C. Kohn, and D.D. Roberts, Suramin inhibits laminin- and thrombospondin-mediated melanoma cell adhesion and migration and binding of these adhesive proteins to sulfatide. *Cancer Res.* 50: 5937 (1990).

24. J. Folkman, What is the evidence that tumors are angiogenesis dependent? *J Natl Cancer Inst.* 82: 4 (1990).

25. J. Folkman, Anti-angiogenesis: New concept for therapy of solid tumors. *Ann Surg.* 175: 409 (1972).

26. J. Folkman and M. Klagsbrun, Angiogenic factors. *Science.* 235: 442 (1987).

27. J.L. Gross, R.S. Morrison, K. Eidsvoog, W.F. Herblin, P.L. Kornblith, and D.L. Dexter, Basic fibroblast growth factor: a potentiate autocrine regulation of human glioma cell growth. *J Neurosci Res.* 27: 689 (1990)

28. T-P.D. Fan and S. Brem, Angiosuppression, *in:* "The Search For New Anticancer Drugs, Cancer Biology Series. Vol 3., pp185-229." M.J. Waring, and B. Ponder, ed., Kluwer Publishers, Lancaster, UK (1992).

29. N. Vaisman, D. Gospodarowicz, and G. Neufeld, Characterization of the receptors for vascular endothelial cell growth factor. *J Biol Chem.* 256: 19461 (1990).

30. M. Horne, C.A. Stein, R.V. La Rocca, and C.E. Myers, Circulating glycosaminoglycan anticoagulant associated with suramin treatment. *Blood.* 71: 273 (1988).

31. C.A. Janis-Dow, G.H. Weiss, and M.J. Merino, Suramin selectively localizes to vascular endothelial cells: a possible basis for the antiangiogenesis activity of suramin (abstr). *Proc Am Assoc Cancer Res.* 31: 60 (1990).

32. R. Sato and D.B. Rifkin, Autocrine activities of basic fibroblast factor: regulation of endothelial cell movement, plasminogen activator synthesis, and DNA synthesis. *J Cell Biol.* 107: 1199 (1988).

33. S. Takano, S. Gately, A.M.C. Tsanaclis, H. Engelhard, J.L. Gross, W.F. Herblin, K. Eidsvoog, and S. Brem, Inhibition of angiogenesis and the proliferation of endothelial and glioma cells by suramin in vitro and in situ in the brain. (Manuscript in preparation).

34. H.I. Scher, D.I. Jodrell, J.M. Iversen, T. Curley, W. Tong, M.J. Egorin, and A. Forrest, Use of adaptive control with feedback to individualize suramin dosing. *Cancer Res.* 52: 64 (1992).

35. M.R. Cooper, R. Lieberman, R.V. La Rocca, P.R. Gernt, M.S. Weinberger, D.J. Headlee, D.R. Kohler, B.R. Goldspiel, C.C. Peck, and C.E. Myers, Adaptive control with feedback strategies for suramin dosing. *Clin Pharmacol Ther.* 52: 11 (1992).

36. K.J. Pienta, W.B. Isaacs, D. Vindivich, and D.S. Coffey, The effects of basic fibroblast growth factor and suramin on cell motility and growth of rat prostate cancer cells. *J Urol.* 145, 199 (1991).

37. X.J. Guo, J. Fantini, R. Roubin, J. Marvaldi, and G. Rougon, Evaluation of the effect of suramin on neural cell growth and N-CAM expression. *Cancer Res.* 30: 5164 (1990).

38. E.M. Berns, A.L. Schuurmans, D.J. Lamb, J.A. Foekens, and E. Mulder, Antiproliferative effects of suramin on androgen responsive tumor cells. *Eur J Cancer.* 26: 470 (1990).

39. I. Tannock, Population kinetics of carcinoma cells, capillary endothelial cells, and fibroblasts in a transplanted mouse mammary tumor. *Cancer Res.* 30: 2470 (1970).

40. K. Forsbeck, K. Bjelkenkrantz, and K. Nilsson, Role of iron in the proliferation of the established human tumor cell lines U-937 and K-562: Effects of suramin and a lipophilic iron chelator (PIH). *Scand J Haematol.* 37: 429 (1986).

41. H, Jindal, C.W. Anderson, R.G. Davis, and J.K. Vishwanatha, Suramin affects DNA synthesis in Hela cells by inhibition of DNA polymerases. *Cancer Res.* 50: 7754 (1990).

42. P. Michel, R.V. Velthoven, M. Petein, S. Gras, C. Etievant, J.L. Pasteels, and R. Kiss, Influence of suramin alone or in combination with DHT and PDGF on the cell proliferation of benign and malignant human prostatic tissues in organ cultures. *Anticancer Res.* 11: 2075 (1991).

43. J.H. Kim, E.R. Sherwood, D.M. Sutkowski, C. Lee, and J.M. Kozlowski, Inhibition of prostatic tumor cell proliferation by suramin: alterations in TGF alpha-mediated autocrine growth regulation and cell cycle distribution. *J. Urol.* 146: 171 (1991).

44. T.M. Walz, A. Abdiu, S. Wingren, S. Smeds, S.E. Larsson, and A. Wasteson, Suramin inhibits growth of human osteosarcoma xenografts in nude mice. *Cancer Res.* 51: 3585 (1991).

45. P.N. Dean, F. Dolbeare, H. Gratzner, G.C. Rice, and J.W. Gray, Cell-cycle analysis using a monoclonal antibody to BrdUrd. *Cell Tissue Kinet.* 17: 427 (1984).

46. L.E. Odekon, Y. Sato, D.B. Rifkin, Urokinase-type plasminogen activator mediates basic fibroblast growth factor-induced bovine endothelial cell migration independent of its proteolytic activity. *J Cell Physiol.* 150: 258 (1992).

47. D.H. Ausprunk, D.R. Knighton, and J. Folkman, Vascularization of normal and neoplastic tissues grafted to the chick chorioallantois: role of host and preexisting graft blood vessels. *Am J Pathol.* 79: 597 (1975).

48. M. Ciomei, E. Pesenti, F. Sola, W. Pastori, M. Mariani, M. Grandi, and F. Spreafico, Antagonistic effect of suramin on bFGF: *In vitro* and *in vivo* results. *Int J Radiat Biol.* 60: 78 (1991).

49. J.W. Wilks, P.S. Scott, L.K. Vrba, and J.M. Cocuzza, Inhibition of angiogenesis with combination treatment of angiostatic steroids and suramin. *Int J Radiat Biol.* 60: 73 (1991).

50. F. San-Galli, P. Vrignaud, J. Robert, J.M. Coindre, and F. Cohadon Assessment of the experimental model of transplanted C6 glioblastoma in wistar rats. *J Neuronc.* 7: 299 (1989).

51. M.T. Giordana, I. Germanol, G. Giaccone, A. Muro, and D. Schiffer, The distribution of laminin in human brain tumors: an immunohistochemical study. *Acta Neuropathol (Berl).* 67: 51 (1985).

52. M. Eriksdotter-Nilsson, H. Bjorklund, and L. Olson, Laminin immunohistochemistry: a simple method to visualize and quantitate vascular structures in the mammalian brain. *J Neurosci Meth.* 17: 275 (1986).

53. S. Baghdiguian, P. Nickel, and J. Fantini, Double screening of suramin derivatives on human colon cancer cells and on neural cells provides new therapeutic agents with reduced toxicity. *Cancer Letters.* 60: 213 (1991).

54. J.P. Friehauf, C.E. Myers, and B.K. Sinha, Synergistic activity of suramin with tumor necrosis factor-α and doxorubicin on human prostate cancer cell lines. *J Natl Cancer Inst.* *82*: 1206 (1990).

55. S. Liu, M.W. Ewing, P. Anglard, E. Trahan, R.V. La Rocca, C.E. Myers, and W.M. Linehan, The effect of suramin, tumor necrosis factor and interferon-γ on human prostate carcinoma. *J Urol.* *145*: 389 (1991).

[125I]IGF I, [125I]IGF II AND [125I]INSULIN RECEPTOR BINDING SITES IN THE RAT HIPPOCAMPAL FORMATION: TOPOGRAPHIC DISTRIBUTION AND RESPONSE TO ENTORHINAL ABLATION

S. Kar,[1] R. Quirion,[1,2] A. Baccichet,[1,2] and J. Poirier[1,2]

[1]Douglas Hospital Research Center, Department of Psychiatry
McGill University, Montreal, Canada H4H 1R3
[2]McGill Center for Studies in Aging, 1650 Cedar Ave, Montreal
Canada H4G 1A4

INTRODUCTION

Lesion-induced synaptogenesis has provided the most convincing evidence that the adult central nervous system (CNS) is capable of new synapse formation and reorganization. One of the best characterized examples of synaptic reorganization in the CNS is that which occurs in the dentate gyrus of the hippocampal formation following damage to the projection from the entorhinal cortex[1-3]. The dentate gyrus receives extrinsic afferents from the entorhinal cortex, the septum and a variety of ascending and descending pathways[4]. Unilateral lesion of the entorhinal cortex induces selective sprouting of septal afferents, the commissural/associational fibers and to lesser extent the contralateral perforant path in the outer 2/3 of the ipsilateral molecular layer, the area normally innervated by the entorhinal cortex afferents[5-7]. While the time-course events of reactive synaptogenesis has been studied extensively, at present relatively little is known regarding its initiation, maintenance and underlying molecular basis. Prospective inducers of the responses include growth factors and cytokines which are considered to play a role in the molecular cascade regulating the reinnervation process that follows the lesion[2,3,8,9]. This is primarily supported by recent evidence that interleukin 1 (IL-1)[8], nerve growth factor (NGF)[10,11] and basic fibroblast growth factor (bFGF)[12]-like immunoreactivity are increased significantly in the outer molecular layer of the dentate gyrus following lesion and are involved in the sprouting of septal afferent fibers. However, as the physiology of a particular population of neurones depends on the actions of multiple trophic factors, it is most likely that other growth factors may be involved in the process of circuitary rearrangement that facilitates neuronal repair and/or survival of neurones.

Insulin-like growth factors I and II (IGF I and II) and insulin are structurally related polypeptide growth factors that are distributed widely throughout the brain[13-15]. However, unlike IGFs, the issue of local synthesis of insulin is equivocal and it is suggested that brain insulin is derived from both local synthesis as well as from the peripheral circulation[13,14]. The physiological actions of these peptides are presumed to be mediated by three distinct cell surface receptors: the two IGF receptors (IGF I and IGF II) and the insulin receptors[13,14,16]. These receptors[17-22] and/or their mRNA[23,24] are widely distributed in the brain, including the hippocampal formation. Functionally, IGFs and insulin, as suggested by *in vitro* studies, are considered to play an important role in neural growth and differentiation possibly involving dendritic maturation, synaptogenesis and myelinization[25-30]. Evidence further supports that in the adult nervous system, these growth factors function in the maintenance of normal brain cytoarchitecture and are also involved in response to pharmacological and surgical injuries[31, 32]. Since ablation of the entorhinal cortex evokes a powerful sprouting of septal afferents as well as other inputs in the outer molecular layer of the dentate gyrus, it is reasonable to contemplate that IGFs and/or insulin may be involved in the process of degeneration and/or subsequent reinnervation that follows the lesion. This seems most appropriate in view of recent evidence suggesting that IGF I induces neurite outgrowth in septal cell cultures[25]. The present study, as a prerequisite to the proposed hypothesis, describes the time-course response of [^{125}I]IGF I, [^{125}I]IGF II and [^{125}I]insulin receptor binding sites in the rat hippocampal formation, particularly in the dentate gyrus i.e, the area of greater synaptogenesis, following lesion of the entorhinal cortex.

MATERIALS AND METHODS

Entorhinal cortex lesion and tissue preparation: Adult Fisher-344 rats (250 - 300 g) obtained from Charles River, St Constant, Canada were subjected to multiple unilateral electrolytic lesions (30 sec) of the entorhinal cortex at three different depths (2, 4 and 6 mm below dura) and in three locations (L: 3.3, B: 0; L: 4.3, B: 0; L: 5, B: 1 mm) as described in detail earlier[33]. Following lesions, rats were decapitated at different time periods (1, 2, 4, 8, 14 and 30 days post-lesion, DPL) and their brains were snap-frozen in 2-methylbutane at -40° C. Four normal control and three to four lesioned brains from each time point were then serially cut through the hippocampal formation, thaw-mounted on gelatin-coated slides and then stored at -80° C until use.

In vitro receptor autoradiography: To determine [^{125}I]IGF I, [^{125}I]IGF II and [^{125}I]insulin receptor binding sites, slide mounted sections from control and lesioned brains were processed as described previously[17,19]. In brief, for [^{125}I]IGF I and [^{125}I]insulin receptor binding sites, sections were incubated for 2 hr at 22° C in N-[2-hydroxyethyl]piperazine-N-[2-ethanesulfonic acid] buffer (HEPES, 10 mM, pH 7.4) containing 0.5% bovine serum albumin, 0.0125% N-ethylmaleimide, 0.025% bacitracin, 100

KIU/ml aprotinin and either 50 pM [^{125}I]IGF I (2000 Ci/mmol, Amersham) or 50 pM [^{125}I]insulin (2000 Ci/mmol, Amersham). The respective non-specific binding was determined in the presence of either 80 nM unlabeled human recombinant IGF I or unlabeled insulin. As for [^{125}I]IGF II binding, the sections were incubated with 25 pM [^{125}I]IGF II (2000 Ci/mmol, Amersham) at 4° C for 16 hr in Tris-HCl buffer (50 mM, pH 7.4). The non-specific labelling was defined in the presence of 70 nM unlabelled human recombinant IGF II. Following incubation, slides from each binding experiment were washed in Tris-HCl buffer (50 mM, pH 7.4), rinsed in cold-water and then apposed to tritium sensitive films i.e, 18-24 hr, 2-3 days and 9-11 days for IGF I, IGF II and insulin binding sites respectively. The autoradiograms were quantified by densitometry using MCID image analysis system (Imaging Research Inc., St Catherines, Ontario). Specific binding was defined as the difference in radioactivity bound in the absence and in the presence of corresponding excess unlabelled peptide. The data were statistically analysed employing the one-way ANOVA test.

RESULTS

Following lesion of the entorhinal cortex, an altered time-dependent response of [^{125}I]IGF I, [^{125}I]IGF II and [^{125}I]insulin receptor binding sites was noted in selective layers of the hippocampal formation. The distributional profile and the time-course alterations of each of the receptor binding sites are shown in Figs 1-4.

IGF I: In normal rat hippocampal formation, a relatively higher density of [^{125}I]IGF I binding sites was noted in the dentate gyrus and the CA3 region than in the CA1 or CA2 regions (Figs 1A, 4). In the dentate gyrus, binding sites were found to be more prominent in the granular than the molecular layer (Fig 4). Within the Ammon's horn, high density of [^{125}I]IGF I binding was evident particularly in the pyramidal layer whereas other layers such as strata oriens, radiatum and lacunosum moleculare showed moderate to low density of specific binding. Unilateral lesion of the entorhinal cortex induced a significant increase ($P < 0.05$) of [^{125}I]IGF I receptor binding sites in the ipsilateral molecular layer of the dentate gyrus at 8 DPL (Figs 1C, 4). The granular layer of the dentate gyrus and CA1-CA3 subfields however, did not exhibit any significant variation between the contralateral vs ipsilateral hippocampal formation at any time periods following lesion (Figs 1B-D, 4). Interestingly, when compared to controls, [^{125}I]IGF I binding sites showed differential responses in CA1-CA3 regions of the Ammon's horn. In the dentate gyrus however, a significant increase ($P < 0.05$) in [^{125}I]IGF I binding sites was noted in the molecular layer between 8-30 DPL whereas in the granular layer a selective decrease ($P < 0.05$) in binding sites was evident only at 1 day following lesion (Figs 1, 4).

IGF II: In control brains, selective [^{125}I]IGF II receptor binding sites were noted primarily in the pyramidal layer of the CA1-CA3 regions and in the granular cell layer of the dentate gyrus. Other layers of the hippocampal formation exhibited relatively low density of binding sites (Figs 2A, 4). Following lesion of the entorhinal cortex, the ipsilateral granular cell layer of the dentate gyrus, compared to its counterpart on the contralateral side, showed a significant decrease ($P < 0.05$) in [^{125}I]IGF II receptor binding sites at 14 DPL (Fig 4). However, compared to controls, a marked bilateral increase ($P < 0.05$) in [^{125}I]IGF II receptor binding sites was evident in the dentate gyrus and in most layers of the Ammon's horn between day 1 and 8 postlesion (Figs 2A-C, 4).

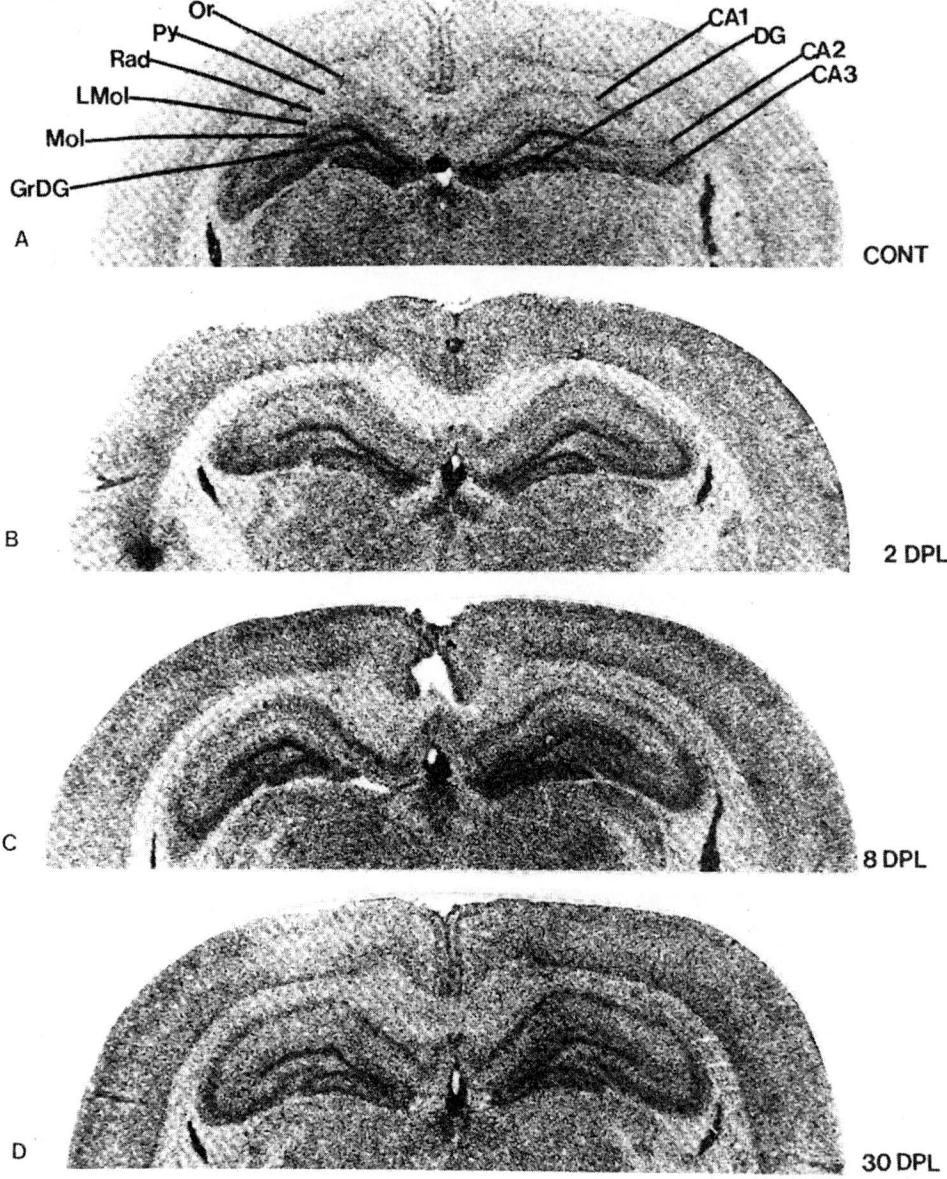

Fig. 1. [^{125}I]IGF I binding sites in the hippocampal formation in control (A), 2 days- (B), 8 days- (C) and 30- days post lesion (D) rats. Right side of the photomicrographs represent the lesioned side. Or, stratum oriens; Py, pyramidal cell layer; Rad, stratum radiatum; LMol, stratum lacunosum moleculare; Mol, molecular layer of the DG; GrDG, granular cell layer of the DG; DG, dentate gyrus.

Insulin: High density of [125I]insulin receptor binding sites was noted in the dentate gyrus and the CA1 region, while the CA2-CA3 subfields exhibited moderate to low density of binding (Figs 3A, 4). In the Ammon's horn as well as in the dentate gyrus, the density of [125I]insulin binding was concentrated primarily in the neuropil layers whereas in the closely packed pyramidal or granular cell layers specific labelling was relatively low. Unilateral lesion of the entorhinal cortex did not induce any significant alteration in [125I]insulin receptor

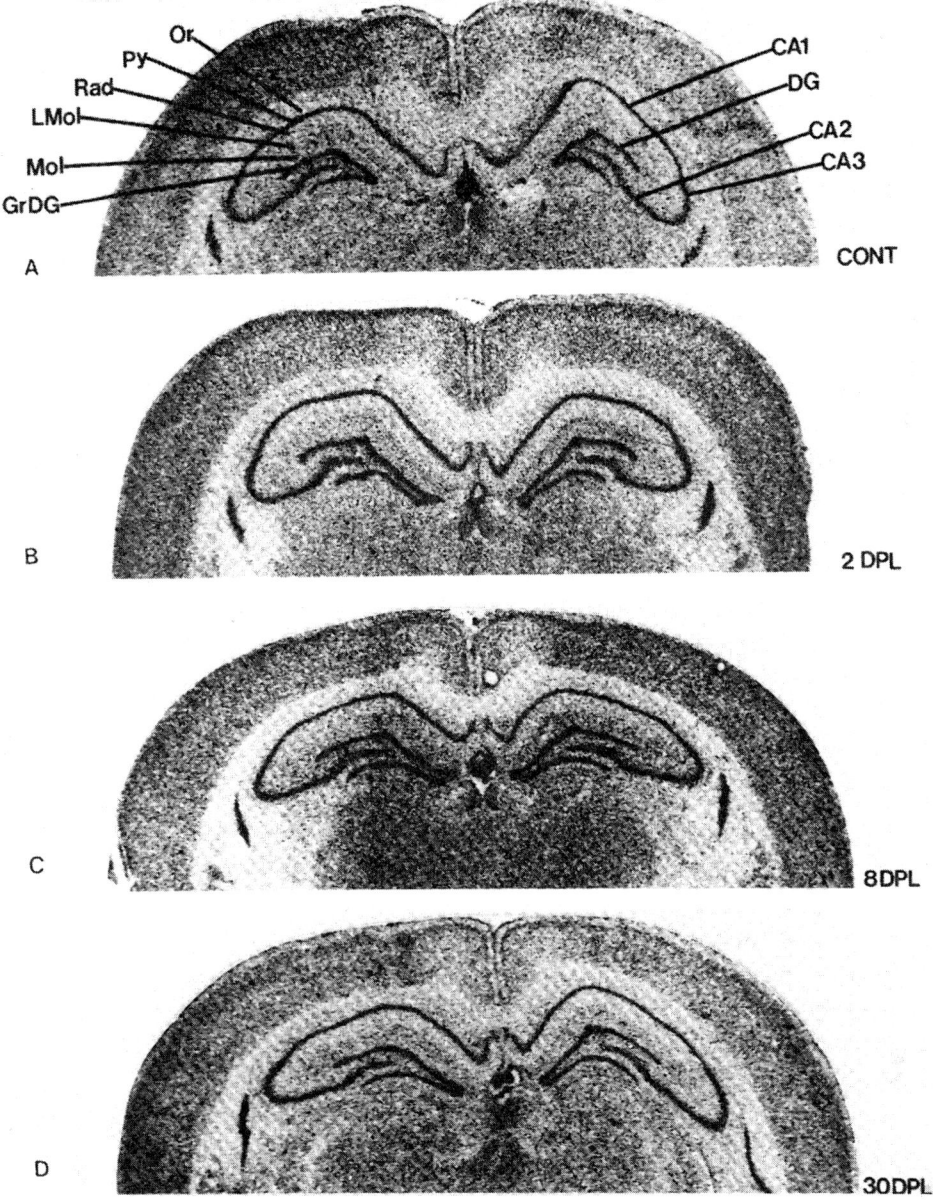

Fig. 2. [125I]IGF II binding sites in the hippocampal formation in control (A), 2 days- (B), 8 days- (C) and 30- days post lesion (D) rats. Right side of the photomicrographs represent the lesioned side. Or, stratum oriens; Py, pyramidal cell layer; Rad, stratum radiatum, LMol, stratum lacunosum moleculare; Mol, molecular layer of the DG; GrDG, granular cell layer of the DG; DG, dentate gyrus.

binding between contra- and ipsilateral hippocampal formation at any time periods (Figs 3B-D, 4). However, compared to controls, a marked bilateral increase ($P < 0.05$) in [^{125}I]insulin binding was noted in the dentate gyrus as well as in CA1-CA3 regions between 2-8 and 30 DPL (Figs 3, 4). In some selected layers of the hippocampal formation significant increase ($P < 0.05$) in [^{125}I]insulin binding was also evident at day 1 and day 14 after lesion (Fig 4).

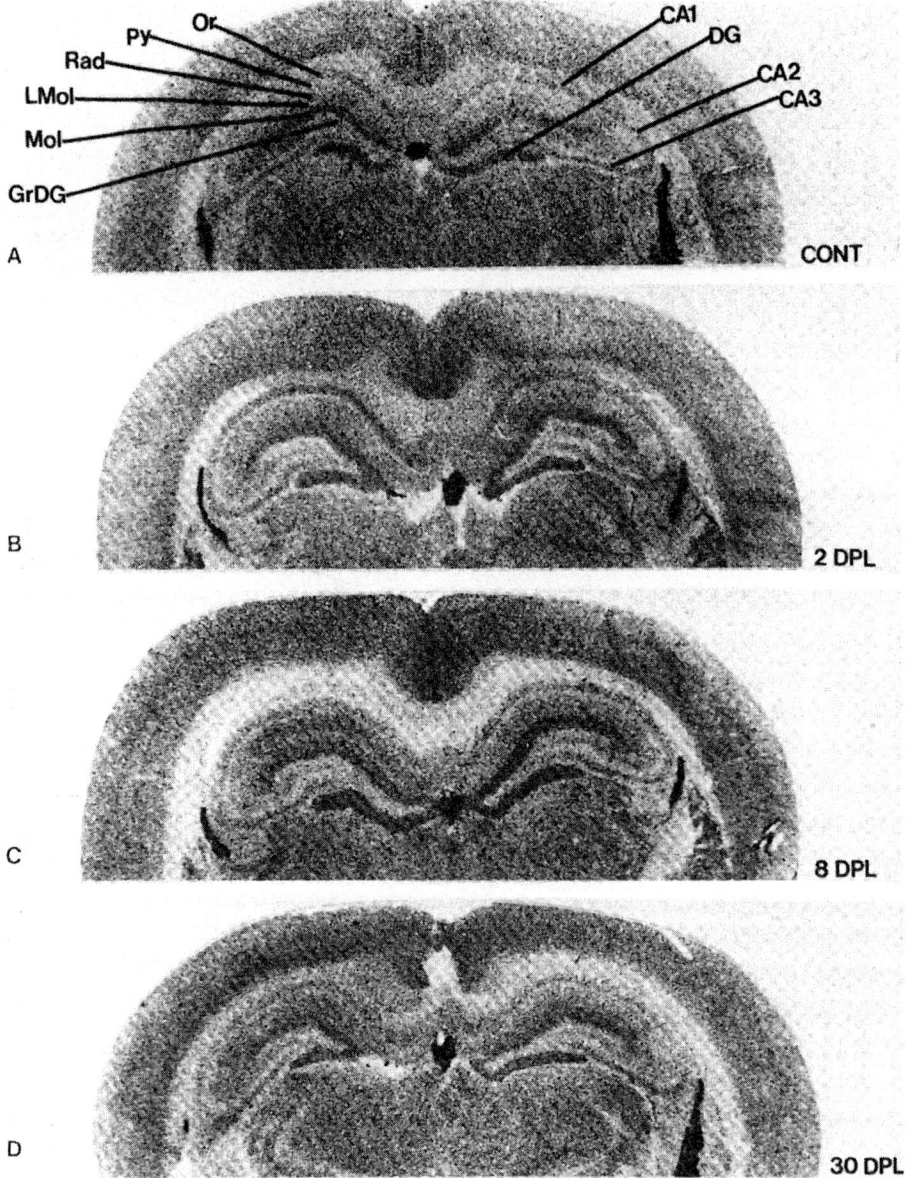

Fig. 3. [^{125}I]insulin binding sites in the hippocampal formation in control (A), 2 days- (B), 8 days- (C) and 30- days post lesion (D) rats. Right side of the photomicrographs represent the lesioned side. Or, stratum oriens; Py, pyramidal cell layer; Rad, stratum radiatum; LMol, stratum lacunosum moleculare; Mol, molecular layer of the DG; GrDG, granular cell layer of the DG; DG, dentate gyrus.

DISCUSSION

The present study showed that $[^{125}I]IGF$ I, $[^{125}I]IGF$ II and $[^{125}I]$insulin receptor binding sites are concentrated in anatomically distinct layers of the hippocampal formation and are selectively affected following lesion of the entorhinal cortex. Taking into account our previous report[19], it is most likely that each radioligand, under the present assay conditions, primarily identifies its own receptor sites and the observed time-course alterations in binding patterns possibly reflect the involvement of IGFs and/or insulin in lesion-induced plasticity.

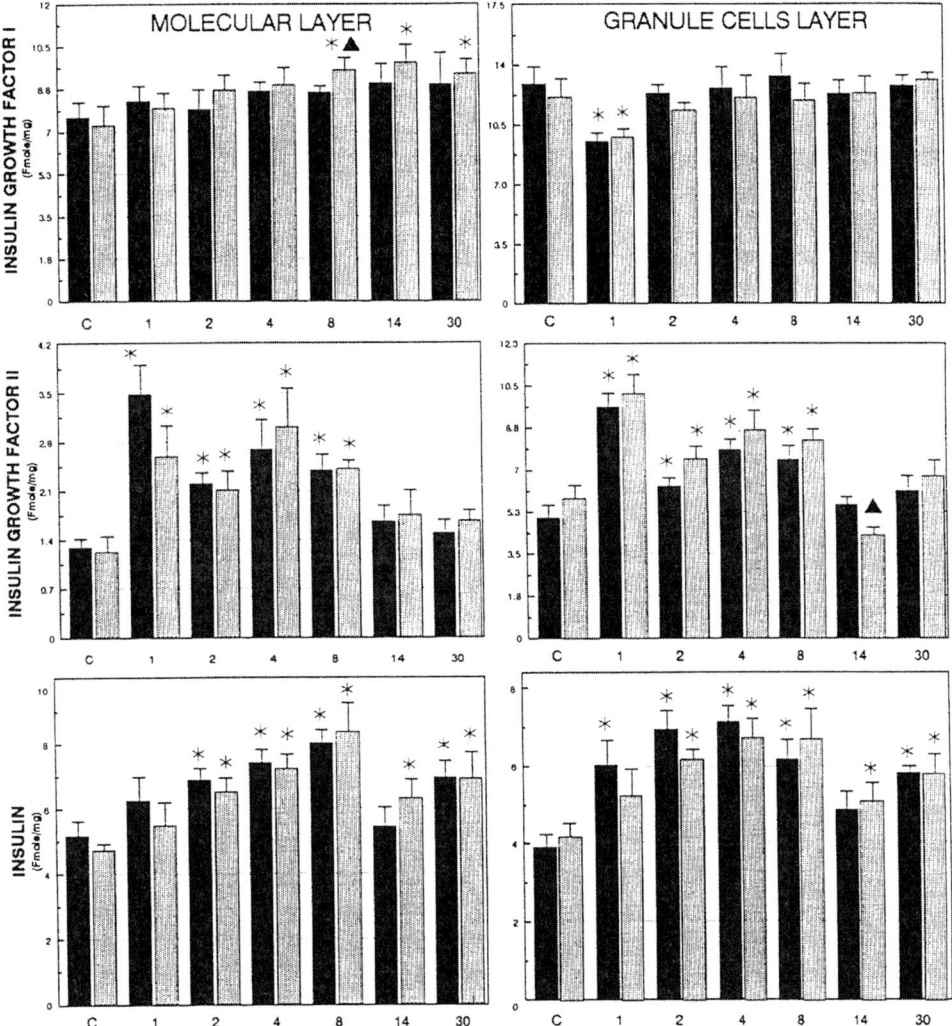

Fig. 4. Time-course alterations in $[^{125}I]IGF$ I, $[^{125}I]IGF$ II and $[^{125}I]$insulin receptor binding sites in molecular and granular cell layers of the dentate gyrus of the hippocampal formation following unilateral lesion of the entorhinal cortex. * represents significant variation (P<0.05) in selective layers of the lesion-induced hippocampal formation compared to unlesioned rat. ▲ represents significant variation (P<0.05) between ipsilateral (stippled bars) and contralateral (solid bars) sides of the lesioned animal. C, control rat; 1,2,4,8,14 and 30 represent post lesion days. Data are presented as the mean percentage ± S.E.M. of 8-10 sections from 3-4 rats.

IGF I: The selective localization of [^{125}I]IGF I receptor binding sites in the rat hippocampal formation correspond well with earlier reports[17-20,22]. However, in contrast to some published results[18,22], we noted relatively high density of [^{125}I]IGF I binding sites in the pyramidal and granular cell layers of the hippocampal formation, which could possibly be ascribed to species variations or assay conditions. Following lesion of the entorhinal cortex, one of the interesting features was the significant increase in [^{125}I]IGF I binding sites in the ipsilateral molecular layer of the dentate gyrus between 8-30 DPL i.e, the time-period parallel to the most rapid phase of sprouting reported in this model of reactive synaptogenesis[1,2,5]. This raises the possibility that IGF I may be involved in the process of triggering growth and/or synaptogenesis that follows the lesion. This is in keeping with *in vitro* studies which have shown that IGF I besides being involved in the proliferation and differentiation of neuronal precursor cells, also promotes neurite formation and outgrowth[27,28,30]. Furthermore, in the periphery the stimulated regeneration of the rat sciatic nerve by IGF I provides an additional support in favour of the possible involvement of IGF I in nerve regeneration and sprouting[34]. It is known that IGF I promotes neurite outgrowth[25], increases choline acetyltransferase enzyme activity[35] in cultured basal forebrain neurones and also regulates acetylcholine release from hippocampal slices[17]. It could therefore be possible that lesion-induced increase in [^{125}I]IGF I binding may be associated with sprouting of septal afferents or alternately may reflect an increase in postsynaptic receptor binding sites at locations of new fiber growth. However, at present the mechanism by which IGF I receptors and the corresponding peptide are involved in the process of sprouting and/or synaptogenesis that follows the lesion remains unclear.

IGF II: Consistent with earlier reports[19-21], high amounts of [^{125}I]IGF II receptor binding sites in normal rat hippocampal formation were noted primarily in the granular and pyramidal cell layers whereas other regions exhibited relatively low density of labelling. Unilateral lesion of the entorhinal cortex induces significant bilateral increase in [^{125}I]IGF II binding in most regions of the hippocampal formation between 1-8 DPL i.e, the time-course parallel to the known macrophage-astrocyte reaction[8,11]. It is known that the terminal degeneration that occurs in the molecular layer of the dentate gyrus following entorhinal cortex lesion activates resident microglia to release IL-1 (and possibly other substances) into the surrounding environment[8]. The IL-1 then induces hypertrophy and proliferation of astrocytes. The reactive astrocytes, as indexed by glial fibrillary acidic protein (GFAP) immunoreactivity, become apparent within two days following lesion and produce neurotrophic factors (i.e, NGF and bFGF) that can subsequently trigger the process of reinnervation and/or promote the survival of neurones[11,36]. Considering these time-course events and the cellular localization of IGF II receptors[37], it is most likely that early increase in [^{125}I]IGF II binding sites is associated with increase number of astrocytes and possibly macrophages observed following lesion. This is further supported by an ischemic lesion study which has shown that reactive astrocytes and macrophages in the vicinity of infarct express IGF II receptor mRNA[31]. The bilateral response in [125I]IGF II binding to unilateral lesion is possibly contributed by a diffusible signal or a threshold response which is triggered initially to about the same degree by degenerating ipsilateral and contralateral cortical afferents[36,38]. As regards to functional significance, evidence supports that IGF II receptors have a role in translocation of enzymes to the lysosome[39,40]. Since macrophages and astrocytes play an important role in scavaging degenerating cell products[2,8,36], it is possible that IGF II receptors participate in recycling phagocytic enzymes. Furthermore, in keeping with the mitogenic role for IGFs[27-30], one should consider the possible involvement of IGF II receptors in the process of proliferation of astrocytes which occurs within first two days following lesion of the entorhinal cortex.

Insulin: The localization of high density of [125I]insulin receptor binding sites in the molecular layer of the dentate gyrus and in the CA1 region of the hippocampal formation correlate well with previous reports[19,20,22]. Following lesion of the entorhinal cortex, a substantial increase in [125I]insulin binding was noted in all regions of the hippocampus between 2-8 and 30 DPL. In some selected areas, significant increase in [125I]insulin binding was also evident at day 14 following lesion. The variation in [125I]insulin binding sites, unlike IGFs, coincide temporally with the period of extensive degeneration (characterized by reactive gliosis) as well as the process of regeneration (i.e, sprouting of the neighbouring fibers) that follows the lesion. Given the evidence that insulin binding sites are localized on both neurones and astrocytes[41-43], it is very likely that increased number of astrocytes as well as the formation of new synapses may account, at least in part, for the dramatic increase of hippocampal [125I]insulin binding sites observed at 2-30 DPL. It is also possible that loss of synaptic density that occurs at the early stage of deafferentation can induce an increase in [125I]insulin binding sites on the postsynaptic sites. The bilateral increase in [125I]insulin binding is consistent with the bilateral response of astrocytes as well as reactive synaptogenesis [36,38]. However, as mentioned earlier, at present it remains unclear whether the bilateral response of [125I]insulin binding sites reflect the action of some diffusable factor or changes in activity produced by the lesion. As for functional significance, it is generally believed that receptors for insulin, but not that of IGFs, are of prime importance in metabolic events such as stimulation of glucose uptake into glial cells, the activation of pyruvate dehydrogenase and the synthesis of proteins and lipids[13,14,16]. If this is the case, then it is very likely that increased [125I]insulin receptor binding sites, evident during reactive gliosis and synaptogenesis, may be associated with enhanced metabolic functions of activated astrocytes and deafferented neurones[14,36,44]. Furthermore, in organotypic culture, it has been shown that insulin, at low concentrations, can influence astroglial morphology and GFAP expression[45]. Since astrocytes become activated following lesion and express high levels of GFAP mRNA[33,36] and its protein[38], it is possible that insulin might be involved in the transition process of astrocytes from a normal to a reactive state.

In summary, following lesion of the entorhinal cortex selective variation in [125I]IGF I, [125I]IGF II and [125I]insulin bindings were noted in specific layers of the hippocampus.

These events, which coincide temporally with the known degeneration and/or reinnervation process suggest that along with other growth factors, IGFs and insulin participate in the neurotrophic cascade that facilitate neuronal repair and/or promotion of neuronal survival.

ACKNOWLEDGEMENTS: This work was supported by Grants from Alzheimer Disease Research Program and MRCC to J. P., and Grants from MRCC and FRSQ to R. Q. J. P. is a recipient of MRC Scholarship; R. Q. is a 'Chercheur-Boursier' and S. K. is a postdoctoral fellow of the FRSQ.

REFERENCES

1. C.W. Cotman and K.J. Anderson, Neural plasticity and regeneration, *in:* Basic Neurochemistry: Molecular, Cellular and Medical Aspects, G.J. Siegel, B.W. Agranoff, R.W. Albers and P.B. Molinoff, eds., pp. 507-522, Raven, New York, (1989).

2. M. Nieto-Sampedro and C.W. Cotman, Growth factor induction and temporal order in central nervous system repair, *in:* Synaptic Plasticity, C.W. Cotman, ed., pp. 407-455, Guilford, New York, (1985).

3. O. Steward, Lesion-induced synapse growth in the hippocampus: in search of cellular and molecular mechanisms, *in:* The Hippocampus, R.L. Isaacson and K.H. Pribram, eds., pp. 65-111. Plenum, New York, (1986).

4. J. O'keefe and L. Nadel, The Hippocampus as a Cognitive Map, Claredon Press, Oxford, (1978).

5. J.W. Geddes, D.T. Monaghan, C.W. Cotman, I.T. Lott, R.C. Kim, and H.C. Chui, Plasticity of hippocampal circuitry in Alzheimer's disease. *Science* 230:1179 (1985).

6. G. Lynch, D.A. Matthews, S. Mosko, T. Parks, and C. Cotman, Induced acetylcholinesterase-rich layer in rat dentate gyrus following entorhinal lesions. *Brain Res.* 42:311 (1972).

7. O. Steward, C.W. Cotman, and G.S. Lynch, Growth of a new fiber projection in the brain: reinnervation of the dentate gyrus by the contralateral entorhinal cortex following ipsilateral entorhinal cortex lesion. *Exp. Brain Res.* 20:45 (1974).

8. A.M. Fagan and F.H. Gage, Cholinergic sprouting in the hippocampus: a proposed role for IL-1. *Exp. Neurol.* 110:105 (1990).

9. M. Nieto-Sampedro and P. Bovolenta, Growth factors and growth factor receptors in the hippocampus. Role in plasticity and response to injury. *Prog. Brain Res.* 83:341 (1990).

10. K.A. Crutcher and F. Collins, Entorhinal lesions result in increased nerve growth factor-like growth-promoting activity in medium conditioned by hippocampal slices. *Brain Res.* 399:383 (1986).

11. F.H. Gage, G. Buzsaki, and D.M. Armstrong, NGF-dependent sprouting and regeneration in the hippocampus. *Prog. Brain Res.* 83:357 (1990).

12. F. Gomez-Pinilla, J.W.K. Lee, and C.W. Cotman, Basic FGF in adult rat brain: cellular distribution and response to entorhinal lesion and fimbria-fornix transection. *J. Neurosci.* 12:345 (1992).

13. M. Adamo, M.K. Raizada, and D. LeRoith, Insulin and insulin-like growth factor receptors in the nervous system. *Mol. Neurobiol.* 3:71 (1989).

14. D.G. Baskin, B.J. Wilcox, D.P. Figlewicz, and D.M. Dorsa, Insulin and insulin-like growth factors in the CNS. *Trends Neurosci.* 11:107 (1988).

15. V.R. Sara and C. Hall, Insulin-like growth factors and their binding proteins. *Physiol. Rev.* 70:591 (1990).

16. S.U. Devaskar, A review of insulin/insulin-like peptide in the central nervous system, *in:* Molecular Biology and Physiology of Insulin-like Growth Factors, M.K. Raizada and D. LeRoith, eds., pp. 385-396. Plenum, New York, (1991).

17. D.M. Araujo, P.A. Lapchak, B. Collier, J.G. Chabot, and R. Quirion, Insulin-like growth factor-I (somatmedin-C) receptors in the rat brain: distribution and interaction with the hippocampal cholinergic system. *Brain Res.* 484:130 (1989).

18. N.J. Bohannon, E.S. Corp, B.J. Wilcox, D.P. Figlewicz, D.M. Dorsa, and D.G. Baskin, Localization of binding sites for insulin-like growth factor-I (IGF-I) in the rat brain by quantitative autoradiography. *Brain Res.* 444:205 (1988).

19. S. Kar, J.G. Chabot, and R. Quirion, Quantitative autoradiographic localization of [^{125}I]insulin-like growth factor I, [^{125}I]insulin-like growth factor II and [^{125}I]insulin receptor binding sites in developing and adult rat brain. (submitted).

20. M.A. Lesniak, J.M. Hill, W. Kiess, M. Rojeski, C.B. Pert, and J. Roth, Receptors for insulin-like growth factors I and II: autoradiographic localization in rat brain and comparison to receptors for insulin. *Endocrinology* 123:2089 (1988).

21. M. Smith, J. Clemens, G.A. Kerchner, and L.G. Mendelsohn, The insulin-like growth factor-II (IGF-II) receptor of rat brain: regional distribution visualized by autoradiography. *Brain Res* 445:241 (1988).

22. G.A. Werther, A. Hogg, B.J. Oldfield, M.J. McKinley, R. Figdor, and F.A.O. Mendelsohn, Localization and charactrization of IGF-I receptors in rat brain and pituitary gland using in vitro autoradiography and computerised densitometry: a distinct distribution from insulin receptors. *J. Neuroendocrinol.* 1:369 (1990).

23. C. Bondy, H. Werner, C.T. Roberts Jr, and D. LeRoith, Cellular pattern of type-I insulin-like growth factor receptor gene expression during maturation of the rat brain: comparison with insulin-like growth factors I and II. *Neuroscience* 46:909 (1992).

24. J.L. Marks, D. Porte, W.L. Stahl, and D.G. Baskin, Localization of insulin receptor mRNA in rat brain by in situ hybridization. *Endocrinology* 127:3234 (1990).

25. M. Faure, C. Desjardins and A. Beaudet, Insulin-like growth factor 1: morphological evidence for preferential action on basal forebrain cholinergic neurones. *XIIth Washington Int. Spring Symp. abst.* 100 (1992).

26. D.G. Puro and E. Agardh, Insulin-mediated regulation of neuronal maturation. *Science* 225:1170 (1984).

27. E. Recio-Pinto, M.M. Rechler, and D.N. Ishii, Effects of insulin, insulin-like growth factor II and nerve growth factor on neurite formation and survival in cultured sympathetic and sensory neurones. *J. Neurosci.* 6:1211 (1986).

28. F.A. McMorris and M. Dubois-Daleq, Insulin-like growth factor I promotes cell proliferation and oligodendroglial commitment in rat glial progenitor cells developing in vitro. *J. Neurosci. Res.* 21:199 (1988).

29. D. Lenoir and P. Honegger, Insulin-like growth factor I (IGF-I) stimulates DNA synthesis in fetal rat brain cell cultures. *Dev. Brain Res.* 7:205 (1983).

30. V.K.M. Han, J.M. Lauder, and A.J. D'Ercole, Characterization of somatomedin/insulin-like growth factor receptors and correlation with biological action in cultured neonatal rat astroglial cells. *J. Neurosci.* 7:501 (1987).

31. W.H. Lee, J.A. Clemens, and C.A. Bondy, Insulin-like growth factors in the response to cerebral ischemia. *Mol. Cell. Neurosci.* 3:36 (1992).

32. J.L. Marks, M.G. King and D.G. Baskin, Localization of insulin and type I IGF receptors in rat brain by in vitro autoradiography and in situ hybridization, *in:* Molecular Biology and Physiology of Insulin-like Growth Factors. M.K. Raizada and D. LeRoith, eds., pp. 459-470. Plenum, New York, (1991).

33. J. Poirier, P.C. May, H.H. Osterburg, J. Geddes, C. Cotman, and C.E. Finch, Selective alterations of RNA in rat hippocampus after entorhinal cortex lesioning. *Proc. Natl. Acad. Sci. USA* 87:303 (1990).

34. M. Kanje, A. Skottner, J. Sjoberg, and G. Lundborg, Insulin-like growth factor I (IGF-I) stimulates regeneration of the rat sciatic nerve. *Brain Res.* 486:396 (1989).

35. B. Knusel, P.P. Michel, J.S. Schwaber, and F. Hefti, Selective and nonselective stimulation of central cholinergic and dopaminergic development in vitro by nerve growth factor, basic fibroblast growth factor, epidermal growth factor, insulin and the insulin-like growth factors I and II. *J. Neurosci.* 10:558 (1990).

36. O. Steward, E.R. Torre, L.L. Phillips, and P.A. Trimmer, The process of reinnervation in the dentate gyrus of adult rats: time course of increases in mRNA for glial fibrillary acidic protein. *J. Neurosci.* 10:2373 (1990).

37. K.L. Valentino, I. Ocrant, and R.G. Rosenfeld, Developmental expression of insulin-like growth factor-II receptor immunoreactivity in the rat central nervous system. *Endocrinology* 126:914 (1990).

38. J. Poirier, M. Hess, P.C. May, and C.E. Finch, Astrocytic apolipoprotein E mRNA and GFAP mRNA in hippocampus after entorhinal cortex lesioning. *Mol. Brain Res.* 11: 97 (1991).

39. C. Polychronakos, H.J. Guyda, U. Janthly, and B.I. Posner, Effects of mannose-6-phosphate on receptor-mediated endocytosis of insulin-like growth factor II. *Endocrinology* 127:1861 (1990).

40. R.A. Roth, Structure of the insulin-like growth factor II: the puzzle amplified. *Science* 239:1269 (1988).

41. F.T. Boyd Jr, D.W. Clarke, T.F. Muther, and M.K. Raizada, Insulin receptors and insulin modulation of norepinephrine uptake in neuronal cultures from rat brain. *J. Biol. Chem.* 260:15880 (1985).

42. D.W. Clarke, F.T. Boyd Jr., M.S. Kappy, and M.K. Raizada, Insulin binds to specific receptors and stimulates 2-deoxy-D-glucose uptake in cultured glial cells from rat brain. *J. Biol. Chem.* 259:11672 (1984).

43. J. Unger, T.H. McNeill, R.T. Moxley III, M. White, A. Moss, and J.N. Livingston, Distribution of insulin receptor-like immunoreactivity in the rat forebrain. *Neuroscience* 31: 143 (1989).

44. R.M. Lindsay, Reactive gliosis, *in:* Astrocytes: Cell Biology ana Pathology of Astrocytes. S. Fedoroff and A. Vernadakis, eds., Vol. 3. pp. 231-261. Academic Press, New York, (1986).

45. C.D. Toran-Allerand, W. Bentham, R.C. Miranda, and J.P. Anderson, Insulin influences astroglial morphology and glial fibrillary acidic protein (GFAP) expression in organotypic cultures. *Brain Res.* 558:296 (1991).

ESTROGEN REGULATION OF THE PITUITARY INSULIN-LIKE GROWTH FACTOR SYSTEM

Kathleen M. Michels[1], Wei-Hua Lee[2], Mark A. Bach[2], Alicia Seltzer[1],
Juan M. Saavedra[1] and Carolyn A. Bondy[2]

[1]Laboratory of Clinical Science, Section on Pharmacology, NIMH
[2]Developmental Endocrinology Branch, National Institute of Child Health
and Human Development, Bethesda, MD 20892

INTRODUCTION

Insulin-like growth factor I (IGF-I) and the closely related IGF-II are homologous to proinsulin and share many of insulin's effects on cell metabolism. In addition they stimulate cellular differentiation and proliferation[1]. Circulating IGF-I, originating mainly from the liver, has classical endocrine actions[2], however, local IGF-I production in many other tissues suggests an autocrine/paracrine role as well[3]. Most biological actions of the IGFs are mediated through the IGF-I receptor (IGFR-I) which has intrinsic tyrosine kinase activity[4,5]. Specific, high-affinity, insulin-like growth factor binding proteins (IGFBPs), present in the plasma and extracellular space, also bind the IGFs, modulating their interaction with the IGF-I receptor and perhaps targeting IGFs to specific loci[6]. To date, six different IGFBPs have been characterized[7], among which IGFBP2 is particularly abundant in the pituitary[8].

The Pituitary IGF System

The first known function of IGF-I was as mediator of growth hormone's (GH's) growth-promoting effects[9]. The production of IGF-I in many tissues is dependent upon GH released from the somatotrophs in the anterior pituitary, and circulating IGF-I in turn feeds back on the pituitary to inhibit GH synthesis and release[10,11]. IGF-I has also been shown to stimulate synthesis of prolactin in anterior pituitary cultures[12] suggesting IGF-I may also have a role in regulating lactotroph function.

In addition to responding to circulating IGF-I, the anterior pituitary itself secretes IGF-I[13] and contains IGF-I mRNA[14,8]. Binding sites for IGF-I are present at a high density in the anterior pituitary[15,16]. These binding sites may represent the IGF-I receptor, mRNA for which is distributed uniformly throughout the pituitary[8]. IGFBPs secreted by pituitary cells[17] may contribute to the binding as well. Possible roles for this intrinsic pituitary IGF system are under investigation.

Growth Factors, Peptides, and Receptors, Edited
by T.W. Moody, Plenum Press, New York, 1993

Estrogen Regulation of the IGF System

Estrogen produces a well-documented increase in anterior pituitary size[18] due largely to lactotroph hyperplasia[19,20]. This hyperplasia may be due to a direct effect of estrogen on pituitary cells or may be mediated by growth factors, such as IGF-I, which are present in the anterior pituitary[21].

The possible involvement of IGF-I in these pituitary effects of estrogen is suggested by analogy to other tissues. Estrogen has been shown to regulate IGF-I gene expression in several tissues including the uterus[22], where estrogen-stimulated uterine proliferation is thought to be mediated by a local increase in IGF-I[23]. IGF-I may also mediate estrogen-stimulated proliferation and growth of mammary gland tumors[24] and IGFBP-2 in these tumors is estrogen-responsive[25].

To determine the effects of estrogen on the rat pituitary IGF system, we coupled quantitative autoradiography with [^{125}I]IGF-I binding and in situ hybridization for IGF-I, the IGF-I receptor and IGF binding protein-2 (IGFBP-2). The animal groups included intact cycling female rats, intact males, and gonadectomized males and females with or without chronic estrogen treatment.

MATERIALS AND METHODS

All experiments were carried out in accordance with the principles and procedures outlined in the NIH Guide for the Care and Use of Laboratory Animals. One week after surgery, ovariectomized females (OVX) were divided into two groups one of which was implanted subcutaneously with cholesterol methyl cellulose pellets containing 17-β estradiol (E2, 0.05 mg/pellet, Innovative Research of America, Toledo, OH) and animals were killed 2 weeks later. E2 levels in the plasma were determined to be 40-50 pg/ml (Hazleton, Vienna, VA), similar to levels described for estrus and diestrus-2[26]. Four groups of cycling females: proestrus, estrus, diestrus-1 (metestrus) and diestrus-2, were used. Castrated males were implanted either with testosterone pellets (T, 25 mg/pellet), or with estrogen pellets (as described for female rats) 3 days postcastration and killed 1 week post-implantation.

Binding of [^{125}I] recombinant insulin-like growth factor-I ([^{125}I]IGF-I, Amersham, Arlington Heights, IL) was carried out on frozen slide-mounted pituitary sections and quantitative autoradiography was carried out as previously described[27].

For in situ hybridization the rat IGF-I[28] IGF-I receptor[29] and IGFBP-2[30] clones were used for synthesis of cRNA probes as previously described[31]. The ^{35}S-labelled cRNA probes were synthesized and used on slide-mounted frozen sections as previously described[32,33].

Statistical analysis was performed using one-way analysis of variance as indicated and the Bonferroni post-test.

RESULTS AND DISCUSSION

We found that estrogen affects several components of the anterior pituitary IGF system in parallel. The number of IGF-I binding sites (Figure 1B and 1C, Figure 2A) and the density of IGFBP-2 mRNA (Figure 1F and 1G, Figure 2B) was significantly increased by estrogen-treatment of ovariectomized females. Similarly, IGF-I mRNA levels were also

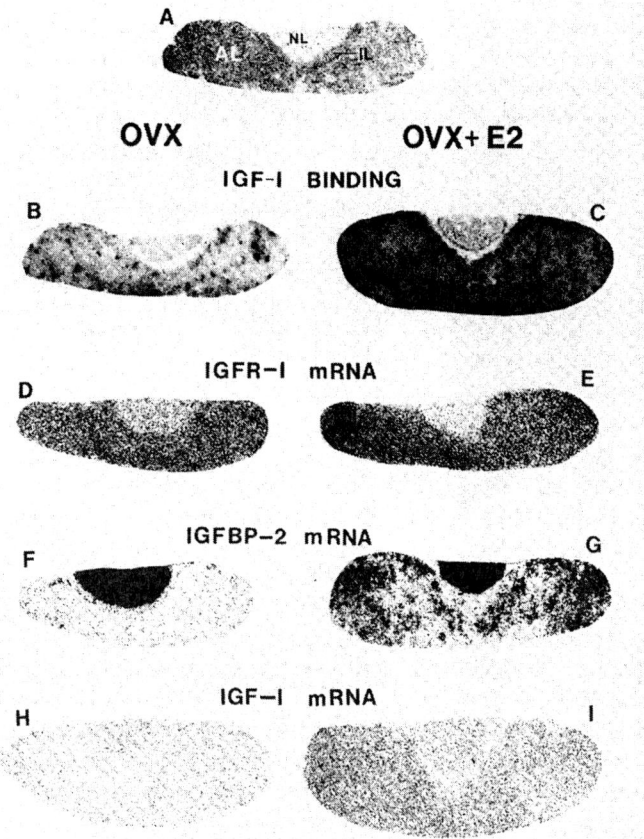

Figure 1. Brightfield autoradiographic images of [^{125}I]IGF-I binding (B and C) and gene expression for IGF-I receptors (IGFR-I, D and E), IGFBP-2 (F and G) and IGF-I (H and I) in representative 16 μm coronal sections of female rat pituitary. A: Pituitary section stained with Toluidine blue for histology. B: OVX incubated in the presence of 0.25 nM [^{125}I]IGF-I (total binding). C: OVX + E2, incubated as in B. Gene expression in OVX: D, F and H. Gene expression in OVX+E2: E, G and I. OVX: ovariectomized, OVX+E2 ovariectomized with estrogen treatment, AL: anterior lobe, NL: neural lobe, IL: intermediate lobe.

increased (Figure 1H and I, Figure 2C). The IGF-I receptor mRNA level, on the other hand, remained stable (Figure 1D and E).

These effects on IGF-I binding, and IGFBP-2 and IGF-I mRNA levels are specifically related to estrogen status, as shown by our findings that IGF-I binding, IGFBP2 mRNA and IGF-I mRNA levels in male anterior pituitary were comparable to those in the OVX female anterior pituitary (Figure 2A, 2B, 2C) and castration of the male had no effect on IGF-I binding or IGFBP2 mRNA (Table I). Furthermore, we found that estrogen increased IGF-I binding, IGFBP2 mRNA and IGF-I mRNA in the castrated male (Table I) while testosterone had no effect. Changes in [^{125}I]IGF-I binding and IGFBP-2 mRNA in the anterior pituitary during the estrous cycle are in agreement with these results. The highest levels of [^{125}I]IGF-I binding and IGFBP-2 mRNA were seen at proestrus (Table II), the phase with the highest level of circulating estrogen[26]. IGF-I mRNA levels were higher at both procstrus and estrus than diestrus (Table II). These results add the anterior pituitary

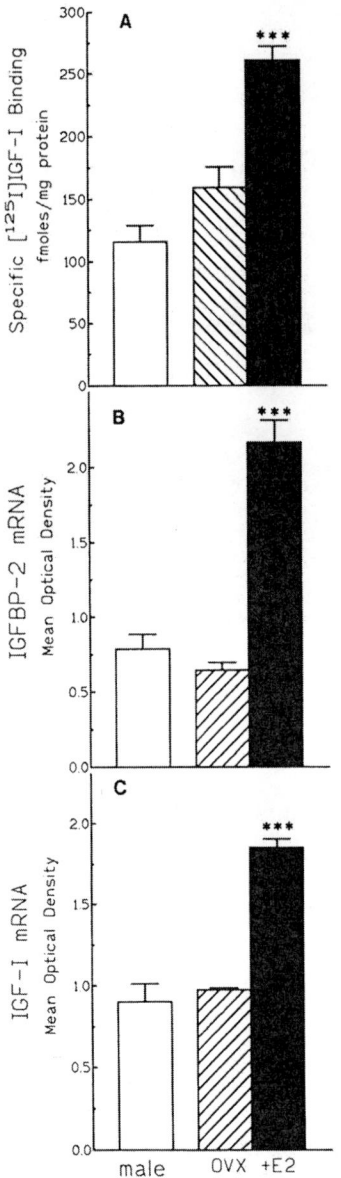

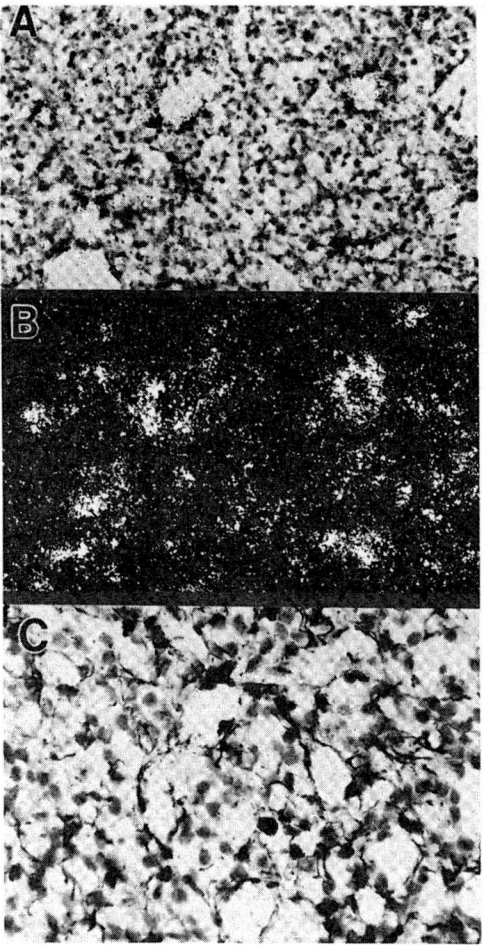

Figure 2. Bar graphs illustrating the results of quantitative autoradiographic analysis of [^{125}I]IGF-I binding (A) and gene expression of iGFBP-2 (B) and IGF-I (C) in anterior pituitary of male and female rats. (A) Results are expressed as the mean ± SEM of specific binding (3-4 sections per animal, n = 6 animals per group). Gene expression (D - I) is expressed as mean optical density ± SEM of mRNA levels. OVX: Ovariectomized. OVX+E2: Ovariectomized treated with estradiol. ***: p<0.001, OVX+E2 compared to OVX and male.

Figure 3. IGFBP-2 mRNA localized in folliculo-stellate cells of the anterior pituitary. Panels A and B are paired bright- and dark field micrographs of a section hybridized to an IGFBP-2 probe. IGFBP-2 mRNA is seen as white grains in the dark-field and is clustered in cells forming the anterior lobe follicles. Panel C shows vimentin-immunostaining of the glial-like folliculo-stellate cells and their processes in a similar section

Table I. The effect of estrogen on [^{125}I]IGF-I binding and gene expression of IGFBP-2 and IGF-I in the anterior pituitary of the male rat.*

	[^{125}I]IGF-I Bound fmoles/mg protein	mRNA optical density	
		IGFBP-2	IGF-I
Male	87 ± 12 (6)	0.22 ± 0.03 (6)	ND
Castrated Male	110 ± 13 (6)	0.31 ± 0.05 (6)	0.07 ± 0.01 (6)
Castrated + E2	240[a] ± 8 (8)	0.60[a] ± 0.02 (8)	0.17[a] ± 0.01 (8)

* Results expressed as mean ± SEM with number of animals in parentheses (2 determinations per animal).
[a] p <.01 castrated +E2 compared to castrated or intact male.
NT: not done.

Table II. Changes in [^{125}I]IGF-I binding and gene expression of IGFBP-2 and IGF-I during the estrous cycle.*

	[^{125}I]IGF-I Bound fmoles/mg protein	mRNA optical density	
		IGFBP-2	IGF-I
Proestrus	511[a] ± 30	1.36[a] ± 0.10	0.18 ± 0.01
Estrus	381 ± 24	0.71 ± 0.05	0.21[b] ± 0.02
Diestrus-1	305 ± 24	0.83 ± 0.04	0.13 ± 0.01
Diestrus-2	326 ± 8	0.72 ± 0.06	0.15 ± 0.02

* Sections were incubated in the presence of 0.25 nM [^{125}I]IGF-I and results are expressed as mean ± SEM.
[a] p<0.01 proestrus different from estrus, diestrus 1, diestrus 2
[b] p<0.01 estrus different from diestrus 1 and diestrus 2.

to the short list of tissues in which estrogen has been shown to regulate IGF-I mRNA. These tissues include uterus and ovarian granulosa cells[34,22] bone[35] and liver[34,36].

IGF-I Distribution and Binding

IGF-I gene expression is distributed in both the ovariectomized and estrogen-treated anterior pituitary in a homogeneous pattern (Figure 1H and 1I). This even distribution makes it unlikely that IGF-I is selectively expressed by a specific subpopulation of endocrine cells such as somato- or lactotrophs, since these are heterogeneously clustered in the dorsolateral wings. Thus it seems likely that IGF-I is expressed either by all endocrine cells equally or by a non-endocrine, widely-distributed support element such as the folliculo-stellate cell[37] (Figure 3C). A previous microscopic study of IGF-I mRNA localization has reported a particulate distribution which is most compatible with the latter possibility[8].

These folliculo-stellate cells perform important trophic functions for the endocrine cells of the anterior pituitary and are the source of other pituitary growth factors such as fibroblast growth factor and vascular endothelial growth factor[37].

The uniform distribution of IGF-I mRNA throughout the anterior pituitary also suggests a widespread synthesis of IGF-I for autocrine or paracrine action. Specificity of IGF-I action in the pituitary, therefore, is probably determined by its interaction with the IGF-I receptor and with IGF binding proteins associated with specific cell populations.

The pattern of IGF-I receptor gene expression in the anterior pituitary is similar to that of IGF-I in its uniformity (Figure 1D and 1E). In contrast, the pattern of [^{125}I]IGF-I binding is heterogeneous, with high-density clusters of binding sites superimposed on a uniform, low-density background (Figure 1C and 1D), a pattern similar to the clustered pattern of IGFBP-2 gene expression (Figure 1G). IGF-I receptor gene expression and [^{125}I]IGF-I binding sites also displayed differential sensitivity to estrogen, since IGF-I receptor mRNA levels did not increase (Figure 1D and 1E) while IGF-I binding sites did increase (Figure 1B and 1C), in response to estrogen treatment. The discrepancy in the pituitary patterns of distribution and regulation of IGF-I receptor mRNA and IGF-I binding sites might be explained if regulation of the IGF-I receptor levels were primarily a function of translation rather than transcription. An alternative, or perhaps additional, explanation, given the parallel regulation of IGF binding and IGFBP-2 mRNA, is that a significant portion of anterior pituitary IGF binding is to IGFBP(s).

Since the IGFBPs bind IGF-I with an affinity equal or greater than that of the IGF-I receptor, but do not bind insulin[6], the pharmacological displacement of IGF-I binding by insulin is used to dissociate IGF binding to the IGF-I receptor from IGF-I binding to IGFBPs. In the SCN of the hypothalamus, for example, insulin displaced all [^{125}I]IGF-I binding, with an IC_{50} approximately 200 fold higher than for IGF-I[27] suggesting all binding was to IGF receptors. However in the present study, insulin displaced only ~ 50% of anterior pituitary IGF-I binding at a concentration almost 10,000 fold higher than the IC_{50} for either IGF-I or IGF-II. The low affinity of insulin for pituitary IGF-I binding sites demonstrated in the present study is in agreement with previous studies on the pharmacology of pituitary IGF-I binding in the male rat[15,16]. This low potency of insulin to displace IGF-I binding in the pituitary suggests that pituitary IGF-I binding sites represent a complex mixture of heterogeneous IGF-I receptor proteins and possibly IGFBPs.

IGF Binding Protein-2

Our findings suggest that IGFBP-2 gene expression is regulated by estrogen in the rat anterior pituitary. Estrogen regulation of IGFBP-2 has been reported previously for rat breast carcinoma cell lines[25] but not for normal rat tissues. However, in the pig uterus, during the estrous cycle, the pattern of change in IGFBP-2 mRNA was found to correlate closely with the pattern of change in circulating estrogen but to be inversely correlated with circulating progesterone[38]. Other IGFBP mRNAs (IGFBP-3,-4,-5) have recently been demonstrated in the male rat pituitary[8]. Whether gene expression of these IGFBPs is also regulated by estrogen remains to be investigated.

The clustering of [^{125}I]IGF-I binding and of IGFBP-2 mRNA within the anterior pituitary is particularly pronounced in intact females and estrogen-treated-gonadectomized males and females (Figure 1). The clustering suggests an anatomical substrate for the targeting of

IGF-I to certain subpopulations of anterior pituitary cells. A recent study[8] shows that in the male rat anterior pituitary, IGFBP-2 mRNA appears to be concentrated in folliculo- stellate cells as shown in Figure 3. Whether IGFBP-2 mRNA has a similar localization in cycling or estrogen-treated female rat pituitaries, where a significantly higher level of IGFBP2 gene expression is seen, remains to be determined. Further studies combining immunocytochemistry with in situ hybridization will be necessary to establish whether estrogen results in an increase in IGFBP-2 in folliculo-stellate cells specifically associated with the somatotroph or lactotroph population, or in the somatotrophs and lactotrophs themselves.

The IGF System in the Neuro-Intermediate Lobe

The high degree of complexity characterizing the regulation of pituitary IGF binding is further demonstrated by our observations on the IGF system of the neural and intermediate lobes of the pituitary. In the neural lobe, IGF binding reflects levels of IGFR-I mRNA (low levels) rather than IGFBP2 mRNA (high levels) and none of these parameters were affected by changes in estrogen status. In the intermediate lobe, on the other hand, where IGF-I receptor gene expression is comparable to IGF-I receptor gene expression in the anterior pituitary, there is little or no IGF-I binding and no detectable IGFBP-2 mRNA or mRNA for the other known IGF binding proteins[8]. It may be that functional IGF-I receptors are not synthesized in the intermediate lobe, or that the most of the autoradiographically detected [^{125}I]IGF-I binding in the anterior pituitary is to IGFBPs, which are not present in the intermediate lobe[8]. These regional differences in pituitary [^{125}I]IGF-I binding, IGF-I receptor and IGFBP-2 gene expression suggest that there is a regional heterogeneity of IGF-I binding characteristics, perhaps related to differences in translation and posttranslational processing for IGFBP-2 as well as the IGF-I receptor.

CONCLUSIONS

In summary, we have shown that estrogen produces an increase in anterior pituitary IGF-I and IGFBP-2 mRNAs and a parallel increase in [^{125}I]IGF-I binding. These results support a possible autocrine/paracrine role for pituitary IGF-I as a mediator of estrogen-induced proliferation and hyperplasia. A number of observations indicate that a significant portion of anterior pituitary IGF binding consists of binding to IGFBP-2 rather than to the IGF-I receptor. The physiological significance of increased production of IGFBP-2 in response to estrogen is not clear. However, we speculate that an increase in IGFBP-2, perhaps binding to cell surfaces or extracellular matrix via its RGD sequence, may result in increased targeting of locally produced or circulating IGF-I to particular pituitary loci. In this way IGFBP-2 could enhance IGF-I's participation in estrogen-induced effects on the anterior pituitary.

ACKNOWLEDGEMENTS

We would like to thank Charles T. Roberts, Jr., Derek LeRoith, Haim Werner, Alexandra Brown, and Matt Rechler for providing the cDNAs used for riboprobe synthesis. K.M. Michels was supported by a National Research Council Associateship and W-H. Lee by a Pharmacology Research Associateship from the National Institute of General Medical Science.

REFERENCES

1. M.M. Rechler and S.P. Nissley, Insulin-like Growth Factors, in: "Peptide Growth Factors and Their Receptors", M.B. Sporn and A.B. Roberts., eds., Springer-Verlag, New York,pp263-367 (1990).
2. E.R. Froesch, C. Schmid, J. Schwander, and J. Zapf, Actions of insulin-like growth factors. Annu. Rev. Physiol. 47:443 (1985).
3. A.J. D'Ercole, A.D. Stiles, and L.E. Underwood, Tissue concentrations of somatomedin C: further evidence for multiple sites of synthesis and paracrine or autocrine mechanisms of action. Proc Natl. Acad. Sci. U. S. A. 81:935 (1984).
4. M.M. Rechler and S.P. Nissley, The nature and regulation of the receptors for insulin-like growth factors. Ann. Rev. Physiol. 47:425 (1985).
5. M.P. Czech, K.T. Yu, R.E. Lewis, R.J. Davis, C. Mottola, R.G. MacDonald, P.C. Necessary, and S. Corvera, Insulin receptor kinase and its mode of signaling membrane components. Diabetes. Metab. Rev. 1:33 (1985).
6. D.R. Clemmons, Insulin-like growth factor binding proteins, in: "Insulin-Like Growth Factors: Molecular and Cellular Aspects", D. LeRoith., ed., CRC Press, Boca Raton,pp151-180 (1991).
7. M.M. Rechler and A.L. Brown, Insulin-like growth factor binding proteins: Gene structure and expression. Growth Regulation 2:55 (1992).
8. M.A. Bach and C.A. Bondy, Anatomy of the pituitary insulin-like growth factor system. Endocrinology (1992).(in press)
9. J. Zapf, C. Schmid, and E.R. Froesch, Biological and immunological properties of insulin-like growth factors (IGF) I and II. Clin. Endocrinol. Metab. 13:3 (1984).
10. M. Berelowitz, M. Szabo, L.A. Frohman, S. Firestone, L. Chu, and R.L. Hintz, Somatomedin-C mediates growth hormone negative feedback by effects on both the hypothalamus and the pituitary. Science 212:1279 (1981).
11. S. Yamashita and S. Melmed, Insulin-like growth factor I action on rat anterior pituitary cells: suppression of growth hormone secretion and messenger ribonucleic acid levels. Endocrinology 118:176 (1986).
12. K. Inoue and T. Sakai, Conversion of growth hormone-secreting cells into prolactin-secreting cells and its promotion by insulin and insulin- like growth factor-1 *in vitro*. Exp. Cell Res. 195:53 (1991).
13. M. Binoux, P. Hossenlopp, C. Lassarre, and N. Hardouin, Production of insulin-like growth factors and their carrier by rat pituitary gland and brain explants in culture. FEBS Lett. 124:178 (1981).
14. D. Olchovsky, J.F. Bruno, M.C. Gelato, J. Song, and M. Berelowitz, Pituitary insulin-like growth factor-I content and gene expression in the streptozotocin-diabetic rat: evidence for tissue-specific regulation. Endocrinology 128:923 (1991).
15. K. Matsuo, M. Niwa, M. Kurihara, K. Shigematsu, S. Yamashita, M. Ozaki, and S. Nagataki, Receptor autoradiographic analysis of insulin-like growth factor- I (IGF-I) binding sites in rat forebrain and pituitary gland. Cellular and Molecular Neurobiology 9:357 (1989).

16. C.G. Goodyer, L. De Stephano, W.H. Lai, H.J. Guyda, and B.I. Posner, Characterization of insulin-like growth factor receptors in rat anterior pituitary, hypothalamus, and brain. Endocrinology 114:1187 (1984).

17. R.G. Rosenfeld, H. Pham, Y. Oh, and I. Ocrant, Characterization of insulin-like growth factor-binding proteins in cultured rat pituitary cells. Endocrinology 124:2867 (1989).

18. S. Holtzman, J.P. Stone, and C.J. Shellabarger, Influence of diethylstilbestrol treatment on prolactin cells of female ACI and Sprague-Dawley rats. Cancer Res. 39:779 (1979).

19. R.V. Lloyd, L. Jin, K. Fields, and E. Kulig, Effects of estrogens on pituitary cell and pituitary tumor growth. Pathol. Res. Pract. 187:584 (1991).

20. L. Jin, J.Y. Song, and R.V. Lloyd, Estrogen stimulates both prolactin and growth hormone mRNAs expression in the MtT/F4 transplantable pituitary tumor. Proc Soc. Exp. Biol. Med. 192:225 (1989).

21. S. Ezzat and S. Melmed, The role of growth factors in the pituitary. J. Endocrinol. Invest. 13:691 (1990).

22. L.J. Murphy, Estrogen induction of insulin-like growth factors and myc proto- oncogene expression in the uterus. J. Steroid Biochem. Mol. Biol. 40:223 (1991).

23. L.J. Murphy and A. Ghahary, Uterine insulin-like growth factor-1: regulation of expression and its role in estrogen-induced uterine proliferation. Endocr. Rev. 11:443 (1990).

24. K.K. Huff, C. Knabbe, R. Lindsey, D. Kaufman, D. Bronzert, M.E. Lippman, and R.B. Dickson, Multihormonal regulation of insulin-like growth factor-I-related protein in MCF-7 human breast cancer cells. Mol. Endocrinol. 2:200 (1988).

25. D.R. Clemmons, C. Camacho-Hubner, E. Coronado, and C.K. Osborne, Insulin-like growth factor binding protein secretion by breast carcinoma cell lines: correlation with estrogen receptor status. Endocrinology 127:2679 (1990).

26. R.L. Butcher, W.E. Collins, and N.W. Fugo, Plasma concentration of LH, FSH, prolactin, progesterone and estradiol-17beta throughout the 4-day estrous cycle of the rat. Endocrinology 94:1704 (1974).

27. K.M. Michels and J.M. Saavedra, Differential development of insulin-like growth factor I binding in the suprachiasmatic nucleus and median eminence of the rat hypothalamus. Neuroendocrinology 54:504 (1991).

28. W.L. Lowe,Jr., S.R. Lasky, D. LeRoith, and C.T. Roberts,Jr., Distribution and regulation of rat insulin-like growth factor I messenger ribonucleic acids encoding alternative carboxyterminal E-peptides: evidence for differential processing and regulation in liver. Mol. Endocrinol. 2:528 (1988).

29. H. Werner, M. Woloschak, M. Adamo, Z. Shen-Orr, C.T. Roberts,Jr., and D. LeRoith, Developmental regulation of the rat insulin-like growth factor I receptor gene. Proc Natl. Acad. Sci. U. S. A. 86:7451 (1989).

30. A.L. Brown, L. Chiariotti, C.C. Orlowski, T. Mehlman, W.H. Burgess, E.J. Ackerman, C.B. Bruni, and M.M. Rechler, Nucleotide sequence and expression of a cDNA clone encoding a fetal rat binding protein for insulin-like growth factors. J. Biol. Chem. 264:5148 (1989).

31. C.A. Bondy, H. Werner, C.T. Roberts,Jr., and D. LeRoith, Cellular pattern of insulin-like growth factor-I (IGF-I) and type I IGF receptor gene expression in early organogenesis: Comparison with IGF-II gene expression. Mol. Endocrinol. 4:1386 (1990).

32. C. Bondy, H. Werner, C.T. Roberts Jr., and D. LeRoith, Cellular pattern of type-I insulin-like growth factor receptor gene expression during maturation of the rat brain: comparison with insulin-like growth factors I and II. Neuroscience 46:909 (1992).

33. W-H Lee, J.A. Clemens, and C.A. Bondy, Insulin-like growth factors in the response to cerebral ischemia. Molecular and Cellular Neurosciences 3:36 (1992).

34. L.J. Murphy, L.C. Murphy, and H.G. Friesen, Estrogen induces insulin-like growth factor-I expression in the rat uterus. Mol. Endocrinol. 1:445 (1987).

35. M. Ernst, J.K. Heath, and G.A. Rodan, Estradiol effects on proliferation, messenger ribonucleic acid for collagen and insulin-like growth factor-I, and parathyroid hormone-stimulated adenylate cyclase activity in osteoblastic cells from calvariae and long bones. Endocrinology 125:825 (1989).

36. L.J. Murphy and H.G. Friesen, Differential effects of estrogen and growth hormone on uterine and hepatic insulin-like growth factor I gene expression in the ovariectomized hypophysectomized rat. Endocrinology 122:325 (1988).

37. W. Allaerts, P. Carmeliet, and C. Denef, New perspectives in the function of pituitary folliculo-stellate cells. Mol. Cell Endocrinol. 71:73 (1990).

38. F.A. Simmen, R.C. Simmen, R.D. Geisert, F. Martinat-Botte, F.W. Bazer, and M. Terqui, Differential expression, during the estrous cycle and pre- and postimplantation conceptus development, of messenger ribonucleic acids encoding components of the pig uterine insulin-like growth factor system. Endocrinology 130:1547 (1992).

EFFECTS OF RADIOFREQUENCY RADIATION ON
GROWTH FACTOR RECEPTORS

A. Makheja[1], E. Albert[1], Q. Balzano[2], L. Cygan[3] and T.W. Moody[1]

[1]Departments of Anatomy, Biochemistry and Molecular Biology
The George Washington University School of Medicine
    and Health Sciences
Washington, D.C. 20037
[2]Motorola Inc.
Ft. Lauderdale, FL 33322
[3]Motorola Inc.
Schaumberg, IL 60196

INTRODUCTION

Swiss 3T3 cells are enriched in peptide and growth factor receptors. Epidermal growth factor (EGF) receptors are present on Swiss 3T3 cells. The 1186 amino acid EGF receptor is characterized by a 621 amino acid extracellular domain which binds EGF, a 23 amino acid transmembrane domain and a 542 amino acid cytoplasmic domain which has tyrosine kinase activity[1]. There is strong sequence homology between the EGF receptor and the v-erb-B oncogene product[2]. Within seconds after the addition of EGF increased transport of amino acids and glucose occurs across the plasma membrane and the intracellular calcium levels increase[3]. Because the EGF receptor functions as a protein kinase, EGF stimulates the phosphorylation of numerous proteins including numerous tyrosine amino acid residues present on the EGF receptor itself[5]. Also, the EGF receptor forms dimers and undergoes internalization[6]. A functional tyrosine kinase is essential for EGF receptor activity and ability to stimulate ^{3}H-thymidine uptake in cells[7].

In addition to EGF receptors, bombesin/gastrin releasing peptide (BN/GRP) receptors (Mr of 75 KDaltons) are present on Swiss 3T3 cells[8,9]. ^{125}I-GRP binds with high affinity (Kd = 0.5 nM) to a single class of sites (Bmax = 100,000/ cell)[10]. BN stimulates phosphatidylinositol (PI) metabolism and PI diphosphate is metabolized by phospholipase C (PLC) to inositol-1,4,5-trisphosphate (IP$_3$) and diacylglycerol (DAG)[11]. Within minutes, the IP$_3$ elevates cytosolic calcium (Ca^{2+}) whereas the diacylglycerol activates protein kinase C (PKC), which can then phosphorylate the EGF receptor[12]. Therefore BN/GRP and EGF receptors cross talk. Within an hour, BN increases expression of the c-fos and c-myc protooncogenes[13]. Within a day, BN stimulates proliferation of Swiss 3T3 cells[14].

Ultraviolet radiation can cause DNA strand breaks impairing DNA replication[15]. In contrast, the effects of electromagnetic field (EMF) radiation on cell transformation are unknown[16]. Previously, EMF radiation was found to alter intracellular Ca^{2+} levels and alter Ca^{2+} flux from cells[17-19]. Thus EMF radiation can alter second messenger production in cells. Because growth factors such as EGF and BN alter cytosolic Ca^{2+} EMF may also alter signal transduction mechanisms induced by growth factors. Here the effects of radiofrequency radiation were investigated on Swiss 3T3 cells which have EGF and BN/GRP receptors.

RECEPTOR BINDING

The effects of pulsed radiofrequency radiation on binding to Swiss 3T3 cells was investigated. Swiss 3T3 cells were treated with and without pulsed RF radiation (840 MHz) for 8 hrs/day for 1, 3 or 7 days. The RF was pulsed at 11 Hz; the pulse average power (4.8 mW/cm2, 2.4 and 1.2 mW/cm2) was 6 times the average power density. The peak instantaneous power density was 5 times the pulse average power density; specific absorption rate measurements and computations gave values of the order of 5-11 microwatts per gram. The Swiss 3T3 cells were grown in DMEM with 10% calf serum. Cells were seeded into 24 well tissue culture plates coated with fibronectin (5 μg/cm^2) and grown to confluence.

When a monolayer of cells formed (3 days), receptor binding studies were performed using 0.2 nM (^{125}I-Tyr^4)BN or ^{125}I-EGF (2000 Ci/mmol). The cells were rinsed four times with DMEM/20 mM HEPES.NaOH (pH 7.4). Routinely, radiolabeled peptide (100,000 cpm/250 μl) was added in DMEM/HEPES containing 0.1% BSA and 100 μg/ml bacitracin in the presence or absence of competitor. After 30 min at 37°C, free peptide was removed and the cells rinsed with 250 μl of cold PBS/0.1% BSA. The cells, which contained bound peptide, were solubilized with 0.2 N NaOH and counted in a gamma counter. The ability of radiofrequency radiation to alter EGF and BN/GRP receptor binding was investigated.

Approximately, 5000 cpm of ^{125}I-EGF bound to the Swiss 3T3 cells (Fig. 1). EGF (0.1 ng/ml) had little effect on binding whereas 1 μg/ml reduced the binding to approximately 600 cpm. EGF inhibited binding in a dose dependent manner. The concentration of EGF required to inhibit half of the specific ^{125}I-EGF binding (IC$_{50}$)

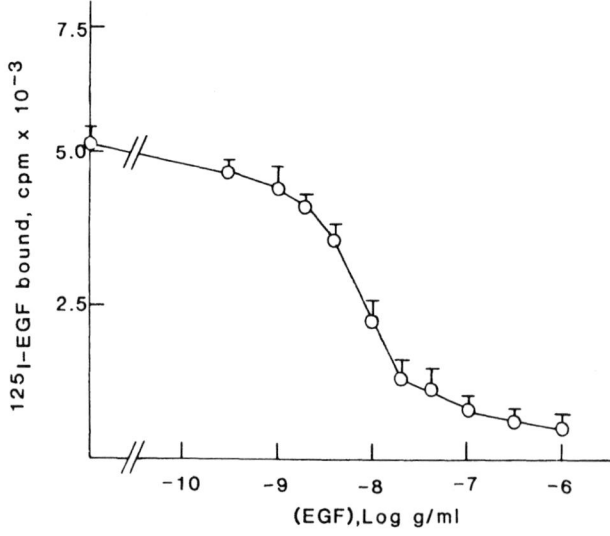

Figure 1. Effect of radiofrequency radiation on EGF receptor binding. The amount of ^{125}I-EGF bound was determined as a function on increasing doses of EGF. The mean value $\pm$ S.E. of 4 determinations is shown.

Table 1. Effect of pulsed radiation on growth factor receptor binding

Growth factor	1 day + hv	1 day - hv	3 day + hv	3 day - hv	control
EGF	12	7	10	10	10
BN	10	8	8	10	14

The IC_{50} value for EGF to inhibit specific ^{125}I-EGF binding (ng/ml) and BN to inhibit specific (^{125}I-Tyr^4)BN binding (nM) was determined. The mean value of 4 determinations is indicated.

was 10 ng/ml in the control plate. Table I shows that in the presence or absence of radiation for 1 day, the IC_{50} value was 12 and 7 ng/ml. In the presence or absence of radiation for 3 days the IC_{50} values were 10 and 10 ng/ml respectively. These data indicate that radiofrequency radiation has little effect on EGF receptor binding.

Approximately, 7000 cpm of (^{125}I-Tyr^4)BN bound to the Swiss 3T3 cells (Fig. 2). BN (1 nM) had little effect on binding whereas 1 μM BN reduced the binding to approximately 800 cpm. BN inhibited binding in a dose dependent manner. The concentration of BN required to inhibit half of the specific (^{125}I-Tyr^4)BN binding (IC_{50}) was 14 nM in the control plate. In the presence or absence of radiation for 1 day, the IC_{50} value was 10 and 8 nM (Table 1). In the presence or absence of radiation for 3 days the IC_{50} values were 8 and 10 ng/ml respectively. These data indicate that radiofrequency radiation has little effect on BN/GRP receptor binding.

Previously it was found that BN/GRP receptors undergo cross talk to EGF receptors in Swiss 3T3 cells. Because BN/GRP stimulates PI turnover the diacylglycerol released may activate protein kinase C. This may cause translocation of PKC from the cytosol to the membrane and phosphorylation of the EGF receptor on the threonine at position 654[20]. Figure 3 shows that BN reduced specific ^{125}I-EGF binding in a dose dependent manner. BN (0.1 nM) had little effect on EGF binding

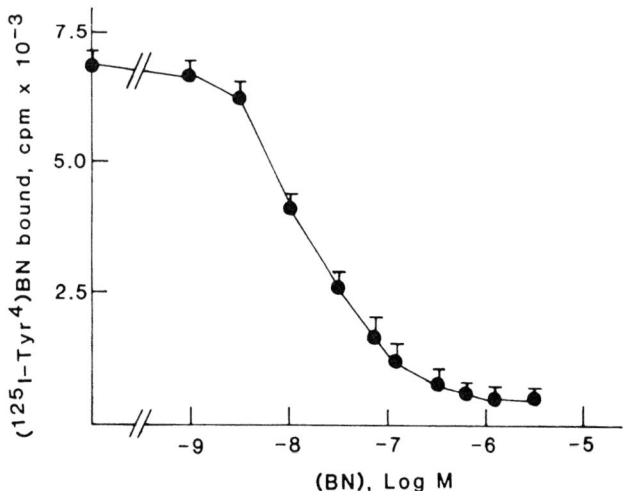

Figure 2. Effect of radiofrequency radiation on BN/GRP receptors. The amount of (^{125}I-Tyr^4)BN bound was determined as a function on increasing doses of BN. The mean value $\pm$ S.E. of 4 determinations is shown.

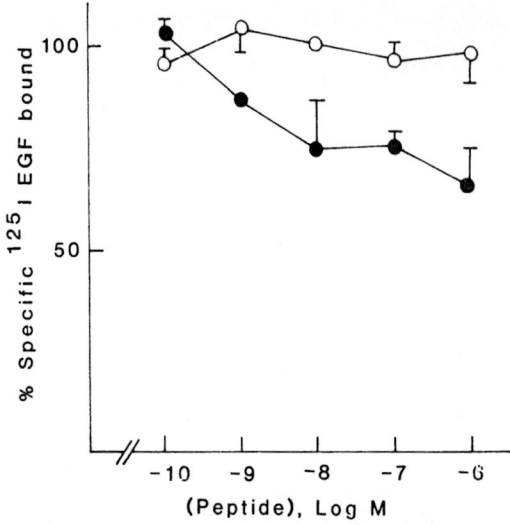

Figure 3. BN/GRP-EGF receptor cross-talk. Specific binding of ^{125}I-EGF was inhibited by BN (●) but not (FA)BN$^{6-13}$ME (o) in a dose dependent manner. The mean value ± S.E. of 4 determinations is shown.

whereas 1 μM BN decreased binding by approximately 35%. The effect was dose dependent in that 1 and 10 nM BN decreased binding by approximately 10 and 20% respectively. The BN/GRP receptor antagonist (D-F-Phe^6, D-Ala^{11})BN$^{6-13}$methyl ester ((FA)BN$^{6-13}$ME) had little effect of ^{125}I-EGF binding. These data indicate that BN/GRP receptor agonists but not antagonists affect EGF receptor binding. Pulsed radiofrequency radiation did not affect the ability of BN to inhibit EGF receptor binding.

Previously BN/GRP receptor antagonists were found to inhibit binding of (^{125}I-Tyr^4)BN, the increase in cytosolic Ca^{2+} caused by BN and the increase in ^{3}H-

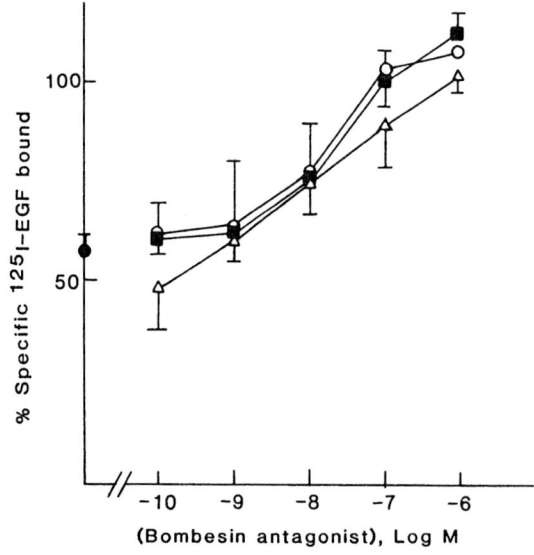

Figure 4. Alteration of receptor cross-talk by BN/GRP receptor antagonist. 10 nM BN reduced ^{125}I-EGF binding to Swiss 3T3 cells (●). The inhibition of EGF receptor binding was reversed by (FA)BN$^{6-13}$ME in a dose dependent manner using control (■) cells and cells treated with (O) and without radiation for 1 day (Δ). The mean value ± S.E. of 4 determinations is indicated.

thymidine uptake caused by BN in Swiss 3T3 cells[21,22]. Because similar data were obtained in SCLC cells, BN/GRP receptor antagonists may be useful in the treatment of SCLC[23,24,25]. Figure 4 shows that 10 nM BN inhibited approximately 40% of the EGF receptor binding and that the inhibition caused by BN is reversed by (FA)BN[6-13]ME in a dose dependent manner. (FA)BN[6-13]ME (0.1 nM) had little effect on the ability of BN to inhibit EGF receptor binding. In contrast, 1 μM (FA)BN[6-13]ME reversed the ability of 10 nM BN to inhibit EGF receptor binding. The effects of (FA)BN[6-13]ME were dose dependent and 18 nM half maximally reversed the effects of BN. These data indicate that when the BN/GRP receptor is blocked by (FA)BN[6-13] ME, BN cannot alter EGF receptor binding. Because BN only inhibits EGF receptor binding at 37 but not 4°C it may indirectly alter the EGF receptor conformation as a result of phosphorylation via kinases resulting in impaired EGF receptor binding. Pulsed radiofrequency radiation had little effect of BN/GRP-EGF receptor cross-talk.

SECOND MESSENGERS

For the cytosolic Ca^{2+} determination, the Swiss 3T3 cells were cultured on fibronectin-treated chambers in DMEM with 10% fetal bovine serum. When the cells were approximately 20% confluent, the serum-supplemented medium was removed and the cells rinsed with buffer (150 mM NaCl, 1 mM $CaCl_2$, 1 mM $MgCl_2$, 10 mM glucose, 5 mM KCl and 20 mM HEPES. NaOH (pH 7.4)) which contained 1% BSA for 5 min. The old buffer was removed and the cells incubated in 1 ml of buffer containing 5 μg/ml of Indo-1 AM for 30 min at 37°C. After loading, the old buffer was removed and the cells washed in incubation buffer. After 5 min the old buffer was removed and new buffer added. The cells were analyzed for cytosolic Ca^{2+} using a Meridian before and after addition of 10 nM BN. The effects of radiation of cytosolic Ca^{2+} were investigated.

Previously it was found that 10 nM BN caused a transient increase in the cytosolic calcium from 150 to 400 nM after 60 sec. The response attenuated and the cytosolic Ca^{2+} returned to basal values after 4 min. The ability of BN to elevate cytosolic Ca^{2+} was investigated. The relative fluorescence intensity varied in individual Swiss 3T3 cells. After addition of 10 nM BN the relative emission ratio increased in approximately half of the cells. The fluorescence intensity was maximal after 60 sec and then slowly decreased after 2.5 min the baseline level was reached. These data indicate the BN transiently increases the cytosolic Ca^{2+}. Similar data were obtained when the Swiss 3T3 cells were treated with radiofrequency radiation for 1 day. These data indicate that pulsed radiofrequency radiation has little effect on the ability of BN to mobilize Ca^{2+}. Similar data were obtained for EGF.

Table 2. Effect of radiofrequency radiation on cytosolic calcium

Condition	BN-induced increase in emission ratio
None	0.60 + 0.05
-hv	0.70 + 0.10
+hv	0.59 + 0.08

The basal emission ratio was determined and subtracted from the maximal emission ratio increase caused by 10 nM BN in 5 different Swiss 3T3 cells. The mean relative increase + S.E. of 3 experiments each performed in duplicate is indicated.

Table 3. Effect of pulsed radiofrequency radiation on ^{3}H-thymidine uptake

Agent added	1 day + hv	1 day - hv	3 days + hv	3 days - hv	7 days + hv	7 days - hv
Control	27982 +6481	34506 +9114	26646 +5394	25999 +4683	0259 +5791	21105 +5206
10% FBS	478832 +28855	519461 +65726	516376 +63664	527699 +39533	561096 +48871	515193 +73318

The mean value $\pm$ S.E. of 10 determinations is indicated.

GROWTH

The ability of radiofrequency radiation to alter ^{3}H-thymidine uptake in Swiss 3T3 cells was investigated. When the Swiss 3T3 cells were confluent, the old media was removed and DMEM containing 0.2% calf serum added. After 24 hr at 37°C, BN and insulin (1 μg/ml) or EGF and ^{3}H-thymidine (1 μCi) added. After 24 hr, the cells were washed three times with cold phosphate buffered saline (PBS), one time with 5% TCA and one time with EtOH/ether (2/1). The cells were solubilized with 0.2 N NaOH, followed by 0.2 N HCl. The samples were placed in a vial, 10 ml of Aquasol added and counted in a ß-counter. Each experiment was repeated in quadruplicate and the mean value $\pm$ S.D. calculated.

Because most Swiss 3T3 cells become arrested in the G_0 growth phase after contact inhibition, they have a very low basal uptake of ^{3}H-thymidine into newly synthesized DNA. The cells can be stimulated to enter S-phase, however, and dramatically increase the uptake of radiolabeled nucleotide into DNA. Fetal bovine serum (FBS) increased ^{3}H-thymidine uptake in Swiss 3T3 cells approximately 20-fold in the presence or absence of radiation for 1, 3 or 7 days (Table 3). Similarly, EGF (100 ng/ml) or BN (10 nM) plus insulin (1 μg/ml) increased ^{3}H-thymidine uptake approximately 10-fold. Radiofrequency radiation did not alter the ability of EGF, BN or FBS to increase proliferation. These data indicate that pulsed radiofrequency radiation has little effects on the growth of Swiss 3T3 cells.

SUMMARY

There is great concern about the effects of EMF radiation stimulating cancer formation. When high intensities of such weak radiation are delivered in a pulsed manner, this can result in thermal effects from increased heating. Under the experimental conditions used here the temperature of the media changed by less than 0.1°C upon exposure to pulsed radiofrequency radiation. Hence the viability of cells treated with radiation was unchanged from the control.

Here the effects of EMF radiation on Swiss 3T3 proteins (EGF and BN/GRP receptors) were investigated. The EGF receptor is a tyrosine kinase which crosses the membrane one time. It can phosphorylate enzymes such as phospholipase Cτ stimulating PI turnover elevating cytosolic Calcium^{2+}[26]. In contrast the BN/GRP receptor contains 384 amino acid residues and spans the plasma membrane 7 times[27,28]. It interacts with G-proteins stimulating PI turnover. The IP produced elevates cytosolic Ca^{2+} and the DAG stimulates PKC. PKC can then phosphorylate the EGF receptor decreasing EGF binding affinity.

EMF radiation had little effect on EGF and BN/GRP receptor binding. Pulsed radiofrequency radiation had little effect on cytosolic Ca^{2+} or the ability of BN to release Ca^{2+} from intracellular pools. Previously EMF radiation decreased calcium flux at low (6-16 hertz) but increased calcium flux at high frequencies (50-450 megahertz)[18]. Also, in HL-60 cells exposure to static magnetic or radiofrequency fields did not significantly alter cytosolic Ca^{2+} but exposure to combined fields increased the cytosolic Ca^{2+} from 120-150 nM. Thus the effects of EMF radiation are dependent upon the conditions and cells utilized.

BN agonists cause receptor cross-talk in Swiss 3T3 cells resulting impaired EGF binding. Here the effects of BN were reversed by BN/GRP receptor antagonists such as $(FA)BN^{6-13}ME$. The BN IGRP receptor may stimulate after agonist activation PKC and the effects if reversed by $(FA)BN^{6-13}ME$. Pulsed radiofrequency radiation had little effect on BN/GRP receptor cross talk. For the EGF receptor tyrosine kinase activation is essential for growth. Because pulsed radiofrequency radiation had little effect of ^{3}H-thymidine uptake, EMF radiation may not affect tyrosine kinase activity.

Thus EMF radiation does not appear to affect BN/GRP or EGF receptors under conditions whereby thermal changes are minimal. It remains to be determined if EMF radiation affects oncogene expression and/or carcinogenesis.

ACKNOWLEDGEMENTS

This manuscript is dedicated to the memory of Dr. E. Albert (deceased). The authors thank Dr. V. Hu for assistance on the calcium measurements.

REFERENCES

1. Y. Yarden and A. Ullrich, Growth factor receptor tyrosine kinases. Ann. Rev. Biochem. 57:443 (1988).
2. T. Yamamoto, S. Ikawa, T. Akiyama, K. Semba, N. Nomura, N.Miyajima, T. Saito and K. Toyoshima. Similarity of protein encoded by the human c-erb-B-2 gene to epidermal growth factor receptor. Nature 319:230 (1986).
3. A. Ullrich and J. Schlessinger. Signal transduction by receptors with tyrosine kinase activity. Cell 61:203 (1990).
4. P. Gosh-Dostidar and C.F. Fox, Epidermal growth factor and epidermal growth factor receptor dependent phosphorylation of a Mr = 34,000 protein substrate for pp60^{src}. J. Biol. Chem. 258:2041 (1983).
5. H. Ushiro and S. Cohen, Identification of phosphotyrosine as a product of epidermal growth factor activated protein kinase in A-431 cell membranes. J. Biol. Chem. 255:8363(1980).
6. J. Schlessinger. Signal transduction by allosteric receptor oligomerization. Trends Biochem. Sci 13:443 (1988).
7. A.M. Honneger, D. Szapary, A. Schmidt, R. Lyall and E. VanObberghen. A mutant epidermal growth factor receptor with defective protein tyrosine kinase is unable to stimulate proto-oncogene expression and DNA synthesis. Mol. Cell Biol., 7:4568 (1987).
8. R. Kris, R. Hazan, J. Villines, T. Moody and J. Schlessinger. Identification of the bombesin receptor on murine and human cells by cross-linking experiments. J. Biol. Chem. 262:11215 (1987).
9. I. Zachary and E. Rozengurt. Identification of a receptor for peptides of the bombesin family in Swiss 3T3 cells by affinity cross-linking. J. Biol. Chem. 262:3947 (1987).
10. I. Zachary and E. Rozengurt. High affinity receptors for peptides of the bombesin family in Swiss 3T3 cells. Proc. Natl. Acad. Sci. USA 82:7616 (1985).
11. S.A. Mendoza, J.A. Schneider, A. Lopez-Rivas, J. Sinnet-Smith and E. Rozengurt. Early events elicited by bombesin and structurally related peptides in quiescent Swiss 3T3 cells. II. Changes in Na^+ and Ca^{2+} fluxes, Na^+,K^+ pump activity and intracellular pH. J. Cell Biology 102:2223 (1986).

12. I. Zachary, J. W. Sinnett-Smith and E. Rozengurt. Early events elicited by bombesin and structurally related peptides in quiescent swiss 3T3 cells. I. Activation of protein kinase C and inhibition of epidermal growth factor binding. J. Cell Biology 102:2211 (1986).

13. A. Palumbo, P. Rossino and P. Comoglio. Stimulation of c-fos expression by bombesin in Swiss 3T3 cells. Expt. Cell Res. 167:276-280; 1986.

14. E. Rozengurt and J. Sinett-Smith, J. Bombesin stimulation of DNA synthesis and cell division in cultures of Swiss 3T3 cells. Proc. Natl. Acad. Sci. U.S.A. 80:2936 (1983).

15. J.A. Reese, R.F. Jostes and M.E. Frazier. Exposure of mammalian cells to 60-Hz magnetic or electric fields: Analysis for DNA single strand breaks. Bioelectromagnetics 9:237 (1988).

16. R.B. Goldberg and W.A. Creasey. A review of cancer induction by extremely low frequency electromagnetic fields: Is there a plausible mechanism? Med. Hypotheses 35:265 (1991).

17. J.J.L. Carlson, F.S. Prato, D.J. Drost, L.D. Diesbourg and S.J. Dixon. Time varying magnetic fields increase cytosolic free Ca^{2+} in HL-60 cells. Am. J. Physiology 259:687 (1990)

18. P. Greengard Effects of electromagnetic radiation on calcium function in the brain. in USAF radiofrequency radiation bioeffects research program A review J.C. Mitchell (Ed) Brooks Air Force Base, Texas, pp264 (1981).

19. E. Albert, C. Blackman, C. and F. Slaby. Calcium dependent secretory protein release and calcium efflux during RF irradiation of rat pancreatic tissue slices. In Proceedings of the URSI conference on electromagnetic waves and biology, Paris, pp325 (1980).

20. C.R. Lin, W.S. Chen, C.W. Lazar, C.D. Carpenter, G.N. Gill, R.M. Evans and M.G. Rosenfeld. Protein kinase C phosphorylation at Thr 654 of the unoccupied EGF receptor and EGF binding regulate functional receptor loss by independent mechanisms. Cell 44:839 (1986).

21. D.H. Coy, P. Heinz-Erian, N.Y. Jiang,. Y. Sasak, J. Taylor, J.P. Moreau, W.T. Wolfrey, J.D. Gardner and R.T. Jensen. Probing the peptide backbone function in bombesin: A reduced peptide bond analogue with potent and specific receptor antagonist activity. J. Biol. Chem. 5056 (1988).

22. D.H. Coy, J.E. Taylor, N.Y. Jiang, S.H. Kim, L.H. Wang, S.C. Huang, J.P. Moreau, J.D. Gardner and R.T. Jensen. Short chain pseudopeptide bombesin receptor antagonists with enhanced binding affinities for pancreatic acinar and Swiss 3T3 cells display strong antimitotic activity. J. Biol. Chem. 264:14691 (1989).

23. S. Mahmoud, J. Staley, J. Taylor, A. Bogden, J.P. Moreau, D. Coy, I. Avis, F. Cuttitta, J. Mulshine and T.W. Moody. (Psi[13,14])bombesin analogues inhibit the growth of small cell lung cancer in vitro and in vivo. Cancer Res. 51:1798(1991).

24. J. Staley, D. Coy, J.E. Taylor, S. Kim and T.W. Moody.
     (Des-met[14])bombesin analogues function as small cell lung cancer bombesin receptor antagonists. Peptides 12:145(1991).

25. T.P. Davis, S. Crowell, J.Taylor, D.L. Clark, D. Coy, J. Staley and T.W. Moody. Metabolic stability and tumor inhibition of bombesin/FRP receptor antagonists. Peptides 13:401 (1992).

26. A.M. Martinelli, L.M. Neri, R.S. Gilmour, P.J. Barker, N.S. Huckisson, F.A. Manzoli and L. Cocco. Temporal changes in intracellular distribution of pkC in Swiss 3T3 cells during mitogenic stimulation with insulin like growth factor I and bombesin: Translocation to the nucleus follows rapid changes in nuclear polyphosphoinositides. Biochem. Biophys. Res. Commun. 177:480 (1991).

27. E.R. Spindel, E. Giladi, T.P. Brehm, R.H. Goodman, and T.P. Segerson. Cloning and functional characterization of a cDNA encoding the murine fibroblast bombesin/GRP receptor. Mol. Endocrinol. 4:1956(1990).

28. J.F. Battey, J.M. Way, M.H. Corjay, H. Shapira, K. Kusano K., R. Harkins, J.M. Wu, T. Slattery, E. Mann and R. Feldman. Molecular analysis of the murine GRP-preferring bombesin receptor. Proc. Natl. Acad. Sci. USA 88:395(1991).

STRUCTURE-FUNCTION ANALYSIS OF FIBROBLAST GROWTH FACTOR-1 (ACIDIC FIBROBLAST GROWTH FACTOR)

Wilson H. Burgess

Department of Molecular Biology
Holland Laboratory
Rockville, MD 20855

INTRODUCTION

Polypeptide growth factors are modulators of cellular proliferation and differentiation *in vitro* and *in vivo*. These functions are mediated, in part, by interaction of the growth factors with relatively high-affinity cell-surface receptors and subsequent alterations in gene expression within responsive cells. The complete cascades of signals initiated by growth factor occupancy of cell-surface receptors responsible for the mitogenic or differentiating responses are not known but are likely to be different for the various growth factor families. The fibroblast growth factor (FGF) family consists of at least seven structurally related polypeptides.[1] Two of these proteins FGF-1 and FGF-2 have been characterized under many different names but are often referred to as acidic FGF and basic FGF or heparin-binding growth factor-1 and heparin-binding growth factor-2. The focus of this report is on the structural basis for the mechanisms of action of FGF-1.

Relatively little is presently known about the precise mechanisms by which FGF-1 elicits various biological responses. These types of studies are complicated by the fact that this growth factor is capable of potentiating a broad range of biological activities both *in vitro* and *in vivo*. These activities will be described in more detail below and include mitogenesis, chemotaxis, neuronal survival, neurite extension, inhibition of terminal differentiation, induction of mesoderm formation, angiogenesis and wound repair. Clearly, the ability to control selected activities of FGF-1 would be of value for understanding the pathways involved in these biological responses and would assist efforts to utilize this protein (and perhaps other members of the FGF family) therapeutically. The scope of this chapter is to provide a description of i) the various structural features of FGF-1 from different species; ii) the functions associated with the protein and what is known about its mechanisms of action, and iii) recent progress in structure-function studies.

STRUCTURES

The "complete" primary structure of bovine FGF-1 was first described as a 140 residue protein.[2] Later studies showed that this isolated protein was an amino-terminal truncation of the full-length protein.[3,4] These studies demonstrated that the full-length forms of both human and bovine FGF-1 consisted of 154 amino acids. In addition, it was established that both proteins contained blocked amino termini as a result of acetylation of their amino terminal alanine residues. The established sequences of full-length human and bovine FGF-1 are in agreement with the protein sequences derived from the human and bovine cDNAs.[5,6] The cDNA sequences of rat and hamster FGF-1 are also available and the derived protein sequences demonstrate that protein obtained from these species also consist of 154 amino acid residues. We have established that chicken FGF-1 also consists of 154 amino acids with acetylation of the amino terminal alanine residue.[7] The established structures and those derived from cDNAs[8,9] are shown in figure 1. There are several features of these structures worth noting. First, it is clear from the figure that the primary structure of FGF-1 is highly conserved (90% or more sequence identity) among mammalian and nonmammalian vertebrates. Included in these conserved residues are cysteine residues at positions 30 and 97 in all species studied to date. In fact, the relative positions of these two cysteines are also conserved in the other six members of the FGF family. A third cysteine is found in most of the FGF-1 sequences either at position 61 or 131. The exception is the avian FGF-1 which contains only the two conserved cysteines. Second, the highest degree of sequence variation between species occurs between residues 130 and 133 where the proteins that contain the cysteine at position 131 have sequences completely different from those that do not, with the exception of lysine 132. The potential significance of this observation is discussed in more detail later in this chapter. Third, it should be noted that several truncated forms of FGF-1 that retain mitogenic activity have been isolated. One was described above and begins with residue 15, another begins with residue 21.[4] Although these amino terminal truncations have relatively little effect on mitogenic activity the amino-terminal 20 residues are as highly conserved amongst the different species as the remainder of the protein. It should be noted that the list of potential biological activities associated with FGF-1 continues to grow and significant differences in the activities of the full-length and truncated forms of the protein may yet be established. These potential differences include differences in the relative stability of the proteins.[10] In summary, the highly conserved sequences imply that multiple functional domains exist within the primary structure of FGF-1.

FUNCTIONS

A major contribution to our current understanding of the fibroblast growth factor family of polypeptides was the observation by Shing et al.[11] that an endothelial cell growth factor (later identified as basic FGF or FGF-2) could be purified to homogeneity using heparin-Sepharose affinity based chromatography. First, this finding provided the basis for the development of rapid and efficient purification schemes for both FGF-1 and FGF-2,[11-13] which in turn provided the quantities of material required for detailed characterization of these polypeptides. Subsequent observations that other structurally related proteins such as the products of the hst/K-fgf and FGF-5 oncogenes and KGF also bind with high affinity to immobilized heparin provided part of the basis for the designation of the heparin-binding or FGF family of polypeptide mitogens.[1,14] Second, the observation provided the impetus for further investigations into the role of heparin or heparin-sulfate proteoglycans in the function of the FGFs. For example, heparin has been shown to potentiate the biological activities of FGF-1 but not FGF-2 (reviewed in 14 and 15). This potentiation is likely to be due in part to the fact that heparin can increase the apparent affinity of FGF-1 for cell surface receptors.[16,17] Heparin also has been shown to protect FGF-1 and FGF-2 from proteolytic degradation[18-20] and from heat and acid inactivation.[18,21] Thus, heparin or heparan sulfate proteoglycans could stabilize extracellular growth factor in the basement membrane or at the cell surface. In addition,

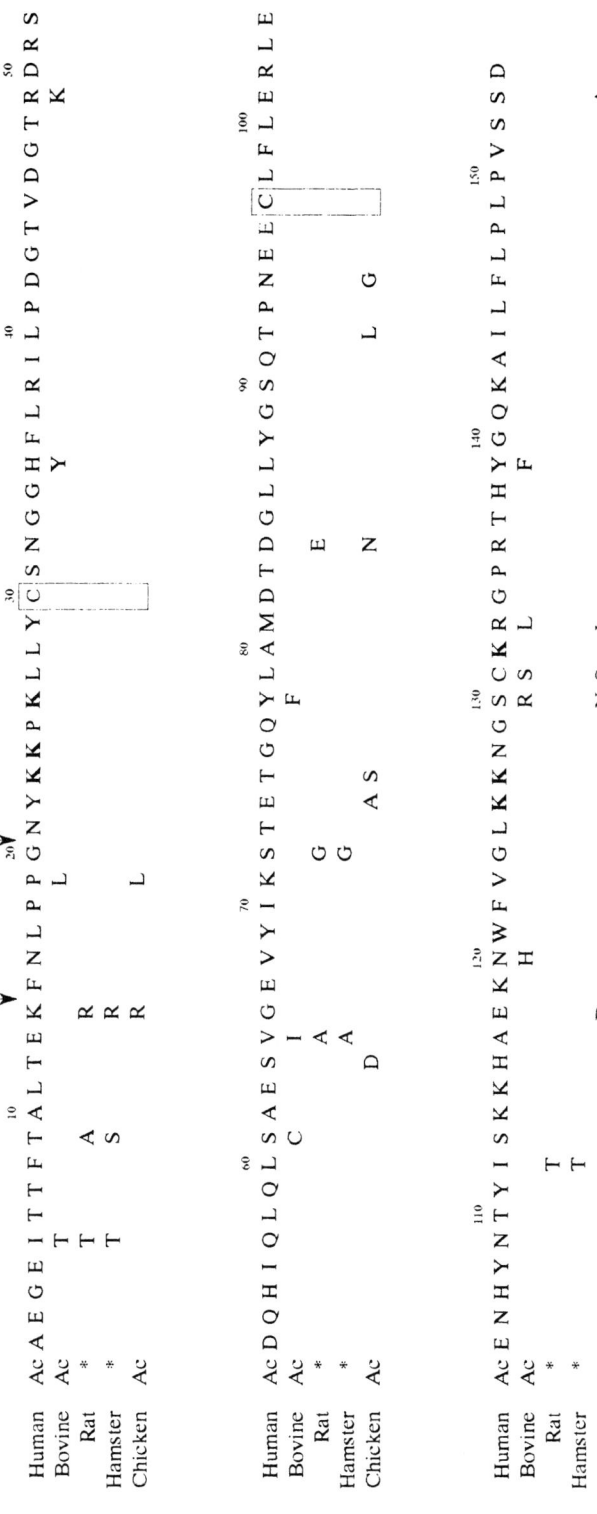

Figure 1. Amino acid sequences of FGF-1 of different species. The complete sequence of the human protein is shown. Only the differences in sequence of other species are indicated. Ac indicates those sequences for which the identity of the amino-terminal blocking group (acetyl) has been established. The bold K indicates those lysine residues that have been studied by site-directed mutagenesis. The boxed C indicates the positions of the two cysteine residues that are conserved in all FGF-1s and other members of the FGF family. The vertical arrows indicate the positions of amino-terminal cleavage products that occur during purification and do not affect mitogenic activity.

recent studies indicate that heparan sulfate proteoglycans are essential for high-affinity binding of FGFs to their tyrosine kinase receptors (see reference 22 for a review).

A family of high-affinity FGF receptors with intrinsic protein tyrosine kinase activities has also been identified.[22] Five distinct receptor gene products have been cloned. In addition, alternate splicing leads to the generation of multiple forms of several of the distinct receptors. This alternate splicing has been shown to affect the ligand binding specificities of the receptors.[22,23] All of the five FGF receptor gene products fall into the class IV receptor tyrosine kinases.[22,24] Common features include the presence of as many as three Ig domains in the extracellular domain, and a 14 residue insert in the cytoplasmic tyrosine kinase domain. The mechanism of action of growth factor binding to receptor tyrosine kinases is not understood but may involve ligand binding followed by receptor oligomerization resulting in trans autophosphorylation of receptor cytoplasmic tyrosines. This in turn allows association of substrates such as phospholipase C-γ, GTPase activating protein, p21^{ras} and others with the activated receptor kinase. It is thought that these receptor mediated events are sufficient to initiate the subsequent mitogenic cascade.[22,24] Although several of the known cellular responses that follow FGF treatment of quiescent cells are consistent with this concept, site-directed mutagenesis studies indicate that additional pathways may be involved (see reference 22 and below).

Addition of FGF-1 to responsive cells clearly leads to an increase in the phosphotyrosine content of FGF receptors and other cellular proteins.[22] As with PDGF and EGF, one of the substrates known to be phosphorylated by activated FGF receptors is phospholipase C-γ (PLC-γ).[25,26] Whereas the role of stimulated phosphoinositide turnover in the signal transduction pathways affected by FGFs remains unclear,[27-29] recent studies demonstrate that a point mutation in one of the FGF receptors can eliminate the FGF-induced phosphorylation and activation of PLC-γ without affecting its ability to transmit a mitogenic signal.[30] The significance of this observation to recent site-directed mutagenesis studies of FGF-1 will be discussed in the next section. Other changes in cellular metabolism in response to FGF have been reported. Both FGF-1 and FGF-2 induce rapid and transient expression of the mRNAs encoding the protooncogene proteins c-*myc*, c-*fos* and c-*jun*.[31-33] In addition, increases in actin mRNA levels,[34] ribosomal gene transcription,[35] PDGF-A chain expression,[33] and synthesis of a number of unidentified cellular proteins[34] have been reported to occur following the addition of FGFs to responsive cells. Whereas the relationship of these changes in gene expression to the cells response to FGFs remain unknown they do provide additional biochemical markers for FGF structure-function studies. To date, the evidence from our work and that of others indicates that many of these events may be necessary but together, not sufficient to complete the FGF-1 induced mitogenic cascade.[36-38] A major challenge in the coming years is to dissect the cellular responses to FGFs and other growth factors to identify the key components in the signal transduction pathways and eliminate those that may be considered "noise" associated with the particular response under study. Although several approaches are possible, site-directed mutagenesis of the ligands and the ligand receptors holds the most promise to understanding the structural basis for FGF function.

STRUCTURE-FUNCTION STUDIES

Our early studies aimed at understanding the structural basis for the various activities of FGF-1 through the use of synthetic peptides or peptides derived from proteolytic cleavage of native protein as potential agonists or antagonists of specific functions. We were not able to identify any peptide or total enzymatic digest that retained any significant biological activity. These results are consistent with the fact that following quantitative reduction and alkylation of the cysteine residues of human recombinant FGF-1 we are unable to detect receptor-binding activity of the modified protein.[39] It is not known whether the results of these experiments are related directly to the cysteine chemistry of FGF-1 or whether they indicate that non-contiguous domains are essential for receptor-binding activity. We were able to detect heparin-binding activity associated with one of the trypsin-derived fragments of FGF-1. A gel overlay procedure was utilized to study the interaction of heparin-binding proteins with ^{125}I-labeled heparin.[40] One peptide corresponding to residues 49 to 71 of human FGF-1 was able to compete for binding of derivatized heparin to full-length FGF-1. This region of FGF-1 is homologous to one of the two regions of heparin-binding activity found in FGF-2.[41] It remains to be established whether peptides that compete with intact FGF-1 or FGF receptors[22] for heparin binding will be useful as antagonists of FGF function. Heparin has been shown to promote the formation of dimers of FGF-2 which in turn may be involved in receptor dimerization and subsequent activation.[42] Subsequent efforts to dissociate the various functions of FGF-1 at the structural level have focused on site-directed mutagenesis of the protein.

The data summarized below focus on two regions of FGF-1 that contain clusters of lysine residues. These two clusters are highlighted in figure 1. The studies on the amino terminal region resulted from a report of mutagenesis of FGF-1 that generated amino-terminal deletions beyond the amino terminus of the des 1-20 form of FGF-1 which is known to be mitogenic.[4,13] The results of this study indicated that the deletion of an additional seven residues reduced the mitogenic activities of the protein significantly without affecting its apparent affinity for immobilized heparin.[38] In addition, it was reported that mitogenic activity was not observed at concentrations of the mutant growth factor that were sufficient to stimulate intracellular kinase activity and induce c-*fos* expression. In contrast to studies summarized below it was not established directly whether tyrosine kinase activity of FGF receptors was activated by this mutant. The deletion of these seven additional residues from the smallest active form of FGF-1 removes a sequence (N-Y-K-K) that fits a consensus sequence for nuclear translocation.[43] We have demonstrated nuclear localization of FGF-1 in NIH 3T3 cells.[37] However, the significance of nuclear localization of FGF-1 or other members of the FGF family has not been established. It is of interest that construction of a chimera protein consisting of the nuclear translocation sequence of yeast histone 2B adjacent to the des 1-27 form of FGF-1 resulted in restoration of full specific mitogenic activity.[38] Although this observation is consistent with a role for nuclear translocation in FGF-1 induced mitogenesis, it is also possible that a cluster of basic amino acids in the amino-terminal domain of FGF-1 is important to its mitogenic capacity. Recently, a consensus for a structural basis for nuclear translocation has been established. Chelsky et al.[43] examined the ability of 18 synthetic peptides to target chicken serum albumin to the nucleus following microinjection into HeLa cells. They derived a consensus sequence where all the peptides that induced full or partial nuclear localization fit the sequence K-K/R-X-K/R where X could be K, R, P, V or A but not N. A search of the protein data base reveals that at best, this sequence can be described as capable but not sufficient for targeting proteins to the nucleus. We utilized site-directed mutagenesis to generate three individual point mutations that change each of the three lysine residues in this region of FGF-1 to a glycine residue.[37] Each of these mutations are predicted to disrupt the consensus sequence for nuclear translocation found in FGF-1. The subcellular distribution of these mutants has not yet been compared to that of wild-type FGF-1. However, their heparin-binding, receptor-binding, mitogenic and neurite-promoting activities are similar to those of the wild-type protein. These results favor the notion that a rare cluster of amino acids in the amino terminal region of an acidic protein may in fact be important to the mitogenic activity of FGF-1. Regardless of mechanism, this deletion mutant study provides evidence that binding of FGF-1 to high-affinity cell surface receptors is not sufficient to initiate the mitogenic cascade. Additional and more convincing evidence is provided by site-directed mutagenesis studies of lysine residues in the carboxy-terminal region of FGF-1 which are summarized below.

Harper and Lobb reported that limited reductive methylation of bovine FGF-1 with formaldehyde and cyanoborohydride resulted in stoichiometric methylation of lysine 132.[44] They reported 90% modification of this residue and showed the modified protein had a reduced apparent affinity for immobilized heparin, a reduction in its ability to stimulate DNA synthesis in NIH 3T3 cells and a similar reduction in its ability to compete with labeled ligand in a radioreceptor assay. A lysine residue is conserved at this position in both FGF-1 and FGF-2 from all species characterized to date. Although these data indicate a critical role for lysine 132 in FGF-1 function, chemical modification data is limited by the fact that non-specific side reactions may introduce modifications of the target protein for which specific assays are either not reported or not possible. These caveats prompted us to address the role of lysine 132 in FGF-1 function by site-directed mutagenesis of this residue in the human sequence to glutamic acid.

We demonstrated that the lysine→glutamic acid mutation at position 132 of human FGF-1 does reduce the apparent affinity of the protein for immobilized heparin (elutes at .5M NaCl vs. 1M NaCl for wild-type).[36] In addition, the mitogenic activity of this mutant for a variety of cell types including NIH 3T3 fibroblasts, Balb MK keritinocytes and human endothelial cells is drastically reduced when compared to wild-type protein. These results are consistent with those obtained by chemical modification studies although the magnitude of the reduction in mitogenic activity of the mutant protein is far greater than for the modified protein (>100 fold). Studies on the binding of the mutant protein to high-affinity cell surface receptors yielded unexpected results. The 132 mutant is able to compete equally well with wild-type protein for receptor binding.[36] In addition, the binding is functional in that the tyrosine kinase activity of the FGF receptor is induced at similar concentration of wild-type and 132 mutant protein. Further, the same pattern of phosphotyrosine containing proteins including the known FGF receptor substrate PLC-γ, are seen in cells treated with either the wild-type or 132 mutant FGF-1.[36]

Another molecular marker associated with growth factor treatment of quiescent cells is the rapid and transient expression of a family of protooncogenes. We have shown that the 132 mutant of FGF-1 promotes the transcription of c-*fos*, c-*jun* and c-*myc* over the same time course and with the same dose response as the wild-type protein.[36,37] Thus, the 132 mutant is able to elicit all of the known early responses associated with occupancy of high-affinity tyrosine kinase receptors as is the wild-type FGF-1 but is not able to complete the mitogenic signal cascade. These results indicate that the concept that receptor binding by growth factors leads to aggregation, transphosphorylation and activation of the receptor and that this activation is sufficient to initiate the mitogenic cascade is probably too simplistic to account for the mitogenic signaling associated with FGF-1. The studies of the lysine 132 mutant FGF-1 demonstrate the concept discussed earlier that in order to fully understand FGF-1 signal transduction, the cellular responses that are necessary must be distinguished from those that are not, and similarly those that are necessary but not sufficient should be identified. The studies summarized above indicate that at best, receptor-binding, tyrosine kinase activation, and induction of protooncogene expression should be considered as necessary but not, as a group, sufficient to sustain the mitogenic effect of FGF-1. Studies summarized earlier[30] demonstrate that tyrosine phosphorylation of PLC-γ falls into the unnecessary category of FGF-1 induced cellular responses with respect to mitogenesis. Activation of this enzyme may be important to other FGF-1 induced activities. Together these results indicate that a key component of the mitogenic signal transduction cascade for FGF-1 has not yet been identified and that once established should prove to be unaffected by the lysine 132 mutant. It is possible that other cell surface receptors exist for FGF-1 that along with the tyrosine-kinase receptors are important to mitogenic signaling by the protein. If these receptors or binding proteins are proteoglycans, then the reduced apparent affinity for heparin characteristic of the lysine 132 mutant could be responsible for its lack of mitogenic activity. It should be noted that such heparan-sulfate proteoglycans would be distinct from those reported to be necessary for high-affinity receptor binding.[45,46] in that the lysine 132 mutant is fully capable of binding and activating the tyrosine-kinase FGF receptors.[36,37] Alternatively, interactions with intracellular targets following receptor-mediated internalization of the growth factor may be important to completion of the mitogenic signal.

The proliferative deficiencies of the lysine 132 FGF-1 mutant can also be observed following transfection and overexpression of wild-type or mutant cDNAs in NIH 3T3 cells.[36,37] Clones that express relatively high and equal levels of wild-type or lysine 132 mutant FGF-1 as judged by Western blotting have been analyzed. The wild-type transfectants rapidly assume a transformed phenotype, exhibit loss of contact inhibition of cell proliferation and will in fact form colonies in soft agar. In contrast, lysine 132 mutant transfected cells exhibit growth properties very similar to those of untransfected 3T3 cells. We have not been able to detect either protein in the conditioned media of the appropriate cells. In addition, we can detect phosphotyrosine-containing proteins in the wild-type transfectants that are not seen in normal or lysine 132 mutant transfected 3T3 cells. These phosphotyrosine containing proteins are distinct from those that are phosphorylated in response to exogenously added growth factor.[47] These results indicate that intracellular tyrosine kinases are present in NIH 3T3 cells and they can be activated by the wild-type but not the lysine 132 mutant FGF-1. Whether these putative kinases play a role in FGF-1 signal transduction has not been established.

Additional evidence for the existence of multiple FGF-1 signal transduction cascades is provided from studies of the differentiation activities of the protein. A role for FGFs and FGF receptors in the formation of ventral mesoderm in amphibian embryogenesis has been established.[48,49] It is of interest that hydrolysis of phosphatidylinositol bisphosphate has been shown to be necessary for the development of ventral mesoderm.[50] This observation provides a potential link between FGF induced activation of PLC-γ and the differentiation related functions of the growth factor. We have examined the ability of the lysine 132 mutant FGF-1 which is capable of activating PLC-γ[36] to induce mesoderm formation in animal cap explants of *Xenopus* embryos.[51] The results of these experiments demonstrate that this mutant FGF-1 is as potent as the wild-type protein in its ability to promote mesoderm formation. This is the only example to date where the mitogenic and differentiation related activities of FGF-1 have been dissociated at the structural level. The results of the studies with the lysine 132 mutant of FGF-1 demonstrate that site-directed mutagenesis can be used to dissect the many functions associated with this growth factor.

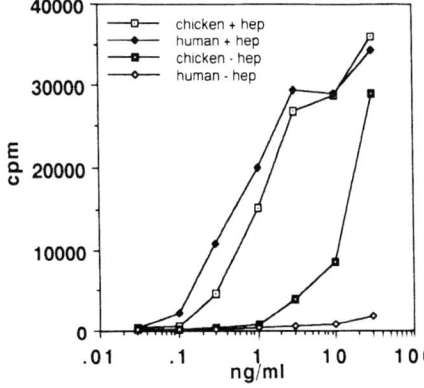

Figure 2. Stimulation of DNA synthesis in Balb MK cells by wild-type human and chicken FGF-1. The assays were performed with the indicated concentrations of FGF-1 in either the presence or absence of 5 units/ml heparin. Values represent the mean of triplicate cultures.

There is some evidence that the mitogenic deficiencies of the lysine 132 mutant may be due in part to events secondary to the single amino acid change. It is known that wild-type human FGF-1 is capable of binding to and activating FGF receptors and promoting protooncogene transcription in the absence of added heparin.[37] However, the human wild-type protein exhibits mitogenic deficiencies in the absence of heparin similar to those displayed by the lysine 132 mutant in its absence (figure 2 and reference 36). The cysteine content of FGF-1s isolated from different species was summarized above. It is of interest that the mitogenic activities of avian FGF-1 and bovine FGF-1 which lack a cysteine residue at position 131 are relatively independent of the presence of exogenously added heparin (figure 2 and reference 52). The conserved cysteines in FGF-1 do not appear to be involved in the formation of an intra-chain disulfide bond.[53] It is possible that the basic cluster surrounding cystine 131 in the human protein plays a role in protecting the growth factor from modification in the presence of heparin. The modification may occur more readily in the lysine 132 mutant. It will be of interest to determine the functional properties of forms of FGF-1 that lack cysteine at position 131 but contain the lysine 132 mutation.

It was noted above that site-directed mutagenesis of any of the three lysine residues at positions 23, 24 and 26 of FGF-1 was without effect on the heparin-binding or mitogenic activities of the protein. Thus, individual residues in this amino-terminal cluster of basic residues do not appear to be critical to FGF-1 function. The region surrounding lysine 132 constitutes another rare cluster of basic residues in FGF-1. We have begun an analysis of the functional properties of mutants lacking the lysine residues found at positions 126 and 127 in the wild-type protein. These residues are conserved in all species of FGF-1 examined to date.

Preliminary studies of these mutants reveal several interesting features. First, mutagenesis of lysine 127 to a glycine residue decreases the apparent affinity of this mutant for heparin-Sepharose to the same extent as the change of lysine 132. Similarly, we are not able to detect significant mitogenic activity associated with the protein. In contrast, mutagenesis of the adjacent lysine 126 to a glycine residue has relatively little effect on the apparent affinity of the mutant FGF-1 for immobilized heparin. This lysine 126 mutant also exhibits some interesting functional properties. As observed for the lysine 127 and 132 mutants, the lysine 126 mutant shows no significant mitogenic activity

Table 1

Ligand	Heparin Binding	Receptor Binding	Mitogenic Potency	Mesoderm Induction
WT	+ + + +	+ + + +	+ + + +	+ + + +
$K_{23} \rightarrow G$	+ + + +	+ + + +	+ + + +	N.D.
$K_{24} \rightarrow G$	+ + + +	+ + + +	+ + + +	N.D.
$K_{26} \rightarrow G$	+ + + +	+ + + +	+ + + +	N.D.
$K_{126} \rightarrow G$	+ + +	N.D.	-/+ + +	+ + + +
$K_{127} \rightarrow G$	+	N.D.	-	N.D.
$K_{132} \rightarrow G$	+	N.D.	-	N.D.
$K_{132} \rightarrow E$	+	+ + + +	-	+ + + +

The relative potencies of wild-type and various mutants of human FGF-1. The mutations and their position in the protein are indicated. K (lysine); G (glycine); E (glutamic acid); N.D. (not determined); -/+ + + (negative for some cell types, positive for others).

for Balb MK keritinocytes. Unlike the 127 and 132 mutants it displays significant mitogenic activity for NIH 3T3 cells. In addition, the 126 mutant is also a potent inducer of mesoderm formation. Further studies are necessary to determine the importance of these residues to the selectivity of FGF-1's action on target cells. A summary of site-directed mutagenesis studies of FGF-1 conducted to date is provided in table 1.

CONCLUSIONS

The results described here provide a basis for future studies of the structural basis for FGF-1 function. There are several conclusions that can be made. First, it may be important to consider functional differences among the amino-terminal truncations of FGF-1 and the full-length protein. The majority of studies related to FGF-1 (i.e. most of those utilizing acidic FGF) have utilized a 14 residue deletion mutant. Second, it appears unlikely that simple linear peptides will be useful as agonists or antagonists of FGF-1 function. A possible exception is that synthetic peptides based on the FGF-1 sequence have been made that have heparin-binding activity. There is, however, no evidence that these peptides have any effect on FGF-1 function. Third, the functional analysis of FGF-1 containing a glutamic acid residue or a glycine residue in place of lysine at position 132 demonstrates that various functions of FGF-1 can be dissociated at the structural level. It is now feasible to generate mutants of FGF-1 that retain certain i.e. mesoderm inducing activity) but not other (i.e. mitogenic activity) biological functions characteristic of the wild-type protein. Fourth, mutagenesis of other lysine residues in the amino-terminal region of FGF-1 indicate that the consensus sequence for nuclear translocation can be disrupted without affecting mitogenic activity. Finally, site-directed mutagenesis of lysine residue 126 indicates that mutant forms of FGF-1 can be generated that retain mitogenic activity for some cell types but not others. The ability to manipulate the various functions of this pleiotropic growth factor at the structural level should aid in the identification of the important components of the cellular responses necessary for the end result (i.e. chemotaxis, proliferation, differentiation, etc.). In addition, these mutants may be useful in identifying the role of post-receptor or intracellular targets of FGF-1.

ACKNOWLEDGEMENTS

This work was supported in part by National Institutes of Health grant HL35762 and a Grant-in-Aid from the American Heart Association (891047) with funds contributed in part by the AHA Maryland Affiliate, Inc. The author thanks Cathryn Wawzinski for her help in the preparation of this manuscript.

REFERENCES

1. A. Baird and M. Klagsbrun, The fibroblast growth factor family an overview, *Annals N.Y. Acad. Sci.* 638:xi (1991).
2. G. Gimenez-Gallego, J. Rodkey, C. Bennet, M. Rios-Candelore, J. DiSalvo and K. Thomas, Brain-derived acidic fibroblast growth factor: complete amino acid sequence and homologies, *Science* 230:1385 (1985).
3. J.W. Crabb, L.G. Armes, S.A. Carr, C.M. Johnson, G.D. Robert, R.S. Bordolic and W.L. McKeehan, Complete primary structure of prostatropin, a prostate epithelial cell growth factor, *Biochemistry* 25:4988 (1986).
4. W.H. Burgess, T. Mehlman, D.R. Marshak, B.A. Fraser and T. Maciag, Structural evidence that endothelial cell growth factor-β is the precursor of both endothelial

cell growth factor-α and acidic fibroblast growth factor, *Proc. Natl. Acad. Sci. USA* 83:7216 (1986).

5. M. Jaye, R. Howk, W.H. Burgess, G.A. Ricca, I.-M. Chiu, M.W. Ravera, S.J. O'Brien, W.S. Modi, T. Maciag and W.N. Drohan, Human endothelial cell growth factor: cloning, nucleotide sequence, and chromosome localization, *Science* 233:541 (1986).

6. C.Y. Halley, Y. Courtois and M. Laurent, Nucleotide sequence of bovine acidic fibroblast growth factor cDNA, *Nucleic Acids Res.* 16:10913 (1988).

7. T. Mehlman and W.H. Burgess (Manuscript in preparation).

8. S.P. Goodrich, G.-C. Yan, K. Bahrenburg and P.-E. Mansson, The nucleotide sequence of rat heparin binding growth factor-1 (HBGF-1), *Nucleic Acids Res.* 17:2867 (1989).

9. J.A. Hall, M.A. Harris, M. Malark, P.-E. Mansson, H. Zhou and S.E. Harris, Characterization of the hamster DDT-1 cell afgf/HBGF-1 gene and cDNA and its modulation by steroids, *J. Cell. Biochem.* 43:17 (1990).

10. W.H. Burgess, Struture-function studies of acidic fibroblast growth factor, *Annals N.Y. Acad. Sci.*, 638:89 (1991).

11. Y. Shing, J. Folkman, R. Sullivan, C. Butterfield, J. Murray and M. Klagsbrun, Heparin affinity: purification of a tumor-derived capillary endothelial cell growth factor, *Science* 223:1296 (1984).

12. R.R. Lobb and J.W. Fett, Purification of two distinct growth factors from bovine neural tissue by heparin affinity chromatography, *Biochemistry* 23:6295 (1984).

13. W.H. Burgess, T. Mehlman, R. Friesel, W. Johnson and T. Maciag, Multiple forms of endothelial cell growth factor, *J. Biol. Chem.* 260:11389 (1985).

14. W.H. Burgess and T. Maciag, The heparin-binding (fibroblast) growth factor family of proteins, *Annu. Rev. Biochem.* 58:575 (1989).

15. M. Klagsbrun, The fibroblast growth factor family: structural and biological properties, *Prog. Growth Factor Res.* 1:207 (1989).

16. A.B. Schreiber, J. Kenney, W.J. Kowalski, R. Friesel, T. Mehlman and T. Maciag, Interaction of endothelial cell growth factor with heparin: characterization by receptor and antibody recognition, *Proc. Natl. Acad. Sci. USA* 82:6138 (1985).

17. J.M. Kaplow, F. Bellot, G. Crumley, C.A. Dionne and M. Jaye, Effect of heparin on the binding affinity of acidic fgf for the cloned human fgf receptors, flg and bek, *Biochem. Biophys. Res. Comm.* 172:107-112 (1990).

18. T.K. Rosengart, W.V. Johnson, R. Friesel, T. Mehlman and T. Maciag, Heparin protects heparin-binding growth factor-1 from proteolytic inactivation *in vitro*, *Biochem. Biophys. Res. Comm.* 152:432 (1988).

19. O. Saksela, D. Moscatelli, A. Sommer and D.B. Rifkin, Endothelial cell-derived heparan sulfate binds basic fibroblast growth factor and protects it from proteolytic degradation, *J. Cell Biol.* 107:743 (1988).

20. R.R. Lobb, Thrombin inactivates acidic fibroblast growth factor but not basic fibroblast growth factor, *Biochemistry* 27:2572 (1988).

21. D. Gospodarowicz and J. Cheng, Heparin protects basic and acidic fgf from inactivation, *J. Cell Physiol.* 128:475 (1986).

22. M. Jaye, J. Schlessinger and C.A. Dionne, Fibroblast growth factor receptor tyrosine kinases: molecular analysis and signal transduction, *Biochem. Biophys. Acta.* 1135:185 (1992).

23. T. Miki, D.P. Bottaro, T.P. Fleming, C.L. Smith, W.H. Burgess, A.M.-L. Chan and S.A. Aaronson, Determination of ligand-binding specificity by alternative splicing: two distinct growth factor receptors encoded by a single gene, *Proc. Natl. Acad. Sci. USA* 89:246 (1992).

24. A. Ullrich and J. Schlessinger, Signal transduction by receptors with tyrosine kinase activity, *Cell* 61:203 (1990).

25. W.H. Burgess, C.A. Dionne, J. Kaplow, R. Mudd, R. Friesel, A. Zilberstein, J. Schlessinger and M. Jaye, Characterization and cDNA cloning of phospholipase C-γ, a major substrate for heparin-binding growth factor-1 (acidic fibroblast growth factor)-activated tyrosine kinase, *Mol. Cell. Biol.* 10:4770 (1990).

26. F. Bellot and M. Jaye (manuscript in preparation).

27. I. Magnaldo, G. L'Allemain, J.C. Chambard, M. Moenner, D. Barritault and J. Pouyssegur, The mitogenic signaling pathway of fibroblast growth factor is not mediated through polyphosphoinositide hydrolysis and protein kinase C activation in hamster fibroblasts, *J. Biol. Chem.* 261:1187 (1986).

28. K. Kaibuchi, T. Tsuda, A. Kikuchi, T. Tanimoto, T. Yamashita and Y. Takai, Possible involvement of protein kinase C and calcium ion in growth factor-induced expression of c-myc oncogene in Swiss 3T3 fibroblasts, *J. Biol. Chem.* 261:1187 (1986).

29. T. Tsuda, K. Kaibuchi, Y. Kawahara, H. Fukuzaki and Y. Takai, Induction of protein kinase C activation and Ca^{2+} mobilization by fibroblast growth factor in Swiss 3T3 cells, *FEBS Lett.* 191:205 (1985).

30. M. Mohammadi, C.A. Dionne, W. Li, N. Li, T. Spivak, A.M. Honegger, M. Jaye and J. Schlessinger, A point mutation at tyr 766 of FGF receptor eliminates FGF induced tyrosine phosphorylation of PLCγ and phosphotidylinositol hydrolysis without affecting mitogenesis, *Nature* (in press 1992).

31. K. Kelly, B.H. Cochran, C.D. Stiles and P. Leder, Cell-specific regulation of the c-*myc* gene by lymphocyte mitogens and platelet-derived growth factor, *Cell* 35:603 (1983).

32. K. Ryder and D. Nathans, Induction of protooncogene c-*jun* by serum growth factors, *Proc. Natl. Acad. Sci. USA* 85:8464 (1988).

33. C.G. Gay and J.A. Winkles, Heparin-binding growth factor-1 stimulation of human endothelial cells induces platelet-derived growth factor A-chain expression, *J. Biol. Chem.* 265:3284 (1990).

34. S.M. Rybak, R.R. Lobb and J.W. Fett, Comparison of the effects of class 1 and class 2 heparin-binding growth factors on protein synthesis and actin mRNA expression in Balb/c-3T3 cells, *J. Cell. Physiol.* 136:312 (1988).

35. G. Bouche, N. Gas, H. Prats, V. Baldin, J.-P. Tauber, J. Tessié and F. Amalric, Basic fibroblast growth factor enters the nucleolus and stimulates the transcription of ribosomal genes in ABAE cells undergoing $G_o \rightarrow G$ transition, *Proc. Natl. Acad. Sci. USA* 84:6770 (1987).

36. W.H. Burgess, A.M. Shaheen, M. Ravera, M. Jaye, P.J. Donohue and J.A. Winkles, Possible dissociation of the heparin-binding and mitogenic activities of heparin-binding (acidic fibroblast) growth factor-1 from its receptor-binding activities by site-directed mutagenesis of a single lysine residue, *J. Cell Biol.* 111:2129 (1990).

37. W.H. Burgess, A.M. Shaheen, B. Hampton, P.J. Donohue and J.A. Winkles, Structure-function studies of heparin-binding (acidic fibroblast) growth factor-1 using site-directed mutagenesis, *J. Cell. Biochem.* 45:131 (1991).

38. T. Imamura, K. Engleka, Zhan, Y. Tokita, R. Forough, D. Roeder, A. Jackson, J.A.M. Maier, T. Hla and T. Maciag, Recovery of mitogenic activity of a growth factor mutant with a nuclear translocation sequence, *Science* 249:1567 (1990).

39. M. Jaye, W.H. Burgess, A.B. Shaw and W.N. Drohan, Biological equivalence of natural bovine and recombinant human α-endothelial cell growth factor, *J. Biol. Chem.* 262:16612 (1987).

40. T. Mehlman and W.H. Burgess, Identification and characterization of heparin-binding proteins using a gel overlay procedure, *Anal. Biochem.* 188:159 (1990).

41. A Baird, D. Schubert, N. Ling and R. Guillemin, Receptor and heparin-binding domains of basic fibroblast growth factor, *Proc. Natl. Acad. Sci. USA* 85:2324 (1988).

42. D.M. Ornitz, A. Yayon, J.G. Flanagan, C.M. Svahn, E. Levi and P. Leder, Heparin is required for cell-free binding of basic fibroblast growth factor to a soluble receptor and for mitogenesis in whole cells, *Mol. Cell. Biol.* 12:240 (1992).

43. D. Chesky, R. Ralph and G. Jonak, Sequence requirements for synthetic peptide-mediated translocation to the nucleus, *Mol. Cell. Biol.* 9:2487 (1989).

44. J.W. Harper and R.R. Lobb, Reductive methylation of lysine residues in acidic fibroblast growth factor: effect on mitogenic activity and heparin affinity, *Biochemistry* 27:671 (1988).

45. A. Yayon, M. Klagsbrun, J.D. Esko, P. Leder and D.M. Ornitz, Cell surface, heparin-like molecules are required for binding of basic fibroblast growth factor to its high-affinity receptor, *Cell* 64:841 (1991).

46. B.B. Olwin and A. Rapraeger, Repression of myogenic differentiation by aFGF, bFGF and K-FGF is dependent on cellular heparan sulfate, *J. Cell Biol.* 118:631 (1992).

47. A.M. Shaheen and W.H. Burgess (manuscript in preparation).

48. G.D. Paterno, L.L. Gillepsie, M.S. Dixon, J.M.W. Slack and J.K. Heath, Mesoderm-inducing properties of INT-2 and kFGF: two oncogene-encoded growth factors related to FGF, *Development* 106:79 (1989).

49. E. Amaya, T.J. Music and M.W. Kirschner, Expression of a dominant negative mutant of the FGF receptor disrupts mesoderm formation in *Xenopus* embryos, *Cell* 66:257 (1991).

50. M.J. Berridge, C.R. Downes and M.J. Hanley, Neural and developmental actions of lithium: a unifying hypothesis, *Cell* 59:411 (1989).

51. R.F. Friesel and W.H. Burgess (manuscript in preparation).

52. A.B. Schreiber, J. Kenney, J. Kowalski, K.A. Thomas, G. Gimenez-Gallego, M. Rios-Candelore, J. DiSalvo, D. Barritault, J. Courty, Y. Courtois, M. Moenner, C. Loret, W.H. Burgess, T. Mehlman, R. Friesel, W. Johnson and T. Maciag, A unique family of endothelial cell polypeptide mitogens: the antigenic and receptor cross-reactivity of bovine endothelial cell growth factor, brain-derived acidic fibroblast growth factor, and eye-derived growth factor-II, *J. Cell Biol.* 101:1623 (1985).

53. S. Ortega, M.-T. Schaeffer, D. Soderman, J. DiSalvo, D.L. Linemeyer, G. Gimenez-Gallego and K.A. Thomas, Conversion of cysteine to serine residues alters the activity, stability, and heparin dependence of acidic fibroblast growth factor, *J. Biol. Chem.* 266:5842 (1991).

Part V
Second Messengers

MDGF-1: A NEW TYROSINE KINASE-ASSOCIATED GROWTH FACTOR

Junichi Kurebayashi[1,2], Mozeena Bano[2], Barbara Ziff[2],
Michael D. Johnson[2], William R. Kidwell[3], Francis G. Kern[2],
and Robert B. Dickson[2]

[1] Second Department of Surgery, Gunma University School
of Medicine, Maebashi, Gunma 371, Japan
[2] Lombardi Cancer Research Center, Georgetown
University, Washington, D.C. 20007
[3] Cellco Advanced Bio-Reactors, Kensington, M.D. 20895

INTRODUCTION

We have isolated and characterized an apparently new tyrosine kinase-associated growth factor from milk. The factor has been provisionally named mammary-derived growth factor-1 (MDGF-1). The polypeptide has been purified to homogeneity from milk as well as from human primary breast tumors; it appears to be identical from the two sources[1]. The biological characteristics of MDGF-1 are marked by differential effects upon cellular collagen synthesis and proliferation. Type I collagen synthesis in normal fibroblasts is stimulated by the factor in the absence growth stimulation. In contrast, type IV collagen synthesis of normal and malignant mammary epithelial cells is stimulated by the factor in the presence growth stimulation[1,2]. We propose that the factor may play a role in normal mammary growth and development and in the regulation of breast cancer by autocrine and paracrine mechanisms.

In this article we will discuss the possible roles of MDGF-1 as an autocrine and paracrine growth factor in normal and malignant mammary tissues. Brief reviews of growth factors in milk, of autocrine and paracrine growth factors in breast cancer, and of collagen synthesis in breast cancer stroma are also included in this chapter. We will summarize our recent experimental data that suggests that MDGF-1 may be an N-glycosylated protein, and that its receptor may be a member of the family of tyrosine kinase-associated growth factor receptors[3,4].

GROWTH FACTORS IN MILK

Milk is important to developing infants not only as a source of nutrients such as proteins, carbohydrates, lipids, minerals and vitamins, but also as a source of biologically active molecules such as immunoglobulins, transferrins, hormones and growth factors. The neonatal intestine has been reported to be much more permeable than the adult intestine, and to have the ability to transport intact proteins at a much higher rate[5]. It has been demonstrated that IgA from human milk may play an important role in the acquisition of passive immunity of the infant, and that human milk contains growth-promoting factors for bifidobacterium in the infant's intestine that may induce a natural defense against E. coli, bacteroides and clostridium[6,7].

Growth Factors, Peptides, and Receptors, Edited
by T.W. Moody, Plenum Press, New York, 1993

Klagsbrun[8] reported that human milk, as well as human serum, contains cell proliferation-promoting mitogens. The growth-promoting properties of milk may be due in part to the presence of growth factors. A series of studies on the isolation and detection of the growth factors in milk has been performed by many investigators. Sinha[9] isolated a colony stimulating factor (CSF) from human milk using in vitro bone marrow culture. Shing and Klagsbrun[10] purified and biologically characterized three different growth factor-containing fractions from human milk. These were designated human milk growth factors I, II and III (HMGF I, II, and III) and were detected using a DNA synthesis assay in Balb/C 3T3 mouse embryo cells. Comparative biochemical studies suggested that HMGF III is primarily human epidermal growth factor (EGF). EGF appears to be a predominant growth factor accounting for about 75% of the total growth-promoting activity in human milk. Zwiebel[11] partially purified a transforming growth factor-a (TGF-α)-like molecule in human milk using anchorage-independent growth assay of normal rat kidney (NRK) cells and radio-receptor assay for EGF receptors. This factor was initially termed MDGF-II. Connolly[12] also demonstrated the presence of EGF or TGFα-like proteins and immunoreactive TGF-α in human milk using a radio-receptor assay and a radio-immunoassay, respectively. The presence of other growth factors such as insulin-like growth factor-I (IGF-I)[13], platelet-derived growth factor (PDGF)[13], TGF-β[13], an epithelial transforming growth factor activity (TGFe)[14] and nerve growth factor (NGF)[15] in milk have also been demonstrated. Recent cellular localization studies have begun to compliment these earlier investigations. Liscia[15] observed that TGFα and EGF receptor are detected by in situ hybridization in biopsies of proliferating human and rat mammary epithelium during lobuloalveolar development of pregnancy. Snedeker[16] has examined the mouse mammary epithelium for EGF and TGFα ranging from virgin through lactational stages. While TGFα and EGF were both detected in virgin and midpregnant glands, EGF strongly predominated in lactation. TGFα was located in proliferating epithelial cap cells, while EGF was primarily associated with luminal secretory cells.

We have detected in human milk an acidic (pI 4.8) growth factor with a molecular mass of 62-kDa using a collagen synthesis stimulating assay with NRK fibroblasts. The molecule, designated MDGF-1, was highly purified from human milk and human mammary tumors using acid/ethanol extraction followed by isoelectric focusing and gel filtration HPLC. A survey of the biochemical properties of known growth factors indicated that MDGF-1 might be a novel growth factor. A computer search of the N-terminal 18 amino acid sequence of MDGF-1 also indicated this novelty[3]. The biological activities of MDGF-1 have been further investigated. In addition to its effects on stromal fibrosis, the factor stimulates the growth and synthesis of type IV collagen synthesis of normal and malignant human mammary epithelial cells. These activities may be mediated through unique membrane receptors that exhibit rapid phosphorylation on tyrosine of a membrane-associated protein[4]. More detailed characterization of the factor is described later in this chapter.

In summary a dozen growth factors have now been detected in milk. If the intact growth factors or active fragments thereof are absorbed by the gastro-intestinal tract of infants, the growth factors have the potential to play specific physiological and/or developmental roles. However, the significance of the milk-derived growth factors in vivo is speculative at present and will require further studies.

AUTOCRINE AND PARACRINE GROWTH FACTORS IN BREAST CANCER

Endocrine hormones such as estrogens, progestins, glucocorticoids, thyroid hormones and prolactin have been demonstrated to play important roles in the development and growth of mammary glands. Estrogens appear to play one of the most critical roles in the development of breast cancer. Recent progress in the molecular analysis of more locally-acting peptide growth factors indicate that a number of the factors may regulate breast cancer cell growth in an autocrine and paracrine manner[18].

TGF-α-like molecules and IGF-related proteins have been reported to be induced by estrogens in some human breast cancer cell lines[19,20]. Both TGF-α and IGFs (IGF-I and

IGF-II) are mitogenic for some human breast cancer cell lines[21,22]. Both TGF-α receptor (EGF receptor) and IGF receptors (IGF-I, IGF-II and insulin receptors) have been detected in human breast cancer[23,24]. Furthermore, recent studies have shown that antibodies directed against TGF-α or EGF receptor suppress both the anchorage-dependent and -independent growth of the responsive cells, and also that antibodies directed against IGF-I receptor suppress the clonal proliferation of the responsive cells[19,25,26]. These studies suggest that both TGF-α and IGFs may be involved in breast cancer cell proliferation through autocrine and paracrine loops.

Many breast cancer cell lines also produce growth factors of the TGF-β-family[27]. Both estrogen receptor-positive and -negative cell lines contain TGF-β receptors, and are inhibited by both TGF-β1 and -β2[28]. Interestingly, TGF-β secretion by MCF-7 estrogen-responsive human breast cancer cells is inhibited by mitogens such as estrogens and insulin, but stimulated by antiproliferative agents such as antiestrogens and glucocorticoids[27]. TGF-β may be acting as a growth-inhibitory factor of breast cancer cells through autocrine and paracrine loops. Actions of TGF-β *in vivo* are also likely to involve tumor-host interactions. Recent studies have suggested that predominant actions of TGF-β in the nude mouse implanted with human breast cancer cells include immunosuppression, fibrosis, and cachexia[29,30].

IGFs, TGF-β and PDGF, synthesized by breast cancer cell lines[31], are potent mitogens for fibroblasts, smooth muscle cells and other stromal cells. All the three factors have the potential to play a paracrine role in breast cancer development, possibly at the level of the stromal fibrotic process known as desmoplasia.

Recently, neovascularization of human breast cancer has been observed to be correlated with the degree of metastatic spread[32]. The fibroblast growth factor (FGF) family has been thought to play an important role in tumor neovascularization and metastasis[33]. In this regard co-amplification of the genes for FGF-3 (*int*-2) and FGF-4 (*K-fgf, hst*) have been detected in about 20% of human breast cancer[34]. However, these factors do not appear to be expressed at the mRNA and protein level in breast cancer. It has been demonstrated, however, that cultured human breast cancer cells express FGF's[35]. These studies suggest that the FGF family may play an important role in malignant progression and metastasis of human breast cancer in a paracrine manner.

MDGF-1 may represent a different type of paracrine growth factor in the breast, one communicating between epithelial and mesenchymal cells. Since MDGF-1 stimulates not only the growth and synthesis of type IV collagen of human mammary epithelial cells, but also stimulates type I collagen synthesis of normal fibroblasts, the factor may also be involved in the growth and development of breast cancer through both autocrine and paracrine mechanisms.

COLLAGEN SYNTHESIS IN BREAST CANCER STROMA

An abnormal deposition of extracellular matrix in the proximity of tumor nests and cords is often recognized in invasive ductal carcinoma of the breast[36]. Biochemical and immunohistochemical analysis of the stromal component in breast cancer indicates that collagen in the stroma is primarily composed of embryo-fetal type I-trimer collagen, in addition to some regular type I, III and V collagen. Minafra[37] suggested that the deposition of unusual type I collagen can be regarded as a part of the interrelationships between the integrated cell-matrix system during malignant growth. Recently, type IV collagen has also been detected immunologically in interstitial elastosis in breast cancer tissues[38].

Type IV collagen is one of the major components of basement membrane. Type IV collagen synthesis appears to be important for the growth and/or survival of the normal epithelium and of well-differentiated epithelial tumors[39]. It has also been suggested that the enzymatic degradation of type IV collagen, initiated by type IV collagenase may be an important characteristic of invasive neoplasms. This process is closely linked to invasive and metastatic potential in several experimental models[40]. Recently, a computerized quantitative image analysis of laminin and type IV collagen in human breast cancer indicated that a decrease in immunostaining of laminin and type IV collagen correlated with the presence of peritumorous vascular invasion, keratin positive-staining cells in bone marrow (suspected to be micrometastasis) and axillary lymph node involvement of the tumor[41].

MDGF-1 strongly stimulates type I collagen production of normal rat kidney fibroblasts by a stimulation of its synthesis rather than by a decrease of collagen turnover[1]. Interestingly, when the fibroblasts were plated on type IV collagen-coated dishes there was no effect of MDGF-1, whereas the cells which had been plated on type I collagen-coated dishes or on tissue culture plastic dishes differentially increased their production of collagen by 3- and 4-fold, respectively, in response to MDGF-1[1].

This protein also stimulates type IV collagen synthesis of normal mouse mammary epithelial cells, and normal and malignant human mammary epithelial cells. Since human mammary tumors contain MDGF-1, and since immunoreactive MDGF-1 has been observed in the cytoplasm of human mammary tumor cells[42], the factor may be produced by many human mammary tumors. Immunoreactive MDGF-1 has also been detected in the concentrated medium prepared from normal and malignant human mammary epithelial cells[3]. These experimental data suggest that MDGF-1, secreted by normal or malignant epithelial cells, might regulate type IV collagen deposition on the basement membrane between the cells and stroma, and that abnormal secretion of MDGF-1 by malignant cells might stimulate stromal fibroblasts in a paracrine manner to produce excessive type I collagen in the extracellular matrix of breast cancer tissues. Moreover, inappropriate deposition of type IV collagen secreted by malignant cells might modulate the invasive and metastatic ability of the cells.

EFFECTS OF MDGF-1 ON GROWTH OF NORMAL AND MALIGNANT MAMMARY EPITHELIAL CELLS

MDGF-1 was observed to stimulate cell growth and collagen synthesis of normal mouse mammary epithelial cells freshly isolated from mouse mammary ducts and alveoli. Optimal stimulation of both cell division and collagen synthesis was induced at an MDGF-1 concentration of 5 ng/ml. The growth of mammary epithelial cells in culture was very low if the cells were isolated from animals depleted of estrogens by ovariectomy. However, if the cells were derived from ovariectomized animals given estrogen replacement, there was about a 100% greater basal growth rate, and this was further enhanced by about 50% by MDGF-1 addition to the culture. Thus, MDGF-1 may act synergistically with estrogen by some mechanism which is not yet defined[1]. We have begun to study such growth regulation by analysis of cell lines *in vitro*. We have systematically screened MDGF-1 effects on anchorage-dependent growth of normal and malignant human mammary epithelial cells (Fig. 1).

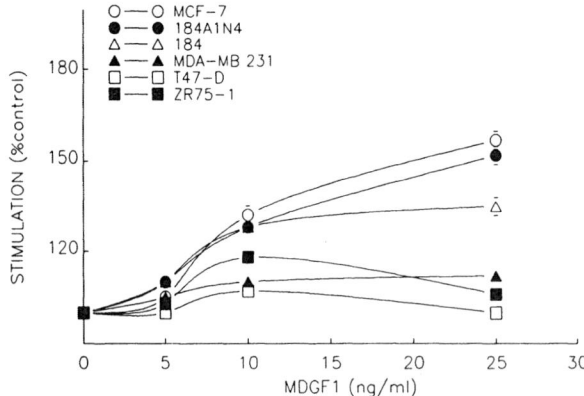

Figure 1. Effect of MDGF-1 on human mammary cell growth. Cells were grown on tissue culture plastic dishes. Cells were incubated with the factor at the indicated concentrations. After 96-144 h cells were trypsinized and cell count was taken using a Coulter counter. Data are mean cell numbers expressed as percent of untreated control cell numbers. (Reproduced from reference 2).

The growth of normal human mammary epithelial cells (non-immortalized 184 cell strain from M. Stampfer derived from reduction mammoplasty) was enhanced by 35% by the addition of MDGF-1, whereas benz[a]pyrene-immortalized non-tumorigenic 184A1N4 human mammary epithelial cells were stimulated by 60-70%. However, transfection of SV40-T, v-Ha-*ras*, and/or v-*mos* oncogenes into 184A1N4 cells desensitized the cells to MDGF-1[2]. Three possible reasons for this decreased sensitivity are that oncogene transfection could result in a loss of function or expression of MDGF-1 receptors or MDGF-1 could be produced in sufficient quantities so that exogenously added MDGF-1 would be ineffective. Alternatively different growth regulatory mechanisms may be operant after transformation. Malignant transformation *per se* does not necessarily result in a loss of sensitivity to MDGF-1. At a concentration of 10-25 ng/ml the factor stimulated the growth of estrogen receptor-positive MCF-7 human breast cancer cells by 50%. It did not have any effects on estrogen receptor-positive ZR75-1 and T47-D human breast cancer cells or on receptor-negative MDA-MB 231 human breast cancer cells. The factor showed a biphasic effect on the estrogen receptor-negative MDA-MB 468 human breast cancer cells at a concentration above 5 ng/ml. Significantly, we observed that the responsive human breast cancer cell lines (MCF-7 and MDA-MB 468) possessed detectable MDGF-1 receptors, whereas the nonresponsive cell lines (ZR75-1, T47-D, and MDA-MB 231) lacked significant numbers of the receptors, as described later.

MDGF-1 PROCESSING

We hypothesized that MDGF-1 might be an autocrine or paracrine growth factor produced by and acting on normal and malignant human breast epithelial cells. As a first step in testing this possibility, we examined whether human breast epithelial cells produce the growth factor. A 62-kDa protein, putative MDGF-1, was immunologically detected in the concentrated conditioned medium and cell lysates prepared from MDA-MB 231 human breast cancer cell line, HBL-100 human milk-derived but nontumorigenic mammary epithelial cell line, and a reduction mammoplasty-derived normal mammary epithelial cell strain designated 184. In contrast, three other cell lines, MCF-7 human breast cancer, MDA-MB 468 human breast cancer, or 184A1N4 immortalized human mammary epithelial strain, did not show any immunologically detectable MDGF-1. Polyclonal antiserum raised against

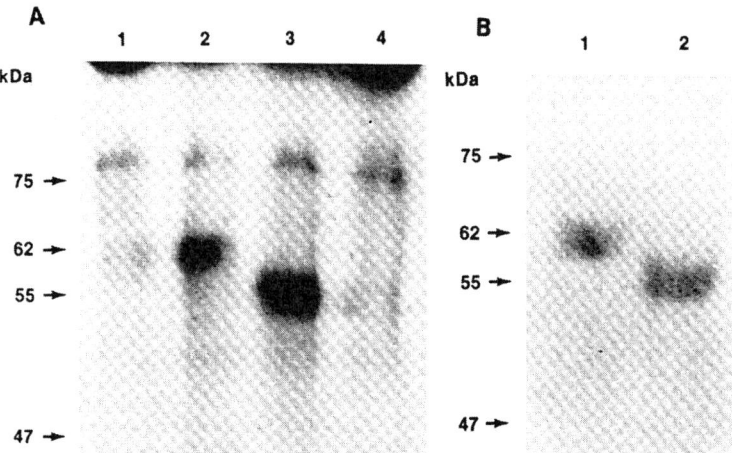

Figure 2.: (A) Tunicamycin treatment. HBL-100 cells were grown to confluency and were treated with tunicamycin for 4 h at 37 °C. Metabolic labeling and immunoprecipitation using anti-MDGF-1 antibody were performed. Immunoprecipitates were analyzed by SDS-PAGE and subsequent fluorography. Lanes 3 and 4 denote tunicamycin treatment and lanes 1 and 2 in the absence of tunicamycin. Results for anti-MDGF-1 sera are shown in lanes 2 and 3 and those for prebled sera in lanes 1 and 4. (B) N-linked glycosylation of 62-kDa MDGF-1. Purified sample was incubated with N-glycanase (lane 2) or buffer only (lane 1) and subjected to Western blot analysis and silver-stained. (Reproduced from reference 3).

the synthetic polypeptide (N-terminal 18 amino acid sequence of MDGF-1) recognizes native milk-derived MDGF-1, absorbs MDGF-1 biological activity from mammary cell-derived conditioned medium, and has been used in these experiments[3].

Following *in vitro* translation of mRNA from the HBL-100 cell line or the MDA-MB 231 cell line immune precipitation with the anti MDGF-1 antisera yielded a protein band with a molecular mass of approximately 55-kDa. After tunicamycin treatment of either cell line metabolically labelled and immunoprecipitated, a 55-kDa protein band was again observed. Treatment of purified MDGF-1 with N-glycanase also led to a reduction in molecular mass to 55-kDa[3] (Fig. 2).

These results indicated that the mature 62-kDa MDGF-1 contains N-linked carbohydrates. The 55-kDa protein may be a precursor of MDGF-1 and may be posttranslationally modified by N-glycosylation.

MDGF-1 RECEPTORS

To determine whether MDGF-1 might have specific receptors in mouse or human mammary epithelial cells, purified MDGF-1 was iodinated, and radio-receptor assays were

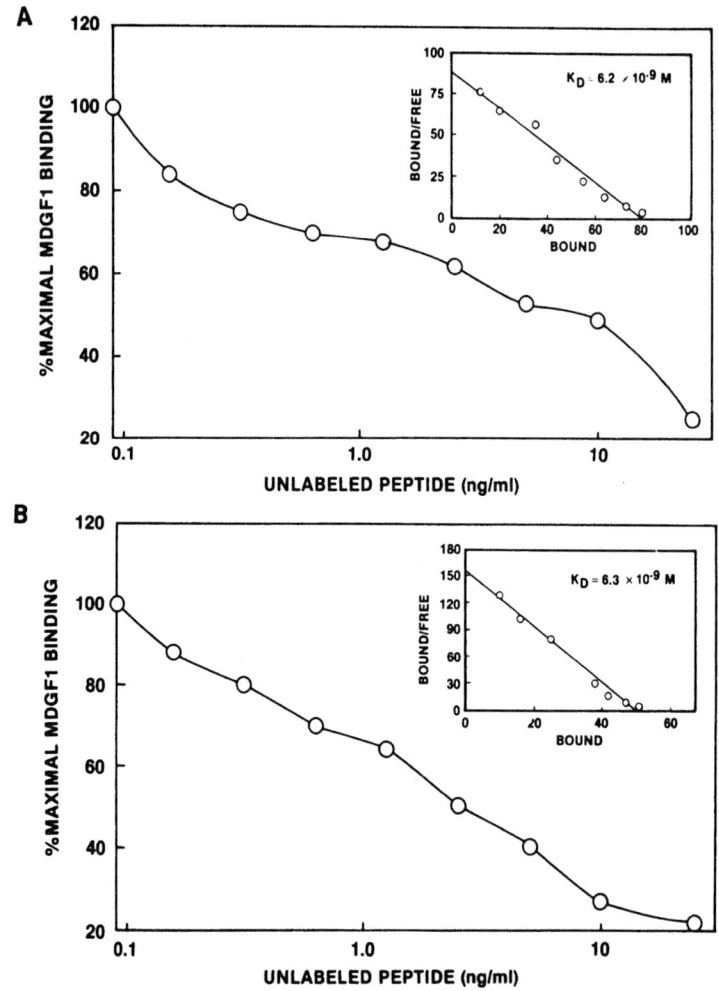

Figure 3. Binding of ^{125}I-MDGF-1 to cells. Aliquots of cells were incubated for 2 h at 37°C with ^{125}I-MDGF-1 in the presence of various concentrations of unlabeled MDGF-1. Nonspecific binding was determined with excess of unlabeled MDGF-1. Data represent the percent maximal binding after nonspecific binding is subtracted. Inset shows the Scatchard analysis. MCF-7 cells in Panel A, and MDA-MB 468 cells in Panel B. (Reproduced from reference 2).

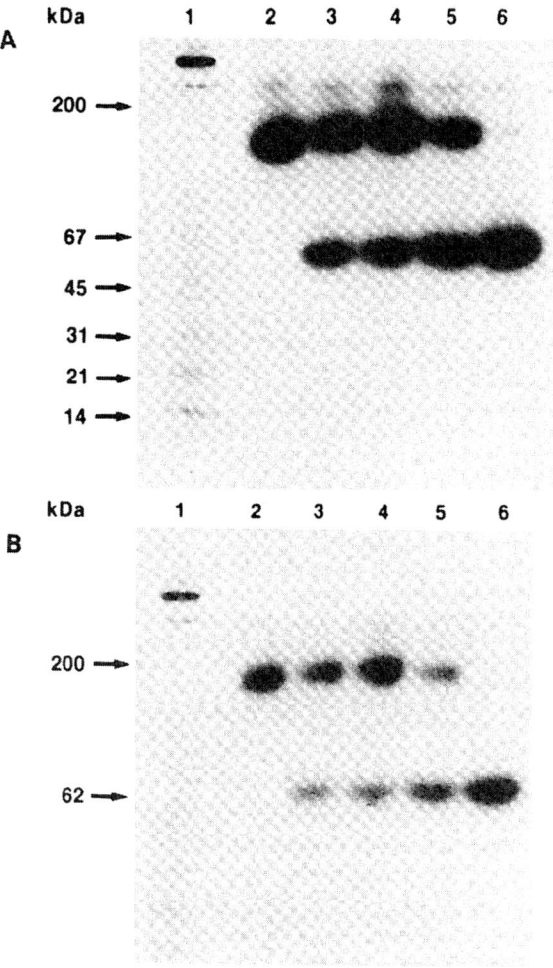

Figure 4. Cross-linking experiments. Cells were incubated with ^{125}I-MDGF-1 without any other additives (lane 2) or in the presence of excess unlabeled MDGF-1 (2.5, 5.0, 10, and 25 ng, lanes 3-6). The cells were then treated with a cross-linking agent DSS. Dissolved cells were subjected to SDS-gel electrophoresis. Lane 1 denotes molecular mass markers. MCF-7 cells in Panel A, and MDA-MB 468 cells in Panel B. (Reproduced from reference 2).

performed. High affinity putative receptors for MDGF-1 (Kd = 2 X 10^{-10} M) were detected on mouse mammary epithelial cells isolated from the ducts and alveoli[1]. High affinity receptors for MDGF-1 (Kd = 6 X 10^{-9} M) were also detected on MCF-7 and MDA-MB 468 human breast cancer cell lines (Fig. 3). Receptors for MDGF-1 (Kd = 4 X10^{-8} M) were also detected on 184 normal human mammary epithelial cell strain, 184A1N4 chemically immortalized cell strain, and the HBL-100 milk-derived human mammary epithelial cell line[2]. Scatchard analysis yielded a linear plot in all cases and indicated that the receptors constituted a single affinity class of binding sites. Some other human breast cancer cell lines (MDA-MB 231, ZR75-1, and T47-D) had no detectable MDGF-1 receptors. The very high affinity receptors for MDGF1 have also been detected on NRK fibroblasts and A431 human epidermoid carcinoma cells[1].

MDGF-1 binding appears to be specific since an excess of EGF, basic FGF, and IGF-I in the radio-receptor assays failed to inhibit the binding of radio-labeled MDGF1 in cell membranes[2].

Further characterization of MDGF-1 receptors was performed using a cross-linking agent disuccinimidyl suberate. The cross-linking experiments indicated that labeled MDGF-1 was associated with a protein migrating with an apparent molecular mass of 130-kDa in MCF-7 and MDA-MB 468 human breast cancer cell lines (Fig. 4), and also in the A431 human epidermoid cancer cell line[2].

MDGF-1 TYROSINE PHOSPHOPROTEIN

To investigate the signal transduction pathways of MDGF-1 receptors, western blot analysis of tyrosine phosphoprotein in the normal and malignant human mammary epithelial cells was performed using a monoclonal anti-phosphotyrosine antibody. The results indicated that MDGF-1 induced the appearance of phosphotyrosine in a 180-185-kDa protein in MDGF-1 receptor-positive cell lines (MCF-7, MDA-MB 468, 184, 184A1N4 and HBL-100) (Fig. 5).

Immunofluorescence staining and membrane fractionation showed that phosphorylation induced by MDGF-1 may be plasma membrane associated. Phosphorylation by MDGF-1 was not blocked by an antibody directed against the binding sites of the EGF receptors. Two dimensional gel electrophoresis and immunoblotting with anti-phosphotyrosine antibody indicated that the tyrosine phosphoprotein induced by MDGF-1 is distinct from that induced by EGF[4].

We hypothesize that the MDGF-1 receptor exists as two components, one for ligand binding and the other a substrate for tyrosine phosphorylation, a situation similar to the tyrosine kinase-associated receptors such as the insulin. However, the nature of the phosphorylation site is under investigation to elucidate if alternatively it might be a cellular substrate of the MDGF-1 receptor.

CONCLUSION

As shown in Table 1, both MDGF-1 and its receptor have been detected in the 184 normal human mammary epithelial cell strain and in the milk-derived HBL-100 human mammary cell line, suggesting the potential for autocrine function in the normal mammary-derived cells. The growth factor, but not the receptor, has been detected in some hormone-independent breast cancer cell lines, while the receptor has been detected in both hormone-dependent and -independent breast cancer cell lines.

Table 1. MDGF-1 production and properties by various cell lines

Cell line	Receptor Binding	Growth	Stimulation of phosphotyrosine	MDGF-1 Production
MCF-7	++	++	++	-
MDA-MB 468	++	+/-	++	-
184A1N4	+	++	+	-
184	+	+	+	++
MDA-MB 231	-	-	-	++
HBL-100	+	++	+	++

Radio-receptor assay and growth stimulation assay were performed[1,2]. Binding studies showed high (++) and low (+) affinity binding sites. for MDGF-1 on different cell lines[2]. For growth assays (++) denotes above 60% stimulation above control and (+) denotes 10-30% stimulation. A biphasic growth effect is depicted by (+/-) which means growth stimulation at low concentration and growth inhibition at high concentration. Western blot of phosphotyrosine in various cell lines were performed after MDGF-1 stimulation using anti-phosphotyrosine monoclonal antibody[4]. Immunoreactive MDGF-1 was detected in the conditioned medium and cell lysates of the three cell lines using Western blot with polyclonal anti-MDGF-1 antibody[3]. (Reproduced from reference 4).

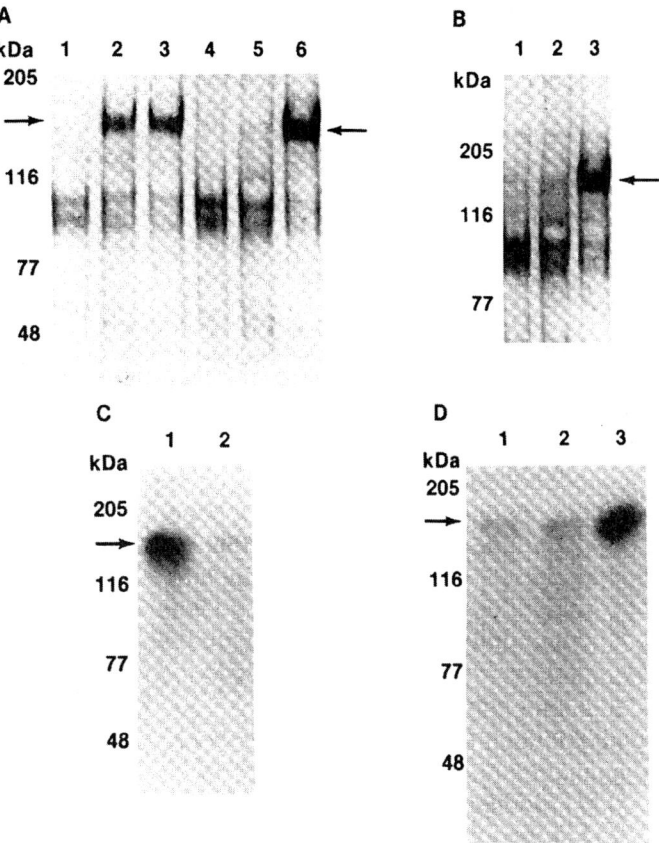

Figure 5. Western blot of phosphotyrosine in MDA-MB 468 cells. Adequate amounts of cells (100,000) were incubated with no addition (*lane 1*), 10 nM purified MDGF-1 (*lane 2*), MDGF-1 in the presence of an excess of anti-EGF receptor monoclonal antibody 528 (*lane 3*), 5 nM EGF (*lane 6*), EGF in the presence of an excess on anti-EGF receptor monoclonal antibody 528 (*lane 5*), and 100 nM monoclonal antibody 528 alone (*lane 4*). Cells were solubilized in sample buffer and subjected to gel electrophoresis (7.5% gels) and Western blot analysis using a commercially available monoclonal anti-phosphoty-rosine antibody (2.5 μg/ml) and visualized with a chromogenic reagent. The *arrow* at *left* indicates a MDGF-1-induced 180-185-kDa size band (*lanes 2* and *3*) and, at *right*, an EGF-stimulated 170-kDa band (*lane 6*), *B*, stimulation of tyrosine phosphorylation of MCF-7 cells by MDGF-1. Confluent MCF-7 cells were treated with 0 (*lane 1*), 10 nM EGF (*lane 2*), or 10 nM MDGF-1 (*lane 3*) for 20 min at 37°C, and then the lysates were subjected to SDS-PAGE, Western blotting, and incubation with monoclonal anti-phosphotyrosine were as above. The *arrow* at *right* indicates a 180-185-kDa size MDGF-1-induced band (*lane 3*). *C*, autoradiogram of phosphorylated proteins in MDA-MB 468 cell lysates. Total cell lysates were analyzed by 7.5% SDS-PAGE gels, followed by immunoblotting with polyclonal anti-phosphotyrosine antibody as explained above. The blots were then processed with ^{125}I-Protein A, dried, and autoradiographed. *Lane 2* denotes control lysate and, *lane 1*, cells treated with 10 nM MDGF-1. The *arrow* indicates the 180-185-kDa band. *D*, detection of the EGF-R autophosphorylation activity in MDA-MB 468 cells. Cultures of untreated (*lane 1*), MDGF-1-treated (10 nM, *lane 2*), or EGF-treated (10 nM, *lane 3*) cells were lysed and then immunoprecipitated with an anti-EGF receptor antibody. The washed immunoprepitates were incubated with 20 mM HEPES. pH 7.4, 0.15 M NaCl, 0.1% Triton X-100, 10% glycerol, 10 nM $MnCl_2$, and 5 μM ATP containing 0.1 μCi of $[\gamma^{32}P]ATP$ (Amersham) for 5 min at 25°C, leading to autophosphorylation. Reactions were stopped in sample buffer, and the samples were electrophoresed on 7.5% PAGE followed by autoradiography. (Reproduced from reference 4).

319

The growth factor thus has the potential for paracrine as well as autocrine function. MDGF-1 may exert a paracrine effect on type I collagen synthesis of stromal fibroblasts. The factor may play an important role as an autocrine and paracrine mediator in epithelial-stromal communication in the normal and malignant mammary tissues. The biochemical studies indicated that MDGF-1 might be a novel N-glycosylated growth factor, and that its receptor may be a member of tyrosine kinase-associated membrane receptors. Current studies are addressing the molecular cloning of cDNA of MDGF-1 to assist in the further characterization of the factor.

ACKNOWLEDGEMENT

The authors would like to thank Marybeth Sabol for excellent technical assistance. This work was partially supported by ACS grant BE 71.

REFERENCES

1. M. Bano.D.S. Salomon, and W.R. Kidwell, Purification of a mammary-derived growth factor from human milk and human mammary tumors, J. Biol. Chem. 260: 5745 (1985).

2. M. Bano, W.R. Kidwell, M.E. Lippman, and R.B. Dickson, Characterization of mammary-derived growth factor 1 receptors and response in human mammary epithelial cell lines, J. Biol. Chem. 265: 1874 (1990).

3. M. Bano, R. Lupu, W.R. Kidwell, M.E. Lippman, and R.B. Dickson, Production and characterization of mammary-derived growth factor 1 in mammary epithelial cell lines, Biochem. 31: 610 (1992).

4. M. Bano, P. Worland, W.R. Kidwell, M.E. Lippman, and R.B. Dickson, Receptor induced phosphorylation by mammary-derived growth factor 1 in mammary epithelial cells, J. Biol. Chem. 267: 10389 (1992).

5. R.J. Grand, J.B. Watkins, and F.M. Torti, Development of the human gatrointestinal tract. A review, Gastroenterology 70: 790 (1976).

6. V.S. Packard, "Human milk and infant formula", Academic Press, New York (1982).

7. H. Beerens, C. Romond, and C. Neut, Influence of breast-feeding on the bifid flora of the newborn intestine, Am. J. Clin. Nutr. 33: 2434 (1980).

8. M. Klagsbrun, Bovine colostrum supports the serum-free proliferation of epithelial cells but not of fibroblasts in long-term culture, J. Cell Biol. 84: 808 (1980).

9. S.K. Sinha and A.A. Yunis, Isolation of colony stimulating factor from human milk, Biochem. Biophys. Res. Commun. 114: 797 (1983).

10. Y.W. Shing and M. Klagsbrun, Human and bovine milk contain different sets of growth factors, Endocrinol. 115: 273 (1984).

11. J.A. Zwiebel, M. Bano, E. Nexo, D.S. Salomon, and W.R. Kidwell, Partial purification of transforming growth factors from human milk, Cancer Res. 46: 933 (1986).

12. J.M. Connolly and D.P. Rose, Epidermal growth factor-like proteins in breast fluid and human milk, Life Science 42: 1751 (1988).

13. J.C. Ness, L. Morgan, H.C. Outzen, and D. Tapper, Specific growth factor activity identifies and predicts murine mammary tumors, J. Surg. Res. 50:6 (1991).

14. D.J. Dunnington, R.G. Scott, M.A. Anzano, and R. Greig, Characterization and partial purification of human epithelial transforming factor, J. Cell. Biochem. 44: 229 (1990).

15. D.S. Licia, G. Merlo, F. Ciardiello, N. Kim, G.H. Smith, R.H. Callahan, and D.S. Salomon, Transforming growth factor α messenger RNA localization in the developing adult rat and human mammary gland by in situ hybridization, Develop. Biol. 140: 123 (1990).

16.   S.M. Snedeker, C.F. Brown, and R.P. DiAugustine, Expression and functional properties of transforming growth factor alpha and epidermal growth factor during mouse gland ductal morphogenesis, Proc. Natl. Acad. Sci. (USA) 88: 276 (1991).

17.   A. Grueters, J. Lakshmanan, R. Tarris, J. Alm, and D.A. Fisher, Nerve growth factor in mouse milk during early lactation: lack of dependency on submandibular salivary glands, Pediatr. Res. 19: 934 (1985).

18.   R.B. Dickson and M.E. Lippman, Growth regulation of the normal and malignant breast epithelium, in <u>The Breast</u>, Bland K. I. and Copeland E.M. (eds.), W.B. Saunders Co. 363 (1990).

19.   S.E. Bates, N.E. Davidson, E. Valverius, C. Freter, R.B. Dickson, J.D. Tam, J.E. Kudlow, M.E. Lippman, and D.S. Salomon, Expression of transforming growth factor a and its mRNA in human breast cancer: its regulation and its possible functional significance, Mol. Endocrinol. 2: 543 (1988).

20.   D. Yee, N. Rosen, R.E. Favoni, and K.J. Cullen, The insulin-like growth factors, their receptors, and their binding proteins in human breast cancer, Cancer Treat. Res. 53: 93 (1991).

21.   S.E. Bates, M.E. McManaway, M.E. Lippman, and R.B. Dickson, Characterization of estrogen responsive transforming activity in human breast cancer cell lines, Cancer Res. 46: 1707 (1986).

22.   K.K. Huff, D. Kaufman, K.H. Gabbay, E.M. Spencer, M.E. Lippman, and R.B. Dickson, Human breast cancer cells secrete an insulin-like growth factor-I related polypeptide, Cancer Res. 46: 4613 (1986).

23.   J.R.C. Sainsbury, A.J. Malcolm, D.R. Appleton, J.R. Farndon, and A.L. Harris, Presence of epidermal growth factor receptor as an indicator of poor prognosis in patients with breast cancer, J. Clin. Pathol. 38:1225 (1985).

24.   R.W. Furlanetto and J.N. DiCarlo, Somatomedin-C receptors and growth effects in human breast cells maintained in long-term tissue culture, Cancer Res. 44:2122 (1984).

25.   S.E. Bates, E.M. Valverius, B.W. Ennis, D.A. Bronzert, J.P. Sheridan, M.R. Stampfer, J.S. Mendelsohn, M.E. Lippman, and R.B. Dickson, Expression of the transforming growth factor α / Epidermal growth factor receptor pathway in normal human breast epithelial cells, Endocrinology 126: 596 (1990).

26.   Q.T. Rohlik, D. Adams, F.C. Kull, and S. Jacobs, An antibody to the receptor for insulin-like growth factor I inhibits the growth of MCF-7 cells in tissue culture, Biochem. Biophys. Res. Commun. 149: 276 (1991).

27.   C. Knabbe, M.E. Lippman, L. Wakefield, K. Flauders, A. Kasid, R. Derynck, and R.B. Dickson, Evidence that TGF β is a hormonally regulated negative growth factor in human breast cancer, Cell 48: 417 (1987).

28.   G. Zugmaier, B.W. Ennis, B. Deschauer, D. Katz, C. Knabbe, G. Wilding, P. Daly, M.E. Lippman, and R.B. Dickson, Transforming growth factors type β1 and β2 are equipotent growth inhibitors of human breast cancer cell lines, J. Cell Physiol. 141: 353 (1989).

29.   S.D. Hurd, M.D. Johnson, J.T. Forbes, T.L. Carty-Dugger, and C.L. Arteaga, Neutralizing TGF - β antibodies increase natural killer (NK) cell activity and inhibit human breast cancer cell tumorigenicity in athymic nude mice, Proceeding of the 1992 meeting of the American Association for Cancer Research (abstract).

30.   G. Zugmaier, S. Paik, G. Wilding, C. Knabbe, M. Bano, R. Lupu, B. Deschauer, S. Simpson, R.B. Dickson, and M.E. Lippman, Transforming growth factor β1 induces cachexia and systemic fibrosis without an antitumor effect in nude mice, Cancer Res. 51: 3590 (1991).

31.　D.A. Bronzert, P. Pantazis, H.N. Antoniades, A. Kasid, N. Davidson, R.B. Dickson, and M.E. Lippman, Synthesis and secretion of PDGF-like growth factor by human breast cancer cell lines, Proc. Natl. Acad. Sci. U.S.A. 84:5763 (1987).

32.　N. Weidner, J.P. Semple, W.R. Welch, and J. Folkman, Tumor angiogenesis and metastasis-correlation in invasive breast carcinoma, N. Engl. J. Med. 324:1 (1991).

33.　J. Folkman, What is the evidence that tumors are angiogenesis dependent? J. Natl. Cancer Inst. 82:4 (1990).

34.　C. Theillet, X.L. Roy, O.D. Lapeyriere, J. Grosgeorges, J. Adnane, S.D. Raynaud, J. Simony-Lafontaine, M. Goldfarb, C. Escot, D. Birnbaum, and P. Gaudray, Amplification of FGF-related genes in human tumors: possible involvement of HST in breast carcinomas, Oncogene 4: 915 (1989).

35.　S. Li and G.D. Shipley, Expression of multiple species of basic fibroblast growth factor mRNA and protein in normal and tumor-derived mammary epithelial cells in culture, Cell Growth and Diff. 2: 195 (1991).

36.　M.S. Al-Adnani, J.A. Kirrane, and J.O. McGee, Inappropriate production of collagen and prolylhydroxylase by human breast cancer cells in vivo, Br. J. Cancer 31: 653 (1975).

37.　S. Minafra, C. Luparello, F. Rallo, and I. Pucci-Minafra, Collagen biosynthesis by a breast carcinoma cell strain and biopsy fragments of the primary tumour, Cell Biol. Int. Rep. 12: 895 (1988).

38.　D. Verhoeven, N. Bourgeois, A. Noel, J-M. Foidart, and N. Buyssens, The presence of a type IV collagen skeleton associated with periductal elastosis in breast cancer, J. Histochem. Cytochem. 38: 245 (1990).

39.　M.S. Wicha, L.A. Liotta, B.K. Vonderhaar, and W.R. Kidwell, Effects of inhibition of basement membrane collagen deposition on rat mammary gland development, Develop. Biol. 80: 253 (1980).

40.　L.A. Liotta, K. Tryggvason, S. Garbisa, I. Hart, C.M. Foltz, and S. Shafie, Metastatic potential correlates with enzymatic degradation of basement membrane collagen. Nature 284: 67 (1980).

41.　C. Charpin, L. Andrac, M. C. Habib, H. Vacheret, M.N. Lavaut, L. Xerri, D. Figarella-Branger, P. Casanova, and M. Toga, Correlation between laminin and type IV collagen distribution in breast carcinomas, and estrogen receptors expression, lymph node and vascular involvement, Med. Oncol. Tumor Pharmacother. 7: 43 (1990).

42.　M. Bano, M.B. Sabol, S. Paik, E. Barker, W.R. Kidwell, M.E. Lippman, and R.B. Dickson, Production and localization of mammary-derived growth factor 1 in human breast cancers, Proceeding of Am. Assoc. for Can. Res. 33: #470, 79 (1991).

INTERLEUKIN-11 MEDIATED SIGNAL TRANSDUCTION
PATHWAYS: COMPARISON WITH THOSE OF
INTERLEUKIN-6

Tinggui Yin[1] and Yu-Chung Yang[1,2,3*]

[1]Department of Medicine (Hematology/Oncology)
[2]Department of Biochemistry and Molecular Biology
[3]Walther Oncology Center
 Indiana University School of Medicine
 Indianapolis, Indiana 46202

INTRODUCTION

Interleukin(IL)-11 was initially identified and molecularly cloned based on its ability to stimulate the proliferation of an IL-6 dependent mouse plasmacytoma cell line, T1165[1]. Subsequent study indicated that IL-11 is a multifunctional cytokine which may play an important role in hematopoiesis. Accumulated data over the past few years have suggested that IL-11, although lacking sequence homology with IL-6, is functionally related to IL-6. For example, IL-11 shares the following biological activities with IL-6: (a) proliferation of an IL-6 dependent mouse plasmacytoma cell line, T1165[1] and an IL-6 dependent mouse B cell hybridoma cell line, B9 (see below); (b) enhancement of IL-3 dependent development of megakaryocyte colonies in both human and mouse bone marrow cultures in vitro [1,2]; (c) stimulation of T cell dependent antigen-specific antibody production in vitro and in vivo in normal mice[3]; (d) acceleration of the regeneration of antigen-specific antibody forming cells in immunosuppressed mice following cyclophosphamide treatment[3]; (e) stimulation of pluripotent stem cells out of G_0 stage of the cell cycle to respond to intermediate or late acting factors such as IL-3 or granulocyte-macrophage colony stimulating factor[4]; (f) induction of acute phase protein synthesis in a rat hepatoma cell line, H-35 and in freshly isolated rat hepatocytes[5] ; (g) inhibition of lipoprotein lipase activity and preadipocyte differentiation[6,7,8]. The similarities in the biological activities between IL-11 and IL-6 led us to speculate that these two cytokines may utilize the same or similar receptor systems.

Recent molecular cloning of the IL-6 receptor has shown that IL-6

---

[*]To whom correspondence should be addressed.

Growth Factors, Peptides, and Receptors, Edited
by T.W. Moody, Plenum Press, New York, 1993

receptor complex composes of a binding subunit gp80[9] and a signal transducing subunit gp130[10,11], which lacks sequence homology to any known protein kinases and produces unknown second messengers. Structural information on the IL-11 receptor system is not yet available. However, our recent studies suggested that IL-11 binding to its receptor can not be competed by adding 1000 fold excess of IL-6 in 3T3-L1 preadipocytes. In T10 cells, which are dependent on either IL-11 or IL-6 for growth, no cross competition in the ligand binding was observed between IL-11 and IL-6 (data not shown), suggesting that IL-11 binding proteins are different from those of IL-6. Affinity cross-linking studies demonstrated that IL-11 binds to a 151 kilodalton (KDa) protein which is distinct from the IL-6 binding protein, gp80 [7,9]. The similarities in biological activities and the difference in the sizes of binding proteins between IL-11 and IL-6 led to the hypothesis that IL-11 and IL-6 utilize its own specific binding protein, but may share the same signal transducer, gp130, which mediates IL-6 biological functions.

The signal transduction mechanisms mediated by IL-11 or IL-6 are not clear at the present time. Our studies on IL-11[7] and others on IL-6[12,13] suggest that IL-11 and IL-6 mediated signals are closely linked to protein tyrosine kinases which may play pivotal roles in cytokine mediated signal transduction[14,15]. This paper will summarize our attempts to understand the signal transduction pathways mediated by IL-6 or IL-11 in several cell lines known to respond to both cytokines.

GROWTH ACTIVITY OF IL-11 ON B9-TY1 CELLS

B9 cells were originally derived from an IL-6 dependent mouse B cell hybridoma[16]. The growth of established B9 cells are dependent on either human or mouse IL-6. Since IL-11 was discovered as a mitogenic factor for an IL-6 dependent mouse plasmacytoma cell line, T1165[1], we examined whether IL-11 can also stimulate the proliferation of IL-6 dependent mouse B9 cells. Our initial studies showed that only a small percentage of B9 cells are responsive to IL-11. An IL-11 dependent cell line designated as B9-TY1 was subsequently established from B9 cells by semi-solid cell cloning

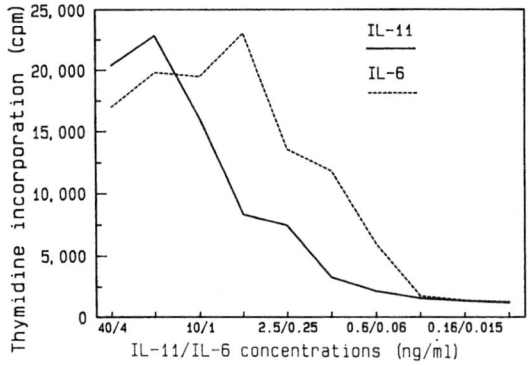

Figure 1. IL-11 and IL-6 induced B9-TY1 cell proliferation. B9-TY1 cells (2 X 10³ cells/well in 96 well plates) were cultured in the presence of different concentrations of IL-11 or IL-6 at 37°C, 5% CO_2 for 48 hours. Cells were pulsed with 0.5 μCi/well of ³H-thymidine for 4 hours, cells were then harvested and radioactivity was counted by Beckman scintillation counter. The results are expressed as cpm. Standard deviation is less than 10%.

techniques in the presence of IL-11. The established B9-TY1 cells are dependent on either IL-11 or IL-6 for growth. As shown in Figure 1, growth of B9-TY1 cells is more sensitive to IL-6 when compared with IL-11. These results may be due to the difference in the number and affinity of IL-11 or IL-6 receptors on the cell surface. Scatchard plot analysis from specific binding data revealed the existence of a single class of IL-11 receptor with a Kd of 9.9×10^{-10} M and a receptor density of 75 sites/cells on B9 cells while there are 1460 sites/cell of IL-6 receptor with a Kd of $1.4 \times 10^{-10}M^{17}$ on the surface of these cells.

EFFECT OF IL-11 ON PROTEIN TYROSINE PHOSPHORYLATION

Many growth factor or cytokine receptors stimulate protein tyrosine phosphorylation following ligand binding. This event is thought to be an essential step of signal transduction mechanisms that mediate the cellular responses[14,15,18,19]. Our initial data indicated that IL-11 mediated signal transduction in 3T3-L1 preadipocytes is closely linked to protein tyrosine phosphorylation[7]. We also examined whether IL-11 can induce protein tyrosine phosphorylation in different IL-11 responsive cell lines such as B9-TY1 cells and H-35 hepatoma cells. As shown in Figure 2A, IL-11 strongly induced tyrosine phosphorylation of a 97/95 KDa protein in B9-TY1 cells. The kinetic studies showed that protein tyrosine phosphorylation appeared at 1 min following IL-11 stimulation, peaked by 15 to 30 min and persisted for at least 150 min (data not shown). Interestingly, both 97 and 95 KDa tyrosine phosphorylated proteins appeared at 1 min and the intensity of 95 KDa tyrosine phosphorylated protein is stronger when compared with 97 KDa protein. The 95 KDa band disappeared by 15 min while 97 KDa band become much stronger when compared with that at 1 min. We speculated that the 95 and 97 KDa tyrosine phosphorylated proteins are the same protein with different levels of tyrosine phosphorylation. The behavior of 97/95 KDa tyrosine phosphorylated proteins is also seen in other cell types such as 3T3-L1 preadipocytes and H-35 hepatoma cells (data not shown).

In contrast to IL-11, IL-6 induced multiple tyrosine phosphorylated proteins with apparent molecular masses of 97/95, 130, 150, and 165 KDa in B9-TY1 cells (Fig. 2B). The kinetics of 97/95 KDa protein tyrosine phosphorylation induced by IL-6 is very similar to that of IL-11 (Figure 2 A and B). These results suggest that there are similarities and differences between IL-11 and IL-6 in the first step of signaling pathways in B9-TY1 cells. The protein tyrosine phosphorylation induced by either IL-11 or IL-6 can be inhibited by a specific protein tyrosine kinase inhibitor genistein, but not by a general protein kinase inhibitor H7 (a strong inhibitor for protein kinase C, cAMP or cGMP dependent protein kinases) (Fig. 2C). In H-35 hepatoma cells, IL-11 and IL-6 induced the same protein tyrosine phosphorylation with molecular masses of 97/95 and 44 KDa. The different pattern of tyrosine phosphorylation induced by IL-11 or IL-6 in different cell types implicated that there are cell type specific protein tyrosine kinases triggered by these two factors . It is critical to further identify these tyrosine kinases and their respective substrates to elucidate the role of protein tyrosine phosphorylation in IL-11 or IL-6 mediated signal transduction pathways.

Although IL-11 induces different patterns of protein tyrosine phosphorylation in various cell types, the 97/95 KDa tyrosine phos-

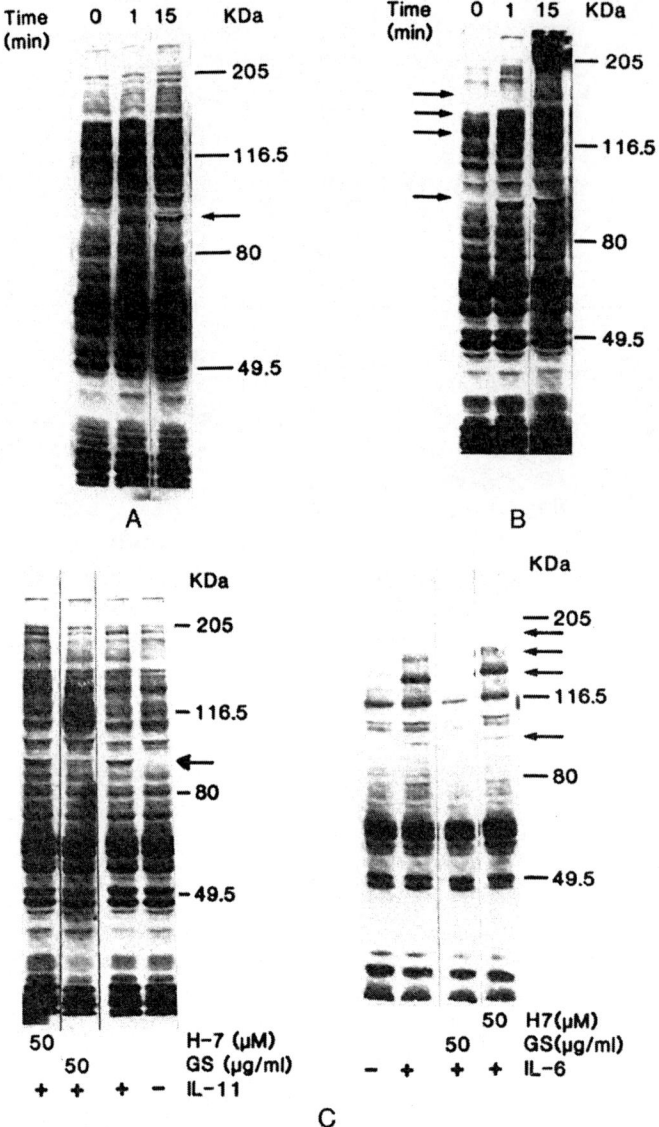

Figure 2. Protein tyrosine phosphorylation induced by IL-11 or IL-6 in B9-TY1 cells. B9-TY1 cells were washed three times with serum free RPMI 1640 medium. After factor starvation for 16 hours, cells were preincubated without (A and B) or with (C) indicated concentrations of genistein (GS) or H7 for 30 minutes. Cells were then stimulated with 250 ng/ml of IL-11 (A) or IL-6 (B) for indicated time (A and B) or for 20 minutes (C). Tyrosine phosphorylated proteins were determined by immunoblotting with anti-phosphotyrosine monoclonal antibody.

326

Table 1. Protein tyrosine phosphorylation induced by IL-11 or IL-6 in various cell lines

Cell lines	IL-11	IL-6	Tyrosine phosphorylated proteins (KDa)					
			165	150	130	97/95	47	44
3T3-L1	+	-	-	+	-	+	+	+
	-	+	-	-	-	-	-	-
B9-TY1	+	-	-	-	-	+	-	-
	-	+	+	+	+	+	-	-
H-35	+	-	-	-	-	+	-	+
	-	+	-	-	-	+	-	+
TF-1	+	-	-	-	-	+	-	-
	-	+	-	-	-	+	-	-
C3H10t1/2	+	-	-	+	-	+	+	+

phorylated protein was present in all the cell lines tested including 3T3-L1 preadipocytes, H-35 hepatoma cells, B9-TY1 cells C3H10t1/2 fibroblasts, and TF-1 leukemia cells[20]. The tyrosine phosphorylated proteins induced by IL-11 or IL-6 in different cell lines are summarized in Table 1.

EFFECT OF TYROSINE KINASE INHIBITOR GENISTEIN AND PROTEIN KINASE INHIBITOR H7 ON IL-11 OR IL-6 INDUCED B9-TY1 CELL PROLIFERATION

It has been increasingly evident that protein tyrosine kinases may be important in growth factor or cytokine mediated cell growth signals[18,19,21,22,23,24]. As demonstrated above, tyrosine kinase inhibitor genistein, but not H7 inhibited protein tyrosine phosphorylation induced by IL-11 or IL-6 in B9-TY1 cells. If protein tyrosine phosphorylation is the key regulatory mechanism in the IL-11 or IL-6 mediated growth signals, we would expect that tyrosine kinase inhibitor genistein, but not H7, will inhibit IL-11 or IL-6 induced B9-TY1 cell proliferation. We tested this possibility by measuring the mitogenic activities of IL-11 and IL-6 on B9-TY1 cells in the presence of genistein or H7. The results demonstrated that genistein inhibited IL-11 or IL-6 induced B9-TY1 cell proliferation in a dose dependent manner. Surprisingly, IL-11 or IL-6 induced B9-TY1 cell proliferation was also inhibited by H7, which did not inhibit tyrosine phosphorylation induced by either of these cytokines. These results not only suggest that tyrosine phosphorylation is the essential step for IL-11 or IL-6 mediated growth signals, but also imply that H7 sensitive protein kinases may play critical roles in IL-11 and IL-6 triggered signal transduction.

ACTIVATION OF TIS8, TIS11, TIS21, AND JUNB PRIMARY RESPONSE GENES BY IL-11 OR IL-6 IN DIFFERENT CELL TYPES

It has become more clear now that the products of primary response genes or oncogenes may play important roles in controlling cell growth and/or differentiation[18,19,25], most likely as a signal transducer or as a transcriptional regulator. It has been shown that IL-6 can activate tis11 and JunB gene expression in mouse B cell hybridoma MH60.BSF-2[12] and in

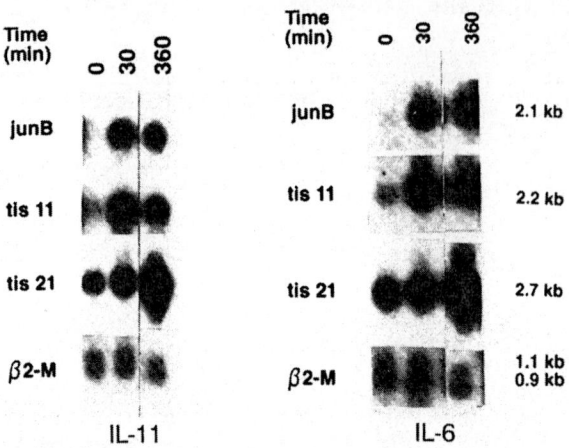

Figure 3. Effects of IL-11 or IL-6 on primary response gene expression in B9-TY1 cells. After factor starvation for 16 hours, cells (5 X 10[6] cells/ml) were stimulated with 500 ng/ml of IL-11 or IL-6 for indicated time. Total RNA was isolated and analyzed by Northern blot with tis11, tis21, and junB gene probes

mouse myeloid leukemia cell line, M1[13]. Because of the overlapping biological activities between IL-11 and IL-6, we examined whether IL-11 can activate the same primary response genes as those of IL-6 in various cell types including B9-TY1 cells, 3T3-L1 preadipocytes, TF-1 leukemia cells, and H-35 hepatoma cells. We tested the expression of eleven primary response genes including tis1, tis7, tis8, tis10, tis11, tis21[26], Fra-1[27], c-fos, c-myc, junB and NGFI-b[28] in these cells. Total RNAs were isolated[29] from factor-starved B9-TY1 cells following stimulation with IL-6 or IL-11 for up to 6 hours and analyzed by Northern or slot blot analysis. The results showed that three (tis11, tis21, and junB) out of eleven primary response genes tested were strongly activated by IL-11 or IL-6 stimulation. The kinetic studies showed that junB and tis11 transcripts appeared by 15 min following IL-11 or IL-6 stimulation, peaked by 30 to 60 min, and persisted for at least 360 min. It was found that tis21 gene is constitutively expressed in B9-TY1 cells and can be further induced following IL-11 or IL-6 stimulation (Fig. 3). In addition to junB and tis11, tis8 gene, which was not expressed in B9-TY1 cells, was rapidly and transiently activated by either IL-11 or IL-6 in H-35 hepatoma cells and 3T3-L1 preadipocytes. Our results revealed that the patterns of primary response gene expression induced by IL-11 or IL-6 are different in different cell types, suggesting that there are cell type specific primary response genes activated by IL-11 or IL-6. It is possible that tis21 may be a specific primary response gene for B9-TY1 cells while tis8 may only be expressed in H-35 and 3T3-L1 preadipocytes. This phenomenon may in part account for the pleiotropic activities of these cytokines. Our results also demonstrate that IL-11 and IL-6 can trigger the same primary response gene expression in all the cell lines tested, suggesting that IL-11 and IL-6 may share common signal transduction pathways as those of leukemia inhibitory factor (LIF) and IL-6[13].

We also examined the mechanisms by which tis8, tis11, tis21, and junB genes are activated by IL-11 or IL-6. Nuclear run on assays were performed to assess whether the activation of these primary response genes is at the transcriptional level. Serum and growth factor-starved B9-TY1 cells were stimulated with or without IL-11 or IL-6 for 60 min and nuclei were isolated for run on transcription assays[13]. The radiolabeled RNA probes hybridized with tis11, tis21, and junB plasmids, indicating that IL-11 and IL-6 strongly induced the transcription of tis11, tis21 and junB. The transcription of these genes was not blocked by pretreatment of cells with cycloheximide for 30 min before stimulation with IL-11 or IL-6, suggesting that the post-translational mechanisms may account for tis11, tis21, and junB gene transcription triggered by IL-6 or IL-11.

Tis21 is one of the immediate early response genes induced by a variety of growth factors[25,30]. The available tis21 sequence suggested that it may encode a transcriptional regulatory protein[30] although the exact role of tis21 is not clear at the present time. Tis11 gene encodes a novel proline, serine and glycine rich nuclear protein[25,31,32] which has a novel zinc finger structure and binds to zinc. Tis11 gene expression can be induced by growth factors or mitogens in a variety of tissues, especially in fetal tissues or growing liver cells[31,33]. This characteristics of tis11 gene expression strongly suggested that the product of tis11 gene may be involved in cellular growth or differentiation. The gene product of junB is a well-characterized transcriptional regulator[34,35] which belongs to one of the jun/AP-1 family of nuclear proto-oncogene proteins. In fact, the protein encoded by junB gene can form heterodimer with fos protein which is involved in transcriptional regulation and binds with increased affinity to AP-1 sites or closely similar sequences[36,37]. Our findings that IL-11 and IL-6 can activate the same primary response genes strongly suggest that these two factors may utilize the same or similar signal transduction pathways.

EFFECTS OF KNOWN PROTEIN KINASE ACTIVATORS OR INHIBITORS ON IL-11 OR IL-6 INDUCED EXPRESSION OF PRIMARY RESPONSE GENES

Our findings that tyrosine kinase inhibitor genistein inhibited both tyrosine phosphorylation and cell proliferation induced by IL-6 or IL-11 in B9-TY1 cells demonstrated that tyrosine kinases may be essential for the growth signals induced by these two growth factors. In addition, a general protein kinase inhibitor, H7, also inhibited B9-TY1 cell proliferation induced by IL-11 or IL-6, suggesting that H7 sensitive protein kinases are also required in IL-11 and IL-6 triggered signal transduction. The effects of genistein and H7 on primary response gene expression induced by IL-6 or IL-11 were also tested in IL-11 responsive cell lines. The results demonstrated that tyrosine kinase inhibitor, genistein, significantly inhibited IL-11 and IL-6 induced primary response gene expression in a dose dependent manner in all the cell lines tested. Interestingly, H7, which can not inhibit tyrosine phosphorylation, completely blocked IL-11 and IL-6 induced primary response gene expression (data not shown).

Since H7 is a general inhibitor for protein kinases, we examined whether known protein kinase activators such as protein kinase C activator 12-o-tetradecanoyl phorbol-13-acetate (TPA), cAMP dependent protein kinase activator N-6,2-o-dibutyryl cyclic AMP (dBcAMP), cGMP dependent protein kinase activator 8-bromo-cyclic GMP (8BrcGMP), cholera toxin (GTP binding protein activator), and calcium ionophore A23187 (calcium/calmodulin dependent protein kinase activator) can induce the same primary response genes as those of IL-11 or IL-6. Among the activators tested, only TPA can induce tis8, tis11, tis21, and junB gene expression in B9-TY1 cells and H-35 hepatoma cells. To understand the me- chanisms of TPA action, the cells were depleted of protein kinase C by TPA treatment for 48 hours. Depletion of protein kinase C did not affect IL-11 or IL-6 induced primary response gene expression, but completely blocked TPA induced primary response gene expression, indicating that protein kinase C is not involved in primary response gene induction by IL-11 or IL-6. Furthermore, neither protein kinase C inhibitor sphingosine nor calcium/ calmodulin protein kinase inhibitor, W7, can inhibit the expression of tis8, tis11, tis21, and junB genes stimulated by IL-11 or IL-6 .

These results are consistent with our hypothesis that H7 sensitive protein kinases which are essential in IL-11 and IL-6 triggered signal transduction are neither protein kinase C nor cAMP, cGMP or calcium/ calmodulin dependent protein kinases. Further identification of novel protein kinases sensitive to H7 inhibitor is critical to the understanding of the mechanisms of IL-11 and IL-6 mediated signal transduction.

POSSIBLE INVOLVEMENT OF IL-6 SIGNAL TRANSDUCER, GP130 IN IL-11 TRIGGERED SIGNALS

It is inconceivable that the two cytokines, such as IL-6 and IL-11, have no sequence homology yet share many of the biological activities. Interestingly, recent computer analysis has suggested the possible common tertiary structures between the two cytokines (M. Walter, personal communication). The similarity in the predicted tertiary structures raises the possibility that IL-11 may share the same or similar receptor system with that of IL-6 which is consisted of a binding subunit (gp80) and a signal transducer subunit (gp130)[9,10,11]. Our initial biochemical studies have ruled out the possibility that IL-11 may bind to the same protein as that of IL-6[7,9]. Our findings suggested that the downstream events such as the expression of primary response genes triggered by IL-11, are very similar to those of IL-6 in all the IL-11 responsive cell lines tested. These results strongly suggested that IL-11 may use the same signal transducer gp130 as that of IL-6. To directly demonstrate that gp130 is involved in IL-11 mediated signal transduction, we have utilized anti-human gp130 mono- clonal antibodies which can block IL-6 induced biological activities. The studies with a multifactor responsive leukemia cells, TF-1, showed that anti-human gp130 antibodies blocked IL-11 or IL-6 induced cell proliferation, but had no effects on granulocyte-macrophage colony stimulating factor or erythropoietin induced TF-1 cell growth. Anti-human gp130 antibodies also inhibited IL-11 or IL-6 induced protein tyrosine phosphorylation and junB gene expression. Furthermore, anti-human IL-6 receptor antibody significantly inhibited IL-6 induced TF-1 cell proliferation, tyrosine phosphorylation, and junB gene expression, but had no effect on the same biological functions mediated by IL-11. These results demonstrated that IL-6 and IL-11 utilize their respective binding proteins and IL-6 signal trans- ducer, gp130 is involved in IL-11 mediated signaling. In addition to IL-6 and IL-11, gp130 has recently been shown to be involved in the signaling pathways mediated by LIF[13,17], oncostatin M[17], and ciliary neurotrophic factor (CNTF)[38,39]. Taken together, these results revealed that gp130 plays pivotal roles in cytokine mediated signal transduction pathways.

SUMMARY AND CONCLUSION

IL-11 is a novel multifunctional cytokine produced by bone marrow stromal fibroblasts. Our studies indicated that most of the biological activities of IL-11 overlap with those of IL-6 in various system although the primary structure between IL-11 and IL-6 is quite different. These observations led to the speculation that IL-11 may share common receptor systems and/or use common signal transduction pathways as those of IL-6. Studies on signal transduction suggested that IL-11 triggered different up-stream signals (protein tyrosine phosphorylation) in 3T3-L1 preadipocytes and B9-TY1 cells when compared with those of IL-6. In H-35 hepatoma cells and TF-1 leukemia cells, IL-11 induced identical protein tyrosine phosphorylation as that of IL-6. The results from downstream signals (primary response gene expression) showed that IL-11 and IL-6 activated the same primary response gene transcription in all cell lines tested. These results strongly suggested that IL-11 and IL-6 may share common signal transduction pathways with the possibility of sharing common signal transducer, gp130. The results from anti-human gp130 antibody studies indicated that TF-1 cell proliferation, tyrosine phosphorylation and primary response gene expression induced by IL-11 or IL-6 can be blocked by anti-human gp130 antibodies, demonstrating that IL-11 can utilize IL-6 signal transducer, gp130. Our results together with other studies on LIF, oncostatin M, and CNTF which use gp130 as a signal transducer, strongly implicated that gp130 is a important signal transducer in a complicated cytokine network. Further identification of protein tyrosine kinases and H7 sensitive protein kinases associated gp130 will be essential to elucidate the mechanisms by which gp130 transduces signals from extracellular stimulation to intracellular events.

ACKNOWLEDGMENTS

We would like to thank Dr. Paul Schendel at Genetics Instititue for providing purified recombinant human IL-11; Dr. H. Herschman for tis1, tis7, tis8, tis11, andtis21 plasmids; Dr. T. Curran for Fra-1 plasmid, Dr. T. Milbrandt for NGFI-b plasmid, and Dr. T. Taga for anti-human gp130 and anti-IL-6 receptor antibodies. Y.-C. Yang is the Scholar of the Leukemia Society of America. T.Y. is supported in part by fellowship from American Heart Association, Indiana Affiniated Inc.

REFERENCES

1. S.R. Paul, F. Bennett, J.A. Calvetti, K. Kelleher, C.P. Wood, R.M. O'Hara,Jr., A.C. Leary, B. Sibley, S.C. Clark, D.A. Williams, and Y.-C. Yang, Molecular cloning of a cDNA encoding interleukin-11, a stromal cell-derived lymphopoietic and hematopoietic cytokine, *Proc. Natl. Acad. Sci. USA* 87:7512 (1990).
2. E. Bruno, R.A. Briddell, R.J. Cooper, and R. Hoffman, Effects of recombinant interleukin-11 on human megakaryocyte progenitor cells, *Exp. Hematol.* 19:378 (1991).
3. T.Yin, P. Schendel, and Y.-C. Yang, Enhancement of in vitro and in vivo antigen specific antibody responses by interleukin-11,*J. Exp. Med. 175:211 (1992).*
4. M. Musashi, Y.-C. Yang, S.R. Paul, S.C. Clark, T. Sudo, and M. Ogawa, Direct and synergistic effects of interleukin-11 on murine hemopoiesis in culture,*Proc. Natl. Acad. Sci. USA* 88:765 (1991).

5. H. Baumann, and P. Schendel, Interleukin-11 regulates the hepatic expression of the same plasma protein genes as interleukin-6, *J. Biol. Chem.* **266**:20424 (1991).

6. I. Kawashima, J. Ohsumi, K. Mita-Honjo, K. Shimoda-Takano, H. Ishikawa, S. Sakakibara, K. Miyadai, and Y. Takiguchi, Molecular cloning of a cDNA encoding adipogeniesis inhibitory faaactor and identity with interleukin-11, *FEBS Letters* **283**:199 (1991).

7. T. Yin, K. Miyazawa, and Y.-C. Yang, Characterization of interleukin-11 receptor and protein tyrosine phosphorylation induced by interleukin-11 in mouse 3T3-L1 cells, *J. Biol. Chem.* **267**:8347 (1992).

8. J. Ohsumi, K. Miyadai, I. Kawashima, H. Ishikawa-Ohsumi, S. Sakakibara, K. Mita-Honjo, and Y. Takiguchi, Adipogenisis inhibitory factor: a novel inhibitory regulator of adipose conversion in bone marrow, *FEBS Letters* **288**:13 (1991).

9. K. Yamasaki, T. Taga, Y. Hirato, H. Yawata, Y. Kawarishi, B. Seed, T. Taniguchi, T. Hirano, and T. Kishimoto, Cloning and expression of the human interleukin-6 receptor, *Science* **241**:825 (1988).

10. T. Taga, M. Hibi, Y. Hirata, K. Yamasaki, K. Yasukawa, T. Matsuda, T. Hirano, and T. Kishimoto, Interleukin-6 triggers the association of its receptor with a possible signal transducer, gp130, *Cell* **58**:573 (1989).

11. M. Hibi, M. Murakami, M. Saito, T. Hirano, T. Taga, and T. Kishimoto, Molecular cloning and expression of an IL-6 signal transducer, gp130, *Cell* **63**:1149 (1990).

12. K. Nakajima, and R. Wall, Interleukin-6 signals activating junB and tis11 gene transcription in a B cell hybridoma, *Mol. Cell. Biol.* **11**:1409 (1991).

13. K.A. Lord, A. Abdollahi, S.M. Thomas, M. Demarco, J.S. Brugge, B.Foffman-Liebermann, and D.A. Liebermann, Leukemia inhibitory factor and interleukin-6 trigger the same immediate early response including tyrosine phosphorylation, upon induction of myeloid leukemial differentiation, *Mol. Cell. Biol.* **11**:4371 (1991).

14. J.B. Bolen, Signal transduction by the SRC family of tyrosine protein kinases in hemopoietic cells, *Cell Growth Differentiation* **2**:409 (1991).

15. A. Ullrich, and J. Schlessinger, Signal transduction by receptors with tyrosine kinase activity, *Cell* **61**:2035 (1990).

16. L.A. Aarden, E.R. De Groot, O.L. Schaap, and P.M. Lansdorp, Production of hybridoma growth factor by human monocytes, *Eur. J. Immunol.* **17**:1411 (1987).

17. D.P. Gearing, M.R. Comeau, D.J. Friend, S.D. Gimpel, C.J. Thut, J. McGourty, K.K. Brasher, J.A. King, S. Gillis, B. Mosley, S.F. Ziegler, and D. Cosman, The IL-6 signal transducer, gp130: an oncostatin M receptor and affinity converter for the LIF receptor, *Science* **255**:1434 (1992).

18. L.C. Cantley, K.R. Anger, C. Carpenter, B. Duckworth, A. Graxiani, R. Kapeller, S. Soltoff, Oncogenes and signal transduction, *Cell* **64**:281 (1991).

19. M.V. Chao, Growth factor signaling: where is the specificity ? *Cell* **68**:995 (1992).

20. T. Kitamura, T. Tange, T. Terasawa, S. Chiba, T. Kuwaki, K. Miyagawa, Y.-F. Piao, K. Miyazono, A. Urabe, and F. Takaku, Establishment and characterization of a unique human cell line that proliferates dependently on GM-CSF, IL-3 or erythropoietin, *J. Cell. Physiol.* **140**:323 (1989).

21. A. Kuriu, H. Ikada, Y. Kanakura, J.D. Griffin, B. Druker, H. Yagura, H. Kitayama, J. Ishikawa, T. Nishiura, Y. Kanayama, T. Yonezawa, and S. Tarui, Proliferation of human myeloid leukemia cell line associated with the tyrosine phosphorylation and activation of the prot-oncogene c-kit product, *Blood* **78**:2834 (1992).

22. H. Ogawara, T. Akiyama, S. Watanabe, N. Ito, M. Kobori, and Y. Seoda, Inhibition of tyrosine kinase actiivty by synthetic isoflavones and flavones, *J. Antibiot.* **42**:340 (1989).

23. A. Okura, H. Arakawa, H. Oka, T. Yoshinari, and Y. Monden, Effect of genistein on topoisomerase activity and on the growth of [val 12] Ha-ras-transformed NIH 3T3 cells, *Biochem. Biophys. Res. Commun.* **157**:183 (1988).

24. C. Linassier, M. Pierre, J.-B.L. Pecq, and J. Pierre, Mechanisms of action in NIH 3T3 cells of genistein, an inhibitor of EGF receptor tyrosine kinase activity, *Biochem. Pharmacol.* **39**:187 (1990).

25. H.R. Herschman, Primary response genes induced by growth factors and tumor promoters, *Annu. Rev. Biochem.* **60**:281 (1991).

26. R.W. Lim, B.C. Varnum, and H.R. Herschman, Cloning of tetradecanoyl phobol ester induced primary response sequences and their expression in density arrested 3T3 cells and a TPA non-proliferative variant, *Oncogene* **1**:263 (1987).

27. D.R. Cohen, and T. Curran, Fra-1, a serum inducible, cellular immediate early gene that encodes a fos-related antigen, *Mol. Cell. Biol.* **8**:2063 (1988).

28. M.A. Watson, and T. Milbrandt, The NGFI-b gene, a transcriptionally inducible member of the steroid receptor gene superfamily: genomic structure and expression in rat brain after seizure induction, *Mol. Cell. Biol.* 9:4213 (1989).

29. P. Chomczynski, and N. Sacchi, Single step method of RNA isolation by acid guanidinium thiocyanate phoenol chloroform extract, *Analyt. Biochem.* 162:156 (1987).

30. B.S. Fletcher, R.W. Lim, B.C. Varnum, D.A. Kujubu, R.A. Koski, and H.R. Herschman, Structure and expression of tis21, a primary response gene induced by growth factors and tumor promotors, *J. Biol. Chem.* 266:14511 (1991).

31. R.N. Dubois, M.W. Mclane, K. Ryder, L.F. Lau, and D. Nathans, A growth factor inducible nuclear protein with a novel cysteine/histidine repetitive sequence, *J. Biol. Chem. 265:19185 (1990).*

32. W.S. Lai, D.J. Stumpo, and P.J. Blackshear, Rapid insulin-stimulated accumulation of an mRNA encoding a proline-rich protein, *J. Biol.Chem.* 265:16556 (1990).

33. R.W. Lim, B.C. Varnum, T.G. O'Berien, and H.R. Herschman, Induction of tumor promotor inducible genes in murine 3T3 variants can occur through protein kinase C dependent and independent pathways, *Mol. Cell. Biol.* 9:1790 (1989).

34. K. Ryder, L.F. Laaaau, and D. Nathans, A gene activted by growth factors is related to the oncogene v-jun, *Proc. Natl. Acad. Sci. USA* 85:1487 (1988).

35. Y. Nakabeppu, K. Ryder, and D. Nathans, DNA binding activities of three murine jun proteins: stimulation by fos, *Cell* 55:907 (1988).

36. R. Chiu, W.J. Boyle, J. Meek, T. Smeal, T. Hunter, and M. Karin, The c-fos protein interacts with c-jun/AP-1 to stimulate transcription of AP-1 responsive genes, *Cell* 54:541 (1988).

37. P. Sassone-Corsi, W.W. Lamph, M. Kamps, and I.M. Verma, Fos-associated cellular p39 is related to nuclear transcription factor AP-1, *Cell* 54:553 (1988).

38. S. Davis, T.H. Aldrich, D.M. Valenzuela, V. Wong, M.E. Furth, S.P. Squinto, G.D. Yancopoulos, The receptor for cilliary neurotrophic factor, *Science* 243:59 (1991).

39. N.Y. Ip, S.H. Nye, T.G. Boulton, S. Davis, T. Taga, Y.Li, S.J. Birren, K. Yasukawa, T. Kishimoto, D.J. Anderson, N. Stahl, and G.D. Yancopoulos, CNTF and LIF act on neuronal cells via shared signaling pathways that involve the IL-6 signal transducing receptor component gp130, *Cell* 69:1121 (1992).

REGULATION OF A MEMBRANE TYROSINE PHOSPHATASE BY SOMATOSTATIN

Jean Pierre Esteve[1], El Mostapha Zeggari[1], Chantal Cambillau[1], Isabelle Rauly[1], Louis Buscail[1], François Thomas[2], Nicole Vaysse[1], Christiane Susini[1]

[1]INSERM U 151 CHU Rangueil, 31054 Toulouse, FRANCE
[2]Ipsen Biotech, rue Cambronne, Paris, FRANCE

INTRODUCTION

Somatostatin is a negative regulator of various cellular functions including secretion and cell proliferation[1]. Somatostatin induces its biological effects by interacting with different receptors subtypes which have been shown to belong to family of G protein-linked receptors[2] and are coupled to a variety of signal transduction pathways[3] including adenylate cyclase, ionic conductance channels[4] and protein dephosphorylation on serine/threonine[5] and tyrosine residues[6].

In pancreatic tumoral AR42J cells, we previously demonstrated that somatostatin analog, SMS 201-995 (SMS), antagonizes the mitogenic effect of growth factors acting on tyrosine kinase receptors such as EGF (Epidermal growth factor)[7] and bFGF (Basic fibroblast growth factor)[8]. SMS exerts this effect by interacting with specific somatostatin receptors negatively coupled to adenylate cyclase. However, this cellular effector system seems to be not involved in the growth inhibiting effect of somatostatin[7].

Here we tested the hypothesis that somatostatin initiates negative growth signal-transduction pathway by activating a membrane tyrosine phosphatase which migh be expected to counteract tyrosine kinase activities mediating growth factor-induced mitogenic responses.

SOMATOSTATIN STIMULATES A MEMBRANE TYROSINE PHOSPHATASE

The effect of somatostatin on membrane tyrosine phosphatase (PTP) activity was investigated by incubating rat pancreatic acini in the presence or not of different concentrations of somatostatin analogs, SMS, BIM 23014 (BIM), RC 160 (generous gift from Dr A.V. Schally). After homogenization, we measured PTP activity on crude membrane fractions with the synthetic substrate ^{32}P-labeled poly [Glu,Tyr][9]. Figure 1 shows that treatment of acini with somatostatin analogs resulted in an increase of PTP activity which was dependent on the analog concentration. A same dose-response curve was obtained with the three analogs with maximal stimulation occuring at 0.1 nM SMS, 0.1-1 nM BIM, 1-10 nM RC 160. However, the potencies of the analogs were different. RC 160 was less active than either BIM or SMS, SMS being the most active.

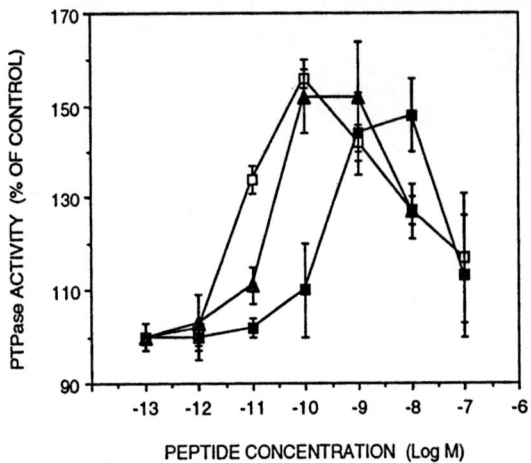

Figure 1. Effect of somatostatin analogs on PTP activity. Acini were incubated for 15 min at 25 °C in the presence or not of different concentrations of SMS (□), BIM (▲), RC 160 (■). Membranes were prepared and assayed for PTP activity using 0.3 μM ^{32}P-Poly [Glu,Tyr]. The reaction was allowed to proceed for 5 min at 30 °C.

To determine the relative affinities of these analogs for somatostatin receptors, we measured their potencies to inhibit somatostatin binding (Figure 2). As observed in Figure 2, RC 160 was again less active than the other analogs whereas SMS was highly effective to inhibit radioligand binding.

A good relationship was observed between the relative potencies of analogs to inhibit somatostatin binding and to stimulate PTP activity (Table 1). Furthermore, similar relative potencies were detected for the inhibitory effect of analogs on AR42J cell growth, BIM being less potent than SMS to inhibit cell proliferation. These results are consistent with the involvement of the somatostatin receptor in mediating the stimulation of tyrosine phosphatase and favor the hypothesis that the enzyme was implicated in the antiproliferative signal activated by somatostatin.

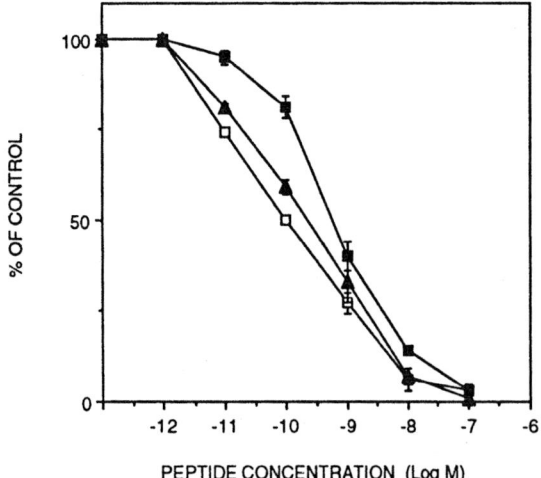

Figure 2. Competition inhibition of somatostatin binding to pancreatic membranes by analogs. Membranes were incubated for 60 min at 25 °C with [^{125}I]somatostatin analog[10] in the presence of increasing concentrations of SMS (□), BIM (▲), RC 160 (■).

Table 1. Receptor binding affinities and PTP-stimulating activities of somatostatin analogs

Analog	Receptor binding IC_{50} (pM)	PTP activity EC_{50} (pM)	Growth inhibition IC_{50} (pM)
SMS	86 ± 12	7 ± 1	24 ± 9
BIM	220 ± 35	37 ± 4	74 ± 6
RC 160	650 ± 125	375 ± 25	ND

The affinities (IC_{50}) of analogs for the somatostatin receptor and their potencies to stimulate PTP activity (EC_{50}) were determined in experiments such as shown in Fig. 1 and 2. Their potencies to inhibit cell growth were determined on AR42J cells as described in [7]. ND:not determined.

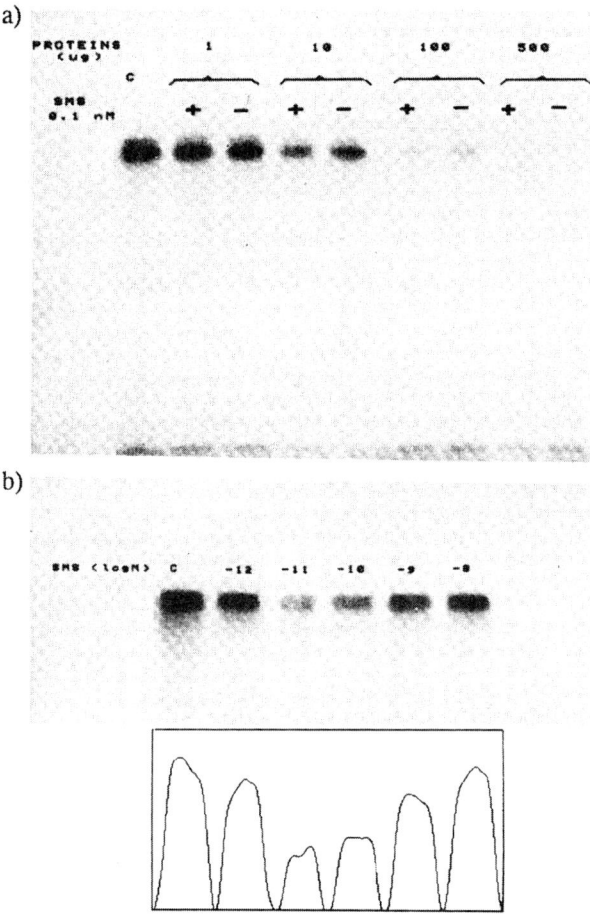

Figure 3. Effect of SMS on dephosphorylation of ^{32}P-EGF receptor induced by tyrosine phosphatase from pancreatic membranes. a) Different concentrations of pancreatic membranes (1-500 μg) were preincubated with (+) or without (-) 0.1 nM SMS for 20 min at 25 °C. b) Pancreatic membranes were incubated with increasing concentrations of SMS (10^{-12}–10^{-8} M). After solubilization, membrane extracts were incubated for 15 min at 30 °C with ^{32}P-labeled A 431 membranes. Proteins were separated by SDS/PAGE, followed by autoradiography.

The effect of somatostatin on PTP activity was also evaluated directly on acinar cell membranes after incubation of membranes with SMS. PTP activity was measured with a physiological substrate, ^{32}P-labeled EGF receptor. EGF receptors from A-431 cell membranes were autophosphorylated[11] and then incubated with solubilized purified pancreatic acinar membranes pretreated or not with SMS. As observed in Figure 3a, the intensity of the major labeled band, corresponding to the 170 kD EGF receptor, was decreased in the presence of increasing concentrations of solubilized membranes suggesting that solubilized membranes contain a PTP activity. Preincubation of membranes with 0.1 nM SMS before solubilization further reduced the intensity of this band.

This demonstrates that dephosphorylation of ^{32}P-EGF receptor is stimulated by SMS. The stimulatory effect was dependent on SMS concentration as shown in Figure 3b with a maximal stimulation occuring at .01-.1 nM. These results indicate that somatostatin can directly stimulate a membrane tyrosine phosphatase after interaction with its receptor.

SOMATOSTATIN RECEPTOR IS ASSOCIATED WITH A TYROSINE PHOSPHATASE

To address further the mechanism of tyrosine phosphatase activation induced by somatostatin , we purified solubilized somatostatin receptors[12] from rat pancreatic membranes preincubated with somatostatin 28, by immunoaffinity using anti-somatostatin 28 antibodies. Analysis by SDS/PAGE and silver staining of the immunopurified proteins revealed a major band of 87 kDa which was specific of somatostatin receptor. Indeed, this band was not observed when immunopurified proteins derived from untreated membranes or when non immun serum was used instead of immun serum.

The immunopurified proteins containing somatostatin receptors exhibited a tyrosine phosphatase activity using ^{32}P-poly [Glu,Tyr] or ^{32}P-EGF receptor as substrates. This activity was inhibited by vanadate, a specific inhibitor of tyrosine phosphatases and stimulated by dithiothreitol in agreement with the essential role of cysteinyl residue of tyrosine phosphatase catalytic unit[13].

When the immunopurified proteins were subjected to Western immunoblotting using anti-tyrosine phosphatase antibodies directed against a conserved domain of tyrosine phosphatase family (kind gift of Dr H.L. Ostergaard), we observed that a band migrating at 50 kD was specifically revealed by the antiserum in the eluted fractions containing somatostatin receptors. This band was not observed in the absence of somatostatin receptor in the eluted fractions.

CONCLUSION

Altogether, these results suggest that a stimulation of a membrane tyrosine phosphatase may be part of the signal transduction pathway promoted by somatostatin receptor occupancy and could be involved in the negative growth signal induced by somatostatin. The molecular mechanism of tyrosine phosphatase stimulation is unknown. It could result from a direct interaction of the enzyme with the C-terminal part of the receptor or an indirect one through a regulatory protein.

ACKNOWLEGEMENTS

This research is supported in part by A European Economic Community grant SC1-CT91-0632 and Association de la Recherche contre le Cancer 6755.

REFERENCES

1. M.J.M. Lewin. The somatostatin receptor in GI tract. Ann Rev. Physiol. 54:455 (1992).

2. Y. Yamada, S.R. Post, K. Wang, H.S. Tager, G.I. Bell, S. Seino. Cloning and functional characterization of a family of human and mouse somatostatin receptors expressed in brain, gastrointestinal tract. Proc. Natl. Acad. Sci. USA 89: 251, (1992).

3. Y.C. Patel, K.K. Murthy, E. Escher, D. Banville, J. Spiess,C.B. Srikant. Mechanism of action of somatostatin: an overview of receptor function and studies of the molecular characterization and purification of somatostatin receptor proteins. Metabolism supp 2, 39:63 (1990).

4. T. Reisine, H.T. He, S. Rens-domanio, J.M. Martin, K. Raynor, S. Borislow, K. Thermos. Biochemical properties of somatostatin receptors. Metabolism supp 2, 39:70 (1990).

5. F. Reyl, M.J.M. Lewin. Intracellular receptor for somatostatin in gastric mucosal cells. Decomposition and reconstitution of somatostatin-stimulated phosphoprotein phosphatases. Proc. Natl. Acad. Sci. USA 79:978 (1982).

6. C. Liebow, C. Reilly, M. Serrano,and A.V. Schally. Somatostatin analogues inhibit growth of pancreatic cancer by stimulating tyrosine phosphatase. Proc. Natl.Acad. Sci. USA 86:2003 (1989).

7. N. Viguerie, N. Tahiri-Jouti, A.M. Ayral, C. Cambillau, J.L. Scemama, M.J. Bastie, J.P. Esteve, L. Pradayrol, C. Susini, N. Vaysse. Direct inhibitory effects of a somatostatin analog,SMS 201-995, on AR42J proliferation via pertussis toxin-sensitivte guanosine triphosphate-binding protein-independent mechanism. Endocrinology 124:1017 (1989).

8. N. Bensaid, N. Tahiri-Jouti, C. Cambillau, N. Viguerie, B. Colas, C. Vidal, J.P. Tauber, J.P. Esteve, C. Susini and N. Vaysse. Int. J. Cancer 50:796 (1992).

9. G. Swarup,and G. Subrahamanyam. Purification and characterization of a protein phosphotyrosine phosphatase from rat spleen which dephosphorylates and inactivates a tyrosine-specific protein kinase. J. Biol. Chem. 264:7801 (1989).

10. S. Knuthsen, J. P. Esteve, B. Bernadet, N. Vaysse and C. Susini. Molecular characterization of the solubilized receptor of somatostatin from rat pancreatic membranes. Biochem. J. 254:641 (1988).

11. P. Vicendo, J. Fauvel, J. Ragab-Thomas, H. Chap. Identification, characterization and purification to near homogeneity of a novel 67 kDa phosphotyrosyl protein phosphatase associated with pig lung annexin extract. Biochem. J. 278:435 (1991).

12. S. Knuthsen, J. P. Esteve, C. Cambillau, B. Colas, C. Susini, N. Vaysse. Solubilization and characterization of active somatostatin receptors from rat pancreas. J. Biol. Chem. 265:1129 (1990).

13. E.H. Fischer, H. Charbonneau, N.K. Tonks. Protein tyrosine phosphatases: a diverse family of intracellular and transmembrane enzymes. Science 253:401 (1991).

CYCLOHEXIMIDE HAS DIFFERING EFFECTS ON IMMEDIATE EARLY GENE EXPRESSION IN IL-3 OR M-CSF STIMULATED HUMAN MONOCYTES

Padmini Rao and R. Allan Mufson

Cell Biology Department
Holland Laboratory
American Red Cross
15601 Crabbs Branch Way
Rockville, MD 20855

INTRODUCTION

The continued survival and differentiation of monocytes and macrophages is dependent of the presence of hematopoietic cytokines such as interleukin-3 (IL-3), granulocyte macrophage colony stimulating factor (GM-CSF), and macrophage colony stimulating factor (M-CSF).[1] Murine monocytes/macrophages require these factors for *in vitro* proliferation as well as optimal differentiation. In contrast, human monocytes do not proliferate in response to cytokine stimulation but require these factors only for optimal survival and differentiation.[2]

The receptors for the cytokines IL-3 and GM-CSF fall into the new family of hematopoietic receptors that do not possess a tyrosine kinase domain.[3] The receptor for M-CSF is the *product of the c-fms* protooncogene, and contains an intrinsic tyrosine kinase domain.[4] All three cytokines, however, stimulate tyrosine phosphorylation after binding to their respective receptors.[5] Few of the other events occurring after CSF stimulation of quiescent human monocytes macrophages have been characterized.

PDGF and other growth factors acting through receptor mediated pathways have been shown to induce the sequential expression of specific genes in quiescent mammalian cells preceding the onset of DNA synthesis and proliferation.[6] The first set of genes induced are rapidly activated (10-60 min) after growth factor addition. This rapid induction of the IER genes is at least partially due to transcriptional activation and occurs even in the presence of inhibitors of protein synthesis, and these genes have thus been called immediate early response genes. This family of genes consists of coordinately expressed genes, the majority of which have not been fully characterized. Among the members that have been identified from this group are genes encoding DNA binding transcription factors of the leucine zipper family of proteins - c-fos, fra-1,2, fos-B, c-jun, jun B,D and the zinc binding proteins EGR-1, zif/268, NGF-1-A, and Krox 20.[7]

The ability of M-CSF and IL-3 to influence survival, proliferation and differentiation in monocyte/macrophage cell types led us to investigate whether these cytokines might also induce these IER genes in postmitotic cells.

CYTOKINE INDUCTION OF IMMEDIATE EARLY RESPONSE GENES

Our laboratory has found that depriving monocytes of human serum and colony stimulating factors for 24-48 hrs reduces the expression of IER genes to undetectable levels. Addition of either M-CSF or IL 3 was associated with the induction of the immediate early gene

Growth Factors, Peptides, and Receptors, Edited
by T.W. Moody, Plenum Press, New York, 1993

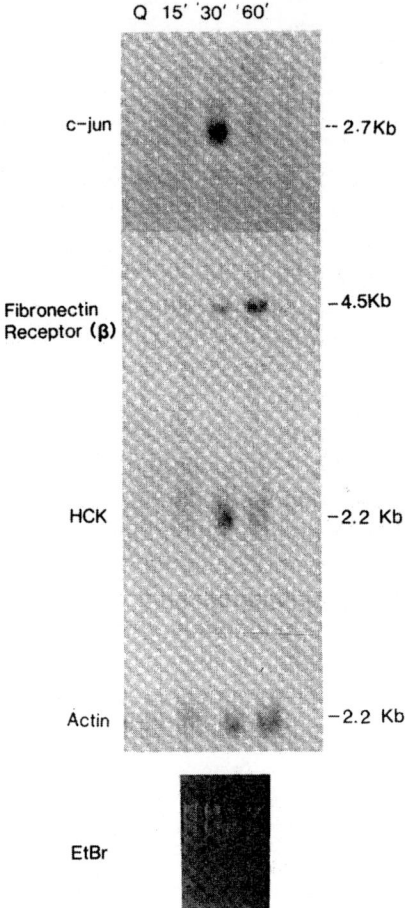

Q 15' '30' '60'

c-jun ---- 2.7Kb

Fibronectin
Receptor (β) --4.5Kb

HCK --2.2 Kb

Actin --2.2 Kb

EtBr

Figure 1. Time course for expression of IER genes by M-CSF in human monocytes. Total RNA was isolated from control quiescent monocytes (Q) or monocytes treated with M-CSF for the indicated time intervals, and subjected to Northern blot analysis. Equal amounts (10 μg) of RNA were loaded in each lane and the blot was sequentially probed for c-jun, fibronectin receptor (β-subunit), hck, and β-actin mRNA. The kb markers indicate the size of each mRNA as determined by comparison to an RNA ladder (BRL, Gaithersburg, MD). A photograph of the ethidium bromide stained agarose gel (*EtBr*) is also shown to demonstrate the equality of RNA loading between the lanes.

response. Within minutes of addition of M-CSF, there occurred a coordinate induction of c-jun, the myelomonocytic specific tyrosine kinase hck, fibronectin receptor and actin mRNA[8] (Fig. 1). IL-3 a more pluripotent cytokine than M-CSF also induced the c-jun protoncogene in factor deprived monocytes. The c-jun gene was rapidly induced within 10 min after M-CSF addition and returns to basal levels 60 min after addition (Fig. 1).

Nuclear run off transcription analysis revealed that the changes in mRNA expression we detected were due to alterations in gene transcription as well as post transcriptional mechanisms.

EFFECTS OF CYCLOHEXIMIDE ON M-CSF AND IL-3 INDUCED C-JUN mRNA

If the induction of the *c-jun* gene in response to cytokines is part of a true immediate early response, then it should be independent of *de novo* protein synthesis. To address this possibility M-CSF induction of IER gene mRNA was performed in the presence of 10 μg/ml of cycloheximide; a concentration that inhibited [³H] leucine incorporation into total cell protein

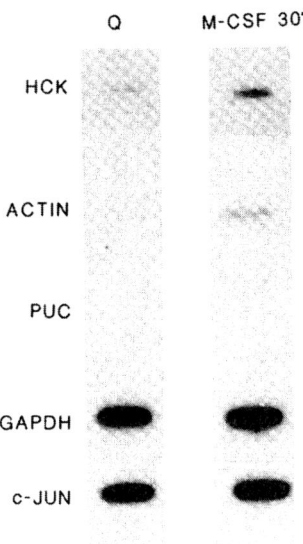

	Q	M-CSF 30'
HCK		
ACTIN		
PUC		
GAPDH		
c-JUN		

Figure 2. Nuclear run on transcription analysis of nuclei isolated from control quiescent and M-CSF treated monocytes. Nuclei were isolated from control quiescent monocytes (Q) or monocytes treated for 30 min with M-CSF. In vitro run on transcription reactions were performed and the [^{32}P] labeled run on transcripts were hybridized to a series of cDNA clones immobilized as slots on nitrocellulose. The hybridized filters were washed and autoradiographed. The pUC plasmid was used as a control for nonspecific hybridization to DNA.

by 95%. This concentration of cycloheximide alone induced the c-jun gene. This concentration of cycloheximide (CHX) also does not block the induction of the IER genes induced by M-CSF and, in fact, cycloheximide appears to synergise with M-CSF, inducing the *hck* and *c-jun* expression above those induced by M-CSF or cycloheximide alone. This synergism between growth factors and cycloheximide has been termed superinduction. Levels of β-actin mRNA however, were not enhanced in the presence of cycloheximide (Fig. 3).

The ability of cycloheximide to induce certain IER genes alone and to superinudce them together with growth factors has generally been ascribed to the inhibition of protein systhesis. Our experiments also reveal that cycloheximide concentrations less than 50ng/ml that do not inhibit protein synthesis also induce the c-jun gene in quiescent monocytes.

Simultaneous exposure of monocytes to cycloheximide and IL-3, surprisingly, did not result in the superinduction of the c-jun gene, despite the ability of either molecule to independently induce the gene. IL-3, in a concentration dependent manner, consistently suppressed the CHX induction of c-jun. Our earlier data[10], as well as those from other labs suggest the activation of serine/threonine protein kinases by IL-3 in monocytes. Since changes in activities of serine/threonine kinases are often related to subsequent changes in serine/threonine phosphatases[11], we investigated the effects of the potent serine/threonine phosphatase inhibitor okadaic acid, on the IL-3 suppression of CHX induced c-jun mRNA expression. These experiments clearly indicated that okadaic acid abrogated the IL-3 suppression of c-jun induction by CHX. The concentration of okadaic acid employed (12.5nM) is consistent with the involvement of a serine/threonine protein phosphatase 1A or 2A in the signal transduction mechanism of IL-3. Okadaic acid neither prevented the induction of c-jun by IL-3, nor did it augment the induction of c-jun mRNA by CHX alone. Since the activation of a protein serine/threonine phosphatase could conceivably inhibit CHX stimulated increases in protein phosphorylation, these results support the conclusion that IL-3 probably activates a serine/threonine phosphatase in monocytes. Transient activation of protein phosphatases by IL-3 could therefore, underlie the differential interactions of IL-3 and M-CSF with CHX, in the induction of the immediate early gene response.

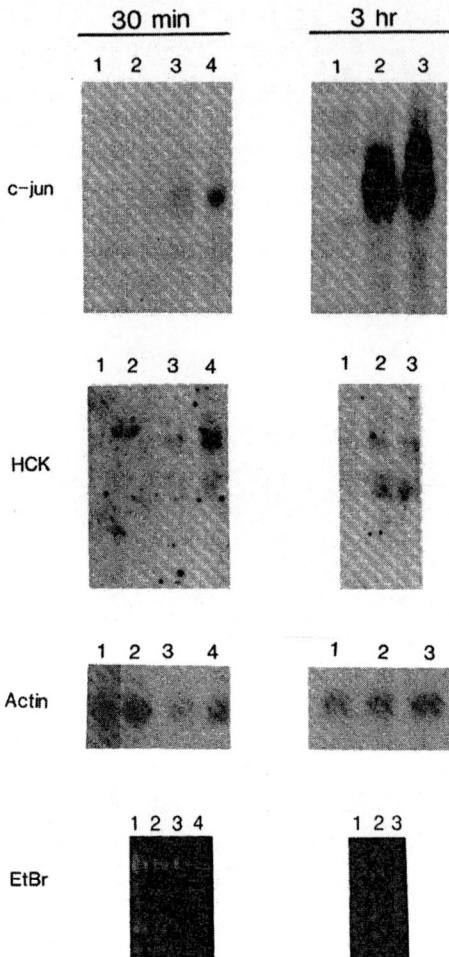

Figure 3. The effect of cycloheximide on IER gene induction. RNA was isolated at 30 min from control quiescent monocytes (1), monocytes treated with 30 ng/ml M-CSF (2) 1 μg/ml cycloheximide or (3) 1 μg.ml cycloheximide and 30 ng.ml m (4), and at 3 h from monocytes treated with 1) M-CSF 30 ng/ml, 2) cycloheximide 1 μg/ml, and 3) M-CSF 30ng/ml and cycloheximide 1 μg/ml, and the RNA was analyzed by Northern blotting. Equal amounts (1 μg) were loaded in each lane and the blots were probed for c-jun, hck, and actin mRNA. A photograph of the ethidium bromide stained gel (*EtBr*) is also shown to demonstrate that equal amounts of RNA were loaded in each lane.

ACKNOWLEDGEMENTS

This work was supported by NIH grant R-01 Ca 53609 and Council for Tobacco Research grant.

REFERENCES

1.   R.J.Tushinski, I.T.Oliver, L.J.Guilbert, W. Tynan, J. Warner and E.R. Stanley.
      Survival of mononuclear phagocytes depends on a lineage-specific growth factor that the differentiated cells selectively destroy. Cell 28:71(1982).

2.   M.K.Warner and P.Ralph. Macrophage growth factor CSF-1 stimulates human monocyte production interferon, tumor necrosis factor, and colony stimulating activity. J.Immunol. 137:2281(1986).

3.   A.Miyajima, T. Kitamura, N.Harada, T.Yokota and K.Arai. Cytokine receptors and signal transduction. Ann. Rev. Immunol. 10:295(1992).

4.   C.J.Sherr, C.W.Rettenmier and M.F.Rossel. Macrophage colony stimulating factor CSF-1 and its protooncogene encoded receptor. Cold Spring Harbor Symp. Quant. Biol. 1:521(1988).

5.   K. Sakamaki, I. Miyajima, Kitamura,T. and A. Miyajima. Critical cytoplasmic domains of the common βsubunit of the human GM-CSF, Il-3 and IL-5 receptors for growth signal transduction and tyrosine phosphorylation. EMBO J. 11:3541(1992).

6.   J.M. Almendral, D. Sommer, H. Mac-Donald-Bravo, J. Benckhardt, J. Perera and R. Bravo. Complexity of the early genetic response to growth factors in mouse fibroblasts. Mol. Cell. Biol. 8:2140(1988).

7.   S.H. Bernstein, S.M. Kharbanda, M.L. Sherman, V. Sukhatme and D.W. Kufe. Post transcriptional regulation of the zinc finger-encoding EGR-1 gene by granulocyte macrophage colony stimulating factor in human U-937 monocytic leukemia cells: Involvement of a pertussis toxin-sensitive Gprotein. Cell Growth and Differentiation. 2:273(1991).

8.   R.A. Mufson. Induction of immediate early response genes by macrophage colony stimulating factor in normal human monocytes. J. Immunol. 145:2333(1990).

9.   L.C. Mahadevan and D.R. Edwards. Signalling and superinduction. Nature 349:747(1991).

10.   R.A. Mufson, J. Szabo and D. Eckert. Human interleukin-3 induction of c-jun in normal monocytes is independant of tyrosine kinase and involves protein kinase C. J. Immunol. 148:1129(1992).

11.   P. Dent, A. Lavoinne, S. Nakielny, F.B. Caudwell, P. Watt and P. Cohen. The molecular mechanism by which insulin stimulates glycogen synthesis Pn mammaliam skeletal muscle. Hature 348:302(1990).

REGULATION OF GASTRIN RELEASING PEPTIDE GENE EXPRESSION

Muriel Draoui[1], James Battey[2], Michael Birrer[3] and Terry W. Moody[1]

[1]Department of Biochemistry and Molecular Biology
The George Washington University School of Medicine and Health Sciences
Washington, D.C. 20037
[2]Laboratory of Neurochemistry
National Institute of Neurological Disorders and Stroke, National Institutes of Health
Bethesda, MD 20892
[3]Biomarkers & Prevention Res. Branch
Division of Cancer Prevention and Control
National Cancer Institute
Kensington, MD 20895

INTRODUCTION

Bombesin/gastrin releasing peptide (BN/GRP) is an autocrine growth factor for some small cell lung cancer (SCLC) cells[1]. The GRP gene, which contains 2 introns and is localized to human chromosome 18q, is expressed in some SCLC cell lines such as NCI-H345 and H209[2-4]. GRP, which is present at high concentrations in classic SCLC cells, is released when the cAMP is elevated[5,6]. GRP binds with high affinity to NCI-H345 cells and stimulates phosphatidylinositol turnover[7,8]. The inositol-1,4,5-trisphosphate released elevates cytosolic calcium and the diacylglycerol released activated protein kinase C[9,10]. Also, GRP or BN stimulate SCLC colony formation and xenograft formation in nude mice[11,12]. SCLC growth in vitro and in vivo is blocked by synthetic BN/GRP receptor antagonists[13,14].

It may also be possible to disrupt the SCLC BN/GRP autocrine growth cycle by altering GRP gene expression. Previously, we found that phorbol esters such as phorbol 12-myristate 13-acetate (PMA), which activate protein kinase C, stimulated GRP gene expression[15]. Here the effects of forskolin, which elevates cAMP levels, were investigated.

INTRACELLULAR cAMP

The effects of forskolin on SCLC cAMP levels were investigated. By radioimmunoassay, the basal cAMP levels were approximately 20 fmol. When 50 μM forskolin was added the cAMP levels increased to 251 fmol. These data indicate that forskolin which interacts with a stimulatory guanine nucleotide binding subune (Gs)

Table 1. NCI-H209 cAMP levels

Addition	cAMP, fmol
None	20 ± 2
Forskolin, 50 μM	251 ± 118
PMA, 1 μM	20 ± 3

NCI-H209 cells (1×10^5) were seeded in 24 well plates. One day after feeding the cells were rinsed twice with SIT medium (RPMI-1640 containing 3×10^{-8} M Na_2SeO_3, 5 μg/ml insulin and 10 μg/ml transferrin. The reaction medium contained 200 μM isobutylmethylxanthine, a phosphodiesterase inhibitor. The compounds were added and after 5 min the reaction was stopped by adding 95% ethanol. The samples were stored at -70°C and the cAMP determined by radioimmunoassay. The mean value $\pm$ S.E. of 4 determinations is indicated.

activating adenylate cyclase increased the intracellular cAMP levels 12-fold. As a control, 1 μM PMA, which activates protein kinase C, has no effect on cAMP levels. Previously, we found that 1 μM vasoactive intestinal peptide (VIP) elevates the cAMP 10-fold in SCLC cells[6]. These data indicate that VIP, which interacts with a cell surface receptor, stimulates adenylate cyclase activity.

The increase in cAMP may activate protein kinase A (PKA). PKA, which phosphorylates serine and threonine amino acid residues on protein substrates, may phosphorylate synapsin I causing exocytosis of granules which contain transmitters and peptides[16]. Thus VIP increases the secretion rate of GRP from SCLC cells in vitro. Also, secretin, which is structurally similar to VIP, increased the plasma levels of GRP in patients with extensive SCLC[6].

C-FOS mRNA

The effects of forskolin on c-fos gene expression were investigated. Previously the c-fos and l-myc oncogenes were found in SCLC cells[16,17]. Figure 1 (top) shows that little c-fos mRNA was expressed at the start of the experiment. C-Fos expression increased dramatically 30 min after the addition of forskolin. The c-fos mRNA was maximal after 1 hr and declined dramatically after 4 hr. As a control, Fig. 1 bottom shows that approximately equal densities of 28 S and 18 S RNA bands were present in each of the samples after ethidium bromide staining. Therefore forskolin transiently increases the density of c-fos mRNA in NCI-H209 cells. It is possible that the SCLC c-fos gene has a cAMP responsive element (CRE) resulting in increased expression[19,20]. Other agents which activated PKA, such as 1 μM VIP, increased c-fos gene expression. Also, PMA, which activates PKC, increased c-fos mRNA. The c-fos gene may therefore have PKC and PMA responsive elements.

EFFECTS ON GRP GENE EXPRESSION

The effects of forskolin on GRP gene expression were investigated. Figure 2 shows that 50 μM forskolin increased GRP gene expression in a time dependent manner. GRP gene expression increased slightly after 1 hour and was maximal after 4-8 hours. In contrast, forskolin had little effect on ß-actin gene expression. Recent

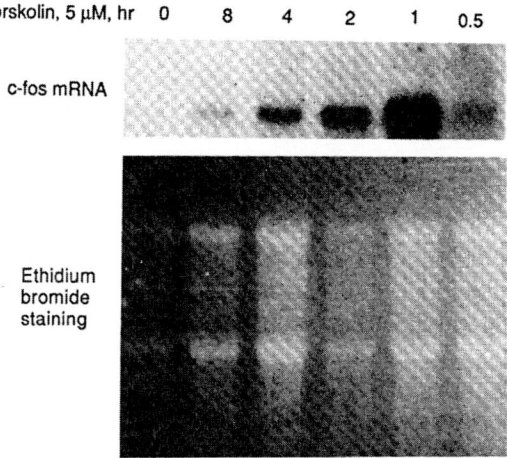

Figure 1. Northern blot analysis for c-fos. NCI-H209 cells were cultured in RPMI-1640 containing 10% fetal bovine serum. Two hours prior to the experiment the cells were placed in RPMI-1640 containing 0.5% fetal bovine serum (R0.5). Then the cells were treated with 5 μM forskolin in R0.5 for the indicated times. The total RNA was isolated using the guanidinium isothiocyanate (GIT) procedure[18] and the c-fos mRNA determined by Northern blot analysis.

data indicate that forskolin increases GRP gene expression as a result of an increase in mRNA synthesis and has no effect on GRP mRNA degradation (M. Draoui, unpublished). Because forskolin elevates the cAMP, a cAMP responsive element (CRE) may be activated on the GRP gene. In this regard a cAMP sequence (GACGTCA) is present in the 5' region 74 nucleotides upstream from the GRP gene. Previously the somatostatin gene was found to have a cAMP responsive element[21].

Table 2 shows that forskolin increased the relative amount of GRP mRNA by 3-fold using cell line NCI-H209. Similarly, PMA increased the relative amount of GRP mRNA by 2.8-fold. Further nuclear run-on analysis indicates that PMA increases GRP mRNA but has no effect on mRNA degradation[15]. Thus in addition to a cAMP responsive element there may be PMA responsive element upstream from the GRP gene. Because PMA increases c-fos mRNA, this may transiently increase c-fos protein. C-fos may form a heterodimer with c-jun and bind to nuclear AP-1 sites, stimulating GRP gene expression[22,23,24].

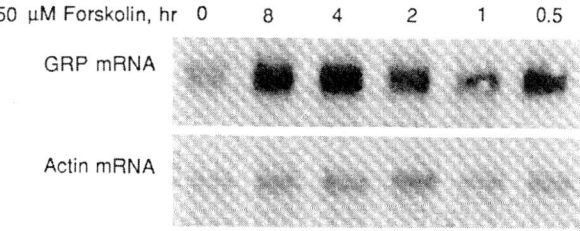

Figure 2. Northern blot analysis for preproGRP. NCI-H209 cells were incubated for the indicated times in R0.5. The cells were treated with 50 μM forskolin in R0.5 for the indicated times. The total RNA was isolated using the GIT method and the preproGRP mRNA determined by Northern analysis.

Table 2. GRP gene expression in SCLC cell line NCI-H209

Agent	Relative GRP mRNA
None	1.0 ± 0.1
PMA 10 M	2.8 ± 0.2**
Forskolin, 50 μM	3.0 ± 1.1*

The relative densities of the preproGRP and ß-actin mRNA were determined and the ratio of preproGRP/ß-actin mRNA calculated. The mean value $\pm$ S.D. of 4 determinations is indicated; $p < 0.01$,**; $p < 0.05$, *.

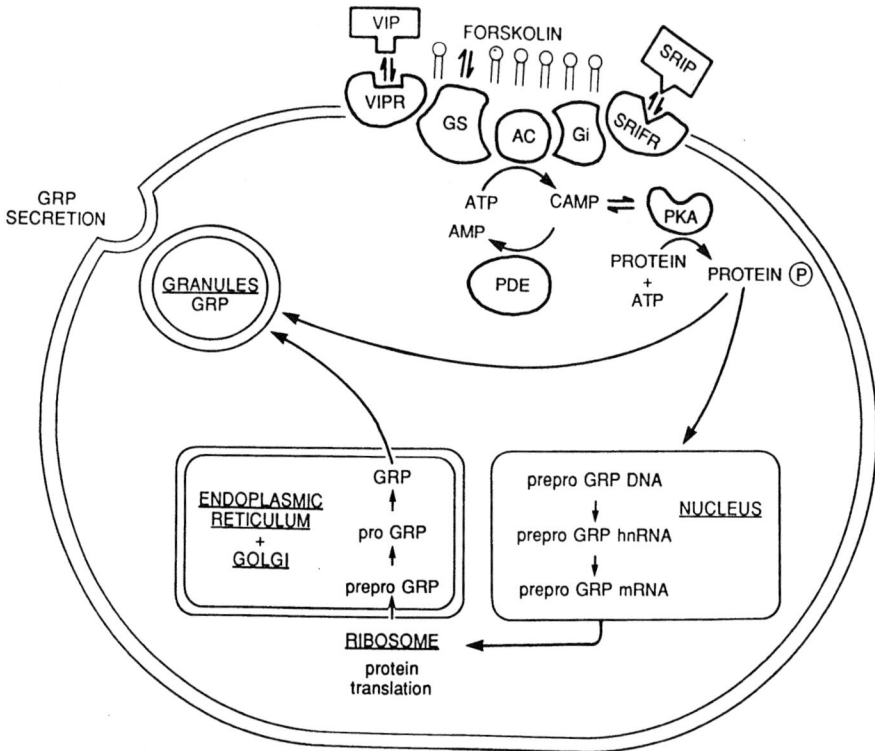

Figure 3. Schematic illustration of cAMP on SCLC cells. VIP binds to its receptor (VIPR) and interact with Gs stimulating adenylate cyclase (AC) and elevating intracellular cAMP. The cAMP is degraded by phosphodiesterase. Somatostatin binds to its receptor (SRIFR) and interacts with Gi inhibiting AC. When the cAMP is elevated PKA is stimulated resulting in phosphorylation of proteins. This resulting in exocytosis of SCLC granules which contain GRP and increased levels of preproGRP mRNA. Upon protein translation the preproGRP is posttranslationally processed to GRP in intracellular organelles.

MECHANISM OF ACTION OF FORSKOLIN

The mechanism by which forskolin may alter GRP gene expression is shown in Figure 3. Forskolin interacts with a stimulatory guanine nucleotide binding protein (Gs) resulting in stimulation of adenylate cyclase and increased intracellular cAMP. Also, VIP cen bind to the SCLC VIP receptors resulting in activation of Gs and adenylate cyclase stimulation. In contrast, somatostatin (SRIF) interacts with SRIF receptors which activate Gi and inhibits the increase in adenylate cyclase caused by VIP. When cAMP is elevated, PKA will phosphorylate substrate protein. Phosphorylation of xynapsin I may stimulate exocytosis of granules which contain GRP. GRP may bind to GRP receptors stimulating SCLC growth. Also, VIP positively and SRIF negatively alters SCLC growth. Due to the secretion of GRP from SCLC, it is then important for the cells to synthesize additional GRP. Some of the proteins phosphorylated by PKA may interact with a nuclear CRE and stimulate preproGRP gene expression. This may result in replication of preproGRP DNA, transcription of propoGRP heavy nuclear RNA and after splicing increased preproGRP mRNA. The preproGRP mRNA may translocate to the cytosol where it is translated into preproGRP protein. Because preproGRP has a signal sequence it is likely translocated into the endoplasmic reticulum where the signal sequence is removed. The preproGRP may then be metabolized by endogenous enzymes in the endoplasmic reticulum and golgi apparatus to GPR. PreproGRP-type I is a 148 amino acid protein with no appreciable biological activity whereas GRP is a 27 amino acid peptide.

The GRP may then be stored in dense core granules and secreted in response to stimuli such as VIP. The secreted GRP binds to cell surface receptors. The GRP receptor contains 384 amino acid residues and 7 hydrophobic domains[25,26]. Thus the GRP receptor has a structure reminiscent of other G-protein coupled receptors such as the substance K receptor. The GRP receptor may then stimulate phosphatidylinositol turnover and growth.

In summary protein kinase A may regulate GRP gene expression in human SCLC. Thus it is possible to regulate SCLC growth using molecular biology approaches which inhibit GRP gene expression.

ACKNOWLEDGEMENTS

This research is supported in part by NCI grant CA-53477.

REFERENCES

1. F. Cuttitta, D.N. Carney, J. Mulshine, J., T.W. Moody, J. Fedorko, A. Fishler and J.D. Minna. Bombesin-like peptides can function as autocrine growth factors in human small cell lung cancer. Nature 316:823(1985).

2. E.R. Spindel, W.W. Chin, J. Price, L.H. Rees, G.M. Besser and J.F. Habener. Cloning and characterization of cDNA encoding human gastrin releasing peptide. Proc. Natl. Acad. Sci. USA 81:5699(1984).

3. E.A. Sausville, A.M. Lebacq-Verheyden, E.R. Spindel, F. Cuttitta, A.F. Gazdar and J.F. Battey. Expression of the gastrin releasing peptide in human small cell lung cancer. J. Biol. Chem. 261:2451(1986).

4. A.M. Lebacq-Verheyden, V. Bertness, I. Kirsch, G.F. Hollis, O.W. McBride and J.Battey. Human gastrin releasing peptide gene maps to chromosome band 18q21. Somatic Cell Mol. Genet. 13:81(1987).

5. T.W. Moody, C.B. Pert, A.F. Gazdar, D.N. Carney and J.D. Minna. High levels of intracellular bombesin characterize humansmall-cell lung carcinoma. Science 214:1246(1981).

6. L.Y. Korman, D.N. Carney, M.L. Citron and T.W. Moody. Secretin/VIP stimulated secretion of BN/GRP from human small cell carcinoma of the lung. Cancer Res. 46:1214(1986).

7. T.W. Moody, D.N. Carney, F. Cuttitta, K. Quattrochi and J.D. Minna. High affinity receptors for BN/GRP-like peptides on human small cell lung cancer cells. Life Sci. 37:105(1985).

8. J.B. Trepel, J.D. Moyer, R. Heikkila and E.A. Sausville. Modulation of bombesin-induced phosphatidylinositol hydrolysis in small cell lung cancer cell lines. Biochem. J. 255:403(1988).

9. T.W. Moody, A. Murphy, S. Mahmoud and G. Fiskum. Bombesin-like peptides elevate cytosolic calcium in small cell lung cancer cells, Biochem. Biophys. Res. Commun. 147:189(1987).

10. E.A. Sausville, J.D. Moyer, R. Heikkila, L.N. Neckers and J.P. Trepel. A correlation of bombesin-responsiveness with c-myc family gene expression in small cell lung carcinoma cell lines. Annals New York Academy of Science 547:310(1988).

11. D.N. Carney, F. Cuttitta, T.W. Moody and J.D. Minna. Selective stimulation of small cell lung cancer clonal growth by bombesin and gastrin releasing peptide, Cancer Res. 47:821(1987).

12. R.W. Alexander, J.R. Upp Jr., G.J. Poston, V.Gupta, C.M. Townsend Jr and J.C. Thompson Effect of bombesin on growth of human small cell lung carcinoma in vitro. Cancer Res. 48:1439 (1988).

13. S. Mahmoud, J. Staley, J. Taylor, A. Bogden, J.P. Moreau, D. Coy, I. Avis, F. Cuttitta, J. Mulshine and T.W. Moody. (Psi^{13,14})bombesin analogues inhibit the growth of small cell lung cancer in vitro and in vivo. Cancer Res. 51:1798(1991).

14. J. Staley, D. Coy, J.E. Taylor, S. Kim and T.W. Moody. (Des-met^{14})bombesin analogues function as small cell lung cancer bombesin receptor antagonists. Peptides 12:145(1991).

15. M. Draoui, T. Moody, Z. Fathi and J. Battey. Alterations of PreproGRP gene expression in small cell lung cancer cells by phorbol esters. J. Cellular Biochem. (1991).

16. J.D. Minna, H. Pass, E. Glatstein AND D.C. Ihde, In "Cancer: Principles and Practice in Oncology," V.T. DeVita Jr., A. Hellman, S.A. Rosenberg (eds) 591, J.B. Lippincott Co., Philadelphia (1989).

17. M.M. Nau, J.B. Brooks, J.F. Battey, E. Sausville, A.F. Gazdar, I.R. Kirsch, O.W. McBride, V. Bertness, G.F. Hollis and J.D. Minna. L-myc, a new myc-related gene amplifies and expressed in human small cell lung cancer. Nature 318:69 (1985).

18. L. Davis, M. Dibner and J.F. Battey. In "BASIC METHODS IN MOLECULAR BIOLOGY" Elsevier, New York (1986).

19. T.M. Fisch, R. Prywes, M.C. Simon and R.G. Roeder. Multiple sequence elements in the c-fos promotor mediate induction by cAMP. Genes & Dev. 3:198(1989).

20. L.A. Berkowitz, K.T. Riabowol and M.Z. Gilman. Multiple sequence elements of a single functional class are required for cAMP responsiveness of the mouse c-fos promotor. Mol. Cell Biol. 9:4272(1989).

21. M.R. Montminy, K.A. Sevarino, J.A. Wagner, G. Mandel and R.H. Goodman. Identification of a cAMP responsive element within the rat somatostatin gene. Proc. Natl. Acad. Sci. USA 83:6682(1986).

22. M. Comb, N.C. Birnberg, A. Seasholtz, E. Herbert and H.M. Goodman. A cAMP and phorbol ester-inducable DNA element. Nature 323:353(1986).

23. P.J. Deutsch, J.P. Hoeffler, J.L. Jameson and J.F. Habener. cAMP and phorbol ester stimulated transcription mediated by similar DNA elements that bind distinct proteins. Proc. Natl. Acad. Sci. USA 85:7922(1988).

24. P. Sassone-Corsi, J.C. Sisson and I.M. Verma. Direct interaction between fos and jun nuclear oncoproteins: Role of the "Leucine zipper" domain. Nature 336:692(1988).

25. E.R. Spindel, E. Giladi, T.P. Brehm, R.H. Goodman, and T.P. Segerson. Cloning and functional characterization of a cDNA encoding the murine fibroblast bombesin/GRP receptor. Mol. Endocrinol. 4:1956(1990).

26. J.F. Battey, J.M. Way, M.H. Corjay, H. Shapira, K. Kusano K., R. Harkins, J.M. Wu, T. Slattery, E. Mann and R. Feldman. Molecular analysis of the murine GRP-preferring bombesin receptor. Proc. Natl. Acad. Sci. USA 88:395(1991).

Part VI
Proliferation

TGFα INDUCED PROLIFERATIVE CHANGES
IN TRANSGENIC MICE

Hitoshi Takagi[1], Chamelli Jhappan[1], Richard Sharp[1],
Hisashi Takayama[1], Gilbert H. Smith[2], and Glenn Merlino[1]

[1]Laboratory of Molecular Biology
[2]Laboratory of Tumor Immunology and Biology
Division of Cancer Biology and Diagnosis and Centers
National Cancer Institute, NIH, Bethesda, MD 20892

INTRODUCTION

Since the discovery of nerve growth factor[1], a multitude of growth factors and their associated receptors have been identified and characterized[2]. Some of these growth factors and receptors have been implicated in oncogenesis. The proto-oncogene c-sis encodes the B chain of platelet derived growth factor (PDGF)[3], the product of c-fms is similar to the colony stimulating factor (CSF)-1 receptor[4], and the c-erb B gene product is highly homologous to the epidermal growth factor (EGF) receptor[5-7].

EGF is a potent mitogen and the founding member of a family of structurally related proteins[8-11], including transforming growth factor alpha (TGFα). Both EGF and TGFα stimulate cellular proliferation by binding to and activating the tyrosine kinase domain of the EGF receptor[12,13].

TGFα is synthesized as a 160 amino acid precursor that is activated by elastase like proteolysis, yielding a mature, secreted 50 amino acid polypeptide[9,14,15] (Figure 1). Introduction of active TGFα expression vectors into cultured cells induces transformation[16-18]. Enhanced production of TGFα and EGF receptor is frequently associated with human cancer and transformed culture cells[19,20]. In addition, elevated levels of TGFα have been detected in the ascites, urine and pleural effusion of cancer patients[21-23].

The transgenic technology, in which foreign genetic information is stably introduced into the germ line, has substantially enhanced our understanding

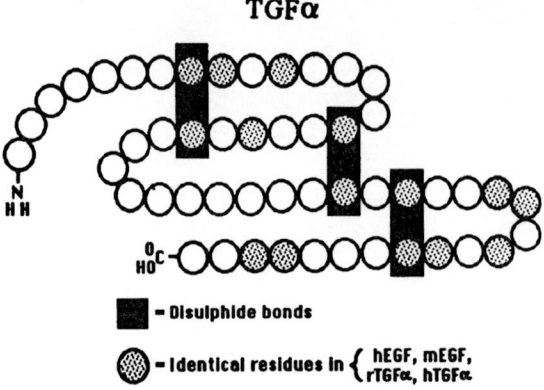

TGFα

■ = Disulphide bonds

◉ = Identical residues in { hEGF, mEGF, rTGFα, hTGFα

Figure 1. Structure of the mature, secreted TGFα protein. hEGF; human EGF, mEGF; mouse EGF, rTGFα; rat TGFα, hTGFα; human TGFα. Location of disulfide bonds marks the position of cysteine residues characteristic of the EGF family of proteins.

of physiologic and pathologic processes[24]. For example, transgenic animals are being used to provide novel insights into gene regulation, mammalian development and the causes and treatment of disease. To determine the consequences of overexpressing TGFα *in vivo*, we have made transgenic mice that express this potent mitogen in a variety of tissues[25]. We found several interesting phenotypic alterations associated with expression of TGFα; this chapter summarizes our current understanding of the multiple pathologic lesions that occur in these mice.

MATERIALS AND METHODS

The method used to generate transgenic mice has been described previously[25]. One-cell mouse embryos from the outbred line CD1 and inbred FVB/N line were microinjected with a DNA fragment consisting of the mouse metallothionein I gene promoter, the human TGFα cDNA, and the polyadenylation signal from the human growth hormone gene (Figure 2). Where appropriate, transgenic and control mice were maintained on drinking water containing 50 mM zinc sulfate (to stimulate the transgenic inducible metallothionein promoter). Transgenic mice were identified by Southern blot

Figure 2. Transgene structure. The MT-TGFα fragment contains 770 bp of the mouse metallothionein I promoter (mMT), 925 bp of the human TGFα cDNA (hTGFα), and 625 bp of the 3' end of the human growth hormone gene polyadenylation signal (hGH)[25]. The arrow indicates the start and direction of transcription, and R1 is EcoR1. Taken with permission from Jhappan et al[25].

hybridization using a TGFα-specific cDNA fragment[25]. Total RNA was prepared as previously described[25]. Typically, mice were injected with 5 mg/kg zinc chloride 4 to 5 hours prior to sacrifice to stimulate transgene expression. Total tissue RNAs were analyzed by either RNase protection or Northern blot hybridization[25]. Tissues to be processed for routine histologic analysis only were fixed in 10% buffered formalin and stained with hematoxylin and eosin (HE). Tissues to be used for immunohistochemical analysis were fixed in Bouin's solution. Immunohistochemistry was performed by standard techniques described elsewhere[26] using rabbit polyclonal antiserum against a rat proTGFα intracellular peptide (residues 137-159)[27].

RESULTS AND DISCUSSION

TGFα Transgenic Mice

Two lines, MT42 (derived from outbred CD1) and MT100 (derived from inbred FVB/N) were examined intensely in this study. These lines contained 2-3 copies and approximately 5 copies of the transgene per haploid genome, respectively. To determine the activity of the TGFα transgene, total RNA was isolated from representative tissues of the F1 progeny of these mice and subjected to RNase protection analysis. Figure 3A demonstrates that human TGFα RNA is highly abundant in liver, kidney, pancreas and testis and moderately abundant in bone, lung, brain, uterus, skeletal muscle, colon and mammary gland. This broad transgene expression pattern conformed to the pattern expected for the endogenous mouse metallothionein gene. Immunohistochemical staining showed high amounts of TGFα in the transgenic liver, pancreas and stomach. In the liver, TGFα was localized membranously around vessels, especially the portal vein (Figure 4B). TGFα can be readily detected in the pancreatic acini (Figure 4D), and in the surface mucosal cells and glandular cells at the base of foveolar crypts in the stomach[26]. Furthermore, TGFα was elevated in the serum and urine of transgenic mice (data not shown).

Liver Tumors in TGFα Transgenic Mice

Macroscopic examination revealed that the young transgenic liver was hypertrophic; however, no tumors developed in mice younger than 6 months of age. Histological examination of livers from male MT42 mice between one and three months of age revealed abnormally large hepatocytes, and transgenic mice over 8 months old possessed dysplastic hepatocytes with enlarged, atypical nuclei[28,29]. Approximately 75% of transgenic male mice over 10 months old developed focal or multifocal liver tumors. Occasionally, macroscopic nodules were detected in male MT42 mice 7 to 8 months of age. In tumor bearing mice,

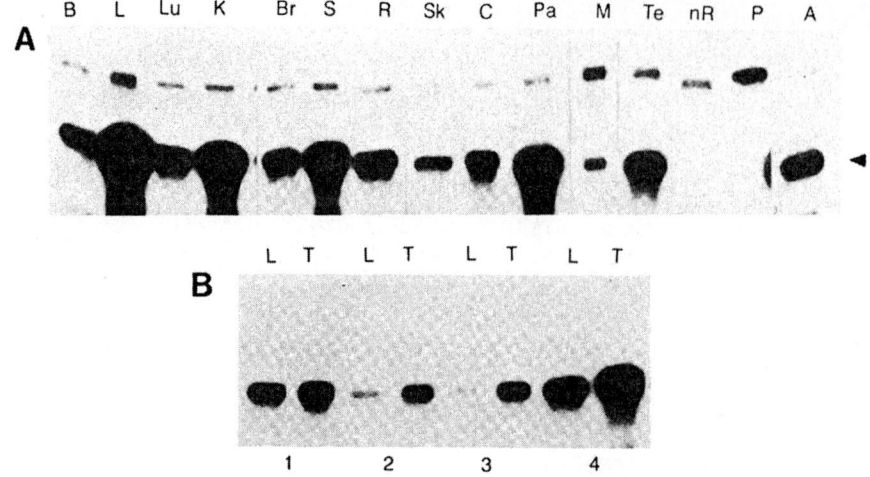

Figure 3. Analysis of expression of the TGFα transgene in multiple tissues by RNase protection. A panel of radiolabeled human TGFα-specific RNAs was protected from RNases A and T1 by 55°C overnight hybridization of a ^{32}P-labeled 385 b riboprobe[25,28] to 15 µg of total RNA. The 310 b labeled digestion products were visualized by autoradiography after electrophoresis on a denaturing 4% polyacrylamide gel. Taken with permission from Jhappan et al[25].

(A) B: bone, L: liver, Lu: lung, K: kidney, Br: brain, S: stomach, R: female reproductive system (uterus and ovary), Sk: skeletal muscle, C: colon, Pa: pancreas, M: mammary gland, Te: testis, nR: nontransgenic female reproductive system (negative control), P: digested riboprobe with no addition of tissue RNA, A: human A431 carcinoma cells.

(B) Transgene expression in four liver tumors and their adjacent nontumorous liver tissues from MT42 male mice. L: adjacent nontumorous liver tissue, T: liver tumor tissue.

about half possessed both hepatocellular carcinomas (HCCs) and adenomas (Figure 5B), while the remaining half had only adenomas. Most of these HCCs were of the well differentiated trabecular type (Figure 5B) and quite similar to human HCCs. In contrast, only about 3% of female transgenic mice developed liver tumors. Although we have not precisely determined the reason for this difference, gonadectomy experiments have suggested that male and female sex hormones stimulate and inhibit tumorigenesis, respectively[28].

Human TGFα RNA levels were previously shown to be high in the livers of MT42 transgenic mice[25]. About 70% of the liver tumors in these transgenic mice exhibited further enhancement in the expression of the TGFα transgene (Figure 3B). These data are in general agreement with results obtained by in situ hybridization of transgenic liver tumors[29]. Endogenous genes whose expression was enhanced in the liver tumors include those encoding c-myc and insulin-like growth factor II (IGF II). In contrast, mutations were not detected in either the *Ha-ras* or *Ki-ras* genes[28].

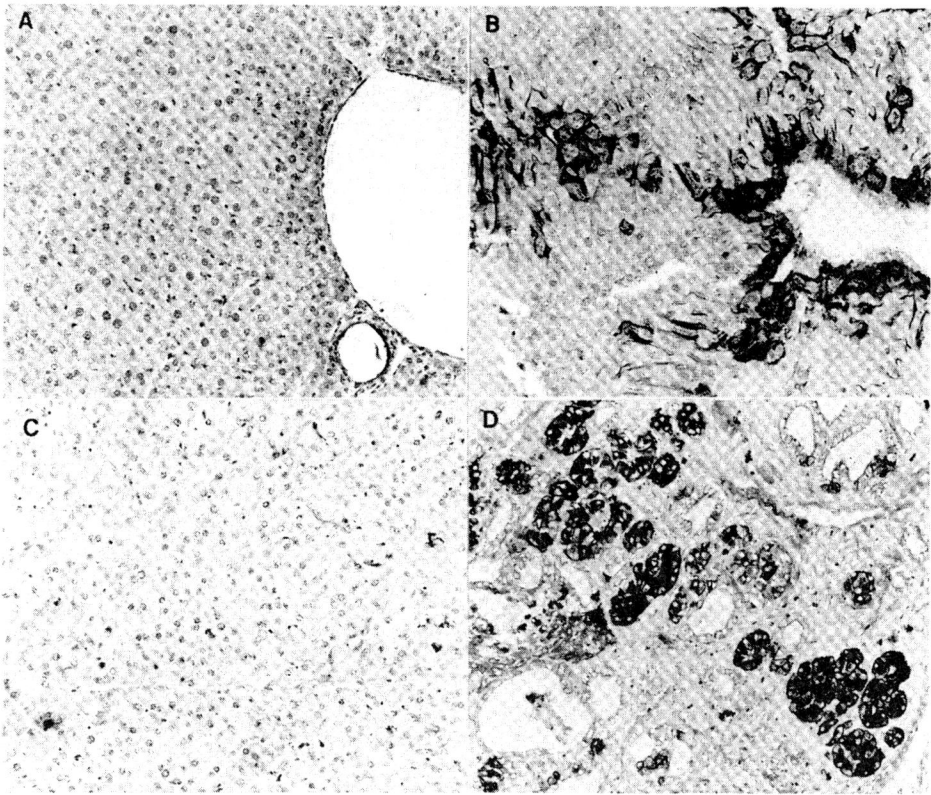

Figure 4. Immunohistochemistry of the liver and pancreas of TGFα transgenic mice. The antibody was made using a peptide sequence found in the intracellular portion of proTGFα[27].

(A) Nontransgenic, control liver shows no obvious staining of endogenous TGFα.

(B) Transgenic hepatocytes membranously stained with TGFα.

(C) Nontransgenic, control pancreas does not stain for TGFα.

(D) TGFα stains the pancreatic acinar cells in transgenic mice, but not the ductal cells or fibrotic tissue. Mag. X200.

Pancreatic Interstitial Fibrosis and Ductular Metaplasia

Human TGFα RNA was highly expressed in pancreas of both MT42 (Figure 3A) and MT100. The pancreas in these mice became progressively fibrotic (Figure 6B). The penetrance of this phenotype was 100%, allowing rapid identification of TGFα mice from negative controls through observation of the grossly rigid and whitish pancreas. The transgenic pancreas became twice the size of its normal counterpart due to accumulation of type I collagen[30]. Histologically these changes reflected intralobular and perilobular fibrosis and a florid ductular metaplasia (Figure 6B) that is well contrasted with the normal pancreas (Figure 6A). Unlike human chronic pancreatitis, inflammatory reactions were not associated with these transgenic pancreatic lesions[25];

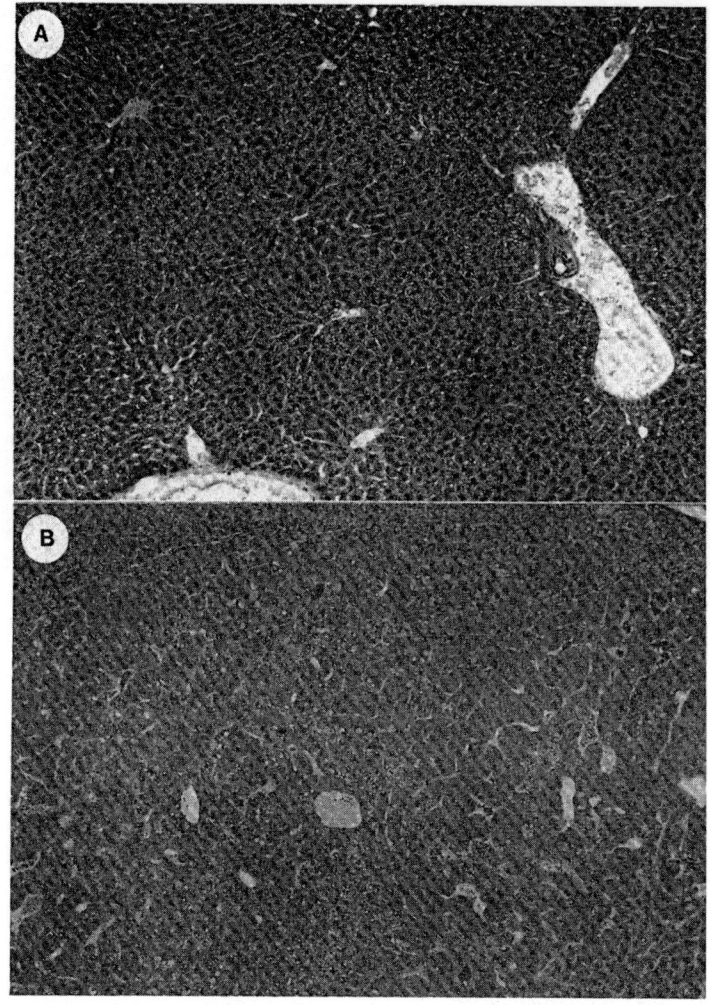

Figure 5. Histology of the liver (HE).

(A) Nontransgenic liver as a normal control.

(B) Transgenic liver shows trabecular structured hepatocellular carcinoma on the right side, and liver cell adenoma on the left side of this picture. Mag. X50.

however, pancreatic cysts with serous contents were occasionally involved. Pancreatic cancer has not been observed, and the transgenic pancreatic tissue was not transplantable. Similar pancreatic lesions have been reported by Sandgren and coworkers[31], who generated transgenic mice overexpressing the rat TGFα gene by using both metallothionein and elastase I gene promoters.

Hypertrophic Gastropathy Resembling Menetrier's Disease

The glandular stomach of MT100 male and female mice developed severe cystic adenomatous hyperplasia (Figure 7B) that resulted in a striking

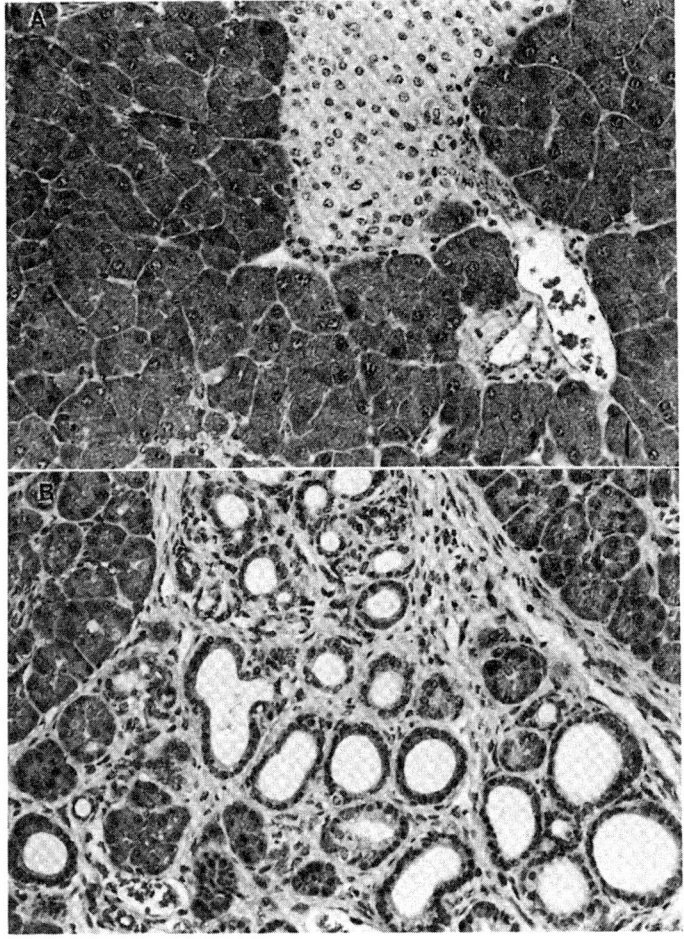

Figure 6. Histology of the pancreas (HE).

(A) Nontransgenic pancreas shows well preserved structure of acinus and islet.

(B) Transgenic pancreas shows interstitial fibrosis with ductular metaplasia. Mag. X250.

nodular thickening of the gastric mucosa[26]. MT42 also exhibited proliferative, cystic mucosal changes of the stomach, though to a lesser degree. These changes began at the age of three weeks and progressed throughout the abbreviated life of the animal, during which time the mass of the transgenic stomach became 3-5 times greater than age-matched controls. Secretions obtained from affected stomachs of mice over 3 months of age contained no detectable gastric acid, suggesting that parietal cell function had been impaired. These findings demonstrate that overproduction of TGFα can stimulate cellular proliferation, suppress acid secretion and perturb organogenesis of the stomach of transgenic mice. These lesions are reminiscent of Menetrier's disease in human patients, which is defined as hypertrophic gastropathy with

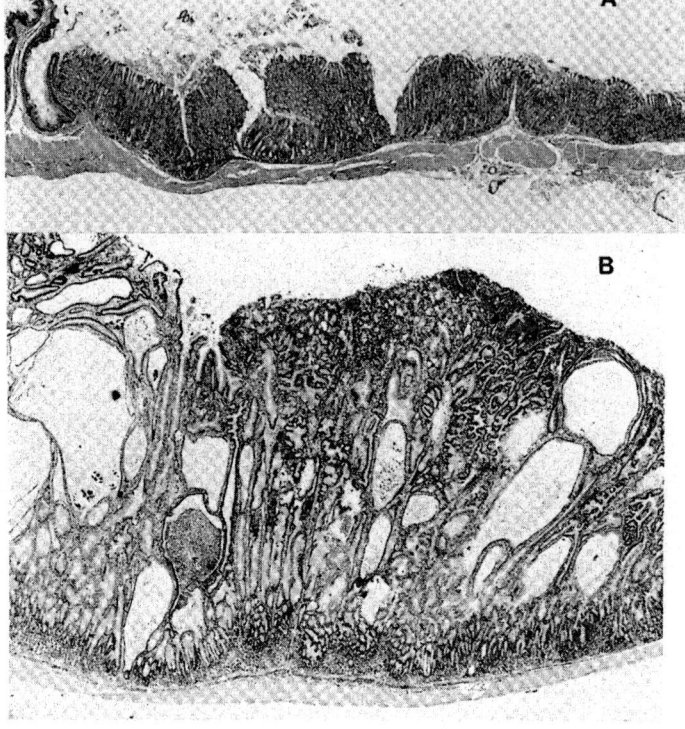

Figure 7. Histology of the stomach (HE).

(A) Nontransgenic stomach as a normal control.

(B) The mucosa of the transgenic stomach is dramatically thicker and hyperplastic, and contains numerous cysts. Some of the cysts contain mucus-laden secretions. Mag. X16.

achlorhydria and protein losing gastropathy[32]. Although we have not confirmed the protein losing gastropathy, these transgenic mice became cachectic and died by 10 months of age. Other characteristics of these transgenic stomachs, including the dramatic pathological changes and low gastric acid secretion, are quite similar to Menetrier's disease in humans. This transgenic mouse model should prove to be valuable for determining the role of TGFα in the development of Menetrier's disease and related hypertrophic gastropathies.

Abnormal Development and Neoplasia of the Breast

TGFα has been implicated as a mediator of mammary gland development and as an epithelial cell mitogen[33,34]. Human TGFα was produced in the transgenic breast and the penetration of the ductal epithelium into the fat pad was impeded at adolescence[25]. Nevertheless, by 12 weeks of age, both the transgenic and nontransgenic fat pads were completely filled.

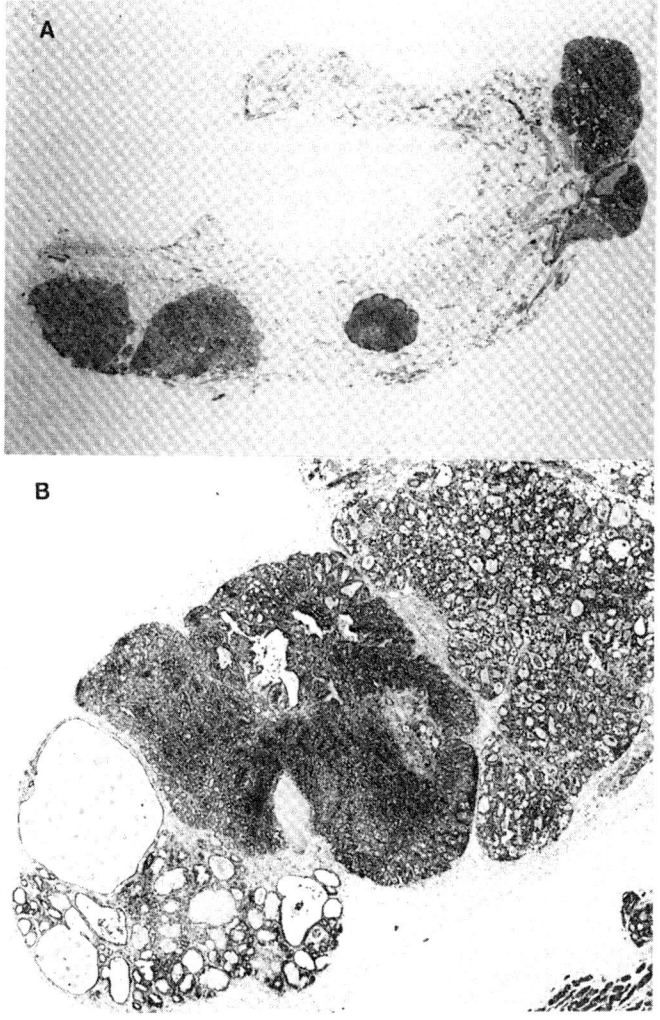

Figure 8. Histology of the mammary gland (HE).

(A) Multiple adenocarcinomas arising from the inguinal mammary fat pad of the multiparous MT100 founder transgenic female mouse. A single lymph node can be seen in the center of the gland. Mag. X4.5.

(B) Secretory adenoma and adenocarcinoma (middle portion) from a multiparous female progeny of the MT100 founder. Mag. X2.5.

Mammary adenomas and adenocarcinomas developed in multiparous female MT100 transgenic mice. Typically three pregnancies were necessary for the development of breast cancer (Figure 8, A and B), suggesting that TGFα can play an important role in mammary oncogenesis, but only in concert with hormones related to pregnancy and/or lactation. Breast cancer has also been observed in mice harboring a rat TGFα transgene[31] and those where TGFα transgene expression is directed by the mouse mammary tumor virus (MMTV)[35]. Breast tissue containing hyperproliferative lesions was successfully transplanted into the mammary fat pads of FVB/N syngeneic females, where cystic adenocarcinomas could form after about 9 months (data not shown).

SUMMARY

We have demonstrated that TGFα overexpression induces proliferative changes in the liver, pancreas, stomach and mammary gland of TGFα transgenic mice. The liver and mammary gland develop malignant tumors, while the pancreas and stomach exhibit consistent proliferative changes but no tumors. These findings demonstrate the pleiotropic effects of TGFα in vivo, and suggest that the establishment of inappropriate autocrine and/or paracrine TGFα stimulation dramatically influences cellular growth and induces the transformed phenotype. This transgenic mouse model can contribute to our understanding of human diseases, including hepatoma, chronic pancreatitis, Menetrier's disease and breast cancer. The study of the development of these lesions in the mouse may lead to the establishment of more efficacious treatment of relevant diseases in humans.

REFERENCES

1. S. Cohen, Purificaton and metabolic effects of a nerve growth-promoting protein from snake venom, *J. Biol. Chem.*.234:1129 (1959).

2. A.S. Goustin, E.B. Leof, G.D. Shipley, and H.L. Moses, Growth factors and cancer, *Cancer Res.* 46:1015 (1986).

3. M.D. Waterfield, G.T. Scrace, N. Whittle, P. Stroobant, A. Johnsson, A. Wasterson, B. Wastermark, C.H. Helden, J.S. Huang, and T.F. Deuel, Platelet derived growth factor is structurally related to the putative transforming protein p28sis of simian sarcoma virus, *Nature* 304:35 1983).

4. C.J. Sherr, C.W. Rettenmier, R. Sacca, M.F. Roussel, A.T. Look, and E.R. Stanley, The c-fms proto-oncogene product is related to the receptor for the mononuclear phagocyte growth factor, CSF-1, *Cell* 41:665 (1985).

5. J. Downward, Y. Yarden, E. Mayes, G. Scarce, N. Totty, P. Stockwell, A. Ullrich, J. Schlessinger and M.D. Waterfield, Close similarity of epidermal growth factor receptor and *v-erb B* oncogene protein sequence, *Nature* 307:521 (1984).

6. G. Carpenter, and S. Cohen, Epidermal growth factor, *J. Biol. Chem.* 265:7709 (1990).

7. G.T. Merlino, Epidermal growth factor receptor regulation and function, *Sem. Cancer Biol.* 1:277 (1990).

8. H. Marquardt, M.W. Hunkapiller, L.E. Hood, D.R. Twardzik, J.E. DeLarco, J. R. Stephenson, and G.J. Todaro, Transforming growth factors produced by retrovirus-transformed rodent fibroblasts and human melanoma cells: amino acid sequence homology with epidermal growth factor, *Proc. Natl. Acad. Sci. U.S.A.* 80:4684 (1983).

9. H. Marquardt, M.W. Hunkapiller, L.E. Hood, and G.J. Todaro, Rat transforming growth factor type I: structure and relation to epidermal growth factor, *Science* 223:1079 (1984).

10. R. Derynck, A.B. Roberts, M.E. Winkler, E.Y. Chen, and D.V. Goeddel, Human transforming growth factor-α: precursor structure and expression in E. coli, *Cell* 38:287 (1984).

11. G.J. Todaro, T.M. Rose, C.E. Spooner, M. Shoyab, and G.D. Plowman, Cellular and viral ligands that interact with the EGF receptor. *Sem. Cancer Biol.* 1:257 (1990).

12. G. Carpenter, Receptors for epidermal growth factor and other polypeptide mitogens, *Ann. Rev. Biochem.* 56:881 (1987).

13. J. Schlessinger, The epidermal growth factor receptor as a multifunctional allosteric protein, *Biochemistry* 27:3119 (1988).

14. R.A. Ignotz, B. Kelly, R.J. Davis, and J. Massague, Biologically active precursor for transforming growth factor type α, released by retrovirally transformed cells, *Proc. Natl. Acad. Sci. U.S.A.* 83:6307 (1986).

15. D.C. Lee, T.M. Rose, N.R. Webb, and G.J.Todaro, Cloning and sequence analysis of a cDNA for rat transforming growth factor-α. *Nature* 313:489 (1985).

16. A. Rosenthal, P.B. Lindquist, T.S. Bringman, D.V. Goeddel, and R. Derynck, Expression in rat fibroblasts of a human transforming growth factor-α cDNA results in transformation, *Cell* 46:301 (1986).

17. S. Watanabe, E. Lazar, and M.B. Sporn, Transformation of normal rat kidney (NRK) cells by an infectious retrovirus carrying a synthetic rat type α transforming growth factor gene, *Proc. Natl. Acad. Sci. U.S.A.* 84:1258 (1987).

18. M.L. McGeady, S. Kerby, V. Shankar, F. Ciardiello, D. Salomon, and M. Seidman, Infection with a TGFα retroviral vector transforms normal mouse mammary epithelial cells but not normal rat fibroblasts, *Oncogene* 4:1375 (1989).

19. D. S. Salomon, N. Kim, T. Saeki, and F. Ciardiello, Transforming growth factor-α: an oncodevelopmental growth factor, *Cancer Cells* 2:389 (1990).

20. R. Derynck, D.V. Goeddel, A. Ullrich, J.U. Gutterman, R.D. Williams, T.S. Bringman, and W.H. Berger, Synthesis of mRNAs for transforming growth factors α and β and the epidermal growth factor receptor by human tumors, *Cancer Res.* 47:707-712 (1987).

21. F. Ciardiello, N. Kim, D.S. Liscia, C. Bianco, R. Lidereau, G. Merlo, R. Callahan, J. Greiner, C. Szpak, W. Kidwell, J. Schlom, and D.S. Salomon, mRNA expression of transforming growth factor alpha in human breast carcinoma and its activity in effusion of breast cancer patients, *J. Natl. Cancer Inst.* 81:1165 (1989).

22. C.L. Artega, A.R. Hanauske, G.M. Clark, C.K. Osborn, R.L. Hazarika, F. Pardue, F. Tio, and D.D. von Hoff, Immunoreactive α transforming growth factor activity in effusions from cancer patients as a marker of tumor burden and patient prognosis, *Cancer Res.* 48:5023 (1988).

23. Y.-C. Yeh, J.-F. Tsai, L.-Y. Chuang, H.-W. Yeh, J.-H. Tsai, D.L. Florine, and J.P. Tam, Elevation of transforming growth factor α and α-fetoprotein levels in patients with hepatocellular carcinoma, *Cancer Res.* 47:896 (1987).

24. G.T. Merlino, Transgenic animals in biomedical research, *FASEB J.* 5:2996 (1991).

25. C. Jhappan, C. Stahle, R.N. Harkins, N. Fausto, G.H. Smith, and G.T. Merlino, TGFα overexpression in transgenic mice induces liver neoplasia and abnormal development of the mammary gland and pancreas,*Cell* 61: 1137 (1990).

26. H. Takagi, R. Sharp, C. Jhappan, G.T. Merlino, Hypertrophic gastropathy resembling Menetrier's disease in TGFα transgenic mice, *J. Clin. Invest* 90:1161 (1992).

27. L.E. Gentry, D.R. Twardzik, G.J. Lim, J.E. Ranchalis, and D.C. Lee, Expression and characterization of transforming growth factor α precursor protein in transfected mammalian cells, *Molec. Cell. Biol.* 7:1585 (1987).

28. H. Takagi, R. Sharp, C. Hammermeister, T. Goodrow, M.O. Bradley, N. Fausto, and G. Merlino, Molecular and genetic analysis of liver oncogenesis in TGFα transgenic mice, *Cancer Res.* 52:5171 (1992).

29. G.H. Lee, G. Merlino, and N. Fausto, The development of liver tumors in TGFα transgenic mice, *Cancer Res.* 52:5162 (1992).

30. D.E. Bockman, and G. Merlino, Cytological changes in the pancreas of transgenic mice overexpressing transforming growth factor alpha, *Gastroenterology*, in press.

31. E.P. Sandgren, N.C. Luetteke, R.D. Palmiter, R.L. Brinster, and D.C. Lee, Overexpression of TGFα in transgenic mice: induction of epithelial hyperplasia, pancreatic metaplasia, and carcinoma of the breast, *Cell* 61:1121 (1990).

32. F. Vilardell, Giant hypertrophy of the gastric mucosa (Menetrier's disease), *in*: "Gastroenterology", H.L. Bockus, ed., pp. 565-571, W.B. Saunders, Philadelphia (1974).

33. C. Neville, and C.W. Daniel. "The Mammary Gland: Development, Regulation and Function", Plenum press, New York (1987).

34. F. Borellini, and T. Oka, Growth control and differentiation in mammary epithelial cells, *Environ. Health Perspect.* 80:85 (1989).

35. Y. Matsui, S.A. Halter, J.T. Holt, B.L.M. Hogan, and R.J. Coffey, Development of mammary hyperplasia and neoplasia in MMTV-TGFα transgenic mice, *Cell* 61:1147 (1990).

SOMATOSTATIN AND HUMAN GASTROINTESTINAL CANCER

Shaun R. Preston, Glenn V. Miller, Linda F. Woodhouse,
Susan M. Farmery and John N. Primrose

Academic Surgical Unit
Department of Clinical Medicine
St. James's University Hospital
Leeds, U.K. LS9 7TF

INTRODUCTION

In the UK colorectal cancer accounts for 11% of malignant neoplasms in men, and 10% in women. It is the second most common cause of cancer death in men, after carcinoma of the lung, and the third in women, after carcinoma of the breast and lung. Colorectal cancer is the commonest of gastrointestinal cancers and has a median 5 year survival rate of approximately 36 %. Although carcinoma of the stomach has a lower incidence (6% in men and 4% in women) it is the fourth most common cause of cancer death in both men and women. It is the second commonest site of gastrointestinal malignancy and has a much poorer prognosis than colorectal cancer with a median five year survival of 11.1% in men and only 9.9% in women (Cancer Research Campaign, 1988-89). The prognosis for both of these conditions is primarily determined by the stage of disease at diagnosis and unfortunately the majority of patients present when the cancer has already invaded the serosal surface of the gut. For this reason it is clear that surgical treatment alone is inadequate.

Recent progress has been made in the adjuvant treatment of colorectal cancer using the immunomodulatory drug levamisole, in combination with 5-fluorouracil. This provides a 41% reduction in cancer recurrence and a 33% reduction in the overall death rate for Dukes' Stage C colon cancer (Moertel et al., 1990). However, there is still scope for alternative treatment modalities to improve the survival of these patients and to help in patients with more advanced disease. Few advances have been made in the treatment of gastric cancer, although adjuvant treatment with H_2-receptor antagonists has been suggested to be of benefit (Tonnesen et al., 1988). Therefore current research has been directed at identification of efficacious alternatives to conventional therapy.

Growth Factors, Peptides, and Receptors, Edited
by T.W. Moody, Plenum Press, New York, 1993

Hormonal Manipulation of Cancer

The concept of hormonal control of carcinogenesis was first suggested by the observations of Beatson (1896). He proposed that the ovary 'held control' over the breast in the absence of 'distinct nervous control'. In support of this he demonstrated a dramatic reduction of breast tumour and axillary node volume in a female patient with advanced breast carcinoma following bilateral salpingo-oophorectomy.

Hormonal manipulation is now widely used in several malignancies:

- breast cancer (tamoxifen and oophorectomy)
- prostatic cancer (oestrogens, orchidectomy and LH-RH agonists)
- thyroid carcinoma (thyroxine)
- neuroendocrine tumours (somatostatin analogues)

This list is far from complete and the passage of time will unveil more tumours being treated with hormonal agonists and antagonists.

Somatostatin

Somatostatin was first isolated from the rat hypothalamus in 1973 (Brazeau et al., 1973) as a cyclic tetra-decapeptide which inhibited the secretion of immunoreactive pituitary growth hormone. Since then somatostatin and its receptors have been identified at a variety of sites both in the brain (Gomez-Pan & Rodriguez-Arnao, 1983) and throughout the body (Reichlin, 1983 Part 1; Patel, 1988).

The relative distribution of somatostatin has been extensively studied in the rat. The gastrointestinal tract accounts for 65% of total body somatostatin, the remainder being distributed between the brain (25%), the pancreas (5%) and 5% at numerous lesser sites (Patel, 1988).

Somatostatin in the gut is present predominantly (approximately 90%) in specialised epithelial cells (D-cells), the remainder being found in the myenteric and submucosal plexuses (Evers et al., 1991). D-cells are present in the gut from the cardia to the anus but are maximally concentrated in the gastric antrum.

The actions of somatostatin in the gastrointestinal tract are to:

- decrease enzyme and electrolyte secretion
- decrease the secretion of all gastrointestinal peptides/hormones
- decrease nutrient absorption
- decrease electrolyte transport
- decrease muscle contraction in the stomach, gall bladder, ileum, and splanchnic vessels
- decrease epithelial cell proliferation

(Reviewed by: Reichlin, 1983 Part 2; Lucey, 1986; Reichlin, 1987; Moreau & DeFeudis, 1987).

The effect of somatostatin on epithelial cell proliferation has led our group, and others, to investigate the potential role of somatostatin in influencing human gastric and colorectal malignancy.

Development of Somatostatin Analogues

The therapeutic use of the native peptide, somatostatin-14, is precluded by its broad spectrum of actions and a short half life in the circulation (approximately 3 minutes (Schally, 1978)). Thus, the development of long acting, stable analogues was essential to the therapeutic application of somatostatin. In addition the development of tissue specific analogues would greatly increase the therapeutic possibilities for the hormone with a reduction in unwanted side effects.

Analysis of the structure of somatostatin revealed that the amino acid sequence essential for biological activity were contained in the beta turn, amino acids 7-10 of somatostatin-14, held in position by the disulphide bridge (Veber et al., 1979). The structure of somatostatin-14 is demonstrated in Figure 1.

This information then paved the way for the same group to synthesise stable cyclical hexapeptide analogues, which upon testing in biological systems were found to be much more potent than somatostatin-14 (Veber et al., 1979; Veber et al., 1981). In addition to their cyclical structure, these analogues contained D-tryptophan residues making them more resistant to degradation.

Bauer et al. (1982) further developed the synthesis of stable D-Trp analogues and produced a number of octapeptide analogues which in addition contained D-Phe, a disulphide bridge and a C-terminal amino alcohol group. One of this group of peptides was octreotide (SMS 201-995). This analogue appeared to be a more selective inhibitor of growth hormone secretion than insulin or glucagon secretion and was therefore initially developed for the treatment of acromegaly.

Solid phase methods have since been used to synthesise a large number of related stable octapeptide analogues which vary in their potencies and biological activities. Two of the best studied analogues are RC-160 (synthesised by Schally and co-workers (Cai et al., 1986)), and somatuline (synthesised by Coy and colleagues (Murphy et al., 1987)).

All subsequent work in our laboratory has concentrated on these three most readily available analogues, the structures of which can be seen in Figure 2.

It is apparent from the structure-sequence information above that all three analogues are cyclic octapeptides with D-amino acids in their backbone. RC-160 and somatuline differ only at the N-terminal amino acid residue. Octreotide differs significantly from RC-160 and somatuline by substitution at positions 3 and 6 within

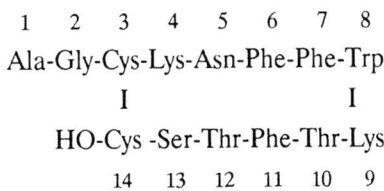

Figure 1. Structure of somatostatin-14

Octreotide / SMS 201-995 (Sandoz, UK)

DPhe-Cys-Phe-DTrp

I   I

Thr(ol)-Cys-Thr- Lys

RC-160 (Debiopharm, Switzerland)

DPhe-Cys-Tyr-DTrp

I   I

NH_2-Trp-Cys-Val- Lys

Somatuline / BIM-23014 (Ipsen, France)

DNal-Cys-Tyr-DTrp

I   I

NH_2-Trp-Cys-Val- Lys

Figure 2. Structure of three somatostatin analogues

the bioactive portion of the molecule and by replacement of the amidated C-terminal amino acid with an amino alcohol group.

Somatostatin Receptor

It has long been suspected that somatostatin exerts its tissue specific effects through a family of somatostatin receptors. Numerous attempts have been made to cross-link and purify the receptor and determine its molecular weight, the resulting protein appearing to vary in size from 27 - 200 kDa (Reviewed in Srikant et al., 1992). The first demonstration of somatostatin receptor subtypes was provided by Reubi in 1984, who revealed the presence of two receptors in rat cerebral cortex. All cortical somatostatin receptors bound somatostatin-14 with a high affinity, but only 75% of the receptors bound the somatostatin analogue octreotide. The octreotide insensitive sub-population appeared specific to rat cerebral cortex as they were not demonstrable in either pituitary or pancreatic beta cells.

More recently somatostatin receptor heterogeneity has been further determined by cross-linking a specific somatostatin photoaffinity probe, EE-581, to the receptors on a variety of rat tissues (Srikant et al., 1992). This revealed the presence of three receptors on the basis of molecular weight. A preponderant 58 kDa protein was found in a variety of tissues, a 32 kDa protein found exclusively in the rat brain, and a larger 80 kDa protein found exclusively in normal pituitary. In addition to their molecular differences there was also a 'functional heterogeneity' in the 58 kDa protein in that its selectivity for somatostatin-14 or somatostatin-28 differs in a tissue specific manner.

Recently two somatostatin receptors were cloned and their functional characteristics determined (Yamada et al., 1992). The two receptors, referred to as SSTR-1 (391 amino acids with a molecular weight of 42.6 kDa) and SSTR-2 (369 amino acids with a molecular weight of 41.3 kDa), were also found to be members of the seven trans-membrane domain receptor superfamily. Both SSTR-1 and SSTR-2 bound somatostatin-14 in preference to somatostatin-28, but differed in their distribution in human tissue. SSTR-1 is present in jejunum, stomach, pancreatic islet cells and at lower levels in colon, colonic cancer, and kidney, whereas SSTR-2 was present in cerebrum and kidney, and at lower levels in jejunum, colon, colonic cancer, liver and hepatoma.

Cloning and Southern blotting analysis further suggest that the somatostatin receptor is a member of a larger family with at least four subtypes, two of which have been sequenced and ligand binding determined (Yamada et al., 1992).

Somatostatin and Cancer

Many human tumours have been shown by both in vitro and in vivo studies to express somatostatin receptors. These are predominantly endocrine tumours including carcinoid, VIPoma, insulinoma, glucagonoma, gastrinoma and small cell lung cancer, but receptors have also been demonstrated on tumours of the brain (Reviewed by: Reubi et al., 1987; Schally, 1988; Vinik, 1990; Evers et al., 1991). The response of these tumours to somatostatin analogues is predominantly one of symptom control, but there have been reports of certain tumours and their metastases actually decreasing in size (Wood et al., 1985; Kraenzlin et al., 1985; Clements & Elias, 1985). In addition there have been numerous reports of somatostatin inhibiting the growth of human solid non-endocrine carcinomata (Reviewed by: Evers et al., 1991; Schally, 1988; Moreau & DeFeudis, 1987). Some examples are given below.

Exocrine Pancreatic Cancer. Experimental studies on rats bearing the acinar tumour DNCP-322 and hamsters bearing pancreatic ductal adenocarcinoma revealed that administration of somatostatin analogues produced a decrease in both tumour weight and tumour volume (Redding & Schally, 1984). Further work revealed that two human pancreatic cancers maintained as xenografts in nude mice could be inhibited by treatment with the somatostatin analogue octreotide (Upp et al., 1988).

Prostatic Cancer. The somatostatin analogues RC-160 and RC-121 have been shown to accentuate the effect of an LH-RH agonist on inhibition of growth of the androgen sensitive prostatic cancer Dunning R-3327H, probably by further inhibiting the hypothalamic-pituitary-gonadal axis (Schally & Redding, 1987).

Breast Cancer. The presence of somatostatin receptors has been demonstrated on a limited number of human breast cancers (Reubi et al., 1987). In vivo studies on cell line xenografts show that octreotide markedly reduces the growth of the oestrogen-dependent MCF-7 cell line and produces a modest prolongation of tumour doubling time in the oestrogen-independent BT-20 cell line (Weber et al., 1989).

Lung Cancer. Xenografts of the human squamous carcinoma cell line H-165 can be inhibited by treatment with the somatostatin agonist somatuline, resulting in reduced tumour volume (Bogden et al., 1990).

Colorectal Cancer. In vitro studies on the human colonic cancer cell line, HT29, and in vivo work on xenografts of CX1 human adenocarcinoma tissue fragments, revealed that treatment with somatostatin-14 produced inhibition of tumour growth (Smith & Solomon, 1988). More recently it has been shown that treatment of xenografts of HT29 cells can be inhibited by treatment with the somatostatin analogue RC-160 (Radulovic et al., 1991).

Gastric Cancer. The presence of somatostatin receptors has been demonstrated on the human gastric cancer cell line HGT-1, and purified to homogeneity (Reyl-Desmars et al., 1990). In addition octreotide has been shown to inhibit the growth of the human gastric cancer cell line MKN45 when established as xenografts in nude mice (Morris et al., 1988).

Mechanism of Action

The exact mechanism by which somatostatin exerts its anti-proliferative effect is not known, but in view of its broad spectrum of biological functions it is unlikely to be by a single effect and may differ between tissue / tumour site. The possible mechanisms are one or a combination of the following: (1) Endocrine:- that somatostatin acts on the hypothalamo-pituitary axis to decrease growth hormone secretion and thus indirectly inhibit cellular proliferation; (2) Paracrine:- somatostatin is known to inhibit secretion of gastrointestinal hormones and growth factors which may act on cells in the immediate vicinity to regulate cell turnover; (3) Autocrine:- somatostatin may act on the target cell to decrease its secretion of growth factors which usually regulate its own growth; (4) Direct:- somatostatin may act at specific receptors on the target tissue to directly influence the intracellular environment and inhibit cell growth; and (5) 'Lumone':- it has occasionally been proposed that somatostatin may act lower down the gastrointestinal tract to influence cellular function after being secreted in the lumen of the gut. This mechanism is not widely accepted.

SOMATOSTATIN RECEPTORS IN GASTRIC AND COLORECTAL CANCER

The possibility that somatostatin may control epithelial cell proliferation directly promoted us to investigate human gastric and colonic adenocarcinomata for the presence of somatostatin receptors. The aims of these studies were to determine a direct or autocrine regulatory role for somatostatin in gastrointestinal cancer; and to assess the therapeutic application of available superactive octapeptide analogues of somatostatin in these prevalent human cancers.

The human tissue used in all assays was obtained fresh following surgical resection for biopsy proven adenocarcinoma of the stomach, colon or rectum. The tumour was sampled, briefly washed in ice cold physiological saline, 'snap' frozen in liquid nitrogen and then stored at -70°C. To determine the presence or absence of somatostatin receptors on these tumours membrane preparations were made by the method of Srikant and Patel (1981). The protein concentration of the resulting membrane suspension was then determined by the method of Bradford (1976) and then used immediately in a standard displacement assay using ^{125}I-Tyr^{11}-somatostatin as the radioligand, displaced with unlabelled somatostatin-14 over the concentration range $1x10^{-11}$M to $1x10^{-5}$M. Data were then processed using the Ligand PC curve fitting program (Munson & Rodbard, 1980).

Using these methods the membrane preparations from 51 surgically resected human gastrointestinal cancers (28 gastric and 23 colorectal) were assayed. These experiments, for the first time, demonstrated the presence of somatostatin receptors on human malignant gastrointestinal tissue (22/28 gastric and 17/23 colorectal cancers). It was interesting to note that only one of the gastric and none of the colorectal cancers expressed receptors with a high affinity (Kd = 0.9nM and Bmax = 0.23 pmol/mg of protein), comparable with those seen in the rat cerebral cortex. The majority of tumours expressed specific low affinity, high capacity receptors. These were found to be similar in both gastric and colorectal cancers (Miller et al., 1991).

At the time of surgical resection for tumour, macroscopically normal mucosa from a site distant from the tumour on the resected specimens, was sampled and stored in the same manner as the cancer. Membrane preparations from this tissue were assayed using standard displacement analysis as above. Specific somatostatin receptors were demonstrated on 21 / 28 samples of gastric mucosa and on 15 / 23 colonic mucosa, all of which were low affinity, high capacity receptors. The results of these studies are shown in Table 1. It is accepted that the normal mucosa contains cells which express high affinity sites, such as the endocrine cells, but these receptors

Table 1. Binding characteristics of human gastrointestinal mucosa and carcinoma. Results are expressed as median (quartiles) for 51 patients (details in text)

		Kd (nM)	Bmax (pmol/mg protein)
Gastric Cancer	n = 21	170 (72-250)	1.2 (0.7-5.7)
	n = 1	0.9	0.23
Gastric Mucosa	n = 22	180 (81-270)	3.9 (1.2-6.6)
Colorectal Cancer	n = 15	140 (89-200)	1.8 (1.2-2.9)
Colorectal Mucosa	n = 15	130 (71-200)	0.7 (0.3 3.5)

are effectively hidden by the preponderant high capacity, low affinity binding sites.

An initial concern was that these low affinity, high capacity receptors were an artefact due to proteolytic cleavage or denaturation of the receptor or ligand. To prevent this the assays were performed in the presence of bacitracin, and the addition of further protease inhibitors made no difference to either dissociation constant or maximum binding capacity. In addition the assays were performed in an identical manner on membrane preparations from rat cerebral cortex to validate the methodology, which revealed the presence of a single class of high affinity somatostatin receptor (median (range) Kd = 0.85 (0.55-0.99) nM and Bmax = 0.22 (0.12-0.32) pmol/mg of protein) consistent with those previously reported (Srikant & Patel, 1981)

Cation Dependence - Further Evidence for Receptor Subtype

The ionic requirements of somatostatin-14 binding to rat brain somatostatin receptors has been investigated (Reubi & Maurer, 1986) revealing that specific binding to the SS_1 (octreotide binding) receptor subtype was enhanced by Mg^{2+} but reduced by 120 mM NaCl, and binding to the SS_2 (octreotide independent) receptor subtype was unchanged by the presence of Mg^{2+} and enhanced by the addition of NaCl. Thus, we proceeded to investigate the effect of NaCl and $MgCl_2$ on somatostatin binding to sites on human gastric mucosa as a representative of our putative gastrointestinal receptor subtype. Rat cerebral cortex was used as control tissue.

The addition of NaCl to the assay system for human gastric mucosa resulted in an increase in specific binding (expressed as a percentage of binding in the absence of NaCl), maximal at 80 mM. In comparison, rat cerebral cortex under identical conditions over the concentration range 60 - 160 mM NaCl resulted in decreased specific binding (Figure 3).

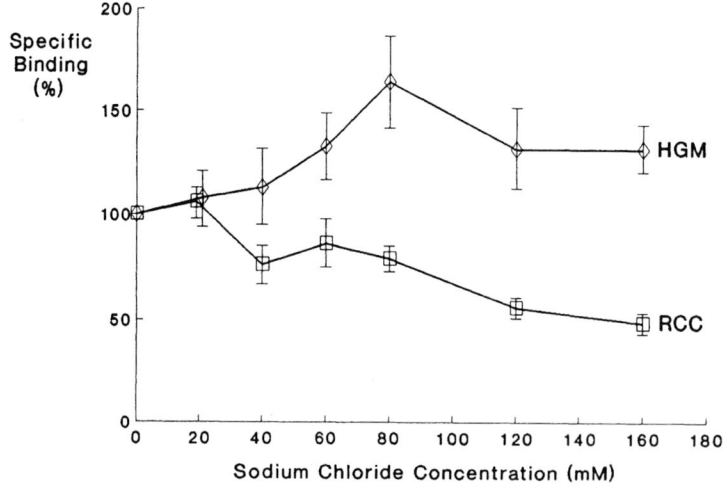

Figure 3. Specific somatostatin-14 binding - dependence upon sodium chloride concentration.

In contrast the addition of MgCl$_2$, in the absence of NaCl, over the concentration range 1.25 - 10 mM had no influence on somatostatin binding to human gastric mucosa, but produced a significant increase in binding to rat cerebral cortex. Thus, the human gastrointestinal somatostatin receptor appears to resemble the SS$_2$ receptor from rat cerebral cortex, but has a lower affinity, and could thus represent a further receptor subtype (Farmery et al., 1991).

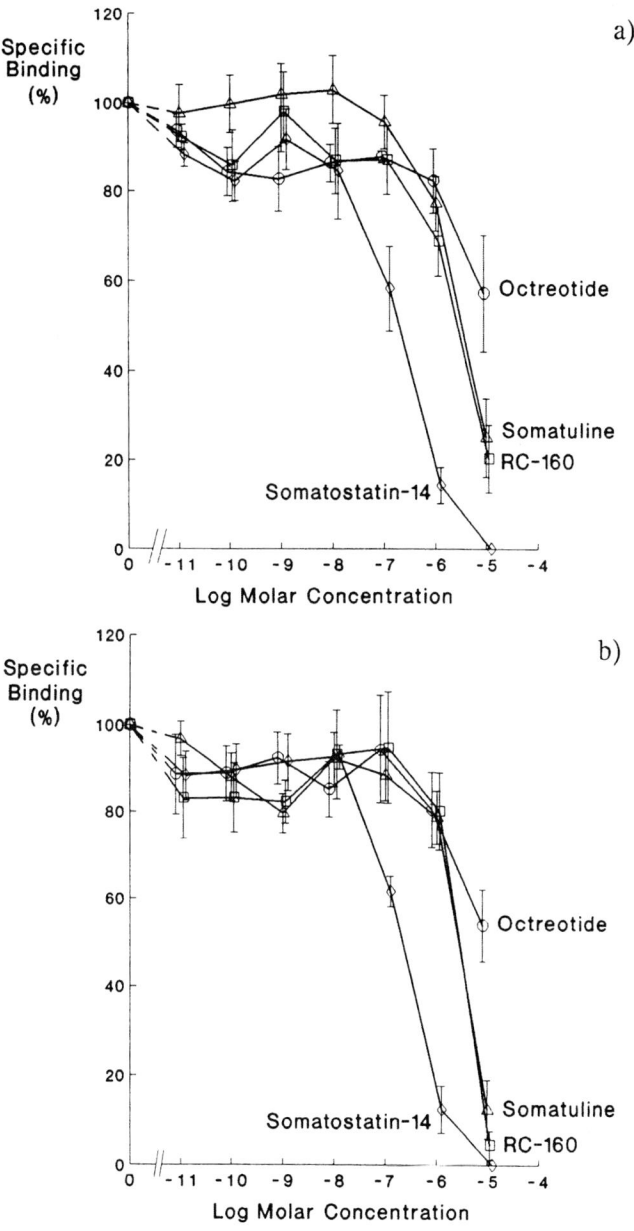

Figure 4. Displacement curve of ^{125}I-Tyr^{11}-somatostatin-14 displaced with somatostatin-14, octreotide, RC-160 and somatuline on a) human gastric cancer and b) human colonic cancer.

Differential Analogue Binding

In view of previous work on rat cerebral cortex, identifying receptor subtypes by the differential binding of the somatostatin analogue octreotide (Reubi, 1984), we attempted to characterise the gastrointestinal receptor subtype using three readily available and well characterised analogues, octreotide, RC-160 and somatuline. These investigations were also performed to identify which analogue, if any, possessed most potential for use in the treatment of gastrointestinal malignancy.

Membrane preparations were again made from human gastric and colonic adenocarcinomata. Displacement assays were performed using ^{125}I-Tyr^{11}-somatostatin-14 as radioligand, displaced in turn with somatostatin and each of the analogues over the concentration range $1\times10^{-11}M$ to $1\times10^{-5}M$. Displacement by the somatostatin analogues was expressed as a percentage of the maximal specific displacement of the radioligand by somatostain-14, over the concentrations tested. From the displacement data the 50% inhibitory concentration (IC_{50}) was calculated for each analogue on each tissue.

Figure 4 demonstrates again the presence of low affinity binding sites, with somatostatin-14 displacing the radioligand with an IC_{50} of 196 nM in human gastric cancer and 197 nM in human colorectal cancer. It can also be seen in both tissue types that RC-160 and somatuline displace the radioligand with an IC_{50} approximately ten fold higher than somatostatin-14, 3.4 uM and 3.6 uM respectively in human gastric cancer. In human colorectal cancer these agonists also displaced with an IC_{50} of 2.5 and 3.0 uM, respectively. Octreotide failed to acheive 50% displacement at any of the concentrations tested on both gastric and colonic cancer.

Conclusion

As a result of work performed in our laboratory we have been able to demonstrate the presence of specific somatostatin binding sites on a high percentage of human gastric and colorectal mucosae and adenocarcinomata. In the vast majority of cases the binding characteristics of these sites are low affinity, high capacity and in this respect differ from those demonstrated elsewhere. However, the presence of high affinity, low capacity binding sites was demonstrated in one gastric cancer, but not on any of the gastric mucosa or colonic tissue, normal or neoplastic, tested. The binding in this tumour, which was not neuroendocrine on the basis of chromogranin and neurone specific enolase activity, serves to illustrate that the methods can demonstrate high affinity sites if they predominate in gastrointestinal tissue.

In addition, by studies of cation dependence and analogue binding we have demonstrated that our proposed gastrointestinal receptor resembles the SS_2 receptor described by Reubi in that its binding of somatostatin-14 is sodium ion dependent and magnesium ion independent (Reubi, 1986), and in the fact that it displays very poor binding of octreotide and binds somatostatin-14 more avidly. The gastrointestinal receptor however differs from the SS_2 receptor subtype in that it has a lower affinity and much larger capacity, as described above.

Thus we propose a further receptor subtype for somatostatin-14 present in the stomach, colon and rectum in both normal mucosa and adenocarcinoma of these tissues. The functional significance of this putative receptor has not been established, but clearly warrants study. If this site proves to be involved in signal transduction it may prove possible to design specific somatostatin analogues with higher affinity for this site which could be exploited in the therapy of both gastric and colorectal neoplasia.

REFERENCES

Bauer, W., Briner, U., Doepfner, W., Haller, R., Huguenin, R., Marbach, P, Petcher, T.J. & Pless, J., 1982, SMS201-995: A very potent and selective octapeptide analogue of somatostatin with prolonged action. Life Sci. **31**, 1133-1140.

Beatson, G.T., 1896, On the treatment of inoperable cases of carcinoma of the mamma. suggestions for a new method of treatment, with illustrated cases. Lancet **ii** 104-107.

Bogden, A.E., Taylor, J.E., Moreau, J.P., Coy, D.H. & LePage, D.J., 1990, Response of human lung tumour xenografts to treatment with a somatostatin analogue (somatuline). Cancer Res. **50**, 4360-4365.

Bradford, M.M., 1976, A rapid and sensitive method for the quantitation of microgram quantities of protein utilising the principle of protein dye binding. Anal. Biochem. **72**, 248-254.

Brazeau, P., Vale, W., Burgus, R., Ling, N., Butcher, M., Rivier, J., Guillemin, R., 1973, Hypothalamic polypeptide that inhibits the secretion of immunoreactive pituitary growth hormone. Science **179**, 77-79.

Cai, R.-Z., Szoke, B., Lu, R., Fu, D., Redding, T.W., Schally, A.V., 1986, Synthesis and biological activity of highly potent octapeptide analogs of somatostatin. Proc. Natl. Acad. Sci. USA. **83**, 1896-1900.

Cancer Research Campaign Factsheets 1.1-9.4, 1988-1990, Cancer Research Campaign, London, UK.

Clements, D. & Elias, E., 1985, Regression of metastatic VIPoma with somatostatin analogue SMS 201-995. Lancet **i**, 874-875.

Evers, B.M., Parekh, D., Townsend, C.M., Thompson, J.C., 1991, Somatostatin and analogues in the treatment of cancer. Ann. Surg. **213**(3), 190-198.

Farmery, S.M., Miller, G.V., Woodhouse, L.F. & Primrose, J.N., 1991, Ionic requirements of [125I]iodo-Tyr[11]-somatostatin-14 binding sites in human gastric mucosa and rat cerebral cortex. Br. J. Pharmacol. **102**(**Suppl.**), 165P.

Gomez-Pan, A. & Rodriguez-Arnao, M.D., 1983, Somatostatin and growth hormone release factor: synthesis, location, metabolism and function. Clin. Endocrinol. Metab. **12**(3), 469-507.

Kraenzlin, M.E., Ch'ng, J.L.C., Wood,S.M., Carr, D.H. & Bloom, S.R., 1985, Long-term treatment of a VIPoma with somatostatin analogue resulting in remission of symptoms and possible shrinkage of metastases.Gastroenterology **88**, 185-187.

Lucey M.R., 1986, Endogenous somatostatin and the gut. Gut **27**, 457-467.

Miller, G.V., Farmery, S.M., Woodhouse, L.F. & Primrose, J.N., 1992, Somatostatin binding in normal and malignant human gastrointestinal mucosa. Br. J. Cancer (In Press).

Moertel, C.G., Fleming, T.R., Macdonald, J.S., Haller, D.G., Laurie, J.A., Goodman, P.J., Ungerleider, J.S., Emerson, W.A., Tormey, D.C., Glick, J.H., Veeder, M.H. & Mailliard, J.A., 1990, Levamisole and fluorouracil for adjuvant therapy of resected colon carcinoma. N. Engl. J. Med. 322(6), 352-358.

Moreau, J.P. & DeFeudis, F.V., 1987, Pharmacological studies of somatostatin and somatostatin-analogues: therapeutic advances and perspectives. Life Sci. 40, 419-437.

Morris, D.L., Watson, S.A., Harrison, J.D. & Durrant L.G., 1988, Somatostatin (SMS201.995) reduces growth of human gastric cancer (MKN45) xenografts [Abstract]. Gut 29, A1477.

Munson, P.J. & Rodbard, D., 1980, Ligand: a versatile computerized approach for characterization of ligand-binding systems. Anal. Biochem. 107, 220-239.

Murphy, W.A., Lance, V.A., Moreau,S., Moreau, J., -P. & Coy,D.H., 1987, Inhibition of rat prostate tumour growth by an octapeptide analog of somatostatin. Life Sci. 40, 2515-2522.

Patel, Y.C., 1988, Articles on Somatostatin & Analogue Peptides. Investigators' Newsletter. Current Medical Literature Ltd. London.

Radulovic, S., Miller,G., Schally, A.V., 1991, Inhibition of growth of HT-29 human colon cancer xenografts in nude mice by treatment with bombesin/gastrin releasing peptide antagonist (RC-3095). Cancer Res. 51, 6006-6009.

Redding, T.W. & Schally, A.V., 1984, Inhibition of growth of pancreatic carcinomas in animal models by analogs of hypothalamic hormones. Proc. Natl. Acad. Sci. USA. 81, 248-252.

Reichlin, S., 1983, Somatostatin (first of two parts). N. Engl. J. Med. 309(24), 1495-1501.

Reichlin, S., 1983, Somatostatin (second of two parts). N. Engl. J. Med. 309(25), 1556-1563.

Reichlin, S., 1987, Secretion of somatostatin and its physiologic function. J. Lab. Clin. Med. 109(3), 320-326.

Reubi, J.C., 1984, Evidence for two somatostatin-14 receptor types in rat brain cortex. Neurosci. Lett. 49, 259-263.

Reubi, J.C. & Maurer, R., 1986, Different ionic requirements for somatostatin receptor subpopulations in the brain., Regul. Pept. 14, 301-311.

Reubi, J.C., Maurer, R., von Werder, K., Torhorst, J., Klijn, J.G.M. & Lamberts, S.W.J., 1987, Somatostatin receptors in human endocrine tumours. Cancer Res. 47, 551-558.

Reyl-Desmars, F., LeRoux, S., Linard, C., Benkouka, F. & Lewin, M.J.M., 1989, Solubilization and immunopurification of a somatostatin receptor from the human gastric tumoral cell line HGT-1. J. Biol. Chem. 264(31), 18789-18795.

Schally, A.V., Coy, D.H., & Meyers, C.A., 1978, Hypothalamic regulatory hormones. Annu. Rev. Biochem. 47, 89-128.

Schally, A.V. & Redding, T.W., 1987, Somatostatin analogs as adjuncts to agonists of luteinizing hormone-releasing hormone in the treatment of experimental prostate cancer. Proc. Natl. Acad. Sci. USA. 84, 7275-7279.

Schally, A.V., 1988, Oncological applications of somatostatin analogues. Cancer Res. 48, 6977-6985.

Smith, J.P. & Solomon, T.E., 1988, Effects of gastrin, proglumide, and somatostatin on growth of human colon cancer. Gastroenterology 95, 1541-1548.

Srikant, C.B. & Patel, Y.C., 1981, Somatostatin receptors: identification and characterisation in rat brain membranes. Proc. Natl. Acad. Sci. USA. 78, 3930-3934.

Srikant, C.B., Murthy, K.K., Escher, E.E. & Patel, Y.C., 1992, Photoaffinity labelling of the somatostatin receptor: identification of molecular subtypes. Endocrinology **130(5)**, 2937-2946.

Srkalovic, G., Cai, R.-Z. & Schally,A.V., 1990, Evaluation of receptors for somatostatin in various tumours using different analogs. J. Clin. Endocrinol. Metab. **70(3)**, 661-669.

Tonnesen, H., Knigge, U., Bulow, S., Damm, P., Fischerman, K., Hesselfeldt, P., Hjortrup, A., Pedersen, I.K., Pedersen, V.M., Siemssen, O.J., Svendsen, L.B. & Christiansen, P.M., 1988, Effect of cimetidine on survival after gastric cancer. Lancet **ii**, 990-992.

Upp Jr., J.R., Olson, D., Poston, G.J., Alexander, R.W., Townsend, C.M. & Thompson, J.C., 1988, Inhibition of growth of two human pancreatic adenocarcinomas in vivo by somatostatin analog SMS 201-995. Am. J. Surg. **155**, 29-35.

Veber, D.F., Holly, F.W., Nutt, R.F., Bergstrand, S.J., Brady,S.F.,Hirschmann,R., Glitzer, M.S. & Saperstein, R., 1979, Highly active cyclic and bicyclic somatostatin analogues of reduced ring size. Nature **280**, 512-514.

Veber, D.F., Freidinger, R.M., Perlow, D.S., Paleveda Jr., W.J., Holly, F.W., Strachan, R.J., Nutt, R.F., Arison, B.H., Homnick, C., Randall, W.C., Glitzer, M.S., Saperstein, R. & Hirschmann, R., 1981, A potent cyclic hexapeptide analogue of somatostatin. Nature **292**, 55-58.

Vinik, A.I., Lloyd,R. & Cho, K., 1990, The use of somatostatin analog in gastroenteropancreatic tumours other than carcinoid. Metabolism **39(9)**, Suppl.2, 156-162.

Weber, C., Merriam, L., Koschitzky, T., Karp, F., Benson, M., Forde, K. & LoGerfo, P., 1989, Inhibition of growth of human breast carcinomas in vivo by somatostatin analogue, SMS 201-995: treatment of nude mouse xenografts. Surgery **106(2)**, 416-422.

Wood, S.M., Kraenzlin, M.E., Adrian, T.E. & Bloom, S.R., 1985, Treatment of patients with pancreatic endocrine tumours using a new long-acting somatostatin analogue symptomatic and peptide responses. Gut **26**, 438-444.

Yamada, Y., Post, S.R., Wang, K., Tager, H.S., Bell, G. & Seino, S., 1992, Cloning and functional characterization of a family of human and mouse somatostatin receptors expressed in brain, gastrointestinal tract, and kidney. Proc. Natl. Acad. Sci. USA. **89**, 251-255.

INHIBITION BY GnRH ANTAGONISTS OF BREAST AND ENDOMETRIAL CANCER CELL GROWTH INDUCED BY ESTROGEN AND INSULIN-LIKE GROWTH FACTORS

Yoav Sharoni and Joseph Levy

Clinical Biochemistry Unit
Soroka Medical Center of Kupat Holim
Faculty of Health Sciences
Ben-Gurion University of the Negev
Beer Sheva, 84105, Israel

DIRECT EFFECT OF GnRH ANALOGUES

Breast and endometrial carcinoma are common female malignancies in Western countries. Estrogen-dependent and independent tumors are present in both diseases. Endocrine therapy, which usually consists of antiestrogens or progestins, is frequently used in the treatment of primary, recurrent, or metastic diseases, but with only limited success. Several recent studies have suggested that synthetic analogues of the hypothalamic hormone, GnRH, have a therapeutic effect on sex steroid-dependent tumors such as carcinoma of the prostate[1] and breast.[2,3] The rationale for this treatment is that continuous administration of GnRH agonists causes inhibition of gonadotropin release from the pituitary and thus reduces steroid production by the ovaries or testes. Although GnRH agonists cause inhibition of pituitary and gonadal function, it recently became clear that an antagonist may be advantageous in the treatment of cancer as compared with the available superagonists. While repeated administration of the GnRH agonist is required to inhibit pituitary and gonadal function, the same effect may be obtained by a single administration of the GnRH antagonist.[4] The inhibition of gonadotropin release by the GnRH antagonist commences immediately after its administration, while agonists frequently cause a transient stimulation of pituitary and gonadal function, resulting in a temporary clinical "flare-up" of the disease.[5]

Conflicting data exist pertaining to the possible direct effect of various GnRH analogues on the growth of mammary tumor cells in culture. Miller et al.[6,7] were the first to propose that GnRH and some of its agonists may inhibit the growth of MCF-7 cells. However, more recent work has not been able to confirm these results.[8] Although GnRH agonists have no significant effect on cell growth, some antagonists do inhibit thymidine incorporation in breast cancer cell lines.[8] New GnRH antagonists have been developed by Dr. A.V. Schally[4] These D-citrulline[6] and D-homocitrulline[6] analogues are very potent antagonists as demonstrated by their ability to inhibit GnRH-induced luteinizing hormone (LH) release, inhibit ovulation in cycling rats, and suppress LH levels in ovariectomized rats. Characteristically, these peptides did not have any edematogenic effects even at high doses.[4]

The direct effect of GnRH agonists and antagonists on the growth of mammary tumors was tested in human estrogen-independent (MDA-MB-231) and estrogen-dependent (MCF-

Growth Factors, Peptides, and Receptors, Edited
by T.W. Moody, Plenum Press, New York, 1993

7) cells in culture. In the MDA-231 cells, we have shown that the new GnRH antagonists inhibit cell growth, but buserelin, a GnRH agonist, has no significant effect.[9] In MCF-7 cells, buserelin, slightly stimulated cells growth in an estrogen-deficient medium, but had no effect in the presence of estrogen.[10] In these cells, there was only slight inhibition of the basal growth rate by the GnRH antagonist, SB-75, but the estrogen-dependent growth was completely abolished.

BINDING OF GnRH ANALOGUES

The apparent dissociation between the effects of agonists and antagonists on mammary tumor cell growth may be explained by the presence of different receptors to these peptides in the tumor cells. To test this hypothesis, we measured GnRH binding sites in DMBA (7,12-dimethylbenz[a]anthracene)-induced rat mammary tumor membranes and on MCF-7 cells. [^{125}I]buserelin, A GnRH super-agonist, was used for these receptor studies. A scatchard plot analysis of buserelin-specific binding revealed a nonlinear plot, suggesting the presence of high affinity and low affinity binding sites. The kinetic parameters of these sites (calculated according to Munson and Rodbard[11]) are shown in Table 1. The binding of [^{125}I]buserelin was displaced equally well by unlabeled buserelin and by GnRH antagonists,[10,12] suggesting that both agonists and antagonists are bound to the same receptor. When parallel experiments were performed with labeled antagonists, the binding was displaced by unlabeled antagonists, but only partially by the unlabeled agonist, buserelin. The results suggest that in mammary tumor cells there is a GnRH antagonist binding site which is not recognizable to GnRH agonists. It is possible that the effect of the antagonists on cell growth is mediated by this low affinity receptor since its dose response on cell growth corresponds to the displacement curve of the low affinity binding sites. Half maximal displacement was achieved at 10^{-6} M and half maximal inhibition of growth was seen at 10^{-6} M and 3×10^{-6} M in endometrial and breast cancer cells, respectively.

ROLE OF INSULIN LIKE GROWTH FACTORS

It has been shown[14] that in the estrogen-dependent mammary tumor cells the cardinal effects of estrogens on growth is mediated by synthesis and secretion of various autocrine growth factors. In the estrogen-independent cells these factors are secreted constitutively in

Table 1. Kinetic parameters of GnRH receptors in MCF-7 cells and in DMBA induced rat mammary tumor membranes

Receptor source	high affinity site		low affinity site	
	Kd (nM)	B$_{max}$ fmol/mg	Kd (µM)	B$_{max}$ pmol/mg
MCF-7 cells	1.4±1.0	36±11	1.3±1.0	42±25
DMBA tumors	2.5±0.8	250±120	1.7±0.3	200±105

MCF-7 cells and DMBA-induced rat mammary tumor membranes were prepared as described.[10,13] Iodination of the GnRH agonist, buserelin, and the conditions for binding to tumor cells and membranes were detailed in recent publications.[10,11,13] Total binding was determined in the absence of unlabeled buserelin and nonspecific binding in the presence of 10^{-5} M buserelin. In regular experiments, the total binding for [^{125}I]buserelin was between 7000 to 10000 cpm which represent, on the average, 6-8% of the total counts present in the tube. The specific binding was 50% to 60% of the total binding. The Kd values and the number of receptors were analyzed as described by Munson and Rodbard[11] using the EBDA and LIGAND programs.

an uncontrolled and excessive manner. The fact that GnRH antagonists mainly inhibit estrogen-stimulated growth of MCF-7 mammary cancer cells and inhibit estrogen-independent growth of MDA-231 cells suggests that these peptide analogues intervene with the autocrine mechanism of growth in the two cell lines

Insulin and insulin-like growth factors (IGFs) are among the important regulators of breast and endometrial cancer growth. Physiological concentrations of insulin, IGF-I[15-17] and IGF-II[18] stimulate cell proliferation in several human breast cancer cell lines. Human breast cancer cells have been shown to specifically bind insulin, IGF-I, and IGF-II with high affinity[15,16,19].

It has been shown that human breast[20] and endometrial[21] cancer cells have the capacity to synthesize and secrete a variety of growth factor-like substances, including the IGFs. Using a RIA and a RRA (after removal of IGF-binding proteins), IGF-II-like activity was detected in conditioned medium of several breast cancer cell lines and IGF-II mRNA transcripts were identified by Northern blot analysis and by RNase protection assay.[18] The level of IGF-II mRNA was increased by estrogen. The ability of these cells to express IGF-II mRNA and secrete IGF-II into the culture medium further supports the hypothesis that IGF-II may have autocrine growth regulatory functions in human breast cancer. We, therefore, investigated whether GnRH analogs interfere with autocrine or paracrine loops of insulin-like growth factors (I and II) and whether inhibition of growth of mammary and endometrial cancer cells by the GnRH antagonist involves inhibition of the release and activity of IGFs.

We compared cancer cell growth in response to estradiol, IGFs and insulin. Maximal cell growth was higher in the presence of growth factors than in the presence of estradiol (Fig. 1), and the lag before initiation of growth was shorter by one to two days (submitted), suggesting that autocrine secretion of IGFs may mediate the effect of estradiol. The relative potency of growth factors on breast and endometrial cancer cell growth is shown in Table 2 and is similar to that reported for breast cancer cells.[18,22] The apparent $K_{1/2}$ shown also corresponds to the relative affinities of the binding of these peptides to the IGF-I receptor, suggesting that IGF-II and insulin effects are mediated by the IGF-I receptor.

Table 2. Relative potency of IGFs and insulin
for induction of mammary and endometrial
cancer cell growth

Growth factor	$K_{1/2}$ (nM)	
	MCF-7	Ishikawa
IGF-I	2±1	3±2
IGF-II	10±4	6±3
insulin	90±35	160±55

Estrogen withdrawn cells were prepared according to Vignon.[23] The cells were grown for six days in 75 cm[2] flasks in Dulbecco Modified Eagle's medium without Phenol red and with only 3% steroid stripped fetal calf serum. Cells were then seeded into 96 multiwell plates (12000-20000 cells per well) and grown for one day. The medium was then replaced with fresh medium containing only 1% steroid stripped serum, with or without the indicated peptides at concentrations between 10^{-10} - 10^{-6} M. After three days the number of cells was estimated by the MTT method as described previously.[10] In typical experiments, the control cell number was 18000 and the stimulated was 42000 cells/well for MCF-7 cells. For Ishikawa cells, the numbers were 30000 and 42000 respectively. Apparent K1/2 (±SEM) is the concentration needed for half maximal activation of cell growth, calculated from the results of three to five different experiments, each done in ten replicates for each concentration.

MODULATION OF IGF INDUCED GROWTH BY GnRH ANALOGUES

We studied whether the proliferation induced by the growth factors is affected by GnRH analogues. The results for MCF-7 cells (Fig. 1) compare the effect of the peptides on basal, estrogen and IGF-I induced growth. The two agonists (buserelin and D-Trp[6]-GnRH) produce a small increase in cell number under basal growth conditions, but do not affect cell growth in the presence of estradiol or IGF-I. The two antagonists (SB-75 and SB-88) reduce the number of cells under basal growth conditions. The reduction in cell number of estradiol and IGF-I treated cells is larger than that of unstimulated cells. Thus, the growth stimulatory effect of IGF-I is also affected by GnRH antagonists. Similar effects were found with IGF-II- induced growth, although higher concentrations were needed to achieve the same activation as is clear from the relative potency of the two growth factors (Table 2).

The effect of GnRH antagonists on IGF induced growth can be explained by reduction of number or affinity of IGF-I receptors (arrow 3 in Fig. 2) or by inhibition of the signal transduction mechanisms. In preliminary results we found that SB-75 reduces tyrosine phosphorylation of the IGF-I receptor which is one of the earliest event in its signal transduction pathway (arrow 4 in Fig. 2).

INHIBITION OF IGF-II RELEASE BY GnRH ANTAGONISTS

IGF-II is an autocrine growth factor which is secreted in response to estrogen treatment. Thus, the inhibition of the IGF induced cell growth by the GnRH antagonists may provide an explanation for the inhibition of estrogen-induced cell growth by these peptides. However, it is also possible that the antagonist has an additional effect on IGF secretion. Thus, we measured IGF-I and IGF-II under conditioned growth medium of mammary and endometrial cancer cells. IGF release was measured by specific RIA following acid extraction and chromatography so as to avoid the interference of IGF-binding proteins. The endometrial and breast cancer cells secrete IGF-II, but not IGF-I. The amount of IGF-II in MCF-7 conditioned medium was increased three-fold by two days treatment with 10^{-9} M 17β-estradiol. At this time, no effect of estradiol on cell growth was evident, showing that the induction of IGF-II release by estrogen precedes the steroid's effect on cell growth.

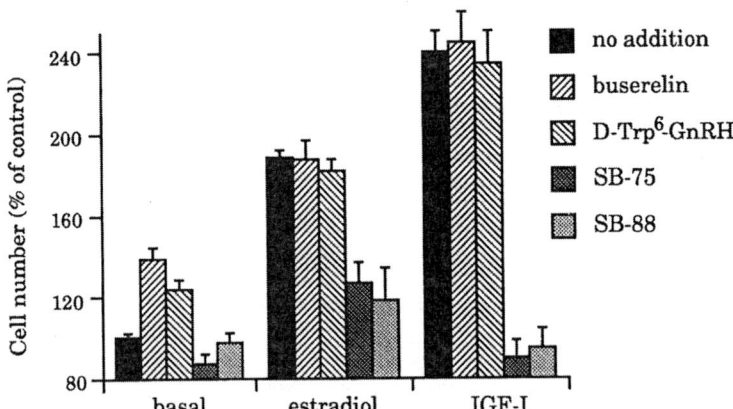

Fig. 1. Effect of GnRH agonists and antagonists on MCF-7 mammary cancer cell growth. Estrogen withdrawn cells were prepared and cell number was measured as described in the legend to Table 2. Cells were grown for three days in the presence of the estrogen (10^{-9} M) or IGF-I (3×10^{-8} M). GnRH analogues were used at 10^{-5} M. The results are expressed as percent (±SEM) of the cell number (18000-25000 cells/well) under basal growth conditions without addition of hormones or GnRH analogues. Results are the average of three to five experiments each done in ten replicates.

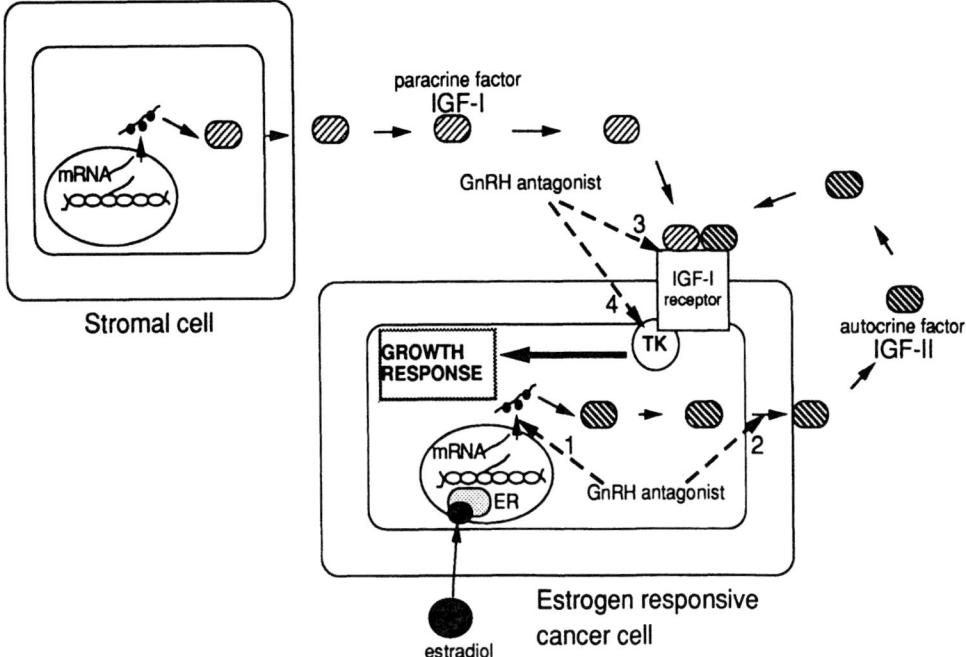

Fig. 2. Possible sites for GnRH antagonist intervention in cancer cell growth induced by estrogen and IGFs. For details see text.

Treatment with a combination of estradiol and SB-75, the GnRH antagonist, reduces the secretion of IGF-II to almost undectable levels, much below that found in control, untreated cells. This reduction is evident after two days of treatment, before any inhibition of cell growth is seen (submitted). Dose response curves of SB-75 effect on IGF-II release and Ishikawa cell growth show that inhibition of IGF-II release occurs at a lower concentration than that needed by the antagonist for reduction in cell number. These results support the notion that inhibition of IGF-II release is one of the earliest events during the treatment with the GnRH antagonist.

As suggested here, and by others,[18] IGF-I is not secreted by mammary cancer cells. However, its secretion from adjacent stromal cells may play a role in proliferation of these cancer cells.[24,25]

CONCLUSIONS

The inhibition of estrogen-induced cancer cell growth by GnRH antagonists of the SB series may be explained by a combination of two processes:

1. Inhibition of IGF-II synthesis or autocrine secretion (arrows 1 and 2 in Fig 2).

2. Inhibition of IGF induced cell growth (arrows 3 and 4 in Fig 2). The two alternative mechanisms are especially important in *in vivo* situations where paracrine secretion of IGF-I by adjacent cells is important for tumor growth.

ACKNOWLEDGEMENTS

We are grateful to Drs. J. Sandow and W. Rechenberg (Hoechst AG, Frankfurt, Germany) for supplying buserelin, Drs. R. Ghraf and Schroder (Ferring GmBH, Kiel, Germany) for supplying D-Trp[6]-GnRH and Dr. A.V. Schally for supplying GnRH antagonists. We are indebt to Dr. A.V. Schally (Tulane University, New Orleans, LA) and to

Dr. D. LeRoith (NIH, Bethesda, MD) for helpful discussions and shring unpublished data. This work was supported by the Israel Ministry of Health and by the Israel Cancer Research Fund (New York).

ABBREVIATIONS

GnRH, Gonadotropin-releasing hormone; SB-88, [Ac-D-Nal(2)[1], D-Phe(4Cl)[2], D-Pal(3)[3], D-Hci[6], D-Ala[10]]-GnRH; SB-75, [Ac-D-Nal(2)[1], D-Phe(4Cl)[2], D-Pal(3)[3], D-Cit[6], D-Ala[10]]-GnRH; Buserelin, [D-Ser(t-Bu)[6],des-Gly[10]-ethylamide]-GnRH; Cit, citruline; Hci, homocitrulline; MTT, 3-[4,5-dimethylthiazol-2-yl]-2,5-diphenyltetrazolium bromide; DMBA, 7,12-dimethylbenz[a]anthracene.

REFERENCES

1. G. Tolis, D. Ackman, A. Stellos, A. Mehta, F. Labrie, A.T.A. Fazekas, A.M. Comaru-Schally and A.V. Schally, Tumor growth inhibition in patients with prostatic carcinoma treated with luteinizing hormone-releasing hormone agonists., *Proc. Natl. Acad. Sci. USA.* 79: 1658 (1982).
2. J.G.M. Klijn, Long-term LHRH-agonist treatment in metastatic breast cancer as a single treatment and in combination with other additive endocrine treatments., *Med. Oncol. Tumor Pharmacol.* 1: 123 (1984).
3. A. Manni, R. Santen, H. Harvey, A. Lipton and D. Max, Treatment of breast cancer with gonadotropin-releasing hormone., *Endocrine Rev.* 7: 89 (1986).
4. S. Bajusz, M. Kovacs, M. Gazdag, L. Bokser, T. Karashima, V.J. Czernus, T. Janaky, J. Guoth and A.V. Schally, Highly potent antagonists of luteinizing hormone-releasing hormone free of edematogenic effects., *Proc. Natl. Acad. Sci. USA.* 85: 1637 (1988).
5. A.V. Schally, T.W. Redding, R.Z. Cai, J.I. Paz, M. Ben-David and A.M. Comaru-Schally, Somatostatin analogues in the treatment of various experimental tumors., *in*: "International Symposium on Hormonal Manipulation of Cancer: Peptides, Growth Factors and New (anti) steroidal agents.", J.G.M. Klijn eds., pp. 431, Raven Press, New York, (1987).
6. W.R. Miller, W.M. Scott, R. Morris, H.M. Fraser and R.M. Sharpe, Growth of human breast cancer cells inhibited by a luteinizing hormone-releasing hormone agonist, *Nature.* 313: 231 (1985).
7. P. Mullen, W.N. Scott and W.R. Miller, Growth Inhibition Observed Following Administration of an LHRH Agonist to a Clonal Variant of the MCF-7 Breast Cancer Cell Line Is Accompanied by an Accumulation of Cells in the G0/G1 Phase of the Cell Cycle, *British Journal of Cancer.* 63: 930 (1991).
8. K.A. Eidne, C.A. Flanagan, N.S. Harris and R.P. Millar, Gonadotropin-releasing hormone (GnRH)-binding sites in human breast cancer cell lines and inhibitory effects of GnRH antagonists., *J. Clin. Endocrinol. & Metabol.* 64: 425 (1987).
9. Y. Sharoni, E. Bosin, A. Miinster, J. Levy and A.V. Schally, Inhibition of growth of human mammary tumor cells by potent antagonists of luteinizing hormone-releasing hormone., *Proc. Natl. Acad. Sci. USA.* 86: 1648 (1989).
10. T. Segal-Abramson, H. Kitroser, J. Levy, A.V. Schally and Y. Sharoni, Direct Effects of Luteinizing Hormone-Releasing Hormone Agonists and Antagonists on MCF-7 Mammary Cancer Cells, *Proc. Natl. Acad. Sci. USA.* 89: 2336 (1992).
11. P.J. Munson and D. Rodbard, A versatile computerized approach for the characterization of ligand biding systems., *Anal. Biochem.* 107: 220 (1980).
12. Y. Sharoni, T. Segal-Abramson, Y. Giat, H. Kitroser, E. Bosin, A. Miinster, B. Feldman, A.V. Schally and J. Levy, Direct effects of GnRH agonists and antagonists on mammary cancer cells., *in*: "The Current Status of GnRH Analogues", B. Lunenfeld and V. Insler eds., pp. In press, The Parthenon Publishing Group Ltd., Lancs, UK, (1992).
13. T. Segal-Abramson, J. Giat, J. Levy and Y. Sharoni, Guanin nucleotides modulation of high affinity GnRH receptors in rar mammary tumors., *Mol. Cell. Endocrinol.* 85: 109 (1992).
14. R.B. Dickson and M.E. Lippman, Estrogenic regulation of growth and polypeptide growth factor secretion in human breast carcinoma, *Endocrine Rev.* 8: 29 (1987).
15. R.W. Furlanetto and J.N. DiCarlo, Somatomedin-C receptors and growth effects in human breast cells maintained in long-term tissue culture., *Canc. Res.* 44: 2122 (1984).
16. K.K. Huff, D. Kaufman, K.H. Gabbay, E.M. Spencer, M.E. Lippman and R.B. Dickson, Human breast cancer cells secrete an insulin-like growth factor-I-related polypeptide., *Cancer Res.* 46: 4613 (1986).
17. C.K. Osborne, G. Bolam, M.E. Monaco and M.E. Lippman, Hormone responsive human breast cancer in long term tissue culture: Effect of insulin, *Proc. Natl. Acad. Sci. USA.* 73: 4536 (1976).

18. C.K. Osborne, E.B. Coronado, L.J. Kitten, C.I. Arteaga, S.A.W. Fuqua, K. Ramasharma, M. Marshall and C.H. Li, Insulin-like growth factor-II (IGF-II): a potential autocrine\paracrine growth factor for human breast cancer acting via the IGF-I receptor., *Mol Endocrinol.* 3: 1701 (1989).

19. Y. Myal, R.P.C. Shiu, B. Bhaumick and M. Bala, Receptor binding and growth factors in human breast cancer cells (T-47D) in culture., *Cancer Res.* 44: 5486 (1984).

20. M.E. Lippman, R.B. Dickson, E.P. Gelmann, N. Rosen, C. Knabbe, S. Bates, D. Bronzert, K. Huff and A. Kasid, Growth regulation of human breast carcinma occurs tgrough regulated growth factor secretion., *J. Cell Biochem.* 35: 1 (1987).

21. L. Murphy and A. Ghahary, Uterine insulin-like growth factor-I: Regulation of expression and its role in estrogen-induced uterine proliferation., *Endocrine Reviews.* 11: 443 (1990).

22. J.A. Fontana, A. Burrowsmezu, D.R. Clemmons and D. Leroith, Retinoid Modulation of Insulin-Like Growth Factor-Binding Proteins and Inhibition of Breast Carcinoma Proliferation, *Endocrinology.* 128: 1115 (1991).

23. F. Vignon, M.-M. Bouton and H. Rochefort, Antiestrogens inhibit the mitogenic effect of growth factors on breast cancer cells in the total absence of estrogens, *Biochem. Biophys. Res. Commun.* 146: 1502 (1987).

24. C.K. Osborne, D.R. Clemmons and C.L. Arteaga, Regulation of Breast Cancer Growth by Insulin-Like Growth Factors, *Journal of Steroid Biochemistry and Molecular Biology.* 37: 805 (1990).

25. K.J. Cullen, H.S. Smith, S. Hill, N. Rosen and M.E. Lippman, Growth Factor Messenger RNA Expression by Human Breast Fibroblasts from Benign and Malignant Lesions, *Cancer Research.* 51: 4978 (1991).

β-ENDORPHIN IS METABOLIZED *IN VITRO* BY HUMAN SMALL CELL LUNG CANCER TO γ-ENDORPHIN WHICH STIMULATES CLONAL GROWTH

Thomas P. Davis[1] and Susan L. Crowell

College of Medicine
Department of Pharmacology
Tucson, AZ 85724 USA

INTRODUCTION

Human small cell lung cancer (SCLC) represents approximately 20-25% of total primary lung carcinomas[1]. It is characterized by a rapid growth rate[1], initial sensitivity to chemo- and radiotherapy[2,3] and a propensity for early metastasis[1]. With the use of defined media, continuous SCLC cell lines were established from 72% of 41 fresh specimens allowing for numerous *in vitro* models of SCLC[4].

SCLC is thought to originate from the Kulchitsky (k) cells found in the tracheobronchial mucosa. K cells are small, basal cells with endocrine properties which share (with SCLC) amine precursor uptake and decarboxylation (AUPD) cell characteristics.[5,6] Small cell lung cancer cells possess many neuroendocrine properties including the production of polypeptide hormones such as: bombesin/GRP,[7,10] calcitonin,[11,12] adrenocorticotrophin (ACTH),[13] neurotensin,[12,14] somatostatin,[12] β-endorphin,[15] and neurotensin.[16]

Certain of these peptides, especially Bombesin/GRP,[17,18] but also neurotensin,[19,20] β-endorphin (βE[20]) and Insulin-like growth factor I[21] have been hypothesized to function as regulatory growth factors in SCLC cells. These growth factors are proposed to stimulate growth via an autocrine mechanism whereby the cells produce and secrete growth factors which feed back to cellular receptors. The ensuing growth factor and receptor interaction is thought to activate a post-receptor signalling pathway which leads to a mitogenic response.[22,23] This endogenous production of growth factors confers on the malignant cell the capacity to survive with fewer exogenous growth factors than the normal cell.[22]

One growth factor, Bombesin/GRP, binds with high affinity to SCLC cells,[24] elevates cytosolic calcium [25-27] and stimulates the growth of SCLC *in vitro*[17,18] and *in vivo*.[28] Moreover, a monoclonal antibody developed against bombesin inhibited the growth of a xenografted SCLC in nude mice and the cloning of the SCLC cell lines in soft agarose.[17] Other neuropeptides (vasopressin, bradykinin, cholecystokinin, galanin, neurotensin and tachykinin peptides) were observed to stimulate transient increases in cytosolic Ca^{2+} in SCLC cell lines as well.[26,29] Furthermore, SCLC tumors have been found to contain somatostatin receptors[6,30] and opioid receptors have been detected in SCLC cell lines.[15,31]

---

[1]Address correspondence to Thomas P. Davis

Growth Factors, Peptides, and Receptors, Edited
by T.W. Moody, Plenum Press, New York, 1993

Opioid peptides are derived from three main precursor proteins: proopiomelanocortin (POMC),[32] proenkephalin A, and proenkephalin B.[33] POMC contains the sequences of adrenocorticotrophin (ACTH), β-endorphin and melanocyte stimulating hormone.[32] β-endorphin has been shown to be metabolized *in vitro* to biologically active peptide fragments. The metabolite γ-endorphin (βE 1-17) in particular, has been shown to be centrally active[34] and to stimulate motility of the canine small intestine.[35]

We previously reported that β-endorphin and neurotensin stimulate growth of SCLC cell lines in soft agarose.[20] Recently we also reported that SCLC cell lines cleave the 13 amino acid peptide neurotensin (NT) into the N-terminal NT 1-8 and the corresponding C-terminal NT 9-13. Furthermore, we found that the C-terminal fragment NT 8-13 and the parent peptide NT 1-13 stimulate clonal growth of SCLC where NT 1-8 did not.[19] Similarly, Staley, *et al.*[36] found that NT 1-13 and NT 8-13 but not the N-terminal NT 1-8 elevated cytosolic calcium in SCLC cell line NCI-H345. Extracellular metabolism of NT may be necessary to activate certain peptide fragment growth factors or it may be that certain metabolites of growth factors retain their biological activity during the degradation process.

In this study, we investigated further the role β-endorphin may play in the regulation of SCLC growth. Intact SCLC cells (NCI-N417 and NCI-H345) were incubated with β-endorphin 1-31 and the metabolic products were assayed for *in vitro* mitogenicity in a soft agarose assay. Data shown here suggests that SCLC cell lines metabolize β-endorphin into γ-endorphin which, along with the parent peptide, stimulates clonal growth of SCLC cell lines. Therefore β-endorphin (as well as γ-endorphin) may be growth factors which SCLC utilizes via an autocrine pathway.

MATERIALS AND METHODS

Cell Lines and Cultures Conditions

Human small cell lung cancer lines utilized include: NCI-N417, NCI-N592, NCI-H345 and NCI-H209. SCLC cells may be classified according to specific characteristics. The first cell line, NCI-N417 is considered to be variant having no L-dopa decarboxylase activity, fast growth with sheet and chain formation in suspension and c-myc oncogene overexpression. The remaining cell lines are classified as classical SCLC cell lines which is defined by high levels of L-dopa decarboxylase, slow growth in tight clumps and no c-myc oncogene overexpression.[11] These four cell lines were cultured in serum-supplemented HITES media (2% serum). HITES medium contained: RPMI 1640 with L-glutamine supplemented with penicillin (100 units/ml), streptomycin (100 μg/ml), insulin (5 μg/ml), human transferrin (10 μg/ml), hydrocortisone (10 nM) and sodium selenite (30 nM). Hydrocortisone, insulin (bovine pancreas), human transferrin, β-estradiol and sodium selenite were purchased from Sigma Chemical Company (St. Louis, MO). RPMI 1640, penicillin, and streptomycin were purchased from JRH Biosciences (Lenexa, KS). Fetal calf serum was obtained from Irvine Scientific (Santa Ana, CA) and heat-inactivated in our laboratory. The cell lines utilized were free of mycoplasma contamination as determined by routine testing (3 month intervals) using a test kit (Gen-Probe Inc., San Diego, CA) as reported previously.[36]

Radioimmunoassay

Pellets of approximately 1×10^7 cells were extracted by sonication in 5 ml acid acetone (acetone, 40: H_2O, 6: HCl, 1). After centrifugation at 10,000 rpm for 20 minutes, the supernatant fluid was dried under N_2. The remaining pellet was analyzed for protein content by the method of Folin and Lowry.[37] These cell extracts were assayed for endogenous human β-endorphin using an RIA kit (Peninsula, Belmont, CA) per the kit insert. The presence of β-endorphin was confirmed by HPLC of combined extracts of NIC-N417 (cell line with highest endogenous β-endorphin by initial RIA). HPLC conditions were similar to those described below but with a modified gradient of 15 to 35% acetonitrile over 45 min versus 100 mM NaH_2PO_4. Collections in 30 second intervals were made in the range of the retention time of a β-endorphin standard. These collections were then radioimmunoassayed for human β-endorphin.

Intact Cell Incubation with β-endorphin

The time-course metabolism of β-endorphin (βE) by intact SCLC cells from cell lines NCI-N417 and NCI-H345 was investigated. SCLC cells were harvested by centrifugation at 1250 X g for 10 minutes then resuspended in sterile RPMI media (containing L-glutamine) to a cell concentration of 2.0 X 10^6 viable cells/ml. Cells (0.4 X 10^6/tube) were incubated with 20 μM βE at 37° C in a time-course fashion (0-180 min). Various control tubes were prepared including no incubation, incubation without cells and an RPMI control without peptide. Following the desired incubation period, enzymatic activity was stopped by the addition of 17.4 M glacial acetic acid and boiling for 5 min; the samples were then centrifuged at 3000 x g for 10 min. The resulting supernatant was analyzed by reversed phase HPLC and the pellet was reserved for protein analysis.[37]

HPLC Analysis

Analysis of the supernatants of similar time-course incubations was described previously.[17] Separations were accomplished as described[38] but modified using a curvilinear gradient of 12-35% acetonitrile (Burdick and Jackson, Chicago, IL) over 80 min versus 100 mM NaH_2PO_4 (Sigma Chemical Co., St. Louis, MO) (pH 2.4). UV absorbance was detected at 210 nm and all glassware used, including autosampler vials were silanized with 5% dimethyldichlorosilane to minimize peptide adsorption loss. Peptide identity was confirmed by amino acid analysis (ninhydrin chemistry) after hydrolysis of collected HPLC peaks.

Clonogenic Assays

Clonogenic assays were performed using a modification of the method of Hamburger and Salmon[39] which has been described previously.[17] Briefly, base layers consisting of 0.4% Sea Plaque agarose (FMC Bioproducts, Rockland, ME) in 2% serum-supplemented HITES media (ss-HITES) (total volume = 1 ml) were prepared in 35 mm x 10 mm plastic petri dishes (Nunc, VWR Scientific, Phoenix, AZ). Cells taken from log phase culture were centrifuged and resuspended in ss-HITES and filtered through a 15 μg nylon mesh to obtain a single cell suspension. Viable cells were counted using trypan blue exclusion and the appropriate concentration was mixed with agarose and ss-HITES in the presence of various nanomolar concentrations of β-endorphin, gamma endorphin (βE 1-17) or βE 18-31 for a final volume of 1.0 ml per plate which was applied to the hardened base layer. β-endorphin was synthesized, purified and composition confirmed in our laboratory using solid phase methodology employing BOC-Glu PAM resin. Gamma endorphin and βE 18-31 were graciously supplied to us by Organon International (Oss, The

Netherlands). Control plates were prepared using media in place of peptide dilution. Peptides were suspended in ss-HITES and filter sterilized through a low-protein binding filter (Acrodisc 13, Gelman Sciences, Inc., Ann Arbor, MI). The NCI-N417 culture plates were seeded at 7.5 X 10^3 cells per plate whereas NCI-H345 were seeded at 2.0 X 10^4 cells per plate. Plates were allowed to gel following the assay and then were placed in humidified incubator with 5% CO_2/95% air at 37° C. Colony counts were performed on day 9 for NCI-N417 and day 14 for NCI-H345 (where day 0= day of experiment). Cell aggregates of 42 μ in diameter were counted using an Omnicon Fas II Image Analysis System (Bausch and Lomb, Inc., Rochester, NY).

Statistical Analysis

Statistical significance was determined by a student's grouped T-test in the pharmacologic calculations system software package. A p value of <0.05 was considered to be significant.

RESULTS

Radioimmunoassay for Human β-Endorphin and HPLC Confirmation

Cell extracts of 4 SCLC cell lines were assayed for human β-endorphin immunoreactivity. β-endorphin immunoreactivity was found to be expressed at levels between 20 and 395 pg/mg protein with NCI-N417 cells expressing the highest level, and NCI-H345, the lowest. Levels of 265 pg/mg protein and 93 pg/mg protein were found in NCI-N592 and NCI-H209 respectively. Extracts from NCI-N417 were combined and injected on an HPLC. Collections were made in the vicinity of the retention time of β-endorphin (based on retention time of standard β-endorphin). β-endorphin immunoreactivity was then measured in these collections by RIA and the resulting radiochromatogram (Figure 1) shows a distinct peak at 31.5 min. (where the β-endorphin standard eluted at 31.41 min). These levels of β-endorphin immunoreactivity are similar to those reported by Maneckjee and Minna of 1.3 to 25 pg/mg protein in six SCLC cell lines.[30]

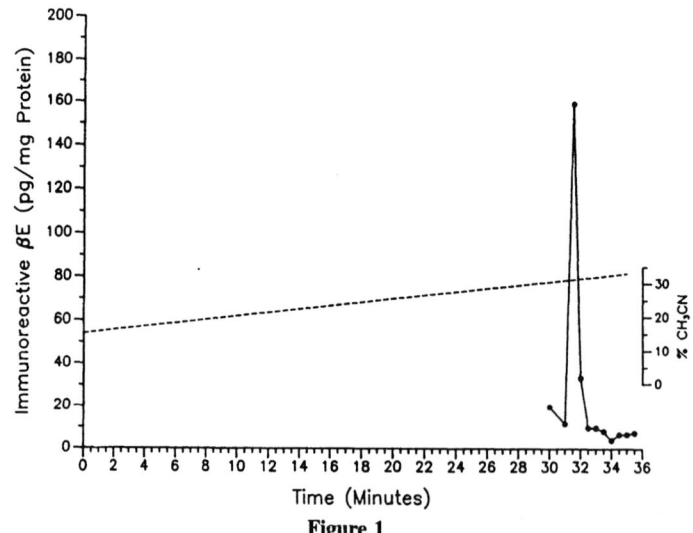

Figure 1

β-endorphin Metabolism by SCLC Cells

The metabolism of β-endorphin by intact SCLC cell lines NCI-N417 and NCI-H345 was investigated. Whole cells were incubated with 20 μM β-endorphin in a time-course fashion. HPLC analysis of the resulting supernatants revealed a major metabolite to be γ-endorphin (βE 1-17) based on comparison of retention time of known standards (Figure 2) and amino acid analysis (Table 1) of collected peaks at the γ-endorphin retention time.

Additional minor amounts of the complementary C-terminal fragment βE 18-31 were also in evidence (Figure 2) which we were unable to confirm by amino acid analysis (Table 1). The metabolic half-life of β-endorphin was found to be 104 min. with NCI-N417 cells and 111 min. for NCI-H345 cells (Table 2). Accumulation of β-endorphin fragments βE 1-17 and βE 18-31 is illustrated in Figure 3.

NCI-N417 cells accumulated roughly four times the quantity of βE 1-17 as NCI-H345 cells (Table 2).

The rate of accumulation of the βE 18-31 fragment, however, was more similar between the two cell lines. The discrepancy between the amounts of accumulated βE 1-17 and βE 18-31 within a cell line (436 ng/mg protein/min for βE 1-17 versus 47 ng/mg protein/min for βE 18-31 in NCI-H345) may be explained by higher concomitant metabolism of the βE 18-31 fragment. Previously, we reported levels

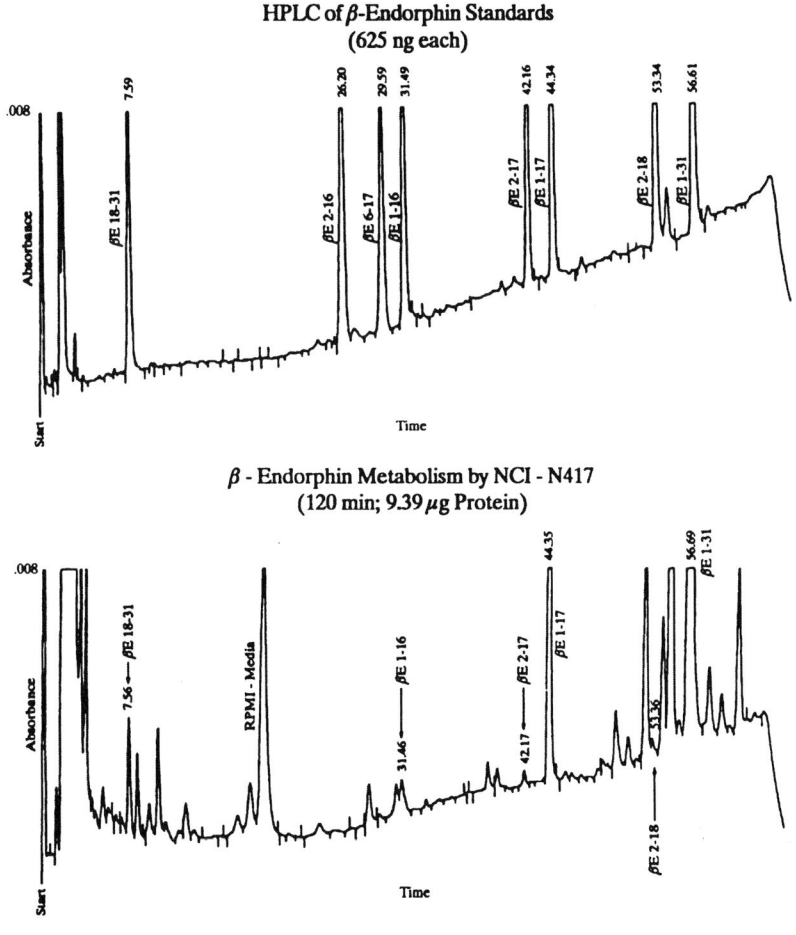

Figure 2

Table 1. Amino acid confirmation of β-endorphin fragment βE 1-17

Amino Acids	βE 1-17 Theoretical	βE 1-17 Observed
Glu	2.00	1.99
Ser	2.00	1.90
Gly	2.00	2.40
Thr	3.00	2.65
Pro	1.00	1.23
Tyr	1.00	0.74
Val	1.00	1.06
Met	1.00	0.77
Leu	2.00	2.11
Phe	1.00	1.03
Lys	1.00	1.10

Intact SCLC cell lines NCI-H345 and NCI-N417 were incubated with β-endorphin in a time course fashion for 0-180 minutes. Enzymatic activity was stopped by the addition of glacial acetic acid and boiling; the samples were then centrifuged and the supernatant, reserved for analysis by HPLC. Following HPLC analysis, the remaining supernatants from several time points were pooled. This pooled sample was injected onto the HPLC and the peak corresponding to the retention time of a βE 1-17 standard was collected from a gradient of 10-40% Acetonitrile (0.1% TFA) versus H_2O (0.1% TFA) over 40 min. The collected sample was lyophilized and later was acid hydrolyzed and amino acids analyzed using ninhydrin chemistry.

of proteolytic enzyme activity in these cell lines.[19] These SCLC cells demonstrate membrane bound trypsin-like activity (61-780 pmoles/mg protein/min), membrane bound metalloendopeptidase (E.C.3.4.24.15) activity (9-17 nmoles/mg protein/min) and carboxypeptidase H (E.C.3.4.17.10) activity (2-6 pmoles/mg protein/ min). However, the endopeptidase enzyme responsible for generating γ-endorphin, gamma-endorphin generating endopeptidase, was not measured but does appear to be present based on the robust generation of βE 1-17 and βE 18-31 (Figure 2 and 3) from β-endorphin.

Table 2. Metabolism of β-endorphin by SCLC cell lines

SCLC Cell Lines	Fragments Accumulated[a] (ng/mg protein/min)		β-endorphin $T_{1/2}$[b]
	βE 1-17	βE 18-31	
NCI-N417	1716.8±59.0	16.9±0.1	104 min.
NCI-H345	436.3±113.3	47.0±3.8	111 min.

Intact SCLC cells (4.0 x 10^5/tube) were incubated with 20 μM β-endorphin in a time course fashion at 37° C. At any given time interval, glacial acetic acid was added and the samples were boiled to stop enzymatic activity. The sample tubes were centrifuged, the supernatants were removed and analyzed by HPLC and the pellets were assayed for protein content. Identity of βE metabolites was based on the retention time of known standards; identity of βE 1-17 was confirmed by amino acid analysis (see Table 1). [a]β-endorphin-related peptide fragments accumulated after 30 min. incubation with intact SCLC cells reported in ng peptide per mg protein per minute. Each point represents the mean of 3 measurements ± S.E.M. [b]Half-life data calculated by linear least square regression analysis of 6 point data.

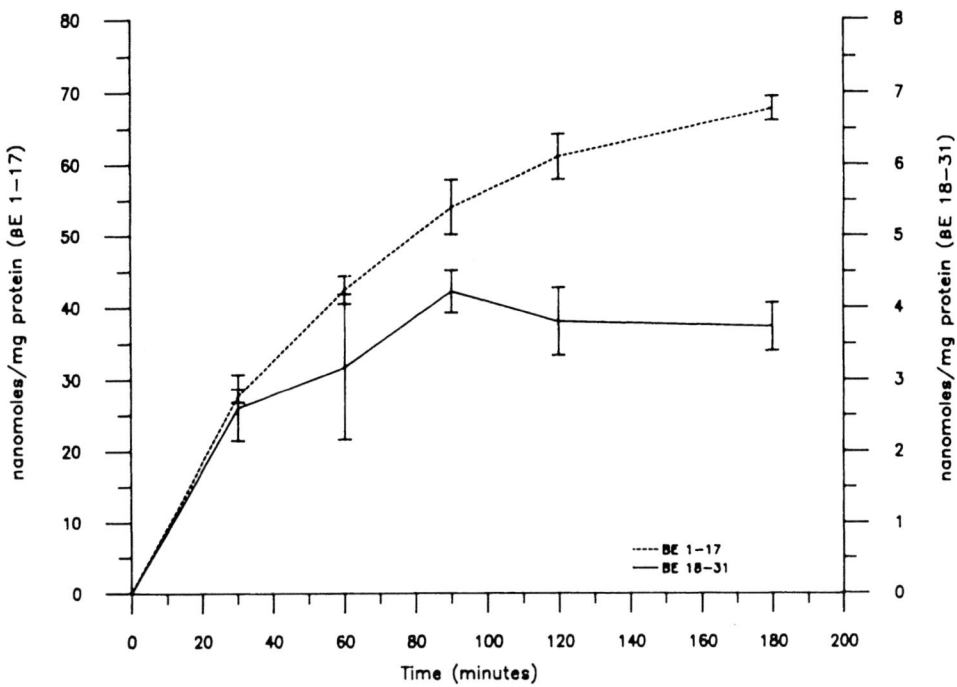

Figure 3 - Formation of β-Endorphin Fragments γ-endorphin (βE 1-17) and βE 18-31 from incubation of β-endorphin with intact SCLC cells

Clonogenic Assays with β-Endorphin and Metabolic Fragments

The potential of β-endorphin and its metabolites βE 1-17 and βE 18-31 to stimulate clonogenic growth as growth factors for SCLC cells was investigated using a soft agarose growth assay and two SCLC cell lines. One of the cell lines, NCI-H345 is considered to be a classic SCLC cell line while the other, NCI-N417, a variant SCLC cell line. Nanomolar concentrations (0 to 50 nM) of the three peptides were added individually in the clonogenic assay with a single cell suspension of NCI-N417 or NCI-H345 cells and the resulting colony formation (after 9-14 days incubation) was compared to that of control plates which received media alone. The response of NCI-H345 cells to β-endorphin and the metabolites γ-endorphin and βE 18-31(0-50 nM) is illustrated in Figure 4.

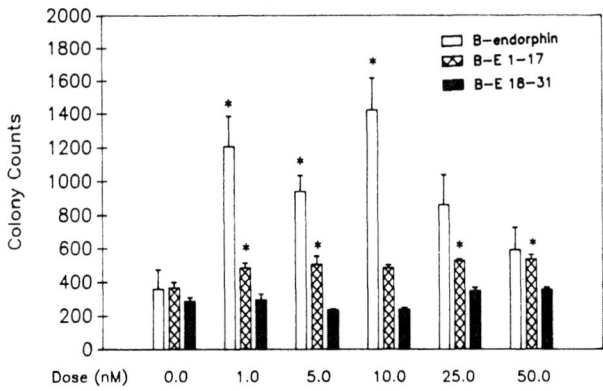

Figure 4 - Clonogenic Assay of NCI-H345 - Dose-Response Effect to β-Endorphin and Metabolites

Experiments with similar colony counts were selected for the sake of comparison. β-endorphin stimulated growth to the highest degree, however, the major metabolite γ-endorphin also significantly stimulated NCI-H345 growth. The carboxy terminal fragment, βE 18-31 on the contrary, did not stimulate HCI-H345 clonal growth. Intact β-endorphin may be initially more mitogenic than γ-endorphin because the cells may cleave βE into γ-endorphin (γ-E) and βE 18-31 resulting in further (but depressed) mitogenic potential through the action of γE. Results of clonogenic assays with these peptides and NCI-N417 as well as NCI-H345 in a dose-response study are shown in Table 3.

Generally, NCI-N417 did not respond to the three peptides to the degree NCI-H345 cells did, however N417 cells were also stimulated by βE and γE and not by βE (18-31).

DISCUSSION

The present study describes the effect of the opioid peptide β-endorphin and its principal metabolites on the autocrine growth of small cell lung cancer. It has been previously shown that SCLC cell lines contain immunoreactive levels of human β-

Table 3 - Effect of β-endorphin and fragments on the <u>in vitro</u> clonal growth of SCLC cell lines NCI-H345 and NCI-N417

Added to Cloning Media		Cell Lines			
Compound	Dose (nM)	NCI-H345[a] colony counts	NCI-H345[b] % of control	NCI-N417[a] colony count	NCI-N417[b] % of control
Control	(0.0)	104.0 ± 30.5	100	443.5 ± 19.6	100
β-endorphin	(1.0)	269.7 ± 27.4^c	259	487.0 ± 17.6	110
	(5.0)	244.0 ± 12.0^c	235	715.7 ± 45.6^d	161
	(10.0)	298.7 ± 56.1^c	287	424.3 ± 34.0	96
	(25.0)	145.5 ± 25.5	140	449.0 ± 41.3	101
Control	(0.0)	364.9 ± 34.0	100	220.1 ± 30.4	100
βE 1-17	(1.0)	484.0 ± 28.7^c	133	283.3 ± 56.7	129
	(5.0)	504.3 ± 48.8^c	138	366.5 ± 39.3^c	167
	(10.0)	484.0 ± 17.6	133	377.5 ± 53.3^c	172
	(25.0)	525.3 ± 10.9^c	144	339.8 ± 38.1^c	154
Control	(0.0)	238.3 ± 27.4	100	443.5 ± 19.6	100
βE 18-31	(1.0)	193.3 ± 10.0	81	437.3 ± 6.1	99
	(5.0)	206.8 ± 24.7	87	515.8 ± 6.2	116
	(10.0)	225.3 ± 19.7	95	250.3 ± 24.5	56
	(25.0)	247.0 ± 7.5	104	418.3 ± 10.0	94

A single cell suspension of NCI-H345 (2.0×10^4 cells/plate) or NCI-N417 (7.5×10^3 cells/plate) (taken from log phase culture) were mixed with agarose, media and nanomolar concentration of peptide (or media for control) and plated in the volume of 1 ml on pre-prepared feeder layers consisting of media and 0.4% agarose. Plates were allowed to gel and were placed in a humidified incubator with 5% CO_2/95% air at 37° C. Colonies $\geq 42\mu$ were counted on day 9-14 using an automated counter (FAS II Image Analysis System). [a]Values are colony counts $\pm$S.E.M. for 3-8 plates. [b]Values are mean growth as % of control. [c]Significantly higher than control, $p<0.05$; [d]$p<0.01$ by students t test. These experiments are representative of several assays which were performed. Overall, a mitogenic response (statistically significant) occurred in 5 of 7 assays of NCI-H345 and β-endorphin and in 2 of 3 assays of NCI-417 and β-endorphin. γ-endorphin was mitogenic in 1 of 2 assays with NCI-H345 and 3 of 3 assays with NCI-N417.

endorphin and related opioid receptors.[15,31] The present study confirms these results and extends work previously reported on opioids and SCLC. Surprisingly, the present results describe a significant stimulation of *in vitro* SCLC clonal growth by β-endorphin and γ-endorphin which differs from the study previously reported by Maneckjee and Minna.[31] They described a negative autocrine loop or tumor suppressing system for opioid peptides. While these authors did not study β-endorphin, they did report inhibition of SCLC growth *in vitro* by studying synthetic selective mu, delta and kappa opioid receptor agonists. Therefore, the difference between the present study and the work of Maneckjee and Minna[31] may be related to our use of endogenous β-endorphin and γ-endorphin as *in vitro* peptide agonists versus selective, but synthetic opioid receptor agonists. Since SCLC cells express opioid receptors and produce endogenous levels of opioid peptides such as β-endorphin, the present study provides strong evidence that β-endorphin is also an autocrine growth factor for SCLC.

Several growth factors have been isolated from SCLC tissue including neurotensin, epidermal growth factor, adrenocorticotrophic hormone, calcitonin, neurokinin A, somatostatin and bombesin/gastrin-releasing peptide.[12-16] However very few of these peptides have been shown to stimulate growth and proliferation of SCLC and even fewer have specific receptors on SCLC cells.[15,24] β-endorphin is a true autocrine growth factor in SCLC because it has been shown to be present in the tumor, is actively secreted and can interact with membrane receptors on the cell surface stimulating growth. Furthermore, we have shown previously that the opioid receptor antagonist naloxone can inhibit the growth response in SCLC cells.[20,37]

β-endorphin also represents a major immunoreactive peptide in the SCLC tumor. Kleber et al[41] demonstrated significant levels in the tumor and in plasma of a patient suffering from an oat-cell carcinoma. However, the ratio of β-lipotropin (the precursor to β-endorphin) to β-endorphin in the plasma of the cancer patient was quite different than in the tumor tissue.[41] This provides evidence that the enzymatic processing of the precursor protein β-lipotropin (β-LPH) may be specific to tumor tissue.

In the present study, we showed that endopeptidase cleavage of β-endorphin can occur with intact SCLC cells leading to the formation of large quantities of γ-endorphin (βE 1-17) and minor quantities of βE 18-31. The amino terminal fragments of β-endorphin such as γ-endorphin and alpha-endorphin (βE 1-16) and methionine enkephalin (βE 1-5) have been shown previously to be centrally and peripherally active in animals.[34,35] We have now demonstrated that γ-endorphin can stimulate *in vitro* growth of both NCI-H345 and NCI-N417 SCLC cells in a dose-dependent manner. However, the carboxy terminal fragment βE 18-31 had no effect on growth. Therefore, the amino terminal extended fragments of β-endorphin are biologically active in a SCLC cell growth assay but the carboxy-extended fragments are not. This is interesting because metabolic endopeptidases responsible for β-endorphin degradation may be leading to the formation of biologically active, receptor specific fragments such as γ-endorphin and not merely to the degradation of the parent peptide. Potential therapies for SCLC focused on metabolic peptide enzymes could therefore have promise if the mechanism of the drug action included inhibition of growth factor processing enzymes or stimulation of growth factor degradative enzymes.

REFERENCES

1. F.A. Greco and R.K. Oldham. Current concepts in cancer: small cell lung cancer. N. Eng. J. Med., 301:355-358, (1979).

2. R.B. Livingston. Current chemotherapy of small cell lung cancer. Chest, 89(suppl): 258S-263S, (1986).

3. D.N. Carney. Recent advances in biology of small cell lung cancer. Chest, 89(suppl):253S-256S, (1986).

4. D.N. Carney, A.F. Gaxder, G. Bepler, J.G. Cuccion, P.J. Marangos, T.W. Moody, M.H. Zweig, and J.D. Minna. Establishment and identification of small cell lung cancer cell lines having classic and variant features. Cancer Res., 45:2913-2923, (1985).

5. G.S. Sidhu. The endodermal origin of digestive and respiratory tract APUD Cells. Am. J. Pathol., 96(1):5-17, (1979).

6. U. Sagman, B. Mullen, K. Kovacs, R. Kerbel, R. Ginsberg, J.-C Reubi. Identification of somatostatin receptors in human small cell lung carcinoma. Cancer, 66:2129-2133, (1990).

7. M.D. Erisman, R.I. Linnoila, O. Hernandez, R.P. Di Augustine, and L.H. Lazarus. Human lung small-cell carcinoma contains bombesin. Proc. Natl. Acad. Sci., 79:2379-2383, (1982).

8. T.W. Moody, V. Bertmess, and D.N. Carney. Bombesin-like peptides and receptors in human tumor cell lines. Peptides, 4:683-686, (1983).

9. T.W. Moody, C.B. Pert, A.F. Gazdar, D.N. Carney, J.D. Minna. High levels of intracellular bombesin characterize human small-cell lung carcinoma. Science, 214:1246-1248, (1981).

10. D.G. Bostwick, K.A. Roth, C.J. Evans, J.D. Barchas, and K.G. Bensch. Gastrin-releasing peptide, a mammalian analog of bombesin, is present in human neuroendocrine lung tumors. Am. J. Pathol., 117:195-200, (1984).

11. R. Yesner. Spectrum of lung cancer and ectopic hormones. Pathol. Ann., 13:207-240, (1978).

12. G. Bepler, M. Rotsch, G. Jacques, M. Haeder, J. Heymann, G. Hartogh, P. Kiefer, and K. Havemann. Peptides and growth factors in small cell lung cancer: production binding sites and growth effects. J. Cancer Res. Clin. Oncol., 114:235-244, (1988).

13. G. Gewritz, and R.S. Yalow. Ectopic ACTH production in carcinoma of the lung. J. Clin. Invest., 53:1022-1032, (1974).

14. T.W. Moody, D.N. Carney, L.Y. Korman, A.F. Gazdar, and J.D. Minna. Neurotensin is produced by and secreted from classic small cell lung cancer cells. Life Sci., 36:1727-1731, (1985).

15. K.A. Roth, and J.D. Barchas. Small cell carcinoma cell lines contain opiod peptides and recrptors. Cancer, 57:769-773, (1986).

16. M. Kapuscinski, A. Shulkes, D. Read, and K.J. Hardy. Expression of neurotensin in endocrine tumors. J. Clin. Endocrin. and Metab., 70(1):100-106, (1990).

17. F. Cuttia, D.N. Carney, T.W. Moody, and J.D. Minna. Selective stimulation of small cell lung cancer clonal growth by bombesin and gastrin-releasing peptide. Cancer Res., 47:821-825, (1987).

18. D.N. Carney, F. Cuttia, T.W. Moody, and J.D. Minna. Selective stimulation of small cell lung cancer clonal growth by bombesin and gastrin-releasing peptide. Cancer Res., 47:821-825, (1987).

19. T.P. Davis, S. Crowell, B. McInturff, R. Louis, and T. Gillespie. Neurotensin may function as a regulatory peptide in small cell lung cancer. Peptides, 12:17-23 (1991).

20. T.P. Davis, H.S. Burgess, S. Crowell, T.W. Moody, A. Culling-Berglund, and R.H. Liu. β-endorphin and neurotensin stimulate in vitro clonal growth of human SCLC cells. Eur. J. Pharmacol., 161:283-285, (1989).

21. V.M. Maculay, M.J. Everard, D. TEale, P.A. Trott, J.J. Van Wyk, I.E. Smith and J.L. Millar. Autocrine function for insulin-like growth factor I in human small cell lung cancer cell lines and fresh tumors. Cancer Res., 50:2511-2517, (1990).

22. M.B. Sporn, A.B. Roberts. Autocrine growth factors and caner. Nature., 313:745-747, (1985).

23. F. Cuttia, D.N. Carney, J. Mulshine, T.W. Moody, J. Fedorko, A. Fischler, and J.D. Minna. Autocrine growth factors in human small cell lung cancer. Cancer Surveys., 4(4):707-727, (1985).

24. T.W. Moody, D.N. Carney, F. Cuttia, K. Quattrocchi, and J.D. Minna. High affinity receptors for bombesin/GRP-like peptides on human small cell lung cancer. Life Sci., 37:105-113, (1985).

25. T.W. Moody, A. Murphy, S. Mahmoud, and G. Fiskum. Bombesin-like peptides elevate cytosolic calcium in small cell lung cancer cells. Biochem. Biophs. Res. Commun., 147:189-195, (1987).

26. N. Takuwa, Y. Takuwa, Y. Ohue, H. Mukai, K. Endoh, K. Yamashita, M. Kumada, and E. Munekata. Stimulation of calcium mobilization but not proliferation by bombesin and tachykinin neuropeptides in human small cell lung cancer cells. Cancer Res., 50:240-244, (1990).

27. R. Heikkila, J.B. Trepel, F. Cuttia, L.M. Neckers, and E.A. Sausville. Bombesin-related peptides induce calcium mobilization in a subset of human small cell lung cancer cell lines. J. Biol. Chem., 262(34):16456-16460, (1987).

28. R.W. Alexander, J.R. Upp, Jr., G.J. Poston, V. Gupta, C.M. Townsend, Jr., and J.C. Thompson. Effects of bombesin on growth of human small cell lung carcinoma in vivo. Cancer Res., 48:1439-1441, (1988).

29. P.J. Woll, E. Rozengurt. Multiple neuropeptides mobilize calcium in small cell lung cancer: effects of vasopressin, bradykinin, cholecystokinin, galanin and neurotensin. Cencer Res., 164(1):66-73, (1989).

30. J.C. Reubi, B. Waser, M. Sheppard, and V. Macaulay. Somatostatin receptors are present in small-cell but not in non-small-cell primary lung carcinomas: relationship to EGF-receptors. Int. J. Cancer, 45:269-272, (1990).

31. R. Maneckjee, and J.D. Minna. Opioid and nicotine receptors affect growth regulation of human lung cell lines. Proc. Natl. Acad. Sci., 87:3294-3298, (1990).

32. B.A. Eipper, R.E. Mains, and E. Herbert. Peptides in the nervous system. Trends Neurosci., 9:463-467, (1986).

33. H. Kakidani, Y. Furutani, H. Takahashi, M. Noda, Y. Morimoto, T. Hirose, M. Asai, S. Inayama, S. Nakanishi and S. Numa. Cloning and sequence analysis of cDNA for porcine β-endorphin/dynorphin precursor. Nature, 198:245-249, 1992).

34. D. De Weid. Pro-opio(melano)cortin and brain homeostasis. Prog. Brain Res., 55:474-482, (1982).

35. T.P. Davis, A.J. Culling, H. Schoemaker, and J.J. Galligan. β-endorphin and its metabolites stimulate motility of the dog small intestine. JPET, 227(2):499-507, (1983).

36. J. Staley, G. Fiskum, T.P. Davis, and T.W. Moody. Neurotensin elevates cystosolic calcium in small cell lung cancer cells. Peptides, 10:1217-1221, (1989).

37. S.L. Crowell, H.S. Burgess, and T.P. Davis. The effect of mycoplasma on the autocrine stimulation of human small cell lung cancer in vitro by bombesin and β-endorphin. Life Sci., 45:2471-2476, (1989).

38. D.H. Lowry, N.J. Rosebrough, A.L. Farr, and R.J. Randall. Protein measurement with the folin phenol reagent. J. Biol. Chem., 193:265-275, (1951).

39. T.P. Davis, and A. Culling-Berglund. High performance liquid chromatographic analysis of in vitro central neuropeptide processing. J. Chromatogr., 327:279-292, (1985).

40. A.W. Hamburger, and S.E. Salmon. Primary bioassay of human tumor stem cells. Science, 197:461-462, (1977).

41. G. Kleber, V. Höllt, W. Del Kers and J.J. Quabbe. Elevated plasma and tissue concentrations of β-endorphin and β-lipotropin assoicated with an ectopic ACTH-producing tumor. Horm. Metab. Res. 12:385-389, (1980).

THE ROLE OF EGF AND ITS RECEPTORS IN THE GROWTH OF HUMAN GASTRIC CARCINOMA CELL LINE (MGC80): EVIDENCE FOR AN AUTOCRINE MECHANISM

Yuan-Fang Chen[1], Zuo-Liang Xiao[1], Guo-Jun Lu[1], Hui-Xin Wang[2] and Liang-Wan Cai[1]

[1]Division of Gastroenterology and Department of Biochemistry
Peking Union Medical College Hospital
Beijing 100739, China
[2]Academy of Military Medical Sciences
Beijing 100851, China

INTRODUCTION

Recent studies demonstrate that epidermal growth factor (EGF) can stimulate the growth of certain tumor cell lines. It may also have a role in experimental carcinogenesis[1-4] and the overexpression of EGF receptors (EGFR) may predict the prognosis of certain tumors[1,3]. In gastric carcinomas, it was reported that patients with synchronous expression of EGF and EGFR have far worse prognosis[2,3,5]. In light of these findings, it is postulated that EGF may play an autocrine regulatory role in the growth and progression of malignant tumors. Most of the results utilized immunohistochemical methods and few studies with gastric cancer cell lines have used receptor assays. One of these reports revealed that the overexpressed EGF receptors in gastric cancer cells are sensitive to EGF in physiological concentrations[1]. More recent studies with molecular probes to EGFR showed that EGFR gene amplification and overexpression are present in gastric cancer cells[6]. EGF may be a very important growth factor for gastric carcinomas.

The present study investigated an autocrine mechanism of the growth-regulating effect of EGF in gastric cancers using growth studies, receptor assays and radioimmunoassays.

MATERIALS AND METHODS

Human gastric carcinoma cell line MGC80 was established by the Department of Biology at Shandong Normal University. Human epidermal growth factor (hEGF) was purchased from Sigma Chemical Co., U.S.A.. Recombinant EGF was produced by the Institute of Basic Medical Sciences at the Chinese Academy of Medical Sciences. EGF antiserum against recombinant hEGF (1-53) was raised in this laboratory. Anti-EGFR monoclonal antibody (mAb) was prepared by the Academy of Military Medical Sciences. RPMI-1640, DMEM and F12 medium was purchased from Sigma Chemical Co. U.S.A. ^{3}H-Thymidine (^{3}H-TdR) was prepared by the Institute of Atomic Energy. Fura-2AM, transferrin, dexamethasone and insulin were purchased from Sigma Chemical Co., U.S.A. Samples were counted in a ß-counter and τ-counter purchased from Beckman Instruments, U.S.A.

Growth assays were done using human gastric carcinoma cells were grown in RPMI-1640 medium supplemented with 10% fetal bovine serum (FBS). After 48-72 hr, the cells were harvested and cultured with RPMI-1640 containing 3% FBS for 24 hr. EGF (10^{-11}-10^{-7} M) with or without EGFR mAb or EGF antiserum was added and cultured for another 48 hr. The cells were pulsed with ^{3}H-TdR for 2 hr (2 μCi/well). Cells were then harvested and the uptake of ^{3}H-TdR determined using a ß-counter. In a separate well, the cell number was determined. The significance was calculated using the Student's t-test.

Binding studies were performed using cells grown in 10% FBS-supplemented RPMI-1640 medium for 48-72 hr. The cells were incubated in serum-free RPMI-1640 for 10 min, digested with 0.1% trypsin, washed three times with HEPES buffer containing 1% BSA and counted. The cells (10^6/ml) were then incubated with ^{125}I-EGF at 4°C for 30 min. For non-specific binding unlabeled EGF was added (10^{-6} M). Then ice-cold 4% BSA-HEPES buffer was added to the cell suspension to stop the reaction and the cells centrifuged and washed with ice-cold 1% BSA-HEPES buffer twice. The pellet was then counted in a τ-counter. The time course of association and dissociation was determined as well as the temperature effect. The displacement curve was done with cells as well as with plasma membranes.

The concentration of EGF in MGC80 cells was determined. The cells were cultured in a serum-free medium (DMEM + F12 medium containing transferrin, dexamethasone and insulin) for 3 days, then harvested and homogenized. The EGF contents of condition medium and cell homogenate were determined by radioimmunoassay.

The effect of EGF on the intracellular Ca^{2+} was determined. Cells were cultured in 10% FBS-RPMI-1640 for 48-72 hr, harvested and washed twice with calcium-free Hank's solution. The cells were resuspended (4×10^5/ml), incubated with 0.5 μM Fura-2AM in a shaking water bath at 37°C for 1 hr with oxygen introduced every 15 min. The cells were washed three times and resuspended in a calcium-free Hank's solution. Three ml of the Fura-2AM loaded cells were added to each spectrofluorometer cuvette. After the basal fluorescence was measured, EGF was added (10^{-7} M) and the change in fluorescence recorded. The Fmax and Fmin were determined by adding 0.1% Triton X-100 and 0.01 M $MnCl_2$ respectively. $(Ca^{2+})i$ was calculated using the formula:

$$(Ca^{2+})i = Kd (F-Fmin)/Fmax-F), \text{ where the } Kd = 224 \text{ nM}$$

RESULTS

The EGF binding assay showed that EGF receptors are expressed on MGC80 cells. The time course indicated that ^{125}I-EGF binding is time dependent reaching a plateau (20.4%) at 30 min. The non-specific binding was around 5% at each time point (Fig. 1).

A study of the effect of temperature on binding indicated that the binding of EGF to its receptors is temperature dependent, with maximum binding (21.4%) at 25°C and 30 min (Fig. 2).

The dissociation curve showed that EGF-receptor binding in MGC80 cells is reversible (Fig. 3).

The displacement curve showed that the binding of ^{125}I-EGF to its cell membrane receptors is competitive and specific. Scatchard analysis yielded a Kd of 1.1×10^{-10} M, and maximal binding sites of 2.5×10^4/cell (Fig. 4).

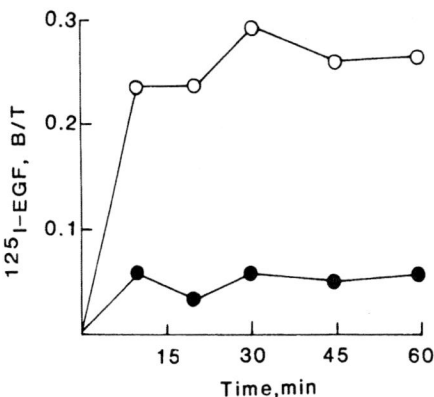

Figure 1. Time course of EGF receptor binding in human gastric carcinoma cell line MGC80. The total binding (O) and non-specific binding (●) is indicated.

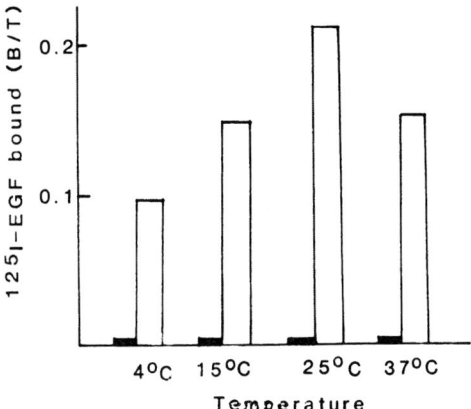

Figure 2. Effect of temperature on the binding of ^{125}I-EGF to MGC80 cells. Total (□) and nonspecific (■) binding is indicated.

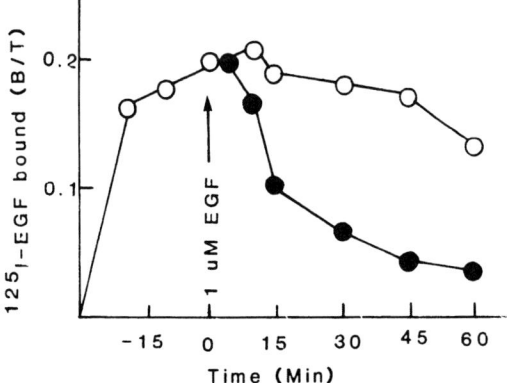

Figure 3. Dissociation curve of EGF receptors at 4°C. At the indicated time the amount of ^{125}I-EGF bound was determined in the presence (O) and absence (●) of 1 μM EGF.

EFFECT OF EGF, EGF ANTISERUM AND EGFR mAb ON THE GROWTH OF MGC80 CELLS

EGF at concentrations of 10^{-11} to 10^{-7} M stimulated the growth of the MGC80 gastric carcinoma cell line. The cell count was $3.8 \pm 0.2 \times 10^5$/well for 10^{-10} M EGF-treated cells as compared to $2.7 \pm 0.1 \times 10^5$/well for controls. ^{3}H-TdR uptake was

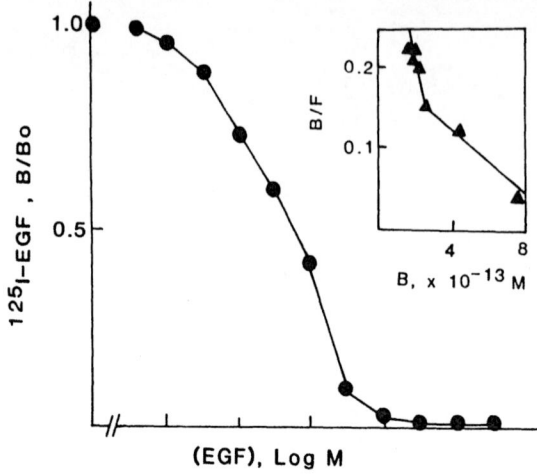

Figure 4. Amount of ^{125}I-EGF bound as a function of EGF concentration. (Insert) Scatchard plot of the specific binding data.

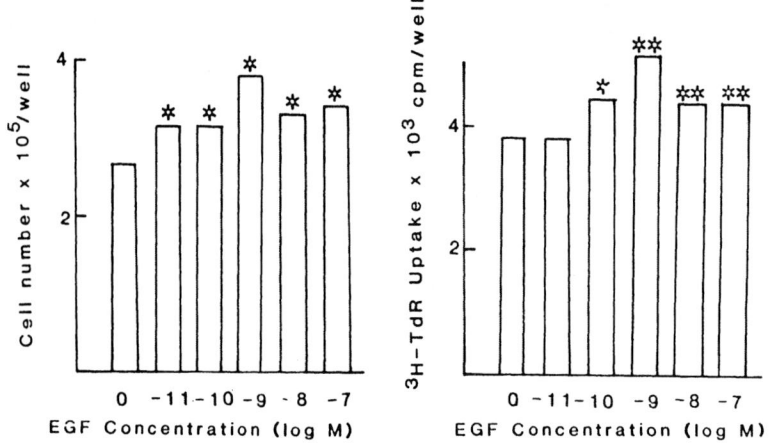

Figure 5. Effect of EGF on growth of human gastric carcinoma cell line MGC80. * p<0.001, ** p<0.01, versus control.

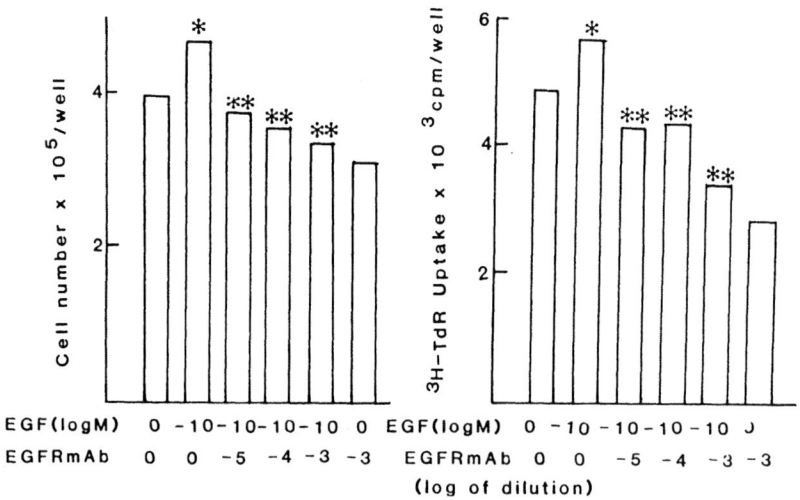

Figure 6. Effect of EGFR mAb on the growth of human gastric carcinoma cell line MGC80. * p <0.001 versus control; ** p<0.01 versus 10^{-10} M EGF.

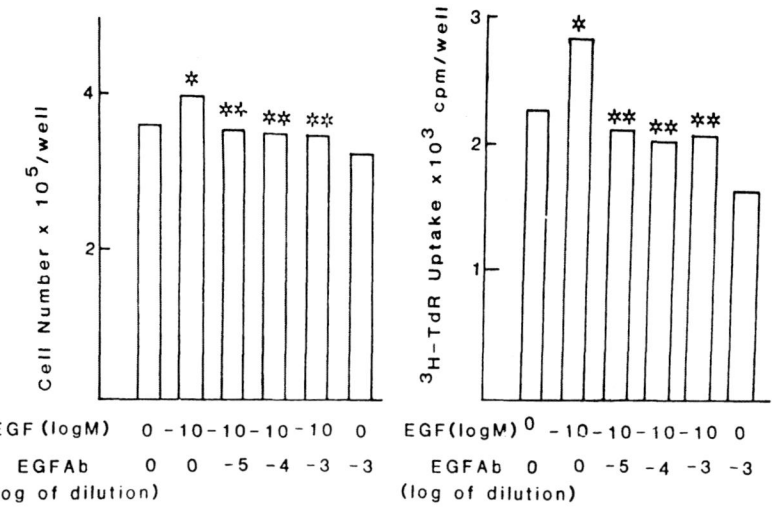

Figure 7. Effect of EGF antiserum on the growth of human gastric carcinoma cell line MGC80. * p <0.001 versus control; ** p<0.01 versus 10^{-10} M EGF.

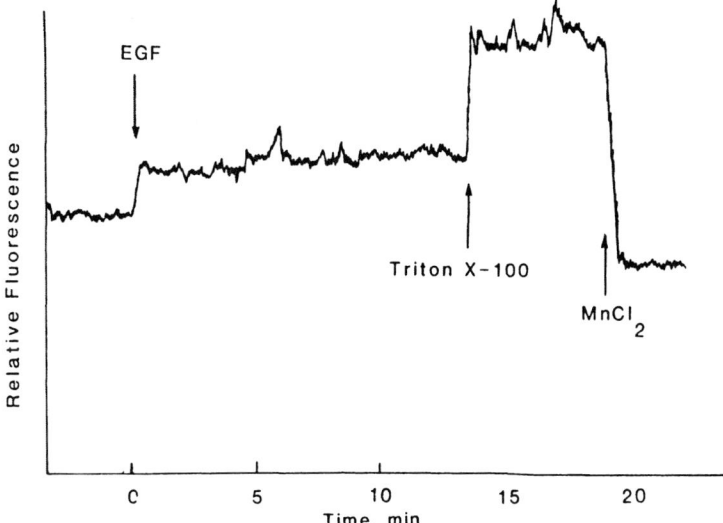

Figure 8. Effect of EGF on cytosolic Ca^{2+} concentration in MGC80 cells. At the indicated times EGF, Triton X-100 and $MnCl_2$ was added.

5097 ± 635 cpm/well in cells treated with 10^{-10} M EGF as compared to 3742 ± 92 cpm/well in controls (Fig. 5).

The growth promoting effect of 10^{-10} M EGF was inhibited in a dose dependent fashion by addition of EGFR mAb (Fig. 6), indicating that the growth stimulation is a receptor-mediated specific effect of EGF. Promotion of growth was also counteracted by EGF antiserum (1:1000 to 1:100,000), further confirming that this growth promotion is a specific effect of EGF (Fig. 7).

EFFECT OF EGF ON INTRACELLULAR Ca^{2+} CONCENTRATION IN MGC80 CELLS

EGF at 10^{-7} M caused a four-fold increase in cytosolic Ca^{2+} concentration from 0.6×10^{-7} M during the control period to 2.3×10^{-7} M after EGF stimulation. These results suggest that EGFR on these cells are biologically active, and that one of the early intracellular events after EGF-EGFR binding is the release of Ca^{2+} from intracellular organelles into the cytosol (Fig. 8).

PRODUCTION AND SECRETION OF EGF BY MGC80 CELLS

The MGC80 cell line can produce and secrete EGF. After cells were cultured in a serum-free medium for 72 hours, the concentration of EGF in the culture medium was 36.4 pg/ml and the EGF in the cell homogenate was 20.6 pg/10^6 cells.

DISCUSSION

Although EGF is recognized as one of the most important growth factors, its effect on the growth of tumor cells is controversial. Some investigators report that EGF has a growth-inhibiting effect and that the effect of EGF on tumor cell growth is dose-dependent[2,7,8]. In the present study, however, we have demonstrated that EGF at most concentrations (10^{-11}-10^{-7} M) can stimulate the growth of human gastric carcinoma cells. This was demonstrated by both ^{3}H-TdR uptake and by cell counts. It was also shown that physiological concentrations (i.e. 10^{-10} M or lower) of EGF can stimulate tumor cell growth. The addition of EGF antiserum to the culture medium in the presence of 10^{-10} EGF significantly suppressed EGF-induced tumor cell growth, further confirming the growth-promoting effect of EGF in MGC80 cells.

As reported by Yoshida et al[3], most gastric carcinomas express EGFR. In our study, radioreceptor assays revealed that MGC80 cells also express EGFR. Furthermore, an anti-EGFR monoclonal antibody (raised against A431 cells) competed for binding of ^{125}I-EGF to MGC80 cells and inhibited their growth in the presence of EGF. These results suggest that EGFR plays a crucial role in the growth-regulatory effect of EGF.

Although transforming growth factor alpha (TGFα), a peptide structurally related to EGF, has been widely recognized as having an autocrine regulatory role in the growth of a number of tumor cell lines[9], the autocrine effect of EGF on tumor cell growth has not yet been universally accepted[2,5]. Previous work in this laboratory showed that a human pancreatic carcinoma cell line established in this hospital produces and secretes EGF and also expresses EGFR[10], suggesting an autocrine growth-regulatory role of EGF. In human gastric carcinomas, most evidence for an autocrine role in growth regulation comes either from immunohistological studies, which showed a co-expression of EGF and EGFR at the protein level[11], or from molecular hybridization studies, which showed a co-expression of EGF and EGFR mRNAs[9]. Using a radioimmunoassay with EGF antiserum raised against recombinant hEGF (1-53), we have shown that in a serum-free medium the gastric carcinoma cell line MGC80 can produce and secrete EGF. This seemingly small amount of EGF has a growth-stimulating effect on MGC80 cells, as evidenced by the inhibition of growth by EGF antiserum added to the culture medium. It has been reported that, unlike the EGFR radioreceptor assay, which cannot distinguish EGF and TGFα because they both bind to the EGFR, the EGF radioimmunoassay using an antibody against EGF (1-53) has little cross-reactivity with TGFα[12]. Our work, therefore, provides direct evidence for the autocrine growth-regulatory role of EGF in human gastric carcinoma cells.

It was reported that co-expression of EGF and EGFR is often associated with a much poorer prognosis in gastric carcinoma[3,11]. The present study lends support to this observation by providing evidence for an autocrine regulatory role of EGF in growth studies of gastric carcinoma cells. In addition, the finding that EGFR mAb can inhibit tumor cell growth may shed some light on the application of EGF mAb in the therapy of gastric carcinoma in the future.

It is recognized that EGF, after binding to EGFRs on the cell surface, may cause dimerization of EGFRs, activate tyrosine kinase in the intracellular domain of the EGFRs, and thus lead to cell proliferation[13]. Our work revealed that EGF may induce the release of free Ca^{2+} from intracellular organelles into the cytosol. This may be one of the early intracellular events of growth factor action after EGF-EGFR binding. The connection between the release of $[Ca^{2+}]i$ and the activation of tyrosine kinase and cell proliferation remains to be further elucidated.

REFERENCES

1. M.H. Werner, P.A. Humphrey, D.D. Bigner and S.H. Bigner. Growth effect of epidermal growth factor (EGF) and a monoclonal antibody against the EGF receptor on four glioma cell lines, Acta Neuropathol. 77:196 (1988).

2. A. Ochiai, A. Takanashi, N. Takekura, K. Yoshida, S. Miyamori, T. Harada, and E. Tahara. Effect of human epidermal growth factor on cell growth and its receptor in human gastric carcinoma cell lines, Jpn. J. Clin. Oncol. 18:15 (1988).

3. K. Yoshida, E. Kyo, T. Tsujino, T. Sano, M. Niimoto, and E. Tahara.Expression of epidermal growth factor, transforming growth factor α and their receptor genes in human gastric carcinomas: Implication for autocrine growth. Jpn. J. Cancer Res. 81:43 (1990).

4. T.J. Velu, Structure, function and transforming potential of the epidermal growth factor receptor. Mol. Cell Endocrinol. 70:205 (1990).

5. W. Yasui, H. Hata, H. Yokozaki, H. Nakatani, A. Ochiai, H. Ito and E. Tahara. Interaction between epidermal growth factor and its receptor in progression of human gastric carcinoma. Int. J. Cancer 41:211 (1988).

6. N.R. Lemoine, S. Jain, F. Silvestre, C. Lopes, C.M. Hughes, E. McLelland, W.J. Gullick and M.I. Filipe. Amplification and overexpression of the EGF receptor and c-erbB-2 proto-oncogenes in human stomach cancer. Br. J. Cancer 64:79 (1991).

7. M. Clementi, A. Gesta, I. Testa, P. Bagnarelli, G. Devescovi, and G. Carloni. Growth-inhibitory effects of epidermal growth factor on human breast cancer and carcinoma of the esophagus transplanted into nude mice. FEBS Lett. 249:297 (1989).

8. Y. Muragama, Expression of high- and low-affinity epidermal growth factor receptors in human hepatoma cell lines. Ann. Surg. 211:263 (1990).

9. A.J. Ekstrand, C.D. James, W.K. Cavenee, B. Seliger, R.F. Pettersson and V.P. Collins. Genes for epidermal growth factor receptor, transforming growth factor α, and epidermal growth factor and their expression in human gliomas in vivo. Cancer Res. 51:2164 (1991).

10. Y. F. Chen, G.Z. Pan, X. Han, T.H. Liu, J. Chen, C. Yanaihara, and N Yanaihara. Epidermal growth factor and its receptors in human pancreatic carcinomas. Pancreas 5:278 (1990).

11. K. Sugiyuma, Y. Yonemura and I. Miyazaki. Immunohistochemical study of epidermal growth factor and epidermal growth factor receptor in gastric carcinoma. Cancer 63:1557 (1989).

12. K. Imanishi, K. Yamaguchi, M. Suzuki, S. Honda, N. Yanaihara and K. Abe. Production of transforming growth factor α in human tumor cell lines. Br. J. Cancer 59:761 (1989).

13. Y. Hirata, M. Uchihashi, M. Nakajima, T. Fujita and S. Matsukura. Immunoreactive human epidermal growth factor in human pancreatic juice. J. Clin. Endocrinol. Metab. 54:1242 (1982).

THE SINGLE-CHAIN IMMUNOTOXINS BR96 sFv-PE40 AND BR96 sFv-PE38: POTENT ANTI-TUMOR AGENTS FOR THE TREATMENT OF HUMAN CANCER

Paula N. Friedman, Dana F. Chace, Susan L. Gawlak, and Clay B. Siegall

Department of Molecular Immunology
Pharmaceutical Research Institute
Bristol-Myers Squibb Co.
3005 First Avenue
Seattle, WA 98121

INTRODUCTION

Immunotoxins are potent agents designed to be selectively cytotoxic against specific populations of cells (1, 2). Such molecules are composed of a binding portion, which is specific for a cell surface antigen or receptor, linked to a cytotoxic agent, such as a protein toxin. *Pseudomonas* exotoxin (PE) is a protein toxin produced by the bacteria *Pseudomonas aeruginosa* and kills cells by ADP-ribosylating elongation factor 2, thereby halting protein synthesis (3). Analysis of the structure of PE revealed there are 3 domains of the protein involved in: 1) binding, 2) processing and translocation, and 3) ADP-ribosylation (4). Forms of PE that lack domain I, the binding domain, are often used in the construction of immunotoxins, as they lack the ability to bind non-specifically to the cell surface receptor.

Immunotoxins can be made either by chemical conjugation (5-8) or by using recombinant DNA technology to produce a fusion protein made expressing two genes joined at the molecular level (9-13). We have constructed two single-chain immunotoxins termed BR96 sFv-PE40 and BR96 sFv-PE38 which contain the variable chains of the carcinoma-reactive antibody BR96 (14) and the binding defective toxins PE40 (15) and PE38 (16). By combining the specificity of the BR96 antibody with the cytotoxic potential of PE, we have produced recombinant fusion proteins that have potential use as anti-tumor agents for the treatment of cancer.

CONSTRUCTION AND EXPRESSION OF BR96 sFv-PE40 AND BR96 sFv-PE38

The plasmid pBW 7.0, which encodes BR96 sFv-PE40, was constructed by fusing the variable chains of BR96 IgG with the gene encoding PE40 (17). An additional

Growth Factors, Peptides, and Receptors, Edited
by T.W. Moody, Plenum Press, New York, 1993

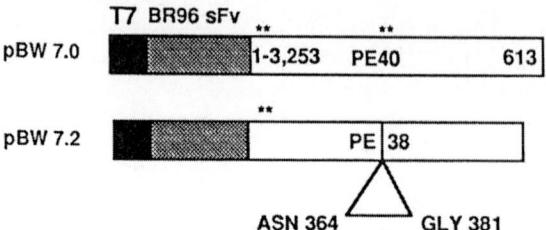

Figure 1: Schematic diagram of plasmids encoding BR96 sFv-PE40 and BR96 sFv-PE38. pBW 7.0 encodes BR96 sFv-PE40 and pBW 7.2 encodes BR96 sFv-PE38, which is a molecule that contains a deletion of 15 amino acids located between domains II and III of *Pseudomonas* toxin. T7 represents the T7 late promoter and BR96 sFv stands for the variable light (V_L) and variable heavy (V_H) chains of BR96 IgG joined by a (Gly 4Ser)3 linker.** represents the disulfide bond location. Note: Diagram not drawn to scale.

construct, pBW 7.2, which encodes the fusion protein BR96 sFv-PE38, was also constructed. This molecule differs from BR96 sFv-PE40 in that it contains an internal deletion of 15 amino acids which removes the disulfide bond formed between cysteine residues at positions 372 and 379 (16). The purpose of such a deletion was to reduce the potential for inappropriate disulfide bond formation and thereby prevent the accumulation of inactive forms of BR96 sFv-PE40 upon renaturation of the protein. A comparitive schematic diagram of these constructs is shown in Figure 1.

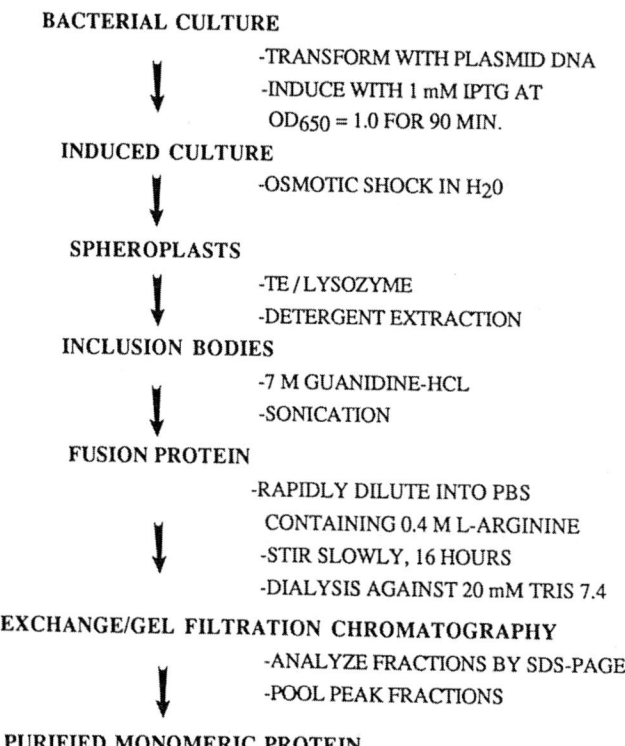

Figure 2: Purification scheme for isolation of the fusion proteins BR96 sFv-PE40 and BR96 sFv-PE38.

In order to produce purified fusion proteins the plasmids pBW 7.0 and 7.2 were transformed into BL21 (λDE3) cells and after induction with IPTG, protein was purified as outlined in Figure 2.

DETERMINATION OF THE CYTOTOXIC POTENTIAL OF BR96 sFv-PE40 AND BR96 sFv-PE38

In order to determine the cytotoxic potential of the purified fusion proteins, as well as their specificity, the two single-chain immunotoxins were tested on a variety of cell lines, namely MCF-7 (breast carcinoma), RCA (colon carcinoma) and KB (epidermoid carcinoma) cells (Figure 3). Both immunotoxins were most active on MCF-7 cells, with BR96 sFv-PE40 being slightly more active than BR96 sFv-PE38, whereas both single-chain immunotoxins were greater than 100-fold more potent than PE40 alone (data not shown). On RCA and KB cells, the activities of the single-chain immunotoxins were almost identical. Hence, the cytotoxic activity of BR96 sFv-PE38 did not seem to be improved over that of the original fusion protein, BR96 sFv-PE40.

To confirm the specificity of the cytotoxic effects of the BR96 immunotoxins, a competition experiment was done with BR96 IgG and, as a control, L6 IgG, an antibody that does not bind to the BR96 antigen. The activity of the fusion protein BR96 sFv-PE40 could be competed by BR96 IgG but not by L6 IgG indicating that the cytotoxic potential of the immunotoxin is due to its specific interaction with the BR96 antigen.

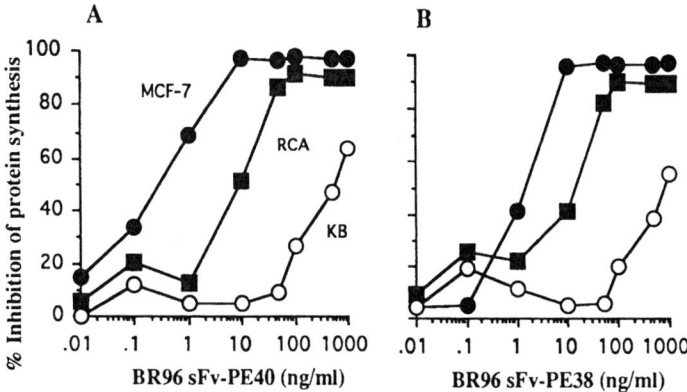

Figure 3: Cytotoxicity of the single-chain immunotoxins on three human cancer lines. The cancer cell lines, MCF-7 (●), RCA (■), and KB (○) were all cultured as monolayers at 37°C in RPMI 1640 supplemented with 10% fetal bovine serum. Tumor cells were plated onto 96-well flat bottom tissue culture plates (1 x 10^4 cells/well) and kept at 37°C for 16 hours. Dilutions of immunotoxin were made in growth media and 0.1 ml added to each well for 20 hours at 37°C. Each dilution was done in triplicate. The cells were pulsed with [3H]-leucine (1 uCi/ well) for an additional 4 hours at 37°C. The cells were lysed by freeze-thawing and harvested using a Tomtec cell harvester (Orange, CT). [3H]-leucine incorporation was determined by a LKB Beta-Plate liquid scintillation counter.

ANTI-TUMOR ACTIVITY OF BR96 sFv-PE40 AGAINST L2987 AND RCA TUMOR XENOGRAFTS

To determine the anti-tumor potential of BR96 sFv-PE40, we treated L2987 (lung) and RCA (colon) carcinoma xenografts with various concentrations of the fusion protein. In the L2987 model, using an administration schedule of q2dx5, we observed complete regressions of the tumor xenografts with doses of BR96 sFv-PE40 from 0.125 to 0.375 mg/kg (Figure 4A). The highest dose was most effective at prolonging tumor regressions. Mice that were untreated or treated with a recombinant immunotoxin that was not specifically toxic to these cells, IL6-PE40 (18), retained their tumor burden and were sacrificed when their tumors reached approximately 1500 cu mm. In the RCA model, using an identical administration schedule and doses of BR96 sFv-PE40, we observed stabilization of tumor growth over a period of 20 days but no regressions (Figure 4B). These results show the effectiveness of the BR96 sFv-PE40 molecule against two different tumor models.

In summary, we have constructed and characterized two single-chain immunotoxins composed of the carcinoma-reactive antibody BR96 and different mutant forms of *Pseudomonas* exotoxin. We have shown that they are effective anti-tumor agents for the treatment of human cancer. The ability of the antibody in sFv form to internalize into solid tumor cells is the key reason that BR96 sFv-PE40 and BR96 sFv-PE38 are effective immunotoxins.

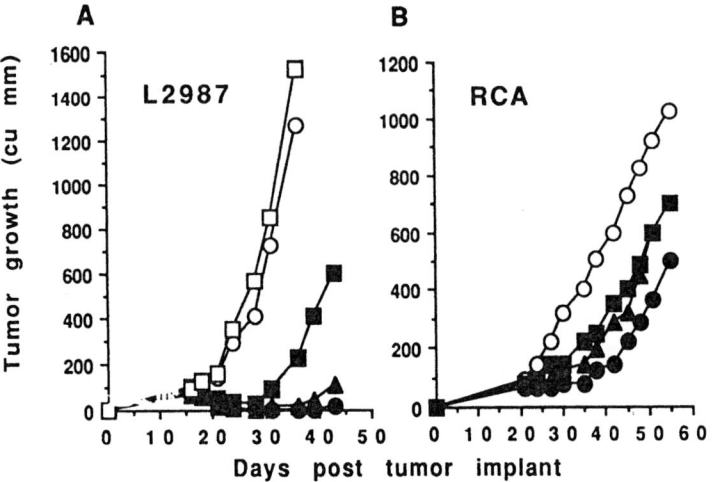

Figure 4: Antitumor activity of BR96 sFv-PE40 against human tumor xenografts in athymic mice. L2987 (A) and RCA (B) tumor fragments were subcutaneously implanted in nude mice. Approximately, two weeks after implantation the animals were randomized and sorted for tumors ranging from 50-100 cubic mm in size. The animals were intravenously injected via the tail vein with the immunotoxin using an administration schedule of q2dx5. Each treatment group consisted of five animals. Doses of BR96 sFv-PE40:0.125 (■), 0.25 (▲), and 0.375 (●) mg/kg were compared with uninjected control animals (○) and, in the L2987 model, with a non-specific immunotoxin, IL6-PE40 (□).

ACKNOWLEDGEMENTS

This research is supported by the Pharmaceutical Research Institute of Bristol-Myers Squibb Company. We thank Drs. K.E. Hellström, P. Fell, S. McAndrew, P.Trail, and J. Brown for helpful discussions.

REFERENCES

1. Pastan, I. and FitzGerald, D. Recombinant toxins for cancer treatment. Science (Washington DC), 254: 1173-1177, 1991.

2. Vitetta, E.S. and Thorpe, P.E. Immunotoxins containing ricin or its A chain. Sem. Cell. Biol., 2: 47-58, 1991.

3. FitzGerald, D. and Pastan, I. Redirecting *Pseudomonas* exotoxin. Sem. Cell. Biol., 2: 31-37, 1991.

4. Hwang, J., FitzGerald, D.J.P., Adhya, S., and Pastan, I. Functional domains of *Pseudomonas* exotoxin identified by deletion analysis of the gene expressed in *E. coli.* Cell, 48: 129-136, 1987.

5. Vitetta, E.S., Krolick, K.A., Miyama-Inaba, M., Cushley, W., and Uhr, J.W. Immunotoxins: A new approach to cancer therapy. Science (Washington DC), 219: 644-650, 1987.

6. Youle, R.J., Murray, G.J., and Neville, D.M. Ricin linked to monophosphopenta-mannose binds to fibroblast lysosomal hydrolase receptors resulting in a cell-type specific toxin. Proc. Natl. Acad. Sci. USA., 71: 5559-5562, 1979.

7. Byers, V.S., Pawluczyk, I.Z., Hooi, D.S., Price, M.R., Carroll, S., Embleton, M.J., Garnett, M.C., Berry, N., Robins, R.A., and Baldwin, R.W. Endocytosis of immunotoxin-791T/36-RTA by tumor cells in relation to its cytotoxic action. Cancer Res., 51: 1990-1995, 1991.

8. Vallera, D.A., Carroll, S.F., Snover, D.C., Carlson, G.J., and Blazer, B.R. Toxicity and efficacy of anti-T-cell ricin toxin A chain immunotoxins in a murine model of established graft-versus-host disease induced across the major histocompatibility barrier. Blood, 77: 182-194, 1991.

9. Chaudhary, V.K., Queen, C., Junghans, J.P., Waldmann, T.A., FitzGerald, D.J. and Pastan, I. A recombinant immunotoxin consisting of two antibody variable domains fused to Peudomonas exotoxin. Nature (Lond.), 339: 394-397, 1989.

10. Chaudhary, V.K., Gallo, M.G., FitzGerald, D.J., and Pastan, I. A recombinant single-chain immunotoxin composed of anti-Tac variable regions and a truncated diphtheria toxin. Proc. Natl. Acad. Sci. USA., 87: 9491-9494, 1990.

11. Brinkmann, U., Pai, L.H., FitzGerald, D.J., Willingham, M., and Pastan, I. B3(Fv)-PE38KDEL, a single-chain immunotoxin that causes complete regression of a human carcinoma in mice. Proc. Natl. Acad. Sci. USA., 88: 8616-8620, 1991.

12. Batra, J.K., FitzGerald, D.J., Chaudhary, V.K., and Pastan, I. Single-chain immunotoxins directed at the human transferrin receptor containing *Pseudomonas* exotoxin A or diphtheria toxin: anti-TFR (Fv). Mol. Cell. Biol., 11: 2200-2205, 1991.

13. O'Hare, M., Brown, A.N., Hussain, K., Gebhardt, A., Watson, G., Roberts, L.M., Vitetta, E.S., Thorpe, P.E., and Lord, J.M. Cytotoxicity of a recombinant ricin A-chain fusion protein containing a proteolytically cleavable spacer sequence. FEBS Lett., 273: 200 204, 1990.

14. Hellstrom, I., Garrigues, H.J., Garrigues, U., and Hellstrom, K.E. Highly tumor-reactive, internalizing, mouse monoclonal antibodies to LeY-related cell-surface antigens. Cancer Res., 50: 2183-2190, 1990.

15. Kondo, T., FitzGerald, D., Chaudhary, V.K., Adhya, S., and Pastan, I. Activity of immunotoxins constructed with modified *Pseudomonas* exotoxin A lacking the cell recognition domain. J. Biol. Chem., 263: 9470-9475, 1988.

16. Siegall, C.B., Chaudhary, V.K., FitzGerald, D.J., and Pastan, I. Functional analysis of domains II, Ib, and III of *Pseudomonas* exotoxin. J. Biol. Chem., 264: 14256-14261, 1989.

17. Friedman, P.N., McAndrew, S.J., Gawlak, S.L., Chace, D., Trail, P.A., Brown, J.P., and Siegall, C.B. BR96 sFv-PE40, a potent single-chain immunotoxin that selectively kills carcinoma cells. 1992 (submitted)

18. Siegall, C.B., Chaudhary, V.K., FitzGerald, D.J., and Pastan, I. Cytotoxic activity of an interleukin 6-*Pseudomonas* exotoxin fusion protein on human myeloma cells. Proc. Natl. Acad. Sci. USA., 85: 9738-9742, 1988.

Part VII

Clinical Correlations

HEMATOPOIETIC CYTOKINES TO SUPPORT REPEATED DOSES OF DOSE-INTENSIVE CHEMOTHERAPY WITH CYCLOPHOSPHAMIDE, ETOPOSIDE, AND CISPLATIN (DICEP)

James A. Neidhart

University of New Mexico Cancer Center
900 Camino de Salud, NE
Albuquerque, NM 87131

INTRODUCTION

The last two decades have witnessed fundamental changes in the practice of clinical oncology. Perhaps the most important is the acknowledgement that partial, or in many cases even complete, clinical remissions do not usually result in substantial survival benefit. The degree of tumor reduction obtainable with present standard therapy regimens is not sufficient to prolong survival in most patients. The importance of dose intensity in achieving a higher response rate[1] and, arguably, survival[2] has encouraged some oncologists, usually in a clinical trials milieu, to accept absolute but reversible granulocytopenia as a justifiable side effect of chemotherapy. This change in attitude has been fostered by the successes of bone marrow transplantation as well as by the development of improved supportive agents and approaches. Concurrently, several hematopoietic cytokines capable of ameliorating hematologic toxicity and the side-effects of dose-intensive therapy have become clinically available. These changes are the basis for a series of trials conducted over the last six years at the University of New Mexico Cancer Center attempting to achieve even greater dose rate intensity and total dose intensity than that used in most transplantation programs. The results of those trials are the substance of this report.

DOSE INTENSITY

Dose-intensity is directly related to response rate with standard chemotherapy regimens given for a number of different types of cancer.[1] Increases of chemotherapy doses sufficient to produce absolute and prolonged granulocytopenia substantially increase the complete response rates in breast cancer[3] and in non-Hodgkin's lymphoma[4,5]. There is no convincing information at present regarding survival benefit from dose-intensive therapy although several on-going randomized trials may fill that void. Neither are there adequate definitions of what constitutes dose-intensity. Modest increases or differences in dose have not yet been shown to improve complete remission rates to a significant degree. In preclinical models, Schabel has shown that doses of drug that are lethal to a high fraction of mice are necessary to obtain cures[6]. The relative benefits or costs, if any, of dose-intensive combination therapy, repeated cycles versus one cycle, the various chemotherapeutic agents or radiation therapy, or pre-intensification induction regimens have yet to be addressed. The need for progenitor replacement is still unproven but is widely assumed. Dose-intensity is a concept in evolution and, in practice, still quite empirical. The interested reader is directed to several reviews [2,3,6,7]. The regimen reported in this study is based on the specific hypothesis that repeated cycles of dose-intensive therapy may be therapeutically better that one cycle and that, at least for certain agents, the bone marrow will recover without progenitor cell replacement.

DOSE-INTENSIVE CYCLOPHOSPHAMIDE, ETOPOSIDE, AND PLATINUM (DICEP)

We first treated patients with the DICEP regimen in January of 1987 and since that time have treated approximately 300 patients with a variety of malignant diseases. The rationale for the regimen is best presented in the report of those original studies[8] but is based on therapeutic synergism, non-overlapping toxicities and the ability to escalate the individual agents at least two-fold above standard doses without reaching limiting non-hematologic toxicities. In addition, Peters had conducted a comparative trial of six different dose-intensive regimens based on the alkylating agents cyclophosphamide and cisplatin and supported by bone marrow transplantation. The analogue of the DICEP regimen had the least hepatic toxicity[9]. We also wished to minimize pulmonary toxicity and prolonged hematologic suppression providing additional reasons for substituting etoposide for carmustine which is used in many transplantation regimens.

The initial chemotherapy dose-finding study explored six dosing levels varying from each other by approximately 25%[8]. All produced absolute granulocytopenia of about nine days duration with the first and second cycle and of about twelve days duration with the third cycle. This study not only demonstrated that the patients would have hematologic recovery without progenitor infusion but also provided important information on patient selection. Six of twelve patients treated with a poor performance status (ECOG 2-4) died with the first cycle of therapy. Four of eighteen with a performance status of 1 and none of twelve with a normal performance status died with the first cycle. Renal function (creatinine clearance below 90 cc/min) and a traditionally "non-responsive" tumor type (e.g. colon or renal cancer or melanoma) were significant predictors of death and/or non-response to DICEP. Dose limiting toxicities were pulmonary, cardiac, and mucositis (esophagitis). Most stomatitis was associated with reactivation of Herpes and was treatable and preventable with acyclovir. The chemotherapy dose chosen for all future studies produced some mucositis in 20% of cycles and grade 2 or greater mucositis in only 5% of cycles. The lessons of this first trial were (1) patient selection and particularly a near normal performance status are critical for safety and success with this regimen, (2) hematologic toxicity is not related to modest dose increases once major escalations are used, (3) non-hematologic toxicities of this regimen are quite tolerable in most patients and are dose-related even at the escalated levels used in this study and (4) complete remissions and prolonged survival are obtainable even in refractory patients.

GRANULOCYTE COLONY-STIMULATING FACTOR (rhG-CSF)

The next trial was one of the first to use a hematopoietic cytokine in support of dose-intensive chemotherapy[10]. Granulocyte colony-stimulating factor (rhG-CSF) was administered starting three days after completion of the chemotherapy using a 30 minute infusion. In retrospect, both the timing and the route of administration are probably less than optimal. A marked and rhG-CSF dose dependent decrease in duration of severe granulocytopenia was demonstrated. At a dose of 60 µg/Kg of rhG-CSF the duration of severe granulocytopenia was shortened from nine days to about 5.5 days. The degree and speed of recovery were, at the time, impressive with granulocyte counts reaching 100,000/ml. and the patients suffering bone pain requiring narcotics for relief. Neither of these effects is necessary nor desirable and cytokines are now stopped when granulocytes reach normal levels. The duration of antibiotic therapy was shortened with rhG-CSF compared to non-randomized controls although we were unable to demonstrate shorter durations of hospitalization. Unfortunately, this study has not been repeated with more appropriate dosing schedules of rhG-CSF, more careful patient selection, and more recent supportive regimens. For those reasons, we cannot compare the use of rhG-CSF and GM-CSF (vide infra) in support of DICEP therapy. It is possible that subcutaneous administration of rhG-CSF started the day following chemotherapy would produce equal or more benefit than GM-CSF in lower doses or decrease the need for hospitalization but in lieu of data we can only speculate.

GRANULOCYTE-MACROPHAGE COLONY-STIMULATING FACTOR (GM-CSF)

Most of our experience with cytokines in support of DICEP therapy has been with yeast derived (glycosylated) GM-CSF and, unless stated differently, the following data refer to that material. Fifty-one patients entered a GM-CSF dose-finding trial starting in October of 1988. The details of that trial have recently been published[11]. That manuscript also outlines in detail the current chemotherapy dosing and supportive regimens for DICEP. Etoposide 500 mg/m^2 is infused every two hours on cycle days 1, 2, 3 followed by cyclophosphamide 2,500 mg/m^2 over one hour on days 1, 2 then cisplatin 50 mg/m^2 over one hour on days 1, 2, 3. Patients are discharged the day following chemotherapy and are rehospitalized for antibiotic therapy if they develop a temperature of 38.5° despite prophylactic acetaminophen. Chemotherapy is administered every 35 days to responding patients for a total of three cycles. The optimal dose of GM-CSF for improving the most clinically important endpoints of duration of hospitalization, antibiotic use and cytopenic fever is between 500 and 750 µg/m^2/day given

TABLE 1

		Days with: mean (range)			
		WBC < 300	Platelets < 10,000	Hospitalization	Antibiotics
Control	(h)	10.2 (5-20)	8.5 (4-21)	22.2 (19-34)	11.0 (7-26)
	(c)	13.2 (9-29)	6.9 (0-20)	15.7 (8-27)	6.3 (3-8)
GM-CSF	(1)	7.6 (6-10)	4.6 (2-8)	9.6 (4-16)	8.5 (5-131)
	(2)	7.6 (5-10)	2.8 (1-6)	9.8 (4-16)	3.0 (0-9)

(1)   GM-CSF 500 µg/m^2/day subcutaneously starting on cycle day 4
(2)   GM-CSF 750 µg/m^2/day subcutaneously starting on cycle day 4

subcutaneously starting the day following chemotherapy. While we divided the daily dose into two injections there is little evidence that improves response. Table 1 presents this data. Two control groups were used and point out that with DICEP alone there is variation in severity of the hematologic toxicity. GM-CSF decreased the duration of severe leukopenia from about 10-13 days in controls to 7.6 days and the duration of hospitalization from 16-22 days to a mean of 9.6 days. The majority of patients, in fact, do not require rehospitalization for cytopenic fever and spend only four days in the hospital for chemotherapy with each cycle. There is a suggestion that the higher dose of GM-CSF can shorten the duration of thrombocytopenia (p <0.05) although not to a clinically important degree. Another interesting finding is that while GM-CSF given intravenously produces a shorter duration of severe leukopenia (WBC <300/ml for 5.9 days) it does not shorten duration of hospitalization. For whatever reason, duration of leukopenia may not be the best marker for the more clinically important endpoints. The second cycle of therapy produces toxicities that are similar to or slightly less than those seen with the first cycle. Recovery from the third cycle is often delayed by several days in terms of WBC and non-hematologic toxicity and by weeks in terms of full platelet recovery although patients usually are transfusion dependent for only 1-2 weeks.

The results of several other studies remain unpublished. Immunologic function, as measured phenotypically or by macrophage killing assays, recovers fully on a timetable similar to granulocyte recovery. There is a rebound in progenitor cells following DICEP therapy with or without cytokine support[12]. GM-CSF given for six days prior to DICEP will increase circulating progenitor cells but does not further shorten the post-chemotherapy duration of cytopenia. E. Coli (non-glycosylated) GM-CSF is more biologically active on a µg-for-µg basis than yeast-derived material with a dose of 3-5 µg/Kg producing hematologic responses equivalent to a dose of 10-18 µg/Kg of yeast-derived material with the DICEP regimen. On the other hand, the E.Coli material was more likely to produce dermatologic and pulmonary side effects...particularly if used at unnecessarily high doses.

ANTITUMOR ACTIVITY

The results of DICEP therapy in Non-Hodgkin's lymphoma (NHL) are similar to those obtained with bone marrow transplantation regimens in terms of continuous disease free survival and perhaps slightly better in terms of complete remission rate and overall survival[5]. The complete remission rate in twenty-three patients with either primarily refractory or refractory relapsed NHL was 52% with an additional 26% attaining PR. Ten patients are alive with a median follow-up of 32 months and six are free of disease. Three of these are in continuous complete remission with no treatment following DICEP. Two had recurrence of unbiopsied lymphadenopathy 5 and 9 months following DICEP and received radiation therapy but remain free of disease at 4.8 and 4.9 years. The other had biopsy proven recurrence of lymphoma at 4.3 years and entered a second CR with two more cycles of DICEP. Two year survival for patients attaining a CR is 75% while no patient failing to attain a CR has lived more than 28 months. This allows selection of those patients who will benefit after the first cycle. Non-responders should not continue therapy. Quality of live is excellent in long-term survivors since the only long-term clinical toxicities to date have been a high frequency hearing loss and slowly resolving neuropathy.

Our early results in pretreated patients with breast cancer were published in summary form and we are currently in the process of up-dating the data.[3] Complete remission rates seem to be in the 40-50% range which is comparable to those seen in bone marrow transplantation programs. We are now treating patients with Stage IV disease and over ten positive nodes at initial presentation. A multi-institutional Phase II trial in breast cancer and lymphoma is currently under way. Our experience in over types of malignancy is with small numbers of patients and is very preliminary. Some encouraging responses have been seen in cervical and ovarian carcinoma, esophageal cancer, gastric cancer, bladder cancer, refractory lymphoblastic leukemia, and even in single patients with pancreatic and prostatic cancer. Rapid and usually complete tumor regressions have been seen in head and neck cancer and in small cell cancer of the lung but the remissions have been of very short duration and probably are not worthwhile unless combined modality treatments can produce more durability. A trial is currently on-going at Ohio State University incorporating DICEP into a multimodality trial for initial treatment of head and neck cancer. (Roy Smith, personal communication). Patients with non-small cell cancer of the lung usually obtain substantial reduction in tumor but, again, regrowth is often rapid. Renal cancer, melanoma, sarcoma, and colon cancer have not been responsive in terms of complete remissions although an occasional patient will have some response. We no longer treat patients with these diseases.

DISCUSSION AND FUTURE DIRECTIONS

DICEP can be given in three repeated cycles without progenitor replacement. The most frequent side effect is fatigue which can interfere with function. Less frequent, but occasionally severe, toxicities are vomiting, esophagitis, diarrhea, and anorexia which are uncomfortable in about 10-20% of patients. Patients have measurable abnormalities in hepatic and pulmonary function and slow recovery to normal hematologic values following DICEP. Quality of life is excellent beginning about 1-2 months after the last treatment. The hematopoietic cytokines and GM-CSF, in particular, shorten the duration of cytopenia and decrease the incidence of infection and the need for hospitalization. Most patients are treated on an outpatient basis after four days of hospitalization for chemotherapy.

Huan has reported a randomized trial of an almost identical chemotherapy regimen given for two cycles with or without bone marrow transfusion.[13] No cytokines were used. While the bone marrow arm had a statistically significant shortening in time to recovery of the absolute granulocyte count (AGC) to 100/ml, the difference was only 1 to 5 days depending on dose and cycle and there were no differences in febrile episodes or rates of documented infections. There were no differences in time to recovery to an AGC of 1000/ml nor in time to platelet recovery. The duration of severe granulocytopenia in that study with bone marrow transplantation was about 4-5 days longer than we obtained with GM-CSF support. The hematopoietic colony stimulating factors and, in our experience, GM-CSF in particular seem to decrease risk and morbidity of DICEP with little cost in terms of added toxicity. Dosing schedules and the type of GM-CSF used are important but incompletely evaluated at present.

We have no doubt that the degree of dose-intensity in the DICEP regimen or most bone marrow supported dose-intensive regimens will produce more complete responses in traditionally responsive tumor types and even in some tumors not usually responsive to chemotherapy. Either type of regimen can produce durable complete remissions in relapsed or refractory lymphoma whereas standard rescue regimens do not. I believe randomized trials will be necessary and are presently appropriate to ferret out the relative benefits or role for each regimen in lymphoma. The data in breast cancer is less mature. Dunphy treated patients with Stage IV, receptor-negative breast cancer with four cycles of adriamycin and cyclophosphamide[14]. Patients with stable or responsive disease then went on to receive two cycles of a regimen almost identical to DICEP with the exception that one half of the patients received bone marrow transplantation and only two cycles of intensive therapy were given[13]. The complete remission rate was 55% with a median progression-free interval of 57 weeks. Approximately 25% of patients were projected to be free of disease at two years. In our early report, four of six patients with non-refractory disease achieved complete responses that were on-going at 3.5 to 10.4 months[3]. We now have more experience in minimally pretreated patients and are currently up-dating our data. Peters has also published data that suggest some patients with advanced breast cancer can achieve durable complete remissions with dose-intensive therapy.[15] There is preliminary data that dose-intensive therapy used in the adjuvant setting can be of substantial benefit in preventing recurrence.(Peters, personal communication) Again, randomized comparisons are appropriate and necessary to determine the relative benefits of these various approaches and standard therapy in breast cancer. Johnson used one or two cycles of a regimen with a dose-intensity approximately 75% of DICEP to treat twenty patients with extensive small cell lung cancer[16]. The complete remission rate was 65% but, despite continued standard therapy, only two patients remained free of disease at the time of publication with a median response duration of 6 months. Our experience, although much more limited, is similar in terms of non-durability of response. While we have no experience with advanced or recurrent testicular cancer, Harstrick has reported a high complete remission rate of 53-72% in 33 patients treated with a slightly less intensive regimen than DICEP with ifosfamide substituted for cyclophosphamide[17].

Where to now? A more precise understanding of the degree of benefit from DICEP and similar regimens in a variety of tumor types is necessary and towards that goal a series of Phase II and III trials is appropriate. We have chosen not to change the chemotherapy regimen as we explored various cytokines but obviously a number of substitutions can be made. Carboplatinum, in particular, may help reduce hearing loss in children. There are a number of newer cytokines such as a GM-CSF/IL-3 hybrid molecule (PIXY), IL-3, IL-6 and various combinations of cytokines that may further reduce hematologic toxicity. We currently are evaluating a combination of rhG-CSF and GM-CSF and will soon start a trial of PIXY. Although some of the chemotherapeutic agents can be escalated to a greater degree, we believe dose limiting mucositis or pulmonary and cardiac toxicity will limit the usefulness of such escalations. Perhaps a more reasonable alternative is to develop a pharmacokinetically driven dosing regimen since it is highly likely that some patients are being undertreated and some overtreated with our present empiric dosing[6]. Perhaps the most interesting and exciting alternative is to develop combined modality approaches based on the ability of dose-intensive regimens to markedly decrease tumor load if not produce cure. Some of the recent innovations and advances in immunotherapy make this alternative particularly attractive. Clearly, the advent of substantial dose escalation as a reasonable treatment alternative for cancer has opened new avenues of research and offered renewed hope to many patients.

REFERENCES

1. W.M. Hryniuk, Average relative dose intensity and the impact on design of clinical trials. Semin in Onc 14(1):56,(1987)

2. E. Frei III, K.Antman, B.Teicher, et al, Bone marrow auto transplantation for solid tumors - prospects. J Clin Oncol 7(4):515(1989).

3. J. Neidhart, Dose intensive treatment of breast cancer supported by granulocyte macrophage colony-stimulating factor. Breast Cancer Treatment 20: S153(1991).

4. J. Armitage, Bone marrow transplantation in the treatment of patients with lymphoma. Blood 73(7):1749(1989).

5. J. Neidhart, R.Kubica, C.Stidley, et al, Multiple cycles of dose-intensive cyclophosphamide, etoposide, and cisplatin (DICEP) produce durable responses in refractory non-Hodgkin's lymphoma. Cancer Invest, In-press.

6. F. Schabel, D. Griswold, T. Corbett, et al, Increasing the therapeutic response rates to anticancer drugs by applying the basic principles of pharmacology. Cancer 54:1160 (1984).

7. J. Henderson, D. Hayes, R. Gelman, Dose-response in the treatment of breast cancer: a critical review. J Clin Oncol 6:1501(1988).

8. J. Neidhart, W. Kohler, C. Stidley, et al, A Phase I study of repeated cycles of high dose cyclophosphamide, etoposide and cisplatin administered without bone marrow transplantation. J Clin Oncol 8:1728(1990).

9. W. Peters, E. Shpall, R. Jones, et al, Critical factors in the design of high-dose combination chemotherapy regimens, in: Advances in Cancer Chemotherapy, High dose therapy and autolougous marrow transplantation. G.P. Herzig, ed., Park Row, Dallas(1987).

10. J. Neidhart, A. Mangalik, W. Kohler, et al, Granulocyte colony stimulating factor (rhg-CSF) stimulates recovery of granulocytes in patients receiving dose-intensive chemotherapy without bone marrow transplantation. J Clin Oncol 7(11):1685(1989).

11. J. Neidhart, A. Mangalik, C. Stidley, et al, Dosing regimen of granulocyte-macrophage colony stimulating factor (GM-CSF) to support dose-intensive chemotherapy. J Clin Oncol (In Press).

12. D. Clark, A. Castillo, J. Neidhart, Myeloid progenitors after high dose chemotherapy and granulocyte-macrophage colony stimulating factor treatment. Proc. Am. Assoc. Cancer Res., April 1990.

13. S. Huan, J. Yau, F. Dunphy, et al, Impact of autolougous bone marrow infusion on hematopoietic recovery after high-dose cyclophosphamide, etoposide, and cisplatin. J Clin Oncol 9:1609(1991).

14. F. Dunphy, G. Spitzer, A. Buzdar,et al, Treatment of estrogen receptor-negative or hormonally refractory breast cancer with double high-dose chemotherapy intensification and bone marrow support. J Clin Oncol 8:1207(1990).

15. W. Peters, E. Shpall, R. Jones, et al, High-dose combination alkylating agents with bone marrow support as initial treatment for metastatic breast cancer. J Clin Oncol 6:1368(1988).

16. D. Johnson, M. DeLeo, K. Hande, et al, High-dose induction chemotherapy with cyclophosphamide, etoposide, and cisplatin for extensive-stage small-cell lung cancer. J Clin Oncol 5:703(1987).

17. A. Harstrick, H. Schmoll, C. Bokemeyer, et al, Cisplatin, etoposide, ifosfamide stepwise escalation with concomitant granulocyte/macrophage-colony-stimulating factor for patients with far-advanced testicular carcinoma. J Can Res Clin Oncol 117:S198(1991).

INTERLEUKIN-2 (IL-2) AUGMENTS THE EXPRESSION OF TRANSFORMING GROWTH FACTOR BETA IN PATIENTS WITH DISSEMINATED CANCER

Pauli Puolakkainen[1,2], Paolo Alberto Paciucci[3], Jane E. Ranchalis[1], Leslie Oleksowicz[3], Daniel R. Twardzik[1,4]

[1]Bristol-Myers Squibb Pharmaceutical Research Institute, Seattle, WA 98121

[2]Second Department of Surgery, Helsinki University Central Hospital, 00250 Helsinki, Finland

[3]The Mount Sinai School of Medicine, New York, New York 10029

[4]Department of Medicine, University of Washington School of Medicine, Seattle, WA 98195

INTRODUCTION

Interleukin-2 (IL-2), originally discovered in the supernatant of lectin or antigen-stimulated T-cell cultures[1-2], possesses a panoply of immunologic effects such as induction of paracrine growth of antigen-stimulated T-lymphocytes and of non-MHC restricted anti-tumor cytotoxic activity[3-5]. The mechanism involved in IL-2-induced signal transduction leading to these effects remains to be resolved[6]. The production of recombinant IL-2 has allowed the exploration of its biological effects as well as its therapeutic potential as an antitumor agent. Clinical Phase I and II studies of IL-2 in patients with disseminated cancer have shown that the lymphokine alone is able to induce often impressive regression of tumors typically insensitive to chemotherapy or radiation therapy[7-10]. Administration of IL-2 at tumoricidal doses, however, causes severe and limiting toxic side effects that often preclude its administration. Thrombocytopenia is universal toxicity of IL-2 therapy : it is not consequent to decreased marrow production but to peripheral sequestration of platelets[11] and it is accompanied, in vitro[12] and in vivo, by platelet degranulation (Paciucci, unpublished data) mediated by the release of thromboxan A-2 from IL-2-activated mononuclear cells in vitro[12] and, perhaps, in vivo. Systemic platelet degranulation, in turn, is responsible for high circulating levels of other factors contained in alpha granules (such as PF-4 and beta thromboglobulin). Interestingly, PF-4 has recently been shown to have in vivo antitumoral effects through inhibition of tumor neovascularization[13,14].

Growth Factors, Peptides, and Receptors, Edited by T.W. Moody, Plenum Press, New York, 1993

Also sequestered in the alpha granule of platelets, albeit in a latent proform complex, is transforming growth factor ß, a potent polypeptide regulator of cell growth and differentiation[15,16]. The mature 24 kD disulfide-linked TGF-ß homodimer inhibits the growth of epithelial, endothelial and haemopoietic cells in vitro[17-19]. In addition, TGF-ß at nanomolar concentrations exhibits antiproliferative effects on a variety of human tumor cell lines and primary cultures of tumor biopsies from solid tumor patients[20]. TGF-ß also exhibits many immunoregulatory properties including inhibition of IL-2 receptor induction[21], IL-1 induced thymocyte proliferation[22], B-cell differentiation and proliferation[21], interferon gamma-induced class II antigen expression[23] and cytotoxic and lymphokine activated killer cell generation[24]. We have previously shown that gamma interferon stimulates the activation of latent recombinant TGF-ß by human monocytes in culture[25]. As such, in situ mechanism(s) of activation will be important in determining in vivo availability of this potent cytokine.

The secondary release of other lymphokines induced by IL-2 in vitro and in vivo has already been shown[26-29]; we herein report that administration of IL-2 to solid tumor patients results in elevation of circulating levels of TGF-ß. Although it is too early to correlate the enhanced expression of TGF-ß with clinical outcome during immunotherapy with IL-2, our data suggests that administration of this cytokine can result in a release of other immunoregulatory factors. The precise definition of the events and of immunoregulatory peptides involved in this "cascade" triggered by IL-2 therapy may allow to further understand the unresolved mechanisms of tumor regression induced by IL-2 therapy.

MATERIALS AND METHODS

Clinical methods

Eleven patients, 6 males and 5 females, aged from 24 to 68 (median age 54), were included in the present study. Patients were treated at the Mount Sinai School of Medicine, New York, in years 1989-90. Three patients each had melanoma and colon cancer, 2 had renal cell carcinoma and 1 each had metastatic transitional bladder carcinoma, adeno-carcinoma of the breast and non-Hodgkin's lymphoma.

Recombinant human interleukin-2 (IL-2) (des-alanyl,serine-125 rIL-2) was generously provided by Chiron Corporation (Emeryville, CA). The lymphokine was administered by constant intravenous infusion, at 20 million IU/m^2 daily. Infusions were given for 6 days a week for 4 consecutive weeks.

At baseline and several times during therapy, heparinized blood specimens were centrifuged within 5 minutes of collection at 160 g for 20 minutes at room temperature. The buffy-coat was platelet-depleted by centrifugation at 800 g for 20 minutes and was stored in aliquots at -80°C.

Biological Assays for TGF-Beta1

TGF-ß was quantitated using growth inhibition assay (GIA) as previously described[30]. Briefly, mink lung epithelial cells (CCL64) were plated in 96-well tissue culture plates at a concentration of 3 x 10³ cells per 50 µl of DMEM containing 10% fetal bovine serum. Samples were added to the test wells in triplicate 5 hours after plating. After incubation at 37°C for 72 hours, ^{125}I-IodoUdr Amersham (IM 355, 1 µCi/ml) was added and the cells were incubated an additional 24 hours. Inhibition of growth was expressed in the percent decrease of ^{125}I-IodoUdr incorporation in treated cells when compared to the incorporation in untreated cells. A standard curve utilizing human platelet TGF-ß1 (0.05-10 ng/ml) was included in each assay.

The clonogenic assay for TGF-ß was done as previously described[31]. NRK cells (clone 49F) were seeded in soft agar in DMEM containing 10% fetal bovine serum. A 1-ml base layer of 0.5% agar was poured into 6-well plates. Samples containing growth factors were resuspended in 750 ml of medium containing 2 x 10⁴ cells/ml and 0.3% agar in the presence or absence of monoclonal antibody (1D11.16) as previously described[32]. Plates were then incubated at 37°C in 5% CO_2 for a period of 1-2 weeks. The wells were scored by counting the number of colonies formed in eight random low-power fields.

Protein Purification

Plasma samples were extracted using acidified ethanol procedure[33]. Recovered proteins were resuspended in 0.1 % trifuoroacetic acid (TFA) and fractionated by high pressure liquid column chromatography (HPLC) on a C-18 µ Bondapak (Waters and Assoc., San Francisco, CA). Following elution with a linear 0-60 % gradient of acetonitrile fractions were assayed in GIA for TGF-ß.

Statistical Analyses

For assessment of statistical significance, t-test analyses were used wherever applicable.

RESULTS

Plasma was randomly pooled from a pool derived from five solid tumor patients at baseline and 5 to 12 days after they received IL-2 infusion and assayed at the dilutions indicated in Figure 1 in a growth inhibition assay on CCL64 mink lung epithelial cells as

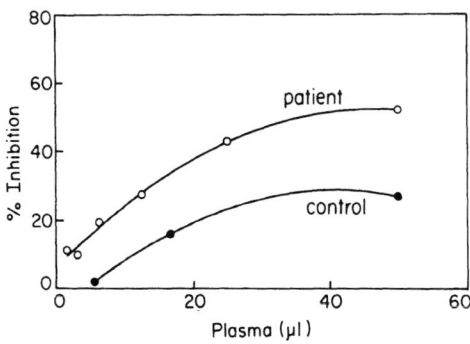

Figure 1. Dilutions of pools of plasma derived from five patients at baseline and 5-12 days after they received IL-2 infusion were assayed in growth inhibition assay on TGF-ß sensitive CCL64 mink lung epithelial cells. Plasma pooled from patients following IL-2 infusion consistently exhibited greater inhibition of DNA synthesis than they did prior to IL-2 treatment. Control (●---●) and plasma from patients after IL-2 infusion (O---O).

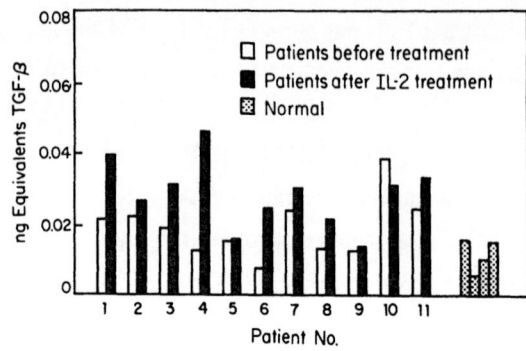

Figure 2. Plasma was tested for inhibition of DNA synthesis on CCL-64 mink lung epithelial cells as described in Materials and Methods. A dilution curve using recombinant simian TGF-ß was used as a standard to convert % inhibition in ^{125}I -IUdr in DNA synthesis into ng equivalents of TGF-ß. Samples were collected 1-13 days post IL-2 infusion and were frozen prior to assay.

described in Materials and Methods. The assay measures inhibition of DNA synthesis (ID$_{50}$ 5.0 pg TGF-ß/ml). Dilution curves of plasma pooled from patients following IL-2 infusion exhibited a 50 % maximal inhibition of DNA synthesis than they did prior to IL-2 treatment, 20 % inhibition.

Following this observation, values of TGF-ß specific activities (ng of TGF-ß per mg of plasma protein) were determined on individual solid tumor patients before and after IL-2 infusion. TGF-ß quantitation was based on triplicate determination of several plasma dilutions; values were extrapolated from a standard curve constructed by measuring inhibition of DNA synthesis in CCL 64 cells by recombinant simian TGF-ß type 1. As shown in Figure 2, TGF-ß-specific activities were elevated in nine out of 11 solid tumor patients following IL-2 infusion. A maximal four-fold increase was seen in one patient (No. 4); although a less exaggerated response was seen in the majority of plasmas tested. Mean (+/- SD) TGF-ß values were 0.0193 +/- 0.009 ng/ mg protein before treatment and 0.0290 +/- 0.01 ng/mg protein) after IL-2 therapy, respectively. This increase was statistically significant ($p < 0.01$, paired t-test). In addition, plasma from about half of the patients prior to IL-2 treatment displayed slightly higher TGF-ß-specific activities relative to plasma derived from healthy donors (mean +/- SD ; 0.0083 +/- 0.006 ng/ mg protein).

Plasma from patient with melanoma was collected at various times following IL-2 treatment. As shown in Figure 3, TGF-ß-specific activity peaked at day 12 (2.7 ng/mg plasma protein) and subsided to basal levels about 30 days following IL-2 treatment.

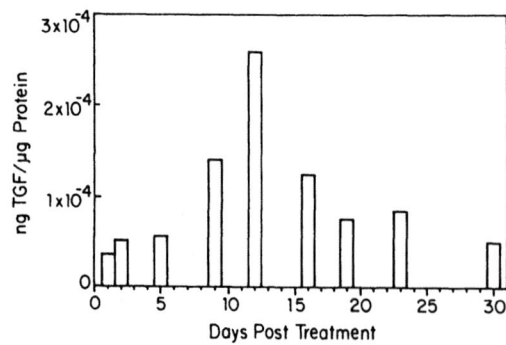

Figure 3. Plasma levels of TGF-ß as a function of time in one solid tumor patient following IL-2 infusion.

Table 1. Neutralization of NRK cell soft agar colony growth induced by plasma from patients after IL-2 infusion

	TGF-ß (ng equivalents/ml)	number of colonies per 8 random low power fields[1]	
		without antibody	with antibody
Purified from platelets	0.00	<10	<10
	0.50	109	31
Patient plasma	0.25	44	<10
	0.50	83	43

[1]Plates of NRK cells were seeded with and without 0.25µg/ml of IgG monoclonal antibody, 1D11.16 which is neutralizing for TGF-ß1, and 2.0 ng epidermal growth factor as a suppliment at the beginning of the experiment.

Experiments were designed to demonstrate that the activity measured in the mink lung growth inhibition assay was TGF-ß. As shown in Table 1, TGF-ß purified from outdated human platelets stimulated normal rat kidney cells to form progressively growing colonies (109) in soft agar in the presence of 2 ng of EGF. Soft agar plates seeded in the presence of a neutralizing monoclonal antibody to TGF-ß type 1 (1D11.16) with the same amount of platelet TGF-ß demonstrated a 28 % reduction in NRK colony growth (31). Likewise, NRK colony growth stimulated by two different doses of TGF-ß activity (0.25 and 0.50 ng/ml), 44 and 83 colonies respectively, derived from patient plasma pooled post-IL-2 infusion is also neutralized by the 1D11.16 monoclonal antibody; approximately 75% and 50%, respectively, relative to control plates not containing neutralizing antibody.

Further, similarities between the growth inhibitory molecule measured in plasma and TGF-ß family of polypeptides was established utilizing HPLC sizing columns run under conditions which favor peptide monomers (high organic soluent, 40 % acetonitrile and low pH, 1.9). As shown in Figure 4, a major peak of TGF-ß growth inhibitory activity was found on analysis of plasma pools derived from solid tumor patients following IL-2 infusion.

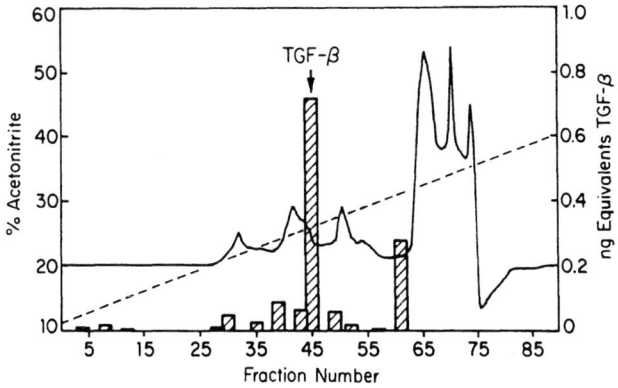

Figure 4. High pressure liquid column chromatography on a C 18 µ Bondapak (Waters and associates) of acid ethanol extracted plasma from a melanoma patient (no. 4) following IL-2 infusion. Time of elution for TGF-ß is indicated with an arrow.

The plasma activity coelutes in the position of authentic TGF-ß purified from human platelets. This elution position corresponds to 24 kD form previously reported for mature TGF-ß.

DISCUSSION

The plasma derived from solid tumor patients 1-2 weeks following IL-2 infusion contains a growth inhibitory activity with many characteristics of the transforming growth factor ß family of polypeptides. In addition to being neutralized by a monoclonal antibody to TGF-ß, this plasma derived activity stimulates anchorage independent growth of rodent fibroblasts in soft agar and coelutes from a HPLC sizing column with a mass identical to mature 24 kD form of platelet TGF-ß. In this study we have shown that the levels of this cytokine were elevated in plasma of nine out of eleven solid tumor patients following IL-2 infusion. Maximal enhancement observed was four-fold, and in one patient examined peak values of the TGF-ß values were found 12 days post IL-2 infusion. Of particular interest is the finding that plasma derived TGF-ß did not require prior acidification for activation. This is unique in that TGF-ß is sequestered in a latent form both in the platelets alpha granule and in supernatants derived from cultured normal and transformed cells.

TGF-ß is synthetized as part of a larger precursor which is then cleaved proteo-lytically to release the biologically active C-terminal domain. The remainder of the protein, latency associated protein (LAP), however, continues to be associated non-covalently with the bioactive domain, blocking its ability to bind to its cell membrane receptor, thus rendering this activity latent. Although this complex of TGF-ß and LAP are all that is required to confer latency, in many situations the LAP protein is bound covalently to a third, genetically unrelated protein called the latent TGF-ß binding protein (LTBP). LTBP may function to stabilize the latent form or confer selectivity to the activation process. TGF-ß as released from degranulating platelets and as secreted from most cells exists in this tertiary complex of TGF-ß/LAP/LTBP[34]. Indeed, in this study we found high molecular weight TGF-ß proforms (55-65 kDa) with Western blot analysis in the plasma of 5 of 5 patients after IL-2 infusion (data not shown).

It is not known what is the source of the TGF-ß activity we are measuring in plasma from these patients. It may be the product of the tumor cells themselves or target cell populations that respond to IL-2 and promote TGF-ß gene expression. Although platelet secretion does not correlate with thrombocytopenia, this mechanism has been proposed for IL-2 -induced thrombocytopenia[11] and platelet degranulation could, therefore, represent the probable source of TGF-ß in patients following IL-2 therapy. This would also correlate with high levels of PF4, which is packaged in alpha granule of platelets along with TGF-ß (Paciucci, unpublished data).

Concomitant with the release of active TGF-ß one would expect regulation of many immunological parameters. TGF-ß has been shown to inhibit IL-2 receptor induction[21] and as such may modulate subsequent IL-2 dependent phenomena. In this regard TGF-ß may antagonize infused IL-2 and thus diminish therapeutic efficacy. In addition, TGF-ß also inhibits IL-1 induced thymocyte proliferation[22], B cell differentiation[21] and interferon-gamma induced class II antigen expression[23]. Downregulation of these activities together with inhibition of cytotoxic and lymphokine activated killer cell generation suggest that IL-2 enhanced production and processing of latent TGF-ß may exaggerate immunosuppression observed in some solid tumor patients.

Direct antiproliferative effects of TGF-ß have been demonstrated on human and rodent tumor cell lines and on a variety of solid tumor biopsied cultured in presence of accessory cells (Celletrex)[20]. In addition, we have previously shown that daily peritumoral administration of TGF-ß in a nude mice model bearing TGF-ß sensitive adenocarcinoma of the lung tumor inhibited tumor growth and promoted a more differentiated phenotypes[35]. However, subsequent attempts to produce similar effects utilizing intravenous route of TGF-ß administration did not succeed. The rapid 5 minute plasma half life of TGF-ß and related delivery problem of this cytokine to appropriate tumor targets has been suggested. As such, TGF-ß plasma levels measured in this study are also most likely diminished by rapid renal clearance and are not representative of absolute TGF-ß plasma values. It is unclear whether plasma levels of TGF-ß following IL-2 infusion may have antitumor effects. It could be postulated that TGF-ß may also excaerbate tumor progression, as discussed above, by downregulating immunofunction, i.e. immunosuppression. Recent findings that metastatic variants of rodent tumors selectively secrete TGF-ß also supports this possibility. Tada et al.[36] showed recently that TGF-ß produced by tumor cells induced deleterious effects on T-cells, especially on the CD4+ Th subset and thus provided an explanation for the previously observed CD4+ Th selective suppression in the tumor bearing state. Further, Perrotti et al.[37] found that the TGF-ß1 gene is expressed in higher levels in highly metastatic variants of Lewis lung carcinoma relative to low levels in variants with less metastatic activity. Also these findings seem to suggest that TGF-ß might have deleterious effects on cancer patients receiving tumor therapy.

However, the patients who later on went on to obtain significant tumor response were those who had higher levels of TGF-ß. This correlation was surprising and indicates that the contention of TGF-ß being immunosuppressive and favor tumor growth may not be so stringent. Lee et al.[38] showed recently, that TGF-ß1 is capable of bifunctional T-cell growth regulation by inhibiting some subsets and inducing others. It is also possible that other peptides released by the platelets (e.g. PF-4, ß-thromboglobulin) may have a role in the regression of tumors.

Although clinical correlates are still lacking, because of the immuno-suppressive effects of TGF-ß and its downregulation of the IL-2 receptor, further study is necessary in patients receiving cancer immunotherapy with IL-2. Also, the production of antagonists to cytokines like TGF-ß may allow one to design adjunct therapeutic strategies to control their production during therapies such as IL-2.

In summary, in the present study we show that administration of IL-2 to cancer patients results in elevated levels of circulating TGF-ß. This suggests that in vivo administration of cytokines i.e. IL-2 may modulate the expression of other biological responsemodifiers and immunoregulators i.e. TGF-ß.

REFERENCES

1. D. A. Morgan, F. W. Ruscetti, and R. C. Gallo. Selective in vitro growth of T-lymphocytes from normal bone marrow. Science 193:1007(1976).
2. J. D. Watson, D. Y. Mochizuki , and S. Gillis. Molecular characterization of interleukin 2. Federation Proc. 42:2747(1983).
3. I. Yron, T. A. Wood, P. J. Spiess, and S. A. Rosenberg. In vitro growth of murine T cells. V. The isolation and growth of lymphoid cells infiltrating syngeneic solid tumors. J. Immunol. 125: 238(1980).
4. E. A. Grimm, A. Mazumder, H. Z. Zhang, and S.A. Rosenberg. Lymphokine-activated killer cell phenomenon. Lysis of natural killer-resistant fresh solid tumor cells by interleukin 2-activated autologous human peripheral blood lymphocytes. J. Exp. Med. 155:1823(1982).
5. W. Domzig, B. M. Stadler, and R. B. Herberman. Interleukin 2 dependence of human natural killer (NK) cell activity. J. Immunol. 130:1970(1983).
6. T. A. Waldman. The interleukin-2 receptor. Minireview. J. Biol. Chem. 266:2681(1991).
7. S. A. Rosenberg, M. T. Lotze, L. M. Muul, A. E. Chang, F. P. Avis, S. Leitman, W. M. Linehan, C. N. Robertson, R. E. Lee, J. T. Rubin, C.A. Seipp, C. G. Simpson, and D. E. White. A progress report on the treatment of 157 patients with advanced cancer using lymphokine-activated killer cells and interleukin-2 or high dose interleukin-2 alone. N. Engl. J. Med. 316:889(1987).
8. W. H. West, K. W. Tauer, J. R. Yannelli, G. D. Marshall, D. W. Orr, G. B. Thurman, and R. K. Oldham Constant-infusion recombinant interleukin-2 in adoptive immunotherapy of advanced cancer. N. Engl. J. Med. 316:898(1987).
9. J. A. Sosman, P. C. Kohler, J. Hank, K. H. Moore, R. Bechhofer, B. Storer, and P. M. Sondel. Repetitive weekly cycles of recombinant human interleukin-2: Responses of renal carcinoma with acceptable toxicity. J. Natl. Cancer Inst. 80:60(1988).
10. P. A. Paciucci, J. F. Holland, O. Glidewell, and R. Oldchimar. Recombinant interleukin-2 by continuous infusion and adoptive transfer of recombinant interleukin-2-activated cells in patients with advanced cancer. J. Clin. Oncol. 7:869(1989).
11. P. A. Paciucci, J. Mandeli, L. Oleksowicz, F. Ameglio, and J.F. Holland. Thrombocytopenia during immunotherapy with interleukin-2 by constant infusion. Am. J. Med. 89:308(1990).
12. L. Oleksowicz, P. A. Paciucci, D. Zuckerman, A. Colorito, J. H. Rand, and J. F. Holland. Alterations of platelet function induced by interleukin-2. J. Immunotherapy 10:363(1991).
13. T. E. Maione, G. S. Gray, J. Petro, A. J. Hunt, A. L. Donner, S. I. Bauer, H. F. Carson, and R. J. Sharpe. Inhibition of angiogenesis by recombinant human platelet factor-4 and related peptides. Science 247:77(1990).
14. R. J. Sharpe, H. R. Byers, C. F. Scott, S. I. Bauer, and T. E. Maione. Growth inhibition of murine melanoma and human colon carcinoma by recombinant human platelet factor 4. J. Natl. Cancer Inst. 82:848(1990).
15. P. Puolakkainen, and D. R. Twardzik. Transforming growth factors alpha and beta. In: Neurotrophic Factors. S. E. Loughlin, and J. H. Fallon , eds., Academic Press (1992) (in press).
16. A. B. Roberts, and M. B. Sporn. Transforming growth factor-ßs. In: Handbook of Experimental Pharmacology 95/I: 419, G.V. R. Born, P. Cuatrecasas, and H. Herken,eds., Springer-Verlag, Berlin, (1990).
17. A. B. Roberts, and M. B. Sporn. Transforming growth factor beta. Adv. Cancer Res. 51:107(1988).
18. R. L. Heimark, D. R. Twardzik, and S. Schwartz. Inhibition of endothelial regeneration by type-beta transforming growth factor from platelets. Science 233:1078(1986).
19. J. Keski-Oja, E. B. Leof, R. M. Lyons, R. J. Jr. Coffey, and H. L. Moses. Transforming growth factors and control of neoplastic cell growth. J. Cell. Biochem 33:95(1987).
20. H. L. Moses, R. F. Tucker, E. B. Leof, R. J. Coffey, J. Halper, and G. D. Shipley. Type-ß transforming growth factor is a growth stimulator and a growth inhibitor. Cancer Cells 3:65(1985).
21. J. H. Kehrl, L. M. Wakefield, A. B. Roberts, S. Jakowlew, M. Alvarez-Mon, R. Derynck, M. B. Sporn, and A. S. Fauci. Production of transforming growth factor ß by human T lymphocytes and its potential role in the regulation of T cell growth. J. Exp. Med. 163:1037(1986).
22. S. M. Wahl, D. A. Hunt, H. L. Wong, S. Dougherty, J. McCartney-Francis, L. M. Wahl, L. Ellingsworth, J. A. Schmidt, G. Hall, A. B. Roberts, and M. B. Sporn. Transforming growth factor-ß is a potent immunosuppressive agent that inhibits IL-1-dependent lymphocyte proliferation. J. Immunol. 140:3026(1988).
23. C. W. Czarniecki, H. H. Chiu, G. H. W. Wong, S. M. McCabe, and M. A. Palladino. Transforming growth factor-ß1 modulates the expression of class II histocompatibility antigens on human cells. J. Immunol. 140:4217(1988).
24. J. J. Mule, S. L. Schwartz, A. B. Roberts, M. B. Sporn, and S. A. Rosenberg. Transforming growth factor-beta inhibits the in vitro generation of lymphokine-activated killer cells and cytotoxic T
25. D. R. Twardzik, J. A. Mikovits, J. E. Ranchalis, A. F. Purchio, L. Ellingsworth, and F. W. Ruscetti. Gamma-interferon-induced activation of latent transforming growth factor-ß by human monocytes. Ann. N. Y. Acad. Sci. 593:276(1990).
26. P. Allavena, G. Scala, J. Y. Djeu, A. D. Procopio, J. J. Oppenheim, R. B. Herberman, and J. R. Ortaldo. Production of multiple cytokines by clones of human large granular lymphocytes. Cancer Immunol. Immunother. 19:121(1985).
27. B. T. Gemlo, M. A. Palladino, H. S. Jaffe, T. P. Espevik, and A. A. Rayner. Circulating cytokines in patients with metastatic cancer treated with recombinant interleukin 2 and lymphokine-activated killer cells. Cancer Res. 48:5864(1988).

28. J. W. Mier, G. Vachino, J. W. Van Deer Meer, R. P. Numerof, S. Adams, J. G. Cannon, H. A. Bernheim, M. B. Atkins, D. R. Parkinson, and C.A. Dinarello. Induction of circulating tumor necrosis factor (TNF alfa) as the mechanism for the febrile response to interleukin-2 (IL-2) in cancer patients. J. Clin. Immunol. 8:426(1988).

29. E. Lopez Hänninen, A. Körfer, M. Hadam, C. Schneekloth, I. Dallman, T. Menzel, H. Kirchner, H. Poliwoda, and J. Atzpodien. Biological monitoring of low-dose Interleukin 2 in humans: soluble interleukin 2 receptors, cytokines and cell surface phenotypes. Cancer Res. 50: 6312(1991).

30. J.E. Ranchalis, J. McPherson, Y. Ogawa, S. M. Seyedin, and D. R. Twardzik. Both forms of cartilage inducing factor inhibit the growth of tumor cells in a similar manner. Biochem. Biophys. Res. Comm. 148:783(1987).

31. D. R. Twardzik, S. A. Sherwin, J. E. Ranchalis, and G. J. Todaro. Transforming growth factors in the urine of normal, pregnant and tumor-bearing humans. J. Natl. Cancer Inst. 69:793(1982).

32. J. R. Dasch, D. R. Pace, W. Waegell, D. Inenaga, and L. Ellingsworth. Monoclonal antibodies recognizing transforming growth factor-ß. J. Immunol. 142:1536(1989).

33. D. R. Twardzik, J. E. Ranchalis, and G.J. Todaro. Mouse embryos contain transforming growth factors related to those isolated from tumor cells. Cancer Res. 42:590(1982).

34. A. B. Roberts, and L. M. Wakefield. Latent TGF-ß: Growth factor in disguise. Scars and Stripes, 2:4 (1992).

35. D. R. Twardzik, J. E. Ranchalis, and P. A. Puolakkainen. Transforming growth factor-ß induces tumor differentiation. In: The status of differentiation therapy of cancer. Vol. II. pp. 133, S. Waxman, G.B. Rossi, and F. Takaku, eds., Raven Press, New York, (1991).

36. T. Tada, S. Ohzeki, K. Utsumi, H. Takiuchi, M. Muramatsu, X.F. Li, J. Shimizu, H. Fujiwara, and T. Hamaoka. Transforming growth factor beta-induced inhibition of T cell function. Susceptibility difference in T cells of various phenotypes and functions and its relevance to immunosuppression in the tumor-bearing state. J. Immunol. 146:1077(1991).

37. D. Perrotti, L. Cimino, S. Ferrari, and A. Sacchi. Differential expression of transforming growth factor ß1 gene in 3LL metastatic variants. Cancer Res. 51:5491(1991).

38. H. M. Lee, and S. Rich. Co-stimulation of T cell proliferation by transforming growth factor beta 1. J. Immunol. 147:1127(1991).

PEPTIDE SIGNALS FOR SATIETY

James Gibbs, Nori Geary, and Gerard P. Smith

Department of Psychiatry, Cornell University Medical College
E.W. Bourne Behavioral Research Laboratory
New York Hospital-Cornell Medical Center
21 Bloomingdale Road
White Plains, NY 10605

INTRODUCTION

Over the past 20 years, basic research aimed at determining the fundamental physiological mechanisms that produce meal-ending satiety has made significant advances. A major element in this progress has been the demonstration of the key role of a small group of gut peptides, including cholecystokinin, glucagon, and bombesin-like peptides. These peptides, which are released peripherally by the ingestion of food, appear to provide an informational link between the gut and the brain. Studies from several laboratories have identified common themes uniting the effects of these preabsorptive signals: (1) When administered systemically they produce potent, dose-related, and behaviorally specific inhibitions of short-term food intake in a wide variety of animal species and in humans. Thus, they appear to elicit or accelerate meal-ending satiety. (2) Their sites of action are peripheral, not central. (3) Afferent nerves relay these initial effects centrally, with the abdominal vagus playing a major role. (4) This information is integrated with other visceral and gustatory information in the dorsal hindbrain. (5) Gastrointestinal peptides interact with each other and with traditional neurotransmitter systems in producing satiety. Finally, (6) the development of highly specific receptor antagonists and antisera directed against these peptides has allowed tests of their physiological relevance, providing demonstrations that endogenous cholecystokinin and glucagon play important roles in controlling food intake.

In this chapter, we review the evidence supporting each of these themes. In addition, we indicate directions for new research to determine the peripheral and central mechanisms by which peptide signals limit meal size and regulate the length of the intermeal interval.

CHOLECYSTOKININ

This classic small intestinal hormone has received the most intensive study, and therefore serves as a model for the brain/gut integration of peptide satiety signals. We demonstrated in a series of studies beginning in the early 1970's that the peripheral administration of cholecystokinin (CCK) produced a specific, potent, dose-related suppression of meal size in rats that was behaviorally and chemically specific[1-3]. Based

Growth Factors, Peptides, and Receptors, Edited
by T.W. Moody, Plenum Press, New York, 1993

on this evidence, we proposed that CCK that is released when ingested food contacts the mucosa of the upper small intestine acts rapidly to inhibit further food intake[1].

An array of evidence has accumulated to support this hypothesis. It is now known that CCK reduces food intake in many animal species under many conditions of food deprivation, diet type, circadian time, etc.[4,5]. Numerous results indicate that this action is not due to non-specific side effects[3]. Finally, several human studies support the conclusions from animal studies. CCK reduced meal size in double-blind, placebo-controlled tests employing slow intravenous infusions in normal-weight and obese humans[6-12]. These reductions were accompanied by discernible side effects in less than 15% of subjects, the side effects which did occur were not correlated with changes in food intake, and volunteers in these tests could not identify days on which they received the peptide.

Thus, extensive evidence points to the conclusion that peripherally-administered CCK can mimic natural satiety at meals. On the other hand, whether this effect of exogenous CCK mimics a physiological function of endogenous CCK has been a continuing source of controversy from the outset. Very recently, this controversy has been settled through the use of the latest generation of highly selective and potent CCK receptor antagonists. Peripheral delivery of devazepide and similar CCK antagonists has been shown to increase test meal size and food intake over periods of hours to days in rats[13-17], mice[18], hamsters[19], pigs[20], and rhesus monkeys[21]; the effect has now been shown in over 25 animal studies. In humans, one study has shown that the oral administration of a CCK receptor antagonist increases reports of subjective hunger[22]. We conclude that the action of endogenous CCK is required for the normal control of food intake. Additional work will be needed to delineate the physiological, nutritional, and environmental contexts in which endogenous CCK plays a role in normal satiety.

In contrast to this extensive evidence for the participation of CCK in satiety, the mechanism of action of CCK is poorly understood. The known facts can be listed briefly: (1) The type-A CCK receptor, which is the more abundant form in gastrointestinal and other peripheral tissue sites, is critical[23,24]. The type-B receptor may be involved under some conditions[25], although this possibility remains controversial. (2) Surgical ablation of the gastric pylorus (an area dense with type-A CCK receptors[26]), attenuates the satiety effect of high doses of peripherally-administered CCK[27]. (3) Capsaicin-sensitive afferent neurons of the abdominal vagus are required for CCK's satiety action[28-30]. (4) Lesion of the hindbrain region which receives these afferents (caudal nucleus tractus solitarius and overlying area postrema) also attenuates the action[31]. (5) Initial integratory processing of the CCK signal also appears to occur in the dorsal hindbrain. For example, in anesthetized rats, intravenous CCK administration potentiates the stimulatory response of neurons in the nucleus tractus solitarius to physical distension of the stomach[32]. These observations strongly suggest that the initial site of action of peripherally-administered CCK is peripheral, rather than central, at least partly gastric, and that afferent vagal neurons relay the satiety signal to receptive areas of the hindbrain, where elaboration of the signal begins.

Central projection areas of dorsal hindbrain are candidates for further processing of peripheral CCK-generated signals, but lesions of known projection areas such as paraventricular nucleus[28,33] and pathways[34,35] have yielded conflicting results. Another layer of complexity is provided by clear evidence that direct central administration of CCK inhibits feeding in a variety of animal species at doses that are too small to inhibit feeding when given peripherally[36-40]. How these central effects relate to the peripheral action of CCK is unknown, because peripherally-administered CCK does not penetrate the blood-brain barrier[41]. CCK and CCK receptors (predominantly but not exclusively type-B) are abundant in brain, and much further work will be required to establish the role of this system in satiety.

Finally, the possibility that CCK or CCK-generated satiety signals may interact with other systems believed to play a role in the control of food intake has been explored. Extensive pharmacological evidence has implicated the indoleamine central neurotransmitter serotonin (5-HT) in satiety processes[42-44]. We have shown that the satiating effect of peripherally-administered CCK was attenuated by metergoline, a relatively non-specific $5-HT_1/5-HT_2$ receptor antagonist[45]. Extending these results, the more selective $5-HT_{1C}/5-HT_2$ receptor antagonist mianserin blocked the action of systemic CCK, but $5-HT_2$ or $5-HT_3$ antagonists did not[46]. The results indicate that $5-HT_{1C}$ receptors are required for CCK-induced satiety; current work is aimed at determining whether the site(s) of interaction of the CCK and 5-HT systems are peripheral or central.

PANCREATIC GLUCAGON

Similar, but less extensive, research supports the idea that pancreatic glucagon, like CCK, is a gut peptide satiety signal contributing to the control of meal size. This is a physiologically plausible hypothesis because plasma glucagon levels often increase during meals in rats[47,48], humans[49-51], and other species. This increase occurs within minutes of meal onset in rats, suggesting that it is due to a preabsorptive action of food, perhaps a cephalic phase reflex[47]. Prandial increases in glucagon levels are accompanied by transient increases in hepatic glycogenolysis and glucose production[52,53], which precede the well-known postprandial decrease in hepatic glycogenolysis due to the absorption of ingested nutrients.

Glucagon administration just before meals by intraperitoneal injection or hepatic portal vein infusion has been reported to inhibit feeding in rats[54-57], dogs[58,59], sheep[60], and other animal species. Exogenous glucagon's inhibitory effect, like CCK's, is rapid and transient, suggesting that its primary action is on meal size. The effect, at least with hepatic portal administration, is also clearly dose-related. Again, similar to CCK and bombesin-like peptides, results from a wide variety of screens of behavioral specificity indicate that glucagon decreases meal size in animals by accelerating normal processes of postprandial satiety[50,61]. These results have been recently extended by the demonstration that intravenous infusion of glucagon during meals reduces meal size in men in the absence of any physical or subjective side effects[9] -- the first test of intravenous glucagon's effects on human food intake. This report is also notable because a low, possibly physiological dose of glucagon was sufficient to inhibit feeding.

Glucagon may interact with CCK to inhibit feeding under some conditions. Although exogenous glucagon failed to inhibit feeding in rats tested with open gastric cannulas that drained ingested liquid food[62], simultaneous injection of glucagon and CCK produces much larger inhibitions that did CCK alone[63,64]. The generality of this interaction has not been established; it has been reported in real feeding rats[64,65], but not in dogs[59] or human volunteers[9].

Unlike CCK and bombesin-like peptides, exogenous glucagon's satiety effect in rats displays a remarkable degree of situational dependence. For example, intraperitoneal glucagon injections inhibit feeding in rats refed at dark onset after 6-18 hr food deprivation, but fail to inhibit feeding at this time after 1-4 hr food deprivation[66-68]. At other times of the day, glucagon does inhibit feeding after brief deprivation[61]. This unusual situational dependence has prompted tests of glucagon's role in the control of spontaneous meals, which are more ecologically and physiologically normal than scheduled test meals. These tests indicate that brief, remotely-controlled hepatic portal infusions of glucagon reduce rats' spontaneous meal sizes both early and late in the dark phase, when rats normally eat most[69]. Further, endogenous glucagon appears to participate in the control of spontaneous meals: this is because similar infusions of specific glucagon antibodies during spontaneous meals dramatically increases meal size[70]. Glucagon antagonism has also increased the size of scheduled test meals[71,72]. These data convincingly indicate peripheral glucagon's relevance in meal-ending satiety.

Many questions remain concerning the mechanism of glucagon's effects on meals size. (1) A comparison of hepatic portal vein and vena cava glucagon administration indicates that glucagon acts in the liver to reduce spontaneous meal size[73]. But whether the signal arises from a metabolic effect such as stimulation of glucose production, a direct action on peripheral nerves, or some other action is unknown. (2) Hepatic vagal fibers, apparently capsaicin-sensitive afferents, appear necessary for glucagon's effects on spontaneous feeding[73], as well as on scheduled test meals under some[68,74,75] but not all[68,76] conditions. The conditions causing recruitment of non-hepatic vagal signals, as well as the identity of these signals, are not understood. Like glucagon's situational dependency, this complexity may provide useful clues in the analysis of the mechanism of glucagon satiety. (3) Lesions of vagal afferent terminals in the dorsal hindbrain attenuate glucagon's inhibitory effect on feeding[77]. Interestingly, the terminal fields involved in glucagon's and CCK's effects do not appear identical[77]. As is the case for CCK, however, initial processing of the signal appears to occur in the nucleus tractus solitarius; for example, multiunit activity of taste-responsive neurons in anesthetized rats is decreased by intravenous glucagon administration[78]. (4) Very little is yet known about the involvement of neural mechanisms rostral to the dorsal hindbrain in glucagon's feeding effects. The ventromedial hypothalamus may be involved in the control of prandial glucagon secretion[47], but electrolytic lesions of the ventromedial hypothalamus did not decrease the inhibitory effect of exogenous glucagon on food intake[79]. There are scattered reports that glucagon administration directly into the cerebral ventricles affects feeding[80,81], but nothing is yet known about the behavioral specificity and physiological relevance of these phenomena, or their relation to the effect of peripheral glucagon.

BOMBESIN-LIKE PEPTIDES

The satiety actions of amphibian bombesin and mammalian bombesin-like peptides show remarkable similarities and some interesting differences from those of CCK and pancreatic glucagon. Our initial demonstration that peripherally-administered bombesin reduced the size of test meals in rats[82] has been confirmed and extended in a number of mammalian and avian species[83-87]. Under double-blind conditions in human volunteers, a slow intravenous infusion of bombesin inhibited food intake at a test meal without producing significant side effects[88]. Gastrin-releasing peptide and the carboxy-terminal decapeptide of gastrin-releasing peptide, which are structurally related to amphibian bombesin and are found in mammalian gastrointestinal tract and brain, also inhibit feeding behavior in rodents and subhuman primates[89-92], but have not been tested in humans. Neuromedin B, another peptide with structural similarities to bombesin, is less effective[93].

As is the case for CCK and pancreatic glucagon, peripheral nerve lesions block the effect of peripherally-administered bombesin on food intake. For bombesin, however, a more extensive nerve lesion is required: Whereas complete abdominal vagotomy failed to affect bombesin-induced satiety, total neural disconnection of gut from brain (produced by the combination of abdominal vagotomy and high thoracic spinal section) completely blocked the action[94]. This blockade strongly indicates, once again, that the initial site of action is peripheral, not central. On the basis of local perfusion studies and the known distribution of bombesin-like peptides in the gut, we have suggested that this initial site is in the upper abdomen, perhaps in the stomach[95,96], but this conclusion is disputed[97].

In addition to its widely confirmed effect in reducing meal size, we have shown that exogenous bombesin has a second satiety function, producing a dose-related extension of the intermeal interval[74,98]. The mechanism of action for this function is different from that mediating the reduction in meal size, because it is not affected by the combined abdominal vagotomy -- spinal visceral disconnection that blocks the meal size effect[94].

In a further parallel to the results with CCK and glucagon, electrolytic lesions of the dorsal hindbrain attenuate the satiety effect of systemically delivered bombesin[99]. Finally, direct hindbrain injections of extremely low doses of bombesin into the fourth cerebral ventricle[100,101] or into the underlying nucleus tractus solitarius[102,103] produce a behaviorally specific inhibition of food intake. Again, the relationship of these effects to the peripheral action of peripherally-administered bombesin is unclear.

Whether endogenous bombesin-like peptides play a role in the physiological control of meal size or intermeal interval is unanswered. Preliminary reports indicate that central[104] or peripheral[105] administration of specific bombesin receptor antagonists increases food intake in rats, but much further work is needed to establish a secure position for this peptide family in the normal regulation of food intake.

CONCLUSION

We have reviewed the status of three candidate preabsorptive peptide signals contributing to the control of meal size. Experimental analyses of the hypothesized satiety actions of these peptides have followed parallel tracks and revealed several commonalities in their actions and mechanisms. Thus, CCK, glucagon, and bombesin-like peptides are localized in the gut in animals and humans and appear to be released by peripheral food stimuli that occur during meals. Exogenous administration of each peptide powerfully and specifically inhibits feeding in animals, and double-blind studies of human volunteers have demonstrated that each of them can produce a similarly specific decrease in human food intake. The development of potent and selective antagonists has allowed tests of the satiety roles of the endogenous peptides; results to date range from compelling proof in the case of CCK to promising initial results for bombesin-like peptides. Each of the peptides' initial actions appears to be peripheral rather than central. Multiple sites appear to be involved, including the liver for glucagon and the stomach and perhaps other sites for CCK and bombesin-like peptides. Finally, the signals arising at these peripheral sites are relayed to the brain over afferent nerves, principally the vagus, and their integration and processing begins in the dorsal hindbrain.

This progress places these peptides in a unique position among presently understood controls of appetite, poising them both for continued experimental analysis using neuroscience's increasingly sophisticated armamentaria of molecular, physiological, and behavioral techniques as well as for basic and clinical studies to assess their roles in normal and disordered human appetite and their therapeutic potentials.

ACKNOWLEDGEMENTS

We thank Mrs. Jane Magnetti for processing this manuscript. Our work is supported by DK33248 (JG), DK32448 (NG), MH40010 and MH00149 (GPS).

REFERENCES

1.   J. Gibbs, R. C. Young and G. P. Smith. Cholecystokinin decreases food intake in rats. *J. Comp. Physiol. Psychol.* 84:488-495(1973).
2.   J. Gibbs, R. C. Young and G. P. Smith. Cholecystokinin elicits satiety in rats with open gastric fistulas. *Nature* 245:323-325(1973).
3.   G. P. Smith, J. Gibbs, and P. J. Kulkosky, Relationships between brain-gut peptides and neurons in the control of food intake. In "The Neural Basis of Feeding and Reward," B.G. Hoebel, D. Novin (eds) Haer Institute, Brunwsick, Maine (1982).

4. G. P. Smith, J. Gibbs, C. Jerome, F. X. Pi-Sunyer, H. R. Kissileff and J. Thornton. The satiety effect of cholecystokinin: a progress report. *Peptides* 2:57-59(1981).

5. G. P. Smith and J. Gibbs, The development and proof of the cholecystokinin hypothesis of satiety. In "Multiple Cholecystokinin Receptors in the CNS," C.T. Dourish, S.J. Cooper, S.D. Iversen, L.L. Iversen (eds) Oxford University Press, Oxford (1991).

6. G. P. Smith and J. Gibbs. The effect of gut peptides on hunger, satiety, and food intake in humans. *Ann. N. Y. Acad. Sci.* 499:132-136(1987).

7. H. R. Kissileff, F. X. Pi-Sunyer, J. Thornton and G. P. Smith. Cholecystokinin-octapeptide (CCK-8) decreases food intake in man. *Amer. J. Clin. Nutr.* 34:154-160(1981).

8. X. Pi-Sunyer, H. R. Kissileff, J. Thornton and G. P. Smith. C-terminal octapeptide of cholecystokinin decreases food intake in obese men. *Physiol. Behav.* 29:627-630(1982).

9. N. Geary, H. R. Kissileff, F. X. Pi-Sunyer and V. Hinton. Individual, but not simultaneous, glucagon and cholecystokinin infusions inhibit feeding in men. *Am. J. Physiol.* 262:R975-R980(1992).

10. N. E. Muurahainen, H. R. Kissileff, J. Lachaussee and F. X. Pi-Sunyer. Effect of a soup preload on reduction of food intake by cholecystokinin in humans. *Am. J. Physiol.* 260:R672-R680(1991).

11. G. Stacher. The effects of cholecystokinin and caerulein on human eating behavior and pain sensation. *Psychoneuroendocrinology* 11:39-48(1986).

12. G. Stacher, H. Bauer and H. Steinringer. Cholecystokinin decreases appetite and activation evoked by stimuli arising from the preparation of a meal in man. *Physiol. Behav.* 23:325-331(1979).

13. C. A. Watson, L. H. Schneider, E. S. Corp, S. C. Weatherford, R. Shindledecker, R. B. Murphy, G. P. Smith and J. Gibbs. The effects of chronic and acute treatment with the potent peripheral cholecystokinin antagonist L-364,718 on food and water intake in the rat. *Soc. Neurosci. Abstr.* 14:1196(1988).

14. R. D. Reidelberger and M. F. O'Rourke. Potent cholecystokinin antagonist L-364,718 stimulates food intake in rats. *Am. J. Physiol.* 257:R1512-R1518(1989).

15. J. Garlicki, P. K. Konturek, J. Majka, N. Kwiecien and S. J. Konturek. Cholecystokinin receptors and vagal nerves in control of food intake in rats. *Am. J. Physiol.* 258:E40-E45(1990).

16. A. Weller, G. P. Smith and J. Gibbs. Endogenous cholecystokinin reduces feeding in young rats. *Science* 247:1589-1591(1990).

17. T. H. Moran, P. J. Ameglio, G. J. Schwartz and P. R. McHugh. Blockade of type A, not type B, CCK receptors attenuates satiety actions of exogenous and endogenous CCK. *Am. J. Physiol.* 262:R46-R50(1992).

18. A. J. Silver, J. F. Flood, A. M. Song and J. E. Morley. Evidence for a physiological role for CCK in the regulation of food intake in mice. *Am. J. Physiol.* 256:R646-R652(1989).

19. T. E. Adrian, A. J. Bilchik, K. A. Zucker and I. M. Modlin. CCK receptor blockade increases hamster body weight and food intake. *FASEB J.* 2:A737(1988).

20. I. S. Ebenezer, C. de la Riva and B. A. Baldwin. Effects of the CCK receptor antagonist MK-329 on food intake in pigs. *Physiol. Behav.* 47:145-148(1990).

21. T. H. Moran, P. J. Ameglio, G. J. Schwartz and P. R. McHugh. Blockade of the actions of endogenous cholecystokinin (CCK) by intragastric administration of the type A CCK receptor antagonist MK-329 significantly increases food intake in rhesus monkeys. *Soc. Neurosci. Abstr.* 17:542(1991).

22. O. M. Wolkowitz, B. Gertz, H. Weingartner, L. Beccaria, K. Thompson and R. A. Liddle. Hunger in humans induced by MK-329, a specific peripheral-type cholecystokinin receptor antagonist. *Biol. Psychiat.* 28:169-173(1990).

23. R. L. Corwin, J. Gibbs and G. P. Smith. Increased food intake after type A but not type B cholecystokinin receptor blockade. *Physiol. Behav.* 50:255-258(1991).

24. J. Miesner, G. P. Smith, J. Gibbs and A. Tyrka. Intravenous infusion of CCKA-receptor antagonist increases food intake in rats. *Am. J. Physiol.* 262:R216-R219(1992).

25. C. T. Dourish, W. Rycroft and S. D. Iversen. Postponement of satiety by blockade of brain cholecystokinin (CCK-8) receptors. *Science* 245:1509-1511(1989).

26. T. H. Moran, P. H. Robinson and P. R. McHugh. Pyloric CCK receptors: site of mediation for satiety? *Ann. N. Y. Acad. Sci.* 448:621-623(1985).

27. T. H. Moran and P. R. McHugh. Gastric and nongastric mechanisms for satiety action of cholecystokinin. *Am. J. Physiol.* 254:R628-R632(1988).

28. G. P. Smith, C. Jerome, B. J. Cushin, R. Eterno and K. J. Simansky. Abdominal vagotomy blocks the satiety effect of cholecystokinin in the rat. *Science* 213:1036-1037(1981).

29. G. P. Smith, C. Jerome and R. Norgren. Afferent axons in abdominal vagus mediate satiety effect of cholecystokinin in rats. *Am. J. Physiol.* 249:R638-R641(1985).

30. E. H. South and R. C. Ritter. Capsaicin application to central or peripheral vagal fibers attenuates CCK satiety. *Peptides* 9:601-612(1988).

31. G. L. Edwards, E. E. Ladenheim and R. C. Ritter. Dorsal hindbrain participation in cholecystokinin-induced satiety. *Am. J. Physiol.* 251:R971-R977(1986).

32. G. J. Schwartz, P. R. McHugh and T. H. Moran. Integration of vagal afferent response to gastric loads and cholecystokinin in rats. *Am. J. Physiol.* 261:R64-R69(1991).

33. J. N. Crawley and J. Z. Kiss. Paraventricular nucleus lesions abolish the inhibition of feeding induced by systemic cholecystokinin. *Peptides* 6:927-935(1985).

34. J. Grill and G. P. Smith. Cholecystokinin decreases sucrose intake in chronic decerebrate rats. *Am. J. Physiol.* 254:R853-R856(1988).

35. J. N. Crawley, J. Z. Kiss and E. Mezey. Bilateral midbrain transections block the behavioral effects of cholecystokinin on feeding and exploration in rats. *Brain Res.* 322:316-321(1984).

36. S. Maddison. Intraperitoneal and intracranial cholecystokinin depress operant responding for food. *Physiol. Behav.* 19:819-824(1977).

37. M. Della-Fera and C. A. Baile. Cholecystokinin octapeptide: continuous picomole injections into the cerebral ventricles of sheep suppress feeding. *Science* 206:471-473(1979).

38. R. R. Schick, T. L. Yaksh and V. L. W. Go. Intracerebroventricular injections of cholecystokinin octapeptide suppress feeding in rats -- pharmacological characterization of this action. *Regul. Peptides* 14:277-291(1986).

39. D. -M. Zhang, W. Bula and E. Stellar. Brain cholecystokinin as a satiety peptide. *Physiol. Behav.* 36:1183-1186(1986).

40. D. P. Figlewicz, A. J. Sipols, P. Green, D. Porte,Jr. and S. C. Woods. IVT CCK-8 is more effective than IV CCK-8 at decreasing meal size in the baboon. *Brain Res. Bull.* 22:849-852(1989).

41. W. H. Oldendorf. Blood-brain barrier permeability to peptides: pitfalls in measurement. *Peptides* 2 (Suppl.2):109-111(1981).

42. J. E. Blundell. Serotonin manipulations and the structure of feeding behavior. *Appetite* 7:39-56(1986).

43. E. Goodall and T. Silverstone, Pharmacological evidence for the involvement of serotonergic mechanisms in human feeding. In "Behavioral Pharmacology of 5-HT," T. Archer, P. Bevan, A.R. Cools (eds) Erlbaum, Hillside, NJ (1989).

44. R. Samanin, Serotonin and feeding. In Pharmacology of 5-HT, T. Archer, P. Bevan, A.R. Cools (eds) Erlbaum, Hillsdale,NJ (1989).

45. D. Stallone, S. Nicolaidis and J. Gibbs. Cholecystokinin-induced anorexia depends on serotoninergic function. *Am. J. Physiol.* 256:R1138-R1141(1989).

46. B. D. Poeschla, J. Gibbs, K. J. Simansky, D. Greenberg and G. P. Smith. Cholecystokinin-induced satiety depends upon activation of 5-HT1C receptors. *Am. J. Physiol.* In press (1992).

47. A. DeJong, J. H. Strubbe and A. B. Steffens. Hypothalamic influecne on insulin and glucagon release in the rat. *Am. J. Physiol.* 233:E380-E388(1977).

48. W. Langhans, K. Pantel, W. Muller-Schell, E. Eggengerger and E. Scharrer. Hepatic handling of pancreatic glucagon and glucose during meals in rats. *Am. J. Physiol.* 247:R827-R832(1984).

49. J. Day, L. K. Johansen, O. P. Ganda, J. S. Soeldner, R. E. Gleason and W. Medgley. Factors governing insulin and glucagon response during normal meals. *Clin. Endocrinol.* 9:443-454(1978).

50. J. J. Holst, Glucagon in obesity, In "Handbook of experimental pharmacology," P.J. Lefebvre (ed) Springer-Verlag, Berlin (1983).

51. W. A. Muller, G. R. Faloona and R. H. Unger. The influence of the antecedent diet upon glucagon and insulin secretion. *N. Engl. J. Med.* 285:1450-1454(1971).

52. H. Denker, P. Hedner, J. Holst and K. G. Tranberg. Pancreatic glucagon response to an ordinary meal. *Scand. J. Gastroenterol.* 10:471-474(1975).

53. W. Langhans, N. Geary and N. Scharrer. Liver glycogen content decreases during meals in rats. *Am. J. Physiol.* 243:R450-R453(1982).

54. J. R. Martin and D. Novin. Decreased feeding in rats following hepatic-portal infusion of glucagon. *Physiol. Behav.* 19:461-466(1977).

55. N. Geary and G. P. Smith. Pancreatic glucagon and postprandial satiety in the rat. *Physiol. Behav.* 28:313-322(1982).

56. N. Geary, W. Langhans and E. Scharrer. Metabolic concomitants of glucagon-induced suppression of feeding in the rat. *Am. J. Physiol.* 241:R330-R335(1981).

57. B. G. Weick and S. Ritter. Dose-related suppression of feeding by intraportal glucagon infusion in the rat. *Am. J. Physiol.* 250:R676-R681(1986).

58. S. Levine, C. E. Siever, B. A. Morley, B. A. Gosnell and S. E. Silvis. Peptidergic regulation of feeding in the dog (Canid familiaries). *Peptides* 5:675-679(1984).

59. T. J. Kalogeris, R. D. Reidelberger, V. E. Mendel and T. E. Solomon. Interaction of cholecystokinin-8 and pancreatic glucagon in control of food intake in dogs. *Am. J. Physiol.* 260:R688-R692(1991).

60. L. E. Deetz and P. J. Wangsness. Influence of intrajugular administration of insulin, glucagon and propionate on voluntary feed intake of sheep. *J. Anim. Sci.* 53:427-433(1981).

61. N. Geary. Pancreatic glucagon signals postprandial satiety. *Neurosci. Biobehav. Rev.* 14:323-338(1990).

62. N. Geary and G. P. Smith. Pancreatic glucagon fails to inhibit sham feeding in the rat. *Peptides* 3:163-166(1982).

63. J. LeSauter and N. Geary. Pancreatic glucagon and cholecystokinin synergistically inhibit sham feeding in rats. *Am. J. Physiol.* 253:R719-R725(1987).

64. J. Le Sauter and N. Geary. Redundant vagal mediation of the synergistic satiety effect of pancreatic glucagon and cholecystokinin in sham feeding rats. *J. Auton. Nerv. Syst.* 30:13-22(1990).

65. V. Hinton, M. Rosofsky, J. Granger and N. Geary. Combined injection potentiates the satiety effects of pancreatic glucagon, cholecystokinin, and bombesin. *Brain Res. Bull.* 17:615-619(1986).

66. N. Geary, N. Farhoody and A. Gersony. Food deprivation dissociates pancreatic glucagon's effects on satiety and hepatic glucose production at dark onset. *Physiol. Behav.* 39:507-511(1987).

67. D. A. VanderWeele, B. L. Macrum and R. L. Oetting. Glucagon, satiety from feeding and liver/pancreatic interactions. *Brain Res. Bull.* 72:539-543(1986).

68. S. C. Weatherford and S. Ritter. Glucagon satiety: diurnal variation after hepatic branch vagotomy or intraportal alloxan. *Brain Res. Bull.* 17:545-549(1986).

69. J. LeSauter and N. Geary. Hepatic portal glucagon infusion decreases spontaneous meal size in rats. *Am. J. Physiol.* 261:R154-R161(1991).

70. J. LeSauter, U. Noh and N. Geary. Hepatic portal infusion of glucagon antibodies increases spontaneous meal size in rats. *Am. J. Physiol.* 261:R162-R165(1991).

71. W. Langhans, U. Zieger, E. Scharrer and N. Geary. Stimulation of feeding in rats by intraperitoneal injection of antibodies to glucagon. *Science* 218:894-896(1982).

72. C. L. McLaughlin, R. L. Gingerich and C. A. Baile. Role of glucagon in the control of food intake in Zucker obese and lean rats. *Brain Res. Bull.* 17:419-426(1986).

73. N. Geary, J. Le Sauter and N. Noh. Glucagon acts in the liver to control spontaneous meal size in rats. *Am. J. Physiol.* (1992). (In Press)

74. S. Mindell, J. A. DiPoala, S. Wiener, J. Gibbs and G. P. Smith. Bombesin increases postprandial intermeal interval. *Soc. Neurosci. Abstr.* 11:38(1985).

75. L. MacIssac and N. Geary. Partial liver denervations dissociate the inhibitory effects of pancreatic glucagon and epinephrine on feeding. *Physiol. Behav.* 35:233-237(1985).

76. L. L. Bellinger and F. E. Williams. Glucagon and epinephrine suppression of food intake in liver denervated rats. *Am. J. Physiol.* 251:R349-R358(1986).

77. S. C. Weatherford and S. Ritter. Lesion of vagal afferent terminals impairs glucagon-induced suppression of food intake. *Physiol. Behav.* 43:645-650(1988).

78. B. L. Giza, T.R. Scott and D.A. VanderWeele. Administration of satiety factors and gustatory responsiveness in the nucleus tractus solitarius of the rat. *Brain Res. Bull.* 28:637-639(1992).

79. N. Geary, G. P. Smith and J. Gibbs. Pancreatic glucagon and bombesin inhibit meal size in ventromedial hypothalamus-lesioned rats. *Regul. Peptides* 15:261-268(1986).

80. A. Inokuchi, Y. Oomura and H. Nishimura. Effect of intracerebroventricularly infused glucagon on feeding behavior. *Physiol. Behav.* 33:397-400(1984).

81. K. Nagai, L. Thibault, K. Nishikawa, A. Hashida, K. Ootani and H. Nakagawa. Effect of glucagon in macronutrient self-selection: Glucagon-enhanced protein intake. *Brain Res. Bull.* 27:409-415(1991).

82. J. Gibbs, D. J. Fauser, E. A. Rowe, E. T. Rolls and S. P. Maddison. Bombesin suppresses feeding in rats. *Nature* 282:208-210(1979).

83. A. Bado, M. J. M. Lewin and M. Dubrasquet. Effects of bombesin on food intake and gastric acid secretion in cats. *Am. J. Physiol.* 256:R181-R186(1989).

84. D. M. Denbow. Centrally and peripherally administered bombesin decreases food intake in turkeys. *Peptides* 10:275-279(1989).

85. M. O. Miceli and C. W. Malsbury. Effects of putative satiety peptides on feeding and drinking behavior in Golden Hamsters. *Behav. Neurosci.* 99:1192-1207(1985).

86. I. L. Taylor, J. Elashoff and R. Garcia. Effects of pancreatic polypeptide, caerulein, and bombesin in obese mice. *Am. J. Physiol.* 248:G277-G280(1985).

87. S. C. Woods, L. J. Stein, D. P. Figlewicz and D. Porte,Jr.. Bombesin stimulates insulin secretion and reduces food intake in the baboon. *Peptides* 4:687-691(1983).

88. N. E. Muurahainen, H. R. Kissileff, J. Thornton and F. X. Pi-Sunyer. Bombesin: another peptide that inhibits feeding in man. *Soc. Neurosci. Abstr.* 9:183(1983).

89. J. A. DiPoala and J. Gibbs. Neuromedin C inhibits food intake in rats. *Soc. Neurosci. Abstr.* 11:38(1985).

90. D. P. Figlewicz, L. J. Stein, S. C. Woods and D. Porte,Jr.. Acute and chronic gastrin-releasing peptide decreases food intake in baboons. *Am. J. Physiol.* 248:R578-R583(1985).

91. L. J. Stein and S. C. Woods. Gastrin releasing peptide reduces meal size in rats. *Peptides* 3:833-835(1982).

92. J. Gibbs, P. J. Kulkosky and G. P. Smith. Effects of peripheral and central bombesin on feeding behavior of rats. *Peptides* 2 (Suppl.2):179-183(1981).

93. J. Gibbs and G. P. Smith. The actions of bombesin-like peptides on food intake. *New York Academy of Sciences* 547:210-216(1988).

94. J. A. Stuckey, J. Gibbs and G. P. Smith. Neural disconnection of gut from brain blocks bombesin-induced satiety. *Peptides* 6:1249-1252(1985).

95. T. C. Kirkham, J. Gibbs and G. P. Smith. The satiating effect of bombesin is mediated by receptors perfused by the coeliac artery. *Am. J. Physiol.* 261:R614-R618(1991).

96. J. Gibbs and G. P. Smith. Gut peptides and food in the gut produce similar satiety effects. *Peptides* 3:553-557(1982).

97. A. M. Hostetler, P. R. McHugh and T. H. Moran. Bombesin affects feeding independent of a gastric mechanism or site of action. *Am. J. Physiol.* 257:R1219-R1224(1989).

98. D. D. Krahn, B. A. Gosnell, A. S. Levine and J. E. Morley. Localization of the effects of corticotropin-releasing factor on feeding. *Soc. Neurosci. Abstr.* 10:300(1984).

99. E. E. Ladenheim and R. C. Ritter. Caudal hindbrain participation in suppression of feeding by central and peripheral bombesin. *Soc. Neurosci. Abstr.* 15:964(1989).

100. F. W. Flynn. Fourth ventricle bombesin injection suppresses ingestive behaviors in rats. *Am. J. Physiol.* 256:R590-R596(1989).

101. E. E. Ladenheim and R. C. Ritter. High dorsal column transection attenuates bombesin-induced suppression of food intake. *Soc. Neurosci. Abstr.* 14:1108(1988).

102. R. deBeaurepaire and C. Suaudeau. Anorectic effect of calcitonin, neurotensin and bombesin infused in the area of the rostral part of the nucleus of the tractus solitarius in the rat. *Peptides* 9:729-733(1988).

103. S. A. Johnston and Z. Merali. Specific neuroanatomical and neurochemical correlates of grooming and satiety effects of bombesin. *Peptides* 9 (Suppl.1):233-244(1988).

104. Z. Merali, T. W. Moody and D. H. Coy. Blockade of the central bombesin (BN) receptors enhances food intake in prefed rats. *Soc. Neurosci. Abstr.* 16:978(1990).

105. R. D. Reidelberger, G. Varga, T. Castonguay and D. H. Coy. Potent bombesin receptor antagonist D-Phe6-bombesin(6-13)-methyl-ester stimulates food intake in rats. *Soc. Neurosci. Abstr.* 17:543(1991).

THE ROLE OF SOMATOSTATIN RECEPTORS IN THE DIAGNOSIS AND THERAPY OF CANCER

J.C. Reubi[1], E. Krenning[2] and S.W. Lamberts[2]

[1]Division of Cell Biology and Experimental Cancer Research, Institute of Pathology, University of Berne, Berne, Switzerland;
[2]Dept. of Internal Medicine, Erasmus University, Rotterdam, The Netherlands

INTRODUCTION

Somatostatin (SS) is a representative member of the family of small regulatory peptides which are often characterised by a wide spectrum of actions in various organs of the human body. One of the most prominent and well-documented roles of SS is its inhibition of hormone secretory processes, in particular in the pituitary (GH, TSH) and in the gastroenteropancreatic (GEP) system (Insulin, Glucagon, ViP, Secretin, etc...)[1]. In addition, SS can modulate the neurotransmission in various brain regions[1]. All these SS actions seem to be mediated through specific SS receptors (SS-R)[2].

In recent years, two observations suggested that SS may also play a significant role in cancer:
1) In numerous animal tumor models and cultured tumoral cell lines, SS and SS analogs were shown to inhibit tumor growth[3].
2) SS receptors were shown to be expressed in a wide variety of primary human tumors and their metastases[4].

The present review will summarize the incidence of SS-R in human tumors and discuss their general characteristics. It will also give experimental evidence for diagnostical and therapeutical implications of tumoral receptors for a peptide, which may be considered paradigmatic among regulatory peptides in the field of cancer.

INCIDENCE OF SS-R IN HUMAN TUMORS

As seen in Table 1, a wide variety of human tumors express SS-R. One of the groups with the highest incidence of SS-R are neuroendocrine tumors[4]; most GH and TSH producing pituitary adenomas, but also several non-secreting adenomas, contain SRIH receptors as does a significant proportion of endocrine hormone-producing gastroenteropancreatic (GEP) tumors (Fig.1); phaeochromocytomas and paragangliomas; some medullary thyroid carcinomas (MTC); several small cell lung

T a b l e 1 .   Incidence of tumors expressing SS-R

Neuroendocrine Tumors	Pituitary Adenomas, Carcinoids, Islet Cell Carcinomas, Paragangliomas, Pheochromocytomas, Medullary Thyroid Carcinomas, Small Cell Lung Carcinomas	High
Tumors of the Nervous System	Astrocytomas, Neuroblastomas, Meningiomas	High
Lymphomas		High
Renal Cell Carcinomas		High
Breast Tumors		~ 50%
Ovarial Carcinomas, Colon Carcinomas		Low
Glioblastomas, Non-Small Cell Lung Carcinomas, Exocrine Pancreatic Carcinomas, Prostate Carcinomas		No Receptors

cancers (SCLC). Usually, these tumors have a high density of SS-R, but there is a great individual variability of receptor density: more than tenfold differences in SS-R density are not rare within individual tumor types. Most of the above mentioned tumors belong to the group of apudomas, i.e. tumors having the APUD cell as common origin.

The second group of human tumors expressing frequently SS-R are tumors of the nervous system[4]. More than 80% of the astrocytomas (Fig.2) as well as a high percentage of neuroblastomas contain SS-R (Table 1). All meningiomas express a high density of SS-R.

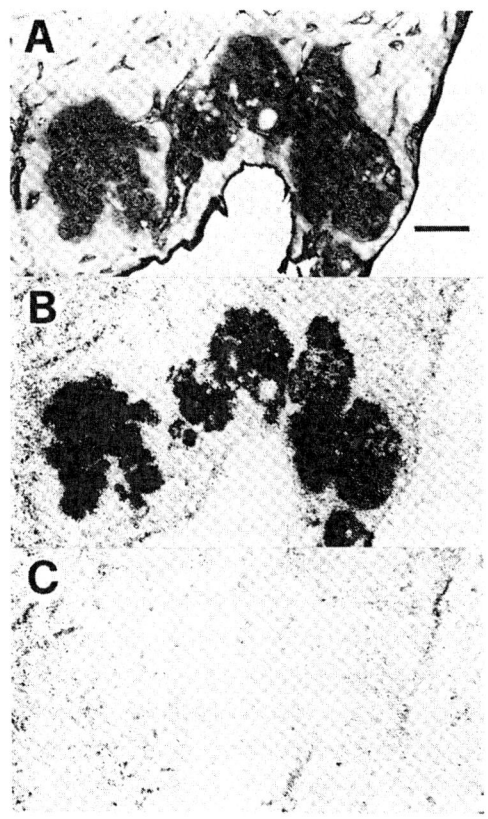

FIG. 1. Somatostatin receptors in a human carcinoid tumor. A: Hematoxylin-eosin stained section. B: Auto-radiogram showing total binding of ^{125}I-[Tyr^3]-octreotide. C: Autoradiogram showing non-specific binding of ^{125}I-[Tyr^3]-octreotide (in presence of 10^{-6} M [Tyr^3]-octreotide). Bar = 1 mm. Notice the selective labelling of tumor tissue.

A third group of tumors having a high incidence of SS-R includes the malignant lymphomas of the Hodgkin and non-Hodgkin type[5]. Recently we identified a fourth group of tumors with a high incidence (72%) but low density of SS-R: the renal cell carcinomas.

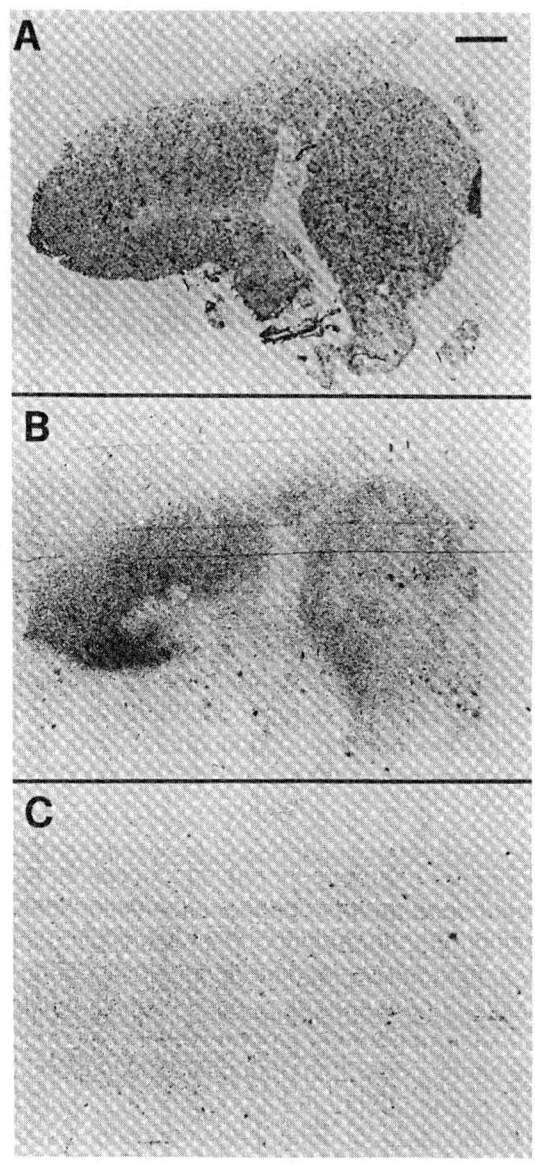

FIG. 2. Somatostatin receptors in a differentiated human astrocytoma. A: Hematoxylin-eosin stained section. B: Autoradiogram showing total binding of ^{125}I-[Tyr^3]-octreotide. C: Non-specific binding. Bar=1 mm.

Apart from the above mentioned tumor types showing usually a high incidence of SS-R, approximately half of the cases of breast cancer were shown to possess SS-R[4,6]. Other tumors, such as colorectal cancers and ovarian tumors, displayed only a low incidence of SS-R (Table 1).

SS-R in the above mentioned tumors have been characterised as high affinity receptors specific for SS, comparable to the SS-R identified in healthy target tissues such as brain, pituitary or pancreas.

However, there is a large group of neoplasms which do not express SS-R: it includes glioblastomas, non-small cell lung cancers (NSCLC), exocrine pancreatic cancers, prostate cancers and most squamous cell carcinomas (Table 1).

In receptor-positive cancers, SS-R are homogeneously distributed in the tumor tissue, whereas the surrounding, healthy tissue is usually lacking SS-R (Fig.1). Interestingly however, there are tumors, in particular of the breast, that often display SS-R only in certain tumor regions, although the whole tumor appears to be histopathologically homogeneous. This observation supports the idea that some breast tumors are composed of numerous different clones with variable biological properties. This has consequences in terms of the precise SS-R status in small samples of large breast tumors samples or of the therapeutic efficacy of SS analogs in such tumors.

GENERAL CHARACTERISTICS OF TUMORS EXPRESSING SS-R

Despite of the fact that human tumors of very different origins are expressing SS-R, many of the SS-R containing tumors seem to have a number of common general features.

Firstly, SS-R positive tumors encompass most of the tumors of neuroendocrine origin. This is true for pituitary adenomas, islet cell carcinomas, carcinoids, phaeochromocytomas, medullary thyroid carcinomas, paragangliomas, neuroblastomas and small cell lung cancer. In addition, a subgroup of breast tumors with neuroendocrine features is usually SS-R positive[7]. The same may apply to the SS-R positive colonic and ovarian carcinomas since it is known that a small percentage of those tumors have also neuroendocrine features.

Secondly, the majority of SS-R containing tumors belong to rather well differentiated tumors. This is true for most of the neuroendocrine tumors. It is worth mentioning that the well-differentiated carcinoids express SS-R in the great majority of the cases, whereas the less differentiated, "atypical carcinoids" often lack SS-R[8]. In a case of a medullary thyroid carcinoma we observed two different areas within the excised tumor sample, a well-differentiated tumor area containing calcitonin and SS-R, and an undifferentiated SS-R negative tumor[9]. Analogous observations have been made in glial tumors, where SS-R are often expressed by astrocytomas I-III, but not by the undifferentiated glioblastomas[4]. At least one exception to this general trend exists, namely the malignant lymphomas, which often contain SS-R, even in high density, in the undifferentiated high grade tumor types[5].

Thirdly, an inverse relationship is frequently seen between the presence of SS-R and of epidermal growth factor(EGF)-R in several types of tumors. SS-R negative NSCLC and glioblastomas are in most instances EGF-R positive, whereas SS-R positive SCLC or astrocytomas usually lack EGF-R[10,11]. A similar tendency is found in breast tumors. It should, however, be noticed that in the vicinity of SS-R positive breast tumor samples EGF-R positive normal breast tissue can be found, while the tumoral tissue remains EGF-R negative[6]. The inverse correlation between SS-R and EGF-R suggests strongly that SS-R positive tumors are usually more differentiated, less aggressive tumors with possibly a better prognosis than their SS-R negative counterparts. Renal cell carcinomas and meningiomas, however, represent an exception since all SS-R positive tumor samples expressed simultaneously EGF-R[11]. Interestingly, in several cases of meningiomas, the two receptors are not necessarily located on identical cellular entities. Indeed, we have observed in these tumors that SS-R are located in high density over the tumor tissue only, whereas necrotic tissues and the necrobiotic tumor area have no SS-R. On the contrary, EGF-R are only weakly expressed in the "resting" tumor area, but intensely in the more aggressive and proliferating part of the tumor near the necrotic zone. Therefore, in some of the meningiomas expressing simultaneously both receptors, SS-R seem preferentially located in the less aggressive region of the tumor[11].

Thus, several common characteristics of SS-R positive tumors can be identified, such as neuroendocrine features, differentiation state or inverse correlation with EGF-R. However, exceptions are regularly observed, such as SS-R positive high grade lymphomas or SS-R and EGF-R positive meningiomas and renal cell carcinomas.

EVIDENCE FOR SS-R SUBTYPES IN SELECTED TUMORS

It has been reported that, in rat and human brain, subpopulations of SS-R exist. For example, pharmacological evidence for such SS-R subpopulations is provided by the differential affinity for the SS analog octreotide (high affinity = SS_1-R; low affinity = SS_2-R) as compared to SS-14 or SS-28, in various brain regions[12,13]; functional correlates for these subtypes exist as observed by the differential response of the adenylate cyclase activity to SS and SS analogs[14] in various brain regions.

The majority of the numerous human tumors tested up to now expressed a SS-R with a high affinity for octreotide (SS_1). Nevertheless, we have observed that a restricted number of SS-R-containing tumors expressed the SS-R subtype SS_2 described above, defined as having a low affinity for octreotide, i.e. at least two orders of magnitude lower than SS-14 or SS-28. This receptor subtype has been observed in a small percentage (<10%) of pituitary adenomas, carcinoids, glial tumors, meningiomas or breast tumors. A higher incidence of SS_2 receptors was found, however, in insulinomas[15]. Moreover, we could demonstrate there that this SS_2 subtype was functional since in those insulinomas cultured in vitro, the insulin release was inhibited by SS-14 and SS-28, but not by octreotide[16]. In addition, these insulinomas were not visualised in vivo using a ^{123}I-[Tyr^3]-octreotide as tracer[16] (see below). Conversely, insulinomas with a SS_1 receptor subtype reacted normally to octreotide in terms of insulin inhibition and were visualised in vivo as expected.

Approximately half of the medullary thyroid carcinomas[9] had also a SS-R subtype with low affinity for octreotide (SS_2). Furthermore, all SS-R positive ovarian tumors displayed this same SS_2 receptor subtype. They can be labelled with SS-14 or SS-28 radioligand but not with [Tyr3]-octre-otide radioligand[17]. Clearly, therefore, some peripheral tumoral tissues have also the potential, as does the healthy brain, to express the SS_2 receptor subtype.

These findings have a number of clinical consequences: first, tumors with only SS_2 receptors will not be visualised **in vivo**, neither with ^{123}I-[Tyr^3]-octreotide nor with ^{111}In-[DTPA,DPhe^1]-octreotide as tracer[16]; secondly, tumors expressing SS_2 receptors will not or only poorly respond to a conventional octreotide therapy[16]; this may be one of the reasons why insulinomas and medullary thyroid carcinomas often react poorly to conventional doses of octreotide[16]. Very high doses of octreotide are required indeed to elicit a reaction in some potentially responsive tumors. This means that for the **in vivo** visualisation or successful therapy of such SS_2 receptor containing tumors other types of SS analogs (based on a structure different from that of octreotide) will have to be developed. Since several SS-R subtypes are likely to exist and to be expressed in SS targets and certain tumors new SS analogs may be found which allow to identify and treat a larger number of human tumors[18].

PUTATIVE FUNCTIONS MEDIATED BY SS-R IN TUMORS

As we could demonstrate in various recent studies, the SS-R detected on human tumors have, in addition to a role as pathobiochemical markers of those tumors, also a specific function. However, the SS function mediated by SS-R may vary depending on the tumor type.

SS-R in human pituitary adenomas and GEP tumors are likely to be functional and mediate primarily SS inhibition of hormone secretion, as shown in four different studies using GH- and TSH-secreting pituitary adenomas as well as carcinoids and islet cell carcinomas[8,16,19,20]. Therefore, SS-R in pituitary and GEP tumors are the likely molecular basis for hormone inhibition by SS and therefore relevant for the therapeutic efficacy of octreotide. To date, there is no conclusive evidence that SS-R present in those human tumors also mediate an antiproliferative action of SS. However, several studies using **in vivo** animal models and **in vitro** cultured cell lines from several tumor types suggest that there is such an effect. One recent study in GEP tumors indicates that SS may play an antiproliferative role in human tumors: Indeed, a SS-R positive human carcinoid, transplanted into a nude mouse and retaining his original carcinoid characteristics, was shown to be inhibited in its growth by 50% after 2 weeks of octreotide therapy[21]. Several other animal and cell culture studies suggest that a direct antiproliferative effect of SS, mediated by SS-R, takes place in certain tumors[3,22]. For instance, both **in vitro** and **in vivo** growth inhibition by the SS analog somatuline was observed in a SS-R positive SCLC cell line[23]. Conversely, in SS-R negative DMBA-induced breast tumors no growth inhibitory effect of SS was observed[24]. Nevertheless, when the growth of a tumor is strongly growth factor dependent, an indirect **in vivo** growth inhibition by SS can take place, despite the absence of tumoral SS-R. In those cases, exogenous SS will act on SS-R located on healthy SS targets (pituitary, pancreas) as shown for the IGF- and insulin-dependent Swarm chondrosarcoma[25].

We have recent evidence that at least in one tumor type, the meningioma, the high density of SS-R present in this neoplasm does <u>not</u> mediate tumor growth inhibition [26]. The SS-R present in meningiomas are functional since SS and SS analogs can inhibit forskolin-stimulated adenylate cyclase activity. In this type of tumor an increase in c-AMP levels induced however a significant tumor growth inhibition, as measured by ^{3}H-thymidine incorporation. Therefore, the addition of SS to cultures of meningiomas resulted in a slight but significant growth stimulation of the tumor cells, but not in growth inhibition[26].

These examples show that SS-R in human tumors are likely to mediate multiple SS actions. There is convincing evidence that they mediate SS inhibition of hormone secretion in pituitary and GEP tumors. They may also mediate antiproliferative effects of SS in certain tumors (breast tumors, SCLC), although a definite proof of cytostatic activity in primary human tumors is still lacking. Finally, they may mediate growth proliferation in selected tumors, such as meningiomas.

The high density of SS-R in many human tumors offers attractive possibilities for their detection **in vivo**. This can be achieved by i.v. injection of ^{123}I-[Tyr^3]-octreotide (^{123}I-[Tyr^3]-SMS 201-995) or ^{111}In-[DTPA-D-Phe^1]-octreotide in patients suspected of having SS-R positive tumors; such tumors are then localised with planar and ECT images obtained with a gamma camera[27-29]. With this method, hot spots representing radioligand binding on SS-R positive tumors are visualised (Fig.3, carcinoid; Fig.4, astrocytoma), as well as healthy organs where the radioligand is circulating or cleared (heart, lung, kidney, liver, gall-bladder, GI tract). Several observations clearly demonstrate that the hot spots identified by this method represent SS-R containing tumor:

In a rat tumor model containing SS-R, systemic injection of non-radioactive octreotide prevents the appearance of the hot spots normally seen on the tumor site after ^{123}I-[Tyr^3]-octreotide injection[22]. There is a highly significant correlation in human tumors between the frequency of SS-R

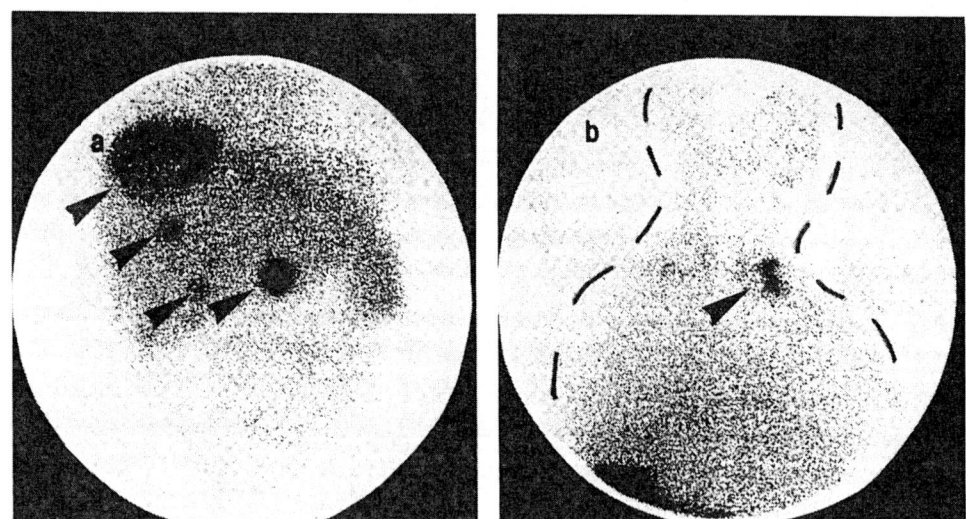

FIG. 3. Gamma-camera pictures 30 min after i.v. administration of ^{123}I-[Tyr^3]-octreotide in a patient with metastatic carcinoid disease. a) liver containing several hot spots (arrows) representing SS-R positive metastases; b) lymph node of the left part of the neck which also contains a metastasis from the carcinoid tumor (arrow).

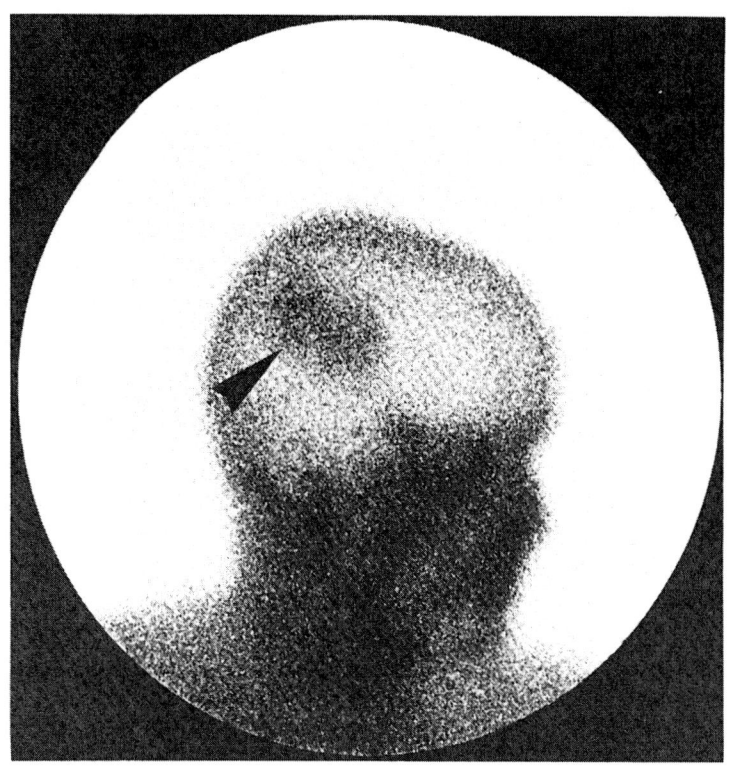

FIG. 4. Grade III astrocytoma on the right side of the brain 4 h after ^{123}I-[Tyr^3]-octreotide administration. Note that the tumor (arrow) but not the somatostatin receptor-positive brain cortex is labelled, suggesting that the blood-brain barrier is permeable to octreotide only in the tumor region but not in the healthy brain.

positivity found by **in vitro** methods and by **in vivo** imaging[22]. Moreover, we have recently confirmed that in all cases of various tumor types with positive scans who later underwent operative biopsies, the hot spots corresponded to SS-R positive tumors as measured with **in vitro** binding techniques. In addition, in two gastrinomas with positive scan, we could show both that they contained SS-R **in vitro** and also that the gastrin release of tumors in culture could be inhibited by octreotide[16,22]. These findings clearly demonstrate the specificity of the **in vivo** imaging method and confirm the **in vitro** finding that several tumors and their metastases contain high densities of SS-R. The **in vivo** labelling of tumors with SS analogs is the first application of a neuropeptide as diagnostic tool in nuclear medicine and opens very promising new avenues for diagnosis and, ultimately, for therapy, for neuropeptides in general.

CLINICAL IMPLICATIONS

The presence of SS-R in a large variety of human tumors has a number of clinical implications. **In vitro** detection of those receptors may be useful for the pathobiochemical characterisation of various tumor types and subtypes, specially since SS-R positive tumors often display neuroendocrine features and a high differentiation grade. **In vivo** detection of these receptors, using radiolabelled octreotide analogs and gamma-camera scintigraphy, is extremely useful for the localisation of SS-R positive tumors, the evaluation of the metastatic disease and, for lymphomas, for the staging of the disease[27-29].

Since these receptors have been shown to be functional, their presence may be predictive for a successful octreotide treatment in those types where an octreotide therapy is indicated (pituitary adenomas; GEP tumors)[8].

Recently, we could demonstrate in two types of human tumors that the SS-R status may also have a predictive value for the prognosis of the tumor. In a retrospective study involving 110 breast cancer patients, we found 17 (15%) of the tumors to be SS-R positive[30]. In this group of patients no significant relationship was observed between the presence of SS-R and lymph node status, or age of the patients. However, the disease-free survival probability for patients with SS-R positive tumors was significantly higher. In the 17 patients with SS-R positive breast tumors the 5 yr disease-free survival was 82%, compared with 46% for the 83 patients with SS-R negative tumors[30]. Furthermore, we have shown in a preliminary study with neuroblastomas that there was in inverse relationship between the presence of SS-R and N-myc oncogene expression[31]. The presence of SS-R in about 50% of tumors seemed to correlate with a favourable prognosis[31].

As a future clinical perspective we should finally mention the possibility of using tumoral SS-R as a radiotherapeutical target. For this purpose, a SS analog linked to an adequate ß-emitting isotope would be needed. However, a careful evaluation of the ratio between the beneficial effects of the radiotherapy in destroying the tumor tissue and the destruction of SS-R containing physiological targets will be a prerequisite.

ACKNOWLEDGEMENTS

We thank Mrs U. Horisberger and B. Waser for excellent technical assistance.

REFERENCES

1. Reichlin S.: Somatostatin. N. Engl. J. Med. 309:1495 (1983).
2. Reubi J.C., Kvols L., Krenning E. and Lamberts S.W.J.: Distribution of somatostatin receptors in normal and tumor tissue. Metabolism 39:78 (1990).
3. Schally A.V.: Oncological applications of somatostatin analogues. Cancer Res. 48:6977 (1988).
4. Reubi J.C., Krenning E., Lamberts S.W.J. and Kvols L.: Somatostatin receptors in malignant tissues. J. Steroid Biochem. Molec. Biol. 37(6):1073 (1990).
5. Reubi J.C., Waser B., van Hagen M., Lamberts S.W.J., Krenning E.P., Gebbers J.O. and Laissue J.: In vitro and in vivo detection of somatostatin receptors in human malignant lymphomas. Int. J. Cancer. 50:895 (1992)
6. Reubi J.C., Waser B., Foekens J., Klijn J., Lamberts S.W.J. and Laissue J.: Somatostatin receptor incidence and distribution in breast cancer using receptor autoradiography: relationship to EGF receptors. Int. J. Cancer 46:416 (1990).
7. Papotti M., Macri L., Bussolati G. and Reubi J.C.: Correlative study on neuroendocrine differentiation and presence of somatostatin receptors in breast carcinomas. Int. J. Cancer 43:365 (1989).
8. Reubi J.C., Kvols L.K., Waser B., Nagorney D.M., Heitz P.U., Charboneau J.W., Reading C.C. and Moertel C.: Detection of somatostatin receptors in surgical and

percutaneous needle biopsy samples of carcinoids and islet cell carcinomas. Cancer Res. 50:5969 (1990).

9. Reubi J.C., Chayvialle J.A., Franc B., Cohen R., Calmettes C. and Modigliani E.: Somatostatin receptors and somatostatin content in medullary thyroid carcinomas. Lab. Invest. 64:567 (1991).

10. Reubi J.C., Waser B., Sheppard M. and Macaulay V.: Somatostatin receptors are present in small cell but not in non-small cell primary lung carcinomas: relationship to EGF-receptors. Int. J. Cancer 45:269 (1990).

11. Reubi J.C., Horisberger U., Lang W., Koper J.W., Braakman R. and Lamberts S.W.J.: Coincidence of EGF receptors and somatostatin receptors in meningiomas but inverse, differentiation-dependent relationship in glial tumors. Am. J. Path. 134:337 (1988).

12. Reubi J.C.: Evidence for two somatostatin-14 receptor types in rat brain cortex. Neurosci. Lett. 49:259 (1984).

13. Reubi J.C., Probst A., Cortes R. and Palacios J.M.: Distinct topographical localisation of two somatostatin receptor subpopulations in the human cortex. Brain Res. 406:391 (1987).

14. Markstein R., Stoeckli K.A. and Reubi J.C.: Differential effects of somatostatin on adenylate cyclase as functional correlates for different brain somatostatin receptor subpopulations. Neurosci. Lett. 104:13 (1989).

15. Reubi J.C., Haecki W.H. and Lamberts S.W.J.: Hormone-producing gastrointestinal tumors contain a high density of somatostatin receptors. J. Clin. Endocr. Metab. 65:1127 (1987).

16. Lamberts S.W.J., Hofland L.J., Koetsveld P. van, Reubi J.C., Bruining H.A., Bakker W.H. and Krenning E.P.: Parallel in vivo and in vitro detection of functional somatostatin receptors in human endocrine pancreatic tumors. Consequences with regard to diagnosis, localisation and therapy. J. Clin. Endocr. Metab. 71:566 (1990).

17. Reubi J.C., Horisberger U., Klijn J.G.M. and Foekens J.A.: Somatostatin receptors in differentiated ovarian tumors. Amer. J. Pathol. 138:1267 (1991).

18. Srkalovic G., Cai R.-Z. and Schally A.V.: Evaluation of receptors for somatostatin in various tumors using different analogs. J. Clin. Endocr. Metab. 70:661 (1990).

19. Reubi J.C. and Landolt A.M.: The growth hormone responses to octreotide in acromegaly correlate with adenoma somatostatin receptor status. J. Clin. Endocr. Metab. 68:844 (1989).

20. Levy A., Eckland D.J.A., Gurney A.M., Reubi J.C., Doshi R. and Lightman S.L.: Somatostatin and thyrotrophin-releasing hormone response and receptor status of a thyrotrophin secreting pituitary adenoma: clinical and in vitro studies. J. Neuroendocr. 1:321 (1989).

21. Evers B.M., Townsend C.M., Upp J.R., Allen E., Hurlbut S.C., Kim S.W., Rajaraman R., Singh P., Reubi J.C. and Thompson J.C.: Establishment and characterization of a human carcinoid in nude mice and effect of various agents on tumor growth. Gastroenterology 101:303 (1991).

22. Lamberts S.W.J., Krenning E.P. and Reubi J.C.: The role of somatostatin and its analogs in the diagnosis and treatment of cancer. Endocr. Rev. 12:450 (1991).

23. Taylor J.E., Bogden A.E., Moreau J.-P. and Coy D.H.: In vitro and in vivo inhibition of human small-cell lung carcinoma (NCI-H69) growth by a somatostatin analogue. Biochem. Biophys. Res. Commun. 153:81 (1988).

24. Bakker G.H., Setyono-Han B., Foekens J.A., Portengen H., van Putten W.L.J., de Jong F.H., Lamberts S.W.J., Reubi J.C. and Klijn J.G.M.: Treatment of rats bearing DMBA-induced mammary tumors with the Sandostatin analog (SMS 201-995). Breast Cancer Res. Treat. 17:23 (1990).

25. Reubi J.C.: A somatostatin analog inhibits chondrosarcoma and insulinoma tumor growth. Acta Endocrin. 109:108 (1985).

26. Koper J.W., Markstein R., Kohler C., Kwekkeboom D.J., Avezaat C.J.J., Lamberts S.W.J. and Reubi J.C.: Somatostatin inhibits activity of adenylate cyclase in cultured human meningioma cells and stimulates their growth. J. Clin. Endocr. Metab. 74:543 (1991).

27. Krenning E.P., Bakker W.H., Breeman W.A.P., Koper J.W., Kooij P.P.M., Ausema L., Lameris J.S., Reubi J.C. and Lamberts S.W.J.: Localization of endocrine-related tumours with radio-iodinated analogue of somatostatin. Lancet i:242 (1989).

28. Lamberts S.W.J., Bakker W.H., Reubi J.C. and Krenning E.P.: The value of somatostatin receptor imaging in the

localization of endocrine and brain tumors. New Engl. J. Med. 323:1246 (1990).

29. Krenning E.P., Bakker W.H., Kooij P.P.M., Breeman W.A.P., Oei H.Y., de Jong M., Reubi J.C., Visser T.J., Bruns C., Kwekkeboom D.J., Reijs A.E.M., van Hagen P.M., Koper J.W. and Lamberts S.W.J.: Somatostatin receptor scintigraphy with [111-In-DTPA-D-Phe-1]-octreotide. J. Nucl. Med. 33:652 (1992).

30. Foekens J.A., Portengen H., Putten W.L.J. van, MacTrapman A., Reubi J.C., Alexieva-Figush J. and Klijn J.G.M.: Prognostic value of receptors for insulin-like growth factor-I, somatostatin and epidermal growth factor in human breast cancer. Cancer Res. 49:7002 (1989).

31. Moertel C.L., Reubi J.C., Scheithauer B., Schaid D.J. and Kvols L.K.: Somatostatin receptors (SS-R) are expressed and correlate with prognosis in childhood neuroblastoma. Proc. Am. Soc. Ped. Res.: Abstr. 306 (1990).

INDEX